ECONOMIC ISSUES, PROBLEMS AND PERSPECTIVES

THE POLITICAL ALGEBRA OF GLOBAL VALUE CHANGE

GENERAL MODELS AND IMPLICATIONS FOR THE MUSLIM WORLD

Economic Issues, Problems and Perspectives

Additional books in this series can be found on Nova's website under the Series tab.

Additional e-books in this series can be found on Nova's website under the e-book tab.

ECONOMIC ISSUES, PROBLEMS AND PERSPECTIVES

THE POLITICAL ALGEBRA OF GLOBAL VALUE CHANGE

GENERAL MODELS AND IMPLICATIONS FOR THE MUSLIM WORLD

ARNO TAUSCH
ALMAS HESHMATI
AND
HICHEM KAROUI

nova publishers
New York

Library of Congress Cataloging-in-Publication Data

ISBN: 978-1-62948-899-8

Published by Nova Science Publishers, Inc. † New York

Contents

ABOUT THIS BOOK

This book should contribute to the homecoming of the *'ilm al umran',* the science of culture, initiated by the Arab classical thinker Ibn Khaldūn (1332-1406) to the countries of the Muslim world, and the Arab world in particular.

The book is intended to provide a new approach to the study of global values and global value change, based on representative international survey data, above all, the *World Values Survey.* This theme is of growing interest not only to the international social science community, but also international policy makers and business and financial executives in the framework of international values and business studies.

The book grew out of lecture series I held at various academic and government education centers in Europe.

I have to admit that the overwhelming force of the empirics which presented themselves to me during the consecutive process of the re-analysis of the *World Values Survey* data base went counter to much of the post-1968 ideological consensus, which I "inhalated" ever since my student days at Salzburg University in Austria from 1969 onwards.

Only now, I began to remember that among my teachers at Salzburg University, there was also the later Nobel laureate in economics, Friedrich August von Hayek, whose entire five semester lecture series I had the chance to attend in the early 1970s. At that time, nobody as yet talked about the Nobel Prize for Hayek, and we were just ten to fifteen students in Hayek's classes at times. I somehow relegated his teachings to my archives, only to re-discover them, 45 years later amidst the powerful statistical effects, which variables, to be associated with his theories, wield for the empirical analysis of *World Values Survey* data and especially for the analysis of realities in the Muslim world and in the Arab world.

Since the book is also designed to serve advanced graduate courses in sociology, political science and economics at Universities, government and business staff training centers, diplomatic and military academies *et cetera,* the book also contains themes for 10-20 minute statements (1000-2000 words) which students and course participants should be able to prepare after attending a course on global values.

Our work continues the very fruitful collaboration between my friend and colleague Professor Almas Heshmati at Sogang University, a Jesuit University in South Korea and Jönköping University, Sweden and myself (Heshmati, Tausch A. and Bajalan 2008; Heshmati and Tausch, 2004, 2007; Tausch and Heshmati, 2004, 2009; Tausch, Heshmati and Brand, 2012).

It also continues my very fruitful collaboration with my friend and colleague, the French intellectual Dr. Hichem Karoui, who was my former graduate student at the Sorbonne in Paris (Tausch and Karoui, 2011). Dr. Karoui's untiring efforts to bring this project to a conclusion must be mentioned in a special way in this brief introductory note.

Any student of the issues of Islam and modernity will have already noticed the path-breaking contributions by the German Jesuit father Christian W. Troll (Troll, 2001a, 2001b, 2003, 2004, 2007; Troll and Bsteh, 1997, Troll and Donohue, 1998; Troll and Vahiduddin, 1986). In this context, we remind our readers also about the important and respectful insights on Islam in world society, which other Roman Catholic clerics and scholars offered to international publics for decades now (Bsteh, 1996; Bsteh and Anawati, 1978; Bsteh and Dupré, 2007; Bsteh and Khoury, 1994; Khoury, 1980, 1991, 1994, 2001, 2002, 2005, 2007a, 2007b, 2008a, 2008b; Khoury et al., 2006). These studies influenced me to a considerable degree. Rethinking the relationship of Islam and modernity in the spirit of Enlightenment should also be the aim of this book.

I would also like to mention here in gratitude two of my late Jesuit academic teachers in Austria, Geraldo de Freitas and Julius Morel, who early on drew my attention to the necessity to study the phenomena of value differences (Karlinger and Freitas 1972; Messner and Freitas, 1988; Morel, 1968, 1972, 1977, 1980, 1986, 1997, 1998, 2003).

I also received great inspiration from the life and work of the six Jesuit liberation theologians massacred on November 16, 1989 at the campus of the Universidad Centroamericana *"José Simeón Cañas"* (UCA) in San Salvador, El Salvador. Their re-thinking of the theology of marriage (which so many in the West now consider as an outdated institution) led them to talk about the union of Adam and Eve (Book of Genesis (Gen.) 2:22) prefiguring the union of God with humanity (Sobrino and Ellacuria, 1993: 228; see also Ellacuria, 1976, 2013; Sobrino, 1990).

I should also mention here the insights which I owe to Professor Bassam Tibi, whose works, especially Tibi, 1990, 1995a, 1995b, 1996, 1997a, 1997b, 1997c, 1998a, 1998b, 2001a, 2001b, 2002, 2002, 2003, 2007 have become an important guiding point in the international debate about Islam, democracy and Enlightenment.

Finally, I also should mention the inspiration received from Dr. Gunther Hauser, senior analyst in the Austrian National Defense Academy, in the framewok of our academic cooperation now lasting for years.

I also mention Mr. Robert Mundigl in Munich and his freely available templates to draw choropleth global statistics maps from Microsoft EXCEL formatted country data. A really interesting system for choropleth maps, which we used in this book! We simply had to enter our data into the freely available EXCEL 2010/Map templates, made available at http://www.clearlyandsimply.com/clearly_and_simply/2009/10/choropleth-maps-with-tableau.html.

To our knowledge, these are the best freely available choropleth global map templates for Microsoft EXCEL data, available today.[1]

[1] We should note with apologies however that Brunei is not contained in the country list of this system. Also data for Macau and Hongkong could not be handled. In addition, values for Taiwan, Province of China, are listed as a separate entitity on the maps. Readers, further interested in drawing such maps of their own with their own data, are kindly invited to use the templates available at https://uibk.academia.edu/ ArnoTausch/Documentation-for-books-and-articles

Chapter 7 develops hypotheses, which were already partly presented by Arno Tausch in his essay (2013) 'A look at recent international survey data about Arab opinion' *Middle East Review of International Affairs,* Vol. 17, No. 3 (Fall 2013), 57-74.

After a long period of joint work, the authors could agree on the final text by around the end of June 2014. The final editing from the authors' side ended on July 31st, 2014.

In thanking my co-authors and also in thanking my family, I hasten to add that the errors and shortcomings of this study which might still exist are all entirely my own.

Arno Tausch
Leopoldsdorf near Vienna, July 31st, 2014

The book presents a very interesting approach to the study of change in global values. This is a very much needed discussion at the particular time. It is remarkably thought provoking for the community of social sciences as well as policy makers.

Ibrahim Sharqieh, Foreign Policy Fellow at the Brookings Institution's Doha Center and an Adjunct Professor in International Conflict Resolution at Georgetown University in Qatar. He is the author of Reconstructing Libya: Stability through National Reconcliation (Brookings Doha Center), and a contributing author to the upcoming book African Renaissance and Afro-Arab Spring (Georgetown University Press).

In this book Arno Tausch and his co-authors consider the very acute and important problem for the future of global value change. They re-discover the positive contribution of religion for society, and this conclusion is based on the creative analysis of numerous data. This wonderful study is able to change our view of the perspectives of the world economy and world politics.

Professor Alexander Dynkin and *Professor Vladimir Pantin,* Institute of World Economy and International Relations, Moscow, Authors of *"Strategic Global Forecast to 2030"* (Magister, 2013), and *"A Peaceful Clash: The U.S. and China",* World Futures, Volume 68, 2012: 506 – 517.

This book unites in a single system such different (but equally significant) dimensions as economic growth, global values, and Islamic civilization. A remarkably original and provocative analysis. This is a bold book, providing rich comparative insight.

Leonid Grinin, Ph.D., Senior Research Professor at the Institute for Oriental Studies of the Russian Academy of Sciences, Vice-President of the International Kondratieff Foundation, Deputy Director of the Eurasian Center for Big History & System Forecasting, and author of *"Social Macroevolution. Genesis and Transformations of the World System"* (Moscow: Russian Academy of Sciences) and *"Macrohistory and Globalization"* (Moscow: Moscow State University).

This work is one of the most remarkable and important studies in cross-national analysis in recent years. The new perspective on the evolutionary potentials of Islamic countries offered by the authors is really stimulating. This work will be of great importance for everyone interested in structural patterns of contemporary global development.

Andrey Korotayev, Ph.D., Head of the Laboratory for Monitoring of Sociopolitical Destabilization Risks of the National Research University, Higher School of Economics, Moscow; Senior Research Professor at the Institute for African Studies of the Russian Academy of Sciences, and author of *"Ancient Yemen"* (Oxford: Oxford University Press) and *"Introduction to Social Macrodynamics"* (Moscow: Russian Academy of Sciences).

This book is a very necessary reconsideration of the entire question of global values, so well-known in the social and economic sciences ever since the path-breaking publications by Hofsteede, Schwartz and Inglehart. Tausch, Heshmati and Karoui challenge hitherto established interpretations by radically introducing the perspective of the shadow economy and the role of family values for the stability of capitalist development. Their optimistic analysis on the long-term perspectives of the Muslim world is based on an analysis of indicators of economic freedom.

Prof. Dr. Dr.h.c.mult. Friedrich Georg Schneider, Department of Economics, Institute of Economic Policy, Johannes Kepler University of Linz, Austria, author of: The Shadow Economy (together with Colin C. Williams), London (UK): The Institute of Economic Affairs, 2013; The Shadow Economy: Theoretical Approaches , Empirical Studies, and Political Implications, together with Dominik Enste, Cambridge (UK):

Cambridge University Press, second edition 2013; Handbook on the Shadow Economy, Edward Elgar Publishing Company, Cheltenham, 2011, and over 200 articles in leading peer-reviewed journals of economics.

BIOGRAPHY OF AUTHORS

Arno Tausch is an Austrian citizen and in his academic functions he is Adjunct Professor (Universitaetsdozent) of Political Science at Innsbruck University, Department of Political Science. Currently, he is also Associate Professor of Economics, Corvinus University, Budapest, and Lecturer of International Development, Vienna University. He served as an Austrian diplomat abroad and was Counselor for Labor and Migration at the Austrian Embassy in Warsaw. His research program is focused on world systems studies, development and dependency studies, European studies, and quantitative world values and peace research. He authored or co-authored 82 articles in peer-reviewed journals and 16 books in English, 2 in French, 8 books in German. His works were published, re-published or are forthcoming in 32 countries. His publications include a number of essays for leading economic and foreign policy think tanks in nine countries, and for the Jean Monnet Institutes of the European Union in three European Union countries.

Almas Heshmati, a Swedish citizen, is currently Professor of Economics at Jönköping University, Sweden, and Sogang University, Soutth Korea. He held similar positions at the Korea University, Seoul National University, and University of Kurdistan Hawler. He was Research Fellow at the World Institute for Development Economics Research (WIDER), during 2001-2004. From 1998 until 2001, he was an Associate Professor at the Stockholm School of Economics. He has a Ph.D. degree from the University of Gothenburg 1994. His research interests include applied microeconomics, globalization, development strategy, efficiency, productivity and growth with application to manufacturing and services. In addition to more than 120 scientific journal articles he has published books on the EU Lisbon Process, Global Inequality, East Asian Manufacturing, Chinese Economy, Technology Transfer, Information Technology, Water Resources, Landmines, Power Generation, Development Economics and Economic Growth. He is member of the Scientific Committee of the International Conference on Panel Data.

Hichem Karoui, a French citizen, is in his academic functions a Scholar, and Writer. Research area and interests: International Relations, Regional studies, and Middle East North Africa (MENA). He holds a Ph.D. in sociology from the Sorbonne in Paris and three M. A. degrees from the same University (Middle-East/Mediterranean studies, English language, literature and civilization, and Arab language, literature and civilization). From 2011 to 2013, he was Research fellow at the *Arab Center for Research and Policy Studies (ACRPS)* in Doha, Qatar, where he served as Coordinator of the Political Unit and academic supervisor of the website. He published widely in English, French and Arabic since the 1980s and was a frequent commentator for leading newspapers in the Arab world:

Books in French: *L'après-Saddam en Irak, les plans, les hommes et les problèmes (Post Saddam Iraq, Plans, Men and Problems):* Paris: L'Harmattan, 2005; *Où va l'Arabie saoudite? (Where Is Saudi Arabia Heading?)* Paris, L'Harmattan; 2006; *Les Musulmans... Un cauchemar ou une force pour l'Europe? (en collaboration avec Arno Tausch).* Paris: L'Harmattan. 2011;

Books in English: *From 9/11 to the Arab Spring: U. S. - Saudi Love-Hate Story.* Middle East Studies/CreateSpace. South Carolina. 2013; *Power Revolving Doors: The Shaping of American Perception of Middle East Studies.* Middle East Studies/CreateSpace. South Carolina. 2013; *Arab Spring: The New Middle East in the Making (Essays).* Middle East Studies/CreateSpace. South Carolina. 2012; *The Bush II Years in the Middle East (2000-2008): A Case Study in the Sociology of International Relations.* Middle East Studies/CreateSpace. South Carolina. 2012.

Books in In Arabic: US Think Tanks and Middle East Studies, *Nama Center, Ryadh, 2014. International Balance, From The Cold War to The Detente. Essay: La Maison arabe du livre. Tunis, 1985; Prolegomena For a Critique of the Arab political Reality.* Maison al nawras. Tunis 1988; *The Seven Pillars of Madness (a novel),* Maison arabe du livre. Tunis. 1984; *Nun (a novel),* Déméter. Tunis 1983.

EXECUTIVE SUMMARY

WHAT WE AIMED TO ACHIEVE

This book is intended to provide a new approach to the study of global values and global value change, based on representative international survey data, above all, the *World Values Survey,* covering the entire population in an ever growing number of countries, now comprising some 90% of the total global population.

Many *World Values Survey* studies were based on the *World Values Survey* waves 1-4, which analyzed global opinion from the early 1980s to 2006. Meanwhile, new and even more complete data have emerged. By and large, our analyses rely especially on the "roll-outs" of the *World Values Survey* data wvs1981_2008_v20090914.sav.

The importance of these data for international politics cannot be overestimated: foreign ministries, international organizations, ministerial planning departments of national governments, national intelligence agencies, military staff colleges, international bankers and investors, global television channels, pension fund managers, global insurance enterprises, organizations of national and international security, NGOs, religious communities, they all can benefit from these freely available data, which indeed will revolutionize our discourse on international politics and political culture.

As we will demonstrate in this book, comparative value studies in the field of international business studies and internatzional business ethics (the research efforts by Hofstede, the subsequent GLOBE project, and Schwartz) basically started by comparing global cultures, but used limited samples across nations – both in terms of the social inclusiveness of the strata suveyed and in terms of the number of persons interviewed. Although we also find in the research programs inspired by Hofstede and Schwartz attempts to use additional country-level data from the *World Values Survey* or the *European Social Survey*, we feel that the time has come to re-analyze the entire question of the underlying dimensions of the *World Values Survey* data.

However much we appreciate the enormous scientific contribution by Professor Ronald Inglehart, who initiated the international data collection of the *World Values Survey*, our re-analysis of the very *World Values Survey* data brought us to question Inglehart's theories, with which he and his associates interpret the mass of the data. Their theoretical approach does not use a sufficiently number of hard-core indicators how global publics view central issues of economic policy and their theories overemphasize a secularistic view of the religious phenomenon in modern society. Their theories predict the gradual waning of the religious phenomena in parallel with the increase of human security, and even cherish at times the tendencies brought about by such a waning of the religious element in advanced democracies. Inglehart spells them out: higher levels of tolerance for abortion, divorce, homosexuality; the erosion of parental authority, the decrease of the importance of family life et cetera. Is that really something to cherish?

Societal and economic development is discontinuous; regional centers of the world economy shift at an enormous speed; and above all, religion and family values can be an important assett in the stability of capitalist development. Economic growth inexorably shifts away from the North Atlantic arena towards new centers of gravitation of the world economy. Alberto Alesina's and Paola Giuliano's new maps of global values (Alesina and Giuliano, 2013) present a real break with the hitherto existing secularistic consensus of global value research. Their maps of family ties, respect for parents *et cetera.* [2] coincide with the global map of economic growth today.

[2] Pages 45 ff in Alesina and Giuliano, 2013, available at http://scholar.harvard.edu/alesina/publications and World Bank data, available at http://data.worldbank.org/ indicator/NY.GDP.MKTP.KD.ZG/ countries? display=map We recommend to our readers to carefully look at Alesina's and Giuliano's maps, one by one, and then to compare these maps with the freely available World Bank maps of global economic growth. The end result will always be the same: economic growth in the world shifts to regions, where *"capitalist family values"* are strong:

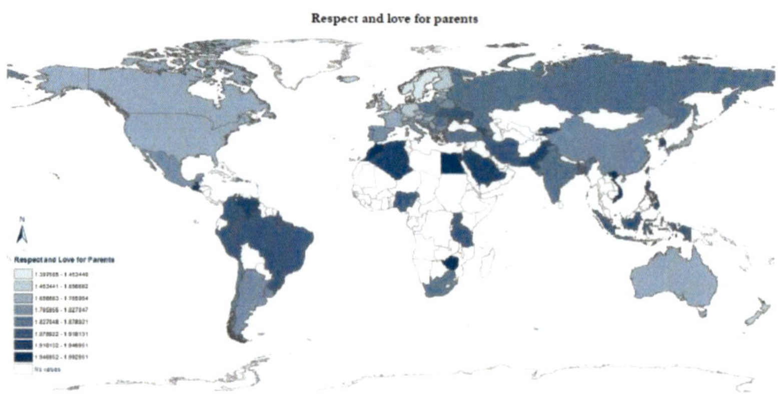

Economic growth, 2009-2013, in % per year.

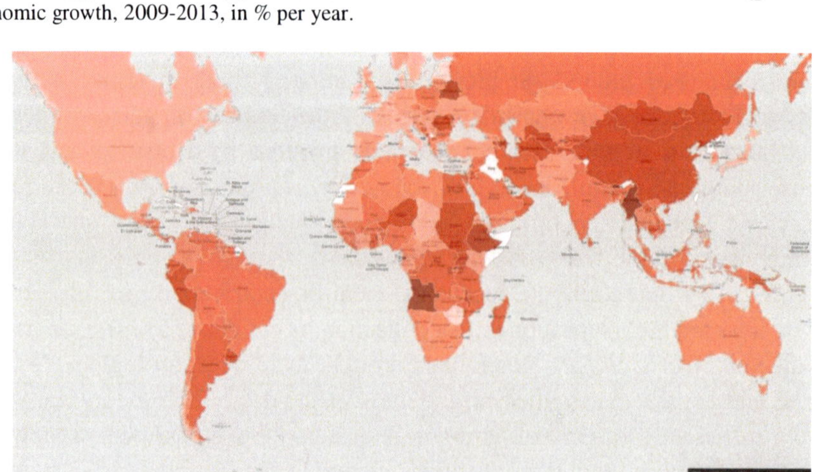

In doing so, we might add that we are researching in good company. Leading representatives of the global economics profession now start to take up the challenge to interpret the mass of the data from the *World Values Survey* project on their own. The essay by Barro and McCleary (2003) was an important beginning and a good example of how today economic research uses data from the *World Values Survey* project to study the relationship between religion, denominations and economic growth.

Alesina (2013); Alesina and Angeletos (2005); Alesina and Fuchs-Schündeln (2007); Alesina and Guiliano (2010, 2011, 2013); Alesina, Cozzi and Mantovan (2012); and Alesina, di Tella, and MacCulloch (2004) all show how the economic discipline can gain hard-core, quantitative and valuable insights from comparative knowledge about such phenomena as generalized trust and social capital, individualism, family ties, morality, attitudes toward work and perception of poverty, and religious practice for economic processes.

In our book, we will attempt to define "cultures" on a global scale largely following Alesina and Guiliano (2013). Although some of our preferred *World Values Survey* indicators are different from those used by Alesina and Guiliano (2013), there is sufficient resemblance between the two approaches, and also there is a high correspondence between their choropleth geographical maps of global values and our own maps.

Leaving behind the omnipresent logic of the confrontation between traditional vs. secular-rational values and survival vs. self-expression values, which is so common for the Inglehart paradigm of global values, opens the way to consider such phenomena as the shadow economy.

We are above all interested in such phenomena as attitudes on competition and free markets, on social expenditures, and on bribery and corruption. Without question, bribery and corruption are one of the main challenges for international business studies nowadays.

From such diverse economic theories as Alesina, Barro and Schumpeter, we re-discover the importance of the data on generalized trust and social capital, family ties, morality, attitudes toward work and religious practice. Democratic and liberal values can correspond to a civilization, characterized by an enlightened religion, in the West and in the Muslim world alike.

In many ways, our investigation puts the large secularistic scientific consensus on the issues under consideration on its head.

We show that the world economic rise of the global South, among them the BRICS countries and the countries of the Arab Gulf, is no coincidence: economic growth in the post-crisis period from 2008 onwards is highly and positively correlated with family values.

All too often, the loss of religion and the rise of the shadow economy go hand in hand, including in leading Western countries. The decay of family values, which are so deeply enshrined in the religious commandments of Judaism, Christianity and Islam, and which are also basic to the other global religious civilizations, goes hand in hand with the decay of economic and social values.

In our study, we used the following variables to arrive at our factor analytical models.

Analysis contained in … of our book	World Values Survey variable
Chapter 5, 6	Age
Chapter 5, 12	Competition good or harmful (harmful)
Chapter 5	Confidence: Armed Forces
Chapter 5	Confidence: The Press
Chapter 6	Education level (recoded)
Chapter 5, 6	Highest educational level attained
Chapter 5, 6, 12	How important is God in your life
Chapter 5, 6, 12	How often do you attend religious services (never)
Chapter 5	Immigrant policy (prevent people from coming)
Chapter 5, 12	Important child qualities: determination and perseverance
Chapter 5, 12	Important child qualities: feeling of responsibility
Chapter 5, 12	Important child qualities: hard work
Chapter 5, 12	Important child qualities: imagination
Chapter 5, 12	Important child qualities: independence
Chapter 5, 12	Important child qualities: obedience
Chapter 5, 6, 12	Important child qualities: religious faith
Chapter 5, 12	Important child qualities: thrift, and saving money and things
Chapter 5, 12	Important child qualities: tolerance and respect for other people
Chapter 5, 12	Important child qualities: unselfishness
Chapter 12	Income equality (large differences needed)
Chapter 6	Jobs scarce: Men should have more right to a job than women (reject)
Chapter 12	Justifiable: abortion
Chapter 5, 12	Justifiable: avoiding a fare on public transport
Chapter 5, 12	Justifiable: cheating on taxes
Chapter 5, 12	Justifiable: claiming government benefits even if one is not entitled to them
Chapter 12	Justifiable: divorce
Chapter 12	Justifiable: euthanasia
Chapter 12	Justifiable: homosexuality
Chapter 12	Justifiable: prostitution
Chapter 5, 12	Justifiable: someone accepting a bribe
Chapter 12	Justifiable: suicide
Chapter 5	Most people can be trusted [highest numerical value: you just can't be too careful])
Chapter 5	Neighbors: Immigrants/foreign workers
Chapter 5	Neighbors: People of a different race
Chapter 5, 12	Political system: Having a democratic political system (reject)

Analysis contained in ... of our book	World Values Survey variable
Chapter 5	Political system: Having the army rule (very bad)
Chapter 5	Private vs. state ownership of business (state)
Chapter 12	Satisfaction with your life
Chapter 5, 12	Self-positioning in political scale (right wing)
Chapter 5, 6	Sex (Gender) [in multivariate analysis: female] (1=male; 2=female)
Chapter 5, 6	University is more important for a boy than for a girl
Chapter 6	Woman as a single parent

Already the great Harvard economist Joseph Alois Schumpeter (8 February 1883 – 8 January 1950) put the decline of family values at the center of his theory about the decline of the capitalist order. Today, Barro goes even further: religion does affect economic outcomes mainly by fostering religious beliefs that influence individual traits such as thrift, work ethic, and honesty. Barro's perspective is largely confirmed in this study.

For Inglehart, phenomena as bribery, corruption, tax evasion, cheating the state to get government benefits for which one wouldn't be entitled practically even do not exist, while the rich data base of the *World Values Survey* itself provides ample evidence about these phenomena. Starting with Schumpeter and his hypothesis about the waning of family values and the capitalist family enterprise as the basis of the crisis of capitalism, we re-discover the positive contribution of religion for society instead of cherishing its decline in the name of "self-expression".

Our book also develops a new perspective on the development potentials of Muslim countries in the light of our new theory. Erich Weede, a well-known follower of Economics Nobel Laureate Friedrich August von Hayek, already observed in 2006 that Islam contains a real chance of promoting capitalism and economic freedom which other religions lack. In principle, Islam provides its own justification for constraining rulers and for limited government (Weede, 2006). Economic Freedom data of the Heritage Foundation are used to show just how important the advance of economic freedom in the Muslim world and the Arab world has become. In particular, our book is optimistic for those countries, which in a way heeded Hayek's lessons in the region, above all the countries of the Arab Gulf.

While our book is fairly optimistic about the long-term economic and poltical tendencies for what we call here "the Indian Ocean arena" and the Muslim world in general, we show in the the concluding Chapter 12 of this book with the latest data from the *World Values Survey*, wave 6, for the years 2010-2014 that today publics in Egypt, Uzbekistan, Kazakhstan, Malaysia, Turkey, Morocco, Yemen, and Tunisia give a higher average importance to democracy than the alleged culture of self-expression in the United States of America, and that the publics in Qatar and Jordan give a higher average importance to democracy than the alleged culture of self-expression in Japan and several European Union member countries. Welcome to the 21st Century!

1. BACKGROUND

So, today, we see a substantial rise in the economic power not only of the so-called BRICS countries (Brazil, Russia, India, China, South Africa), especially China and India, but also of some Arab states and other emerging Muslim economic powers, among them Indonesia and Turkey. Our book is conscious about these important tectonic shifts in the world economy, but it tries to relate them to the loss of many values, important for the good functioning of democracies and market economies in the West.

Of special importance is the legacy of the classical Arab sociologist and historian Ibn Khaldūn (1332-1406), who explained in his most important book *Al-Muqaddimah ("Introduction to history")* that historical change and the succession of dynasties are a function of the interactions between nomadic culture and urban civilization. The core of his sociology is found in the concept of *"Asabiyya"* (social solidarity, or group feeling). Ibn Khaldūn was the first to propose a model of value change within the time span of four generations (i.e., for Khaldūn, 160 years). The first generation retains the desert qualities, desert toughness, and desert savagery. Under the influence of royal authority and a life of ease, the second generation changes from the desert attitude to sedentary culture, from privation to luxury and plenty. Thus, the vigor of group feeling is broken to some extent. People become used to lowliness and obedience. The third generation, then, has (completely) forgotten the period of desert life and toughness, as if it had never existed. Luxury reaches its peak among them, because they are so much given to a life of prosperity and ease. Group feeling disappears completely. When someone comes and demands something from them, they cannot repel this person. The ruler, then, has the need of other, brave people for his/her support. He/she takes many clients and followers. They help the dynasty to some degree, until God permits it to be destroyed, and it goes with everything it stands for (…) In the course of these three generations, the dynasty grows senile and is worn out. Therefore, it is in the fourth generation that (ancestral) prestige is destroyed.

Barro is decidedly speaking in favor of the importance of religion for sound economic growth and long-run economic well-being. Instead of viewing 'Islam' or 'religious beliefs' as an 'impediment' of economic growth, he tends to see them as requirements of a resilient society today. Barro thinks that there are causal influences from religion to economic growth, rather than the reverse. His conjecture is that higher religious beliefs stimulate growth because they help to sustain aspects of individual behavior that enhance productivity. Respect of parents is related in a clear-cut positive manner to economic growth, as well as the belief in hell.

2. THE METHODOLOGY OF THIS PROJECT

This Chapter informs our readers about the methodologies used in this book.

Modern social science research increasingly uses the kind of indices, first designed by the UNDP Human Development Programme. While their use in socio-economic development accounting is already quite common, their use in cross-national value research has been rather limited, and yet there are great potentials for such a type of research. Future research could concentrate for example on such dimensions as avoiding economic permissiveness, avoiding racism and the authoritarian character, and support for the market economy and democracy.

In contrast to the research which still characterizes the reflexion on *World Values Survey* data in current social sciences, we think that the time has come to use techniques which properly allow for stronger relations between the "factors" underlying the correlations between the variables. We think that the Promax Factor Analysis is such an advanced analytical technique, and we use it throughout this book. Global value research would also be unthinkable today without factor analytical index construction. Our research results and the statistical foundations presented here are an invitation for decision makers and researchers to start for themselves developing indices based on the freely available *World Values Survey* data. All they need are competent statisticians, modern statistical software and the firm intention and will to develop projects facilitating the path towards a more mature and encompassing democracy – everywhere around the globe.

The choice of the factor analytical method used to reduce the number of variables of the *World Values Survey* project to its unerlying dimensions is not just a matter for the specialist but it also has many different consequences. Inglehart and most other researchers rely on standard linear factor analysis, which is basically a statistical methodology already developed before the Second World War.

3. The Dark Side of Economics: Why It Is Good from an Economist's Perspective That Sociology Starts to Analyse *World Values Survey* Data on the Shadow Economy and Why Corruption Is a Development Impediment

It is surprising that the massive items available from the *World Values Survey* data base on the shadow economy and core economic values have not yet been hitherto used to a sufficient degree in the dominant theory construction of empirical social science.

Shadow economic activity has been on the rise and causing violations of laws and regulations, lower tax revenue collections, statistical discrepancies, inequality, corruption and public budget deficits and public debt problems for the state and its organizations. The rise of the shadow economy around the world is attributed to the stronger presence of government activity, the increase in tax rates, and the desire to escape taxes and regulatory restrictions.

We have reviewed a large number of recent studies investigating the theoretical and empirical aspects of the shadow economy, in particular the measurement of the shadow economy and its causality with economic growth. In the process, emphasis is made on issues such as corruption, economic freedom, state and institutions capacity and quality, as well as morale and trust. Many of these studies use large data sets allowing for cross country and country groups comparisons.

A negative relationship is found between shadow economic activity and the level of development, democracy and press freedom. Countries with abundant natural resources for export are facing a larger degree of shadow economic activity, corruption, inefficiency, inequality and lower economic growth.

Schneider (2005) considers shadow economic activities a fact of life. Most societies attempt to reduce its magnitude by controlling activities through legal measures such as punishment and persecution or by preventive measures with investment in welfare and education. Despite significant investment in the collection of data on shadow economic activities, it is rather difficult to obtain accurate information about its nature and magnitude. Schneider mentions the existence of a comprehensive literature on particular aspects of the shadow economy, but the subject remains controversial. Furthermore, there is disagreement among researchers about the definition, estimation procedures and their use in economic analysis and policy making.

In a common approach, Schneider (2005) defines the shadow economy to include all market-based legal production of goods and services that are deliberately concealed from public authorities for the following reasons: (i) to avoid payment of income, value added or other taxes, (ii) to avoid payment of social security contributions, (iii) to avoid having to meet certain legal labor market standards, and (iv) to avoid complying with certain administrative procedures. However, this definition does not include economic activities that are illegal and fit the characteristics of classical crime, as well as the informal household economy or tax evasion.

Schneider (2005), in the context of industrialized and transition economies, mentions that the shadow economy is expected to influence the tax system and its structure, the efficiency of resource allocation between sectors, and the official economy in a dynamic sense.

Schneider (2005) concludes that for all countries investigated, the shadow economy as share of GDP has reached a remarkably large size (Africa 33.9-41.2; Americas 34.2-41.5; Asia 20.9-26.3; Transition countries 31.5-37.9 and highly developed OECD countries 13.2-16.8). The average percentage shares of GDP in all cases are increasing over time. The author demonstrates empirically a strong interaction of the shadow economy with government policies and with the official economy. He draws three further conclusions. First, an increasing burden of taxation and social security payments, combined with rising state regulatory activities, are the major driving forces underlying the size and growth of the shadow economy. Second, the shadow economy has a statistically significant and quantitatively important influence on the growth of the official economy. Increases in the shadow economy have a negative effect on the official growth in a developing country, but a positive effect in the developed industrialized and transition countries. People engage in shadow economic activity because of government actions, most notably high levels of taxation and regulation.

Although *World Values Survey* (WVS) data are used by the economics profession in their attempt to estimate the drivers of the shadow economy, contemporary sociological WVS research has failed hitherto to integrate the shadow economy into general theories of values and value change.

The data base of the *World Values Survey* indeed contains very precise items in connection with a neo-liberal interpretation of economic processes, like the acceptancy or rejection of free competition, or various items on the acceptability or desirability of state intervention. In contrast to hitherto published research, we try to integrate these elements into the core-model of global value change. For this good reason, we present here an extensive survey how contemporary economic theory already integrates economic freedom in its empirical and theoretical research. Our survey is intended to be an overview especially for readers who are more familiar with other disciplines of the social sciences and not economics. In the presentation, we use a non-mathematical language.

Muslim societies sometimes already perform very well on indices of economic freedom, irrespective of their development level. As definded by the well-known think tank *"Heritage Foundation"* (http://www.heritage.org/index/book/chapter-5) economic freedom is the condition in which individuals can act with maximum autonomy and minimum obstruction in the pursuit of their economic livelihood and prosperity. As Hayek observed, to be controlled in our economic pursuits would mean to be controlled in everything.

For the Heritage Foundation, a comprehensive view of economic freedom encompasses all liberties and rights of production, distribution, or consumption of goods and services.

The Heritage Foundation Index of Economic Freedom takes a broad and comprehensive view of country performance, measuring 10 separate areas of economic freedom. Each economic freedom is individually scored on a scale of 0 to 100. A country's overall economic freedom score is a simple average of its scores on the 10 individual freedoms. The 10 economic freedoms are grouped into four broad categories:

- Rule of law (property rights, freedom from corruption);
- Government size (fiscal freedom, government spending);
- Regulatory efficiency (business freedom, labor freedom, monetary freedom); and
- Market openness (trade freedom, investment freedom, financial freedom).

The 1991 collapse of the Soviet Union led to creation of 15 new states and a transition from centrally planned economies to market economies for 25 states for the period from 1998 to 2005. These countries experienced heterogeneous growth and development. Pääkkönen (2010) reviewed the political economy of economic growth in the post-communist economies transition to free markets. The focus is on the role of economic policy and institutions in the transition process. The author tested the hypothesis that better institutions, measured in terms of economic freedom, contributed to growth. The empirical results confirm this hypothesis. Increased government consumption has a negative effect on growth suggesting the presence of wasteful spending and hindrance to growth.

Economic research provides very important insights into the conditions of *"effective democracy"*, which cannot depend – as contemporary *World Values Survey* research often contends – on "self expression" values alone. In accordance with *World Values Survey* research, *"effective democracy"* is an indicator, which combines performance in terms of what Western scholarship understands by *'human rights'* and the *rule of law* (concept of the World Bank).

Also, all the dimensions, which were shown to be relevant in our own empirical value research (see Chapter 5) already surfaced in prior economic research, like economic permissiveness, which most strongly affects large parts of Latin America (except Venezuela, Peru, and the countries of the Southern Cone), some but not all nations of Africa, most of the former USSR and some other countries of East Central and Southeastern Europe, Thailand and the Philippines, as one of the main stumbling blocks against effective democracy.

Corruption is widespread in particular among developing countries. A precondition for achieving growth and reducing poverty is to deal with corruption in an effective way. This view is supported by a growing literature, which we review in this Chapter. Potentially effective instruments in rooting out corruption are democratic elections and press freedom.

4. The Sociological and Psychological Theories of Global Values

The major existing sociological theories about global values do not focus as yet sufficiently on these phenomena: Hofstede, Schwartz/Davidov and Inglehart, the three major theoretical approaches guiding the discipline of value research today.

First we mention Geert Hofstede. According to Hofstede, who investigated above all the value systems of the global employees of a particular multinational U.S. company (IBM), there are four to six basic clusters of international value systems, and they are all defined along the scales of how different national societies handle ways of coping with inequality, ways of coping with uncertainty, the relationship of the individual with her or his primary group, and the emotional implications of having been born as a girl or as a boy.

Hofstede's power distance scale is generally lowest in Scandinavian and in the Anglo-Saxon democracies, and also in the social welfare state Austria, and it is, generally speaking, highest in the ex-USSR, in India and South-East Asia and in China, in Mexico and Venezuela and in some countries of Central and Eastern Europe.

The most collectivistic countries are to be found in the Andean region of Latin America, in China and in countries of South and South-East Asia, while the highest individualism is to be found in the United States and in the Anglo-Saxon democracies, with high scores also to be found in the Netherlands, Denmark, and also in Italy and in Hungary.

The highest masculinity scores are to be found in some countries of Central and Eastern Europe, in Japan and in Venezuela, while the countries of Scandinavia and the Baltic region, the Netherlands and, interestingly enough, Chile and Portugal are characterized by high femininity as well.

Uncertainty avoidance is highest in Greece, Portugal and Uruguay, and is generally highest in Roman Catholic and Orthodox cultures, and also in Japan and South Korea, while it is lowest in Protestant cultures, but also in China and in Iran, in South Asia and in South-East Asia.

Hofstede's long-term orientation (LTO) is highest in South Korea, Japan, China, in Russia and the Baltic Republics, and in Germany and some other European countries, while it is especially absent in Latin America, in the Anglo-Saxon democracies and Scandinavia, and in Morocco and Iran; with most of the other European countries and the countries of South and South-East Asia classified in the same ranks.

Hofstede's Indulgence Factor is highest in Venezuela, in countries of the Caribbean and in Mexico, and, interestingly enough, also in Sweden, New Zealand, Australia, the UK, the US and Canada and in some European countries (Denmark, Netherlands, Switzerland and Austria) and in Chile, while Restraint is highest in Pakistan, in most former communist countries, and in China. Most continental European countries seem to be characterized by low indulgence and higher restraint.

Just as Hofstede, the works of the Israeli psychologist Shalom Schwartz received an enormous international reception, especially in the expanding field of international business studies. Schwartz, 2006a, 2009 highlights a famous map of global values. But there is a problem of limited country samples. Schwartz began his investigations not on generalized surveys of the total population, but on relatively small samples of global schoolteachers and college students. Most samples included only between 180 and 280 respondents. Schwartz identifies seven basic cultural orientations and the structure of interrelations among them: West European, English-speaking, Latin American, East European, and South Asian, Confucian influenced, and African and Middle Eastern.

His seven country-level cultural dimensions are

1. Embeddedness
2. Hierarchy
3. Mastery
4. Affective Autonomy
5. Intellectual Autonomy
6. Egalitarianism
7. Harmony

Correlations between the sample means were used in a multidimensional scaling analysis to assess the presence of seven cultural orientations and the relations among them. Muslim societies rank very high on values of embeddedness. Embeddedness combines such values as social order, respect tradition, forgiving, moderateness, obedience, politeness, cleanness, national security, devoutness, wisdom, self-discipline, family security, honoring elders, reciprocation of favors, protecting the public image. In cultures with an emphasis on embeddedness, people are viewed as entities embedded in the collectivity. Meaning in life comes largely through social relationships, through identifying with the group, participating in its shared way of life, and striving toward its shared goals. *"Embedded cultures emphasize maintaining the status quo and restraining actions that might disrupt in-group solidarity or the traditional order. Important values in such cultures are social order, respect for tradition, security, obedience, and wisdom." (Schwartz, 2006a)*

A careful reading of the Schwartz categories and our projections of his scales on real and concrete choropleth maps would suggest that there is a considerable discrepancy of his research results with the interpretation of Muslim civilization by Huntington. The country results for embeddedness, hierarchy and egalitarianism still fit in with the Huntingtonian predictions for the Islamic countries, but the country results for mastery, affective autonomy, intellectual autonomy and longing for harmony suggest a completely different picture of Muslim countries: in contrast to what one would have to expect from Huntington's works, key Muslim countries are even at the forefront of the values of affective autonomy, intellectual autonomy, and the quest for harmony.

Eldad Davidov pays great attention to the statistical-mathematical foundations of comparative value analysis, including the detection of contradictory statements in survey research to reduce the measurement error. He further developed the Schwartz approach. Among his major works one finds Davidov, 2008, 2010; Davidov, Meuleman, Billiet, and Schmidt, 2008; Davidov, Schmidt and Billiet, 2011; and Davidov, Schmidt, and Schwartz, 2008, where Davidov and associates base their analysis on a cascade of human values.

In his 2008 study on the contradictions of the European integration process, Davidov also highlights the close interrelationship between gender prejudice, homophobia, anti-

Semitism, xenophobia, social out-groups (as for example homeless people), Islamophobia and the overall devaluation of newcomers.

Inglehart, in some of his main publications, developed an interpretation of global value change (Inglehart, 1970, 1977, 1990, 1997, 2000), which rests on a well-known two-dimensional scale of global values and global value change. It is based on the statistical technique of factor analysis of up to over twenty key *World Values Survey* variables from the originally more than 900 WVS survey items. These 900 items cover practically all major areas of human concern, from religion to politics, and from economic to social life.

The two Inglehart dimensions are: (1) the Traditional/Secular-Rational dimension and (2) the Survival/Self-expression dimension. In a factor analysis of ten indicators, these two dimensions explain more than 70 percent of the cross-national variance. Each of these dimensions is strongly correlated with scores of other important variables. For Inglehart and Baker, 2000, all of the preindustrial societies show relatively low levels of tolerance for abortion, divorce, and homosexuality; and tend to emphasize male dominance in economic and political life. There is a deference to parental authority, and the importance of family life is of primary concern, and these societies are relatively authoritarian. Most of them place strong emphasis on religion. Advanced industrial societies tend to have the opposite characteristics.

When survival is uncertain, cultural diversity seems threatening. When there isn't "enough to go around," foreigners are seen as dangerous outsiders who may take away one's sustenance. People cling to traditional gender roles and sexual norms, and emphasize absolute rules and family norms in an attempt to maximize predictability in an uncertain world. Conversely, when survival begins to be taken for granted, ethnic and cultural diversity become increasingly acceptable. Beyond a certain point, diversity is not only tolerated, it may be positively valued because it is interesting and stimulating. In advanced industrial societies, people seek out foreign restaurants to taste new cuisine; they pay large sums of money and travel long distances to experience exotic cultures. Changing gender roles and sexual norms no longer seem threatening.

Tanzania, Puerto Rico and Jordan are the least secular countries of the world, while Sweden, Japan and the Czech Republic are the most secular countries of the world.

New Zealand, Australia and the United States are the most self-expression oriented countries of the world, while all of the five most survival oriented and least self-expression oriented countries of the world were of Orthodox Christian cultural heritage: Moldova, Ukraine, Russia, Belarus and Romania.

We also further take the results of Inglehart's calculations at their face value and draw a scatterplot of our own. Since secularization could be regarded as the driver of societal processes and self-expression as the final result, we put secularization on the x-axis and self-expression on the y-axis.

As had to be expected from standard principal components analysis, the bi-variate correlation between the two factors is indeed very small. There is an interesting wave structure in the relationship between secularism and self-expression. With rising secularization, first there is a certain implosion and then a rise in self-expression values. Inglehart considers these self-expression values to be very important for *"effective democracy"*. There is then a rise of self-expression values, in order to implode again. *Only at very high levels of secularization, self-expression rises in a linear fashion.* This wave-structure of modernization, secularization and self-expression is an important qualification of Inglehart's theory.

We discover that the countries of the former USSR and the countries of the Christian Orthodox cultural tradition in general, and not the Muslim countries, present the real problematic cases for *"effective democracy"*: they combine high secularization with low self-expression. While Egypt and its neighbor Israel present residuals of about the same size, most Muslim countries (with the exception of post-Soviet Kyrgyzstan and Azerbaijan) are even ahead of the three Baltic EU-member states Lithuania, Latvia and Estonia. The EU-member country Poland is ranked about equally with Turkey and is even behind several other Muslim countries. Thus prospects for Muslim countries in the strict framework of Inglehart's sociology must be considered as relatively good.

5. TOWARDS A NEW POLITICAL GEOGRAPHY OF HUMAN VALUES

In this Chapter, we present the results of our re-analysis of the entire openly available global *World Values Survey* data base. We use the advanced statistical multivariate analysis technique of the Promax factor analysis, which allows for correlations between factors. It is available to the global public via the IBM-SPSS statistical package XXI. We eliminated missing values by listwise delition. There were 92289 interview partners from around the globe with complete data for all the 30 variables of our research design. We worked with the very best documented *World Values Survey* items.

Our model explains 47.89% of the total variance of all the 30 variables. In Table 5.5 of this Chapter, we highlight the relationships between the original 30 variables and the newly derived factor analytical dimensions:

a) economic permissiveness
b) traditional religion
c) racism
d) higher education for the younger generation (education gap between the generations)
e) distrust of the army and the press
f) authoritarian character
g) tolerance and respect
h) the 'ego' company (i.e., the rejection of obedience and unselfishness as values in education)
i) [predominantly] female rejection of the market economy and democracy

We also look at the trajectory of global society by analyzing the factor scores along the path of the Human Development Indicator of the UNDP *("human security indicator",* also used by Inglehart and his associates):

- Traditional religion diminishes in its societal role, as human security is rising

- Tolerance and respect, distrust of the army and the press, the “ego-company” orientation, and the [predominantly] female rejection of the market economy and democracy all rise with human security
- There is an inverted U-shaped relationship between human security and permissiveness/pessimism, racism, the education gap between the generations, and the authoritarian character

Economic permissiveness clearly captures the dimension of lawlessness, moral-ethical decay and the shadow economy, so prominent in contemporary economic theory of growth. In statistical terms, it is the most important of all the resulting factors.

Traditional religion is linked in a very complex way to the absence of economic permissiveness. We also look at the exceptional performers (“residuals”) which best avoided economic permissiveness on each stage of secularization.

We also present Chropleth maps of human values across the globe, and show the regional implications of our analysis.

Even if traditional religion is weakly and positively associated with the “authoritarian character”, this association is much weaker than expected and leaves 95% of the variance unexplained. Religious or highly religious societies with a remarkable low or medium authoritarian character are

Ethiopia
Bangladesh
Indonesia
Jordan
Thailand
United States
Mexico
Tanzania
Poland
Puerto Rico
Dominican Republic

Relatively secular societies with more than usual authoritarian tendencies are:

Russian Federation
Belarus
Czech Republic
Latvia
Estonia
Ukraine
Armenia
Bulgaria
Lithuania
Hungary
Serbia and Montenegro
Spain

Taiwan
Viet Nam

"Modernity" (which we interpret throughout this work as the natural logarithm of GDP per capita, ln GDP per capita^2) in global society is characterized by

- less traditional religion and
- more tolerance and respect

and the negative phenomena of

- more distrust in the army and the press and
- an increasingly ego-centric culture
- the female rejection of democracy and the market economy

In global society, the paths to "modernity" are indeed not conflict-free, and, at middle levels of development we generally reach a climax of societal

- racism
- the generation gap in education
- economic permissiveness
- the authoritarian character

In 2011, the Arab States reached a Human Development Index of 0.641 on average. In the framework of the general tendencies to be deduced from our theory our analysis would predict that the Arab nations are indeed currently undergoing a critical phase in their development. In the light of our analysis, we would expect a rather optimistic scenario: from the level of UNDP HDI = 0.600 onwards, permissiveness will also decrease in Muslim countries. The weight of religious traditionalism will decline. Racism already reached its climax and will continue to decline. The education gap between the generations will decline dramatically. Tolerance and respect will increase. Egoism will decrease. The trajectory of the authoritarian character and the [predominantly] female rejection of the market economy and the press will not increase sharply. Decision makers and democratic civil society will be confronted however with one important negative phenomenon: from the level of UNDP HDI = 0.600 onwards, distrust of the army and the press will increase with rising human development.

Among the results which have a direct consequence for the explanation of different economic growth rates since the global economic crisis since 2008, we find our factor "authoritarian character". It emerges that with important other statistical predictors of economic growth or social development being constant, the "authoritarian character" is not conducive to economic growth. Our concept of the "authoritarian character" uses some elements from the classical sociological work by Adorno et al. (1950).

We define the authoritarian character by the following five factor loadings equal or above the absolute value of .30:

Most people can be trusted (you can't be too careful)	+0,562
Important child qualities: hard work	+0,437
Important child qualities: obedience	+0,300
Important child qualities: independence	-0,508
Important child qualities: imagination	-0,613

Based on stepwise multiple regressions and the freely available "Corvinus University" macroquantitative data set for world systems analysis (http://economics.uni-corvinus.hu/index.php?id=47854), we can show that market size (the population of a given country as a percentage of world population) is an important driver of contemporary growth. The authoritarian character is one of its main stumbling blocks. Our analysis is based on n = 62 countries with complete data.

Our results hold under due consideration of the non-linear tradeoffs between economic development levels and subsequent growth rates ("convergence effects"). We then go on to show the trajectories of Muslim societies and world society by comparison:

a) Economic permissiveness is highest at low levels of Muslim Human Development, while for global society, it reaches a climax at the current average UNDP Human Development Index level of the Arab world. Thus it can be said that development in the Muslim world promises to be more corruption free and shadow economy free than in the rest of the countries of our globe.
b) Traditionalist religion and racism both in the Muslim world and in global society indeed are evolving in an inverted U-shaped pattern along the trajectory of development, but the climax levels at middle development stages are in fact lower in Muslim countries than in global society.
c) The generational education gap is a problem for Muslim societies too, but still the climax levels are higher in global society.
d) The distrust level concerning the army and the press is indeed a problem for global Muslim development. But still, Muslim distrust levels of the Army and the Press are lower than in global society at high levels of the UNDP Human Development Index.
e) Yes indeed, there is also a problem with the authoritarian character, but in global society the climax of the authoritarian character phenomenon is far worse than in Muslim countries.
f) Societal egoism and the female rejection of democracy and the market economy at high levels of Human Development are relatively lower than in mature non-Muslim societies of our globe.
g) What is a problem, indeed, is the future trajectory of tolerance and respect. Muslim societies – in accordance with the prescriptions of the Noble Quran, have to learn to become open towards immigration and multiculturalism; and to forgo – particularly in the Gulf – the old and useless tradition of "sponsorship", and to be more respectful toward human rights (including the rights of foreign workers).

Our global value development index combines law-abiding and social capital, avoiding racism; trust of the army and the press; no authoritarian character; a high degree of tolerance and respect + post-materialism; and a female acceptance of the market economy and democracy.

The weight, given to each factor, corresponds to the Eigen values listed in Graph 5.4 of this work.

Our country results show that the five best ranked countries of our entire globe are all western democracies with a solid historical anchoring of their societies in the traditions of the Enlightenment – Switzerland, Norway, Sweden, New Zealand, and Australia. But we already find among the next five countries Canada, the two developing countries Vietnam and Tanzania, and the EU-member countries Italy (predominantly Roman Catholic, with a long history of liberal Catholicism since the Second Vatican Council) and Finland (predominantly Protestant). Our global value development index ranks the predominantly Muslim nation of Morocco twelfth – just behind the United States of America – and still ahead the Latin American democracy Uruguay and the EU-country Germany, to be followed by Bosnia and Indonesia.

While in general terms our analysis is quite optimistic about the civil society foundations for a stable democracy for several Muslim countries, including Morocco, Bosnia, Indonesia, Turkey and Jordan, our analysis is fairly pessimistic for the former communist countries and successor states of the former Soviet Union, predominantly Muslim and non-Muslim alike. Our materials also show that the euro-centric assumptions by European decision makers, who pushed European Union Enlargement ahead of democratic consolidation after the fall of Communism in Eastern Europe, are utterly wrong. Table 5.13 of this study provides readers with a first overview of the highest and lowest placed regions in the current European Union members and the EU-accession countries, ranked again by an average Value Development Index score. Finally, we also show country ranks of the Global Value Development Index.

Once more, it emerges that the political geography of global values in the 21st Century is far away from the imaginary of culturalist scientists and populist politicians alike: even for neighboring countries, culturalist explanation mechanisms utterly fail. Just compare Uruguay and Brazil, both predominantly Roman Catholic; Italy and Hungary, both predominantly Roman Catholic, Morocco and Mali, both predominantly Muslim, or for that matter, Morocco and Spain, just separated by the Straits of Gibraltar; Tanzania and Zambia, two neighboring African countries, and Vietnam and Thailand, two Asian neighboring countries. While Uruguay, Italy, Morocco, Tanzania and Vietnam are real frontrunners in value development, we find that Brazil; Hungary, Mali, Zambia, and Thailand are real laggards in global value development. These phenomena hold independently of the attained development level of a country, measured by the Human Development Index of the UNDP. All of a sudden we discover how exceptional countries like Uruguay, Italy, Morocco, Tanzania and Vietnam really are. Global sociological research would do well to focus on factors which contributed towards their performance.

The idea that global values are often distributed in the nations of the world in a highly regionally contradictory pattern is relatively new in the research literature on the subject. It also emerges that even in highly developed democracies, regional value differences are considerable. Differences between the deeply religious "Bible Belt" in the US South and the relatively secular New England are but one example. Secular Western Turkey quickly catches up with other European regions concerning the "Westernization" of values, while Central Anatolia lags behind. Table 5.13 provides us with a first overview of the highest and lowest placed regions in the current European Union member states and the EU-accession countries, ranked again by the average Value Development Index score.

The history of Communism which began to be implemented in Russia in 1917 destroyed the religious fabric of society and left a hyper-authoritarian society in place. Russia's global value performance percentile performance scores, which might be interpreted as a serious question mark about Russia's future trajectory, are the following (see below). The best value performance of Russia is avoiding the distrust of the army and the press, where Russia is only at a standardized rank 61 among a hypothetical 100 nations. Avoiding economic permissiveness, Russia is only at a standardized rank 78 among a hypothetical 100 nations *et cetera, et cetera.* Regaring traditional religion and avoiding the authoritarian character, Russia has the lowest values on earth.

avoiding the distrust of the army and the press	61%
avoiding economic permissiveness	78%
tolerance and respect + post-materialism	79%
avoiding the rejection of the market economy and democracy	87%
traditional religion	100%
avoiding the authoritarian character	100%

Our materials also support the verdict that the member countries of the European Union and the European Commission should carefully weigh the costs and benefits of further enlargements.

6. The Global Analysis of Feminism and Its Regional Implications for the Muslim World. Voices from Quantitative Social Science and from a Religious Muslim Viewpoint

In this Chapter, we re-analyze the question of Islam and feminism, based on an analysis of all respondents from the *World Values Survey*. The Muslim population covered in our survey comprises representatives of 62.6% of the Muslim population of our globe.

The data were based on the following variables:

- Age
- Education level (recoded)
- Highest educational level attained
- How important is God in your life
- How often do you attend religious services (never?)
- Important child qualities: religious faith
- Jobs scarce: Men should have more right to a job than women (reject)
- Sex (Gender)
- University is more important for a boy than for a girl (reject)
- Acceptancy of woman as a single parent

The respondents (all denominations) comprised n = 173231 representative global citizens in 83 countries and territories. After Promax factor analysis, three factors explained 53.8% of total variance. While the distance to religious practice is explained to some 4% by the education level (correlation between the two factors is 0.192), one can say with certainty that there is no real sharp contradiction between religion and feminism on a global scale. And while gender determines feminist convictions, contained in our analysis to some 40%, it is also evident that feminist convictions are not only held by women, but also increasingly by enlightened men, non-Muslims and Muslims alike. Interestingly enough, our data also show that people supporting typical feminist contentions, like female access to tertiary education and jobs even at a time of crisis (Factor 3), are not necessarily too strongly in support of the acceptancy of women as a single parent (factor loading 0.352, i. e., only 12.39% of variance explained). Single parenthood is a form of household organization very common now in Western countries: the argument is that marriage is an outdated institution *et cetera*. Support for single parenthood by women is rather an expression of the distance towards religion around the globe (factor loading of 0.431, i.e., 18.58% of variance explained).

The correlations between the factors suggest that the often described contradiction between feminism and religion hardly exists. Only 1.12% of the variance of feminism is explained by secularism. This is a long distance from the often militant contentions of the critics of religion around the globe. They expect the incompatibility between any form of religion and feminism.

We also present Choropleth maps of global feminism and analyze the regional implications of our analysis for the Muslim countries in the analysis. For the first time in the global scientific literature, we present data on regional aspects of feminism in such countries as Saudi-Arabia.

To our great surprise we learn that instead of talking about gender and politics in "the West" and in "the Muslim Orient" there is a rather wide range of regional variations within the two civilizations, making underdeveloped regions in the West much more similar to the "average Orient", while at the same time showing the already existing advances in feminism and values of gender justice in the more highly developed regions of the Muslim world.

Regarding feminism's concerns, the predominantly Muslim country of Indonesia is not very different from the neighboring predominantly Catholic country of the Philippines. Simple facts of concomitant neighboring geography and not the cultural theories, forwarded by Professor Samuel Huntington explain to us why distance to religion in Bosnia and Herzegovina is at about an equal level as in neighboring Serbia, and Macedonia. Similarily, the distance to religion in former communist countries, like Albania, Kyrgyzstan, Azerbaijan, and Moldova is very similar.

A feminist movement has been active among both local and global Muslim women since many years. This we can infer from the life and works of the Benghali Roquia Sakhawat Hussain (1880 – 1932), the Azerbaijani Hamida Javanshir (1873 –1955), the Lebanese May Ziadeh (1886-1941), the Pakistani Zaib-un-Nissa Hamidullah (1921 – 2000), the Iranian Farideh Mashini (d. 2012), the Egyptian Nawal al-Saadawi, the Moroccan Fatema Mernissi, the French Fadela Amara and Rama Yade, the Sudanese Lubna Ahmed al-Hussein, the American Saleemah Abdul-Ghafur and Amina Wadud, and the Canadian Raheel Raza, to mention just a few examples.

Muslim women movements are not a simple import. Many could be probably considered to be an extension of a social women's movement born in the early 20th Century, whose leaders have been advocating women education and emancipation.

However, while acknowledging a perfectly equal status as human beings for both men and women, and treating them as equals, entitled to equal rights, Islam does, however, differentiate between man and woman with regard to their special functions in life (Kutub, 1984).

Our regional analysis provides global research and global policy-making with insights about the degree of secularism, the rising educational level of the younger generations, and feminism, which have become a reality in the core centers of the Muslim world, and not just in the post-communist Muslim countries of Eastern Europe and the former USSR. Our regional results also allow comparisons with regions in such countries as Australia, Canada, or the United States of America and also poorer regions of the European Union.

7. Human Values and the "Arab Spring"

In this Chapter, we evaluate Arab public opinion with a newly available source, the "*Arab Opinion Index*" by the Arab Center for Research and Policy Studies (ACRPS) in Doha, Qatar, practically neglected by Western scholarship until now. The *Arab Opinion Index* project is currently the largest of its kind in the world. It covers 12 Arab countries, representing 85 percent of the population of the Arab world. It is thus a larger Arab opinion survey project than any other scholarly effort to estimate Arab opinion. The Index compiles the findings of 16,173 face-to-face interviews with subjects who were drawn from a random, representative sampling of the populations of their countries of origin. The questionnaire was prepared in 2010 and the survey was conducted in the first half of 2011. The findings are freely available from the ACRPS website. Indeed, the Arab Center for Research and Policy Studies clearly reflects an "Arab national viewpoint" in international affairs.

For the purpose of this Chapter, the original data of the *Arab Opinion Index* were weighted by UNDP population figures for the year 2010/2011, so that we can arrive at conclusions about the totality of opinions in the Arab states.

On the one hand, the *Arab Opinion Index* indeed shows the overwhelming support for democracy and change in the region. At the same time, also these data may suggest some basic weaknesses of the civil society support for the structures of democracy.

The liberal heritage, which resulted from the movement of Enlightenment and democracy, implying respect of other religions and civilizations, is part of modern Islamic culture. It does not need to be sown anew, but simply supported through institutions facilitating cooperation, interaction and acculturation.

In this new order, emerging economies would not be dependent on the West. The US and the EU, still suffering from economic hardships, lack the resources and the political will to provide global stability. The divergences about the Syrian issue have been preceded by similar divergences about the Libyan issue before Western military intervention. Observers have noted that such divergences were a symptom of a crisis which was largely neglected. The domestic problems may hinder Europe from playing a major political role abroad (Karoui, 2012), which relatively explains the refusal of the British Parliament to support a military strike against the regime of Assad, charged of using chemical weapons against its people.

Support for the separation of religious practices from political and social life is only expressed by 46.6% of the population in the Arab countries, and the separation of religion from politics is only supported by 42.8% of the population. That political freedom and civil liberties are a requirement of democracy is only supported by 36.3%, and that equality and justice among citizens are a requirement of democracy is supported by only 19.5% of the Arab world.

Equally interesting is the extent of Arab rejection of what is denominated in the West as "peace process" between Israel and its Arab neighbors. A resounding 83.7% of Arabs are against the recognition of the State of Israel, and 59.6% support nuclear proliferation in the region to counter the perceived Israeli possession of nuclear weapons.

Today, there is indeed a Western Islam, as well as a Middle-Eastern, African, or Asian Islam. There is Shiite and Sunnite Islam, and many other branches, sections, and segments developing from the same tree, since centuries. These different branches and segments have different views about not only how to relate to non-Muslims, but also how to relate to Muslims holding opposite opinions.

For many experts in Arab-Islamic civilization, this cultural patrimony is to be divided broadly into two distinct tendencies, which we may call today: conservative, and liberal. The first encouraged conservative reproduction of the same. It begat the old fundamentalist tradition (in both the Shiite and Sunnite branches), fighting progress and freedom of thought, always under the pretention of defending the "true and pure" religion. Such a tradition is held responsible for the centuries of darkness and backwardness to which the Muslims have been subjected. They were the servants of Tyrants pretending to serve Islam. The second encouraged the intellectual effort *(Ijtihad),* the creative imagination, progress, scientific research, arts, and opened the minds of people to other cultures and civilizations.

If the Arab world wants positively to confront the 21st Century and become a full and mature democracy, issues like tolerance, sexism, and the overcoming of authoritarian thought patterns become important for the future of the democratic system. While "Islamophobia" in the West remains to be confronted, xenophobia and discrimination against immigrants and minority groups in Muslim countries have also to be faced. Even if a strong emphasis on religion is expressed, the religious doctrines are invited to start to learn from their own Enlightenment traditions, especially of Muslim philosophy, and the great global Enlightenment thinkers of the 18th and the 19th Century.

According to our preliminary hypothesis, one of the main reasons of the trajectory of radicalism, hatred and intolerance, which emerges from our survey, is the still dire state of higher education, especially in the social sciences in the region. The highest ranking Arab University or research institute in the recent very comprehensive SCIMAGO/SIR global University ranking report is Cairo University, and it enjoys only a rank at 526 out of all 3290 ranked institutions.

And even this is rather thanks to Cairo's respectable performance in science and medicine, and not so much in the humanities and the social sciences.

8. Islamic Values, Knowledge, and Morals

In this chapter, we try to find out the common grounds between modern Islam and the rest of the human kind, on the basic assumption that the modern culture is universal. Therefore, Islamic societies cannot stay away from it. We show that the belief that Islam is against modernity, the West, and the values humanity has produced since the 18th Century, is unfounded and even contradictory to the Islamic vision of knowledge and morals.

In a book about global value change, the question about how Muslims perceive the problem of knowledge and morals is crucial. Actually, it is at the basis of all the issues related to universal values, modernity, democracy, and human rights. If we consider that these are global issues, since modern science and technology as well as world trade are blurring the limits between cultures and civilizations more than distinguishing and pushing them to a clash (as believed by Huntington), we may talk of this global era as a universal civilization.

Regardless of how modernity is defined, we can see its effect everywhere in the Arab and Muslim world, in such domains as education, architecture, graphic arts, urban development, political and cultural institutions, *et cetera.* Indeed, the immersion in modernity does not concern exclusively the West. As an Iranian author put it, *"this immersion has expanded to such an extent that presently in the bookstores of most major cities in Iran one encounters the translated works of such thinkers as Arendt, Aron, Carnap, Dewey, Diderot, Habermas, Heidegger, Hume, Jaspers, Kant, Levi-Strauss, Locke, Marcuse, Nietzsche, Pascal, Popper, Rousseau, Bertrand Russell, Sartre, Spinoza, Weber, Wittgenstein, and many others. As the above list indicates, some of the seminal figures of analytic philosophy (logical positivism, linguistic philosophy) and continental philosophy (phenomenology, existentialism, structuralism) have been introduced to Iranian readers" (Boroujerdi, 1997).* And we can surely say the same thing about the bookstores of the major Arab cities.

The belief that Islam is against modernity, against the West, or against the values humanity has produced in its long march since the 18th Century, is naïve and ignorant. Studying critically Western thought does not imply necessarily opposing dogma to science, obscurantism to Enlightenment, irrationality to rationality. After all, medieval Muslim thinkers have dedicated much time and energy to analyzing and commenting texts of Greek philosophers, to see what was adaptable and useful to their own progress. Such a task cannot be achieved without critical thinking. This endeavor is scientifically right. It is proven by the history of philosophy (Hegel, Marx; Frankfort school; M. Foucault, Deleuze, Derrida, etc…) and science.

Muslim thinkers do criticize Western thinkers, in the same measure that Western thinkers criticize each other, not for "religious" reasons, but for inconsistency and logical flaws. Very often, such studies involve interesting questions about the "self" and the "other", science, technology, arts, politics, economy, systems of thought and systems of belief. There is no unanimity among Muslim thinkers concerning the West. If such unanimity were possible, it would rather belong to conspiracy theories, which, since Shakespeare's *Julius Ceasar,* added nothing new to human knowledge. The other side of the coin is likely the absence of unanimity in the West concerning the Islamic world. Thus, more and more Westerners today seek a better, i.e., unbiased – understanding of what is going on in the Islamic world.

Indeed, when we find “Paris” and “London” in the malls of Doha and Dubai; when young Arabs and Muslims dance the hip hop in the streets of Tunis, Manama, or Cairo; when young Saudi or Palestinian musicians play the rock & roll or the rap; when Egyptian movies imitate so many Hollywood productions; when Arab actors play Shakespeare, Becket and Pirandello ... should we call this “clash of civilizations,” globalization, or universal civilization?

Islam has been a precursor and a major force in promoting scientific endeavor. Unlike medieval Catholicism, Islam did not stifle the spirit of scientific enquiry. Although the first attempts to engage into philosophical argumentation have aimed at a better understanding of the Islamic message, soon Muslim scholars adopted new mental attitudes merging philosophical rational questioning with Islamic precepts and directives, and broadening the span of knowledge to new disciplines previously unknown to the Arabs, like chemistry, physics, mathematics, geometry, etc...

Here, it should be emphasized that Islam has a serene view of knowledge in connection to faith and morals, whereas “religion” and Western science seemed to meet in a theatre of conflict for many centuries. This has origins that are centuries old, during the apogee of Christianity, when burning of dissenting individuals, scores of them scholars (Bruno, Huss, Servetus...), was widespread. The Roman Catholic Church banned science but not because of the religious text itself, but as Bucaille explains, science was instead being hampered by those who claimed to be the Church's servants, and who acted on their own initiatives. (Zaimeche, 2002).

Jamal al-Din al-Afghani (1838-97), in his celebrated attack on the 'materialists', i. e., *Haqiqat-i mazhab-i naichiri wa bayan-i hal-i nachiriyan,* translated into Arabic by Muhammad Abduh as al-Radd 'ala'l-dahriyyin, was engaged in a self-proclaimed battle of saving science from the positivists, a battle for which he derived support from both the history of Islamic and modern sciences.

In *Tahdhib al-akhlaq wa-tathir al-a'raq* (refinement of character and purification of dispositions) *“Miskawayh makes religion a foundation for training and refinement; and after the supports of faith are established in the boy's soul, he can study books on ethics, then arithmetic and geometry, or whatever can be deduced or proved by rational proofs.”* (Jamal al-Din, N. 1994). Miskawayh defines moral happiness as *“the happiness enabling the human being to live happily, in accordance with the requirements of virtue. Thus it was a personal happiness which the human being could reach through intellectual effort, and striving to acquire the sciences which would make his thought inclusive of all the areas and all existent beings, and make him free himself from material things so as to reach the degree of wisdom whereby to grasp human perfection.”* (Jamal al-Din, N. 1994).

9. GRETCHEN, PERMISSIVENESS AND ECONOMIC GROWTH

In this Chapter, we maintain that the decay of religious values in the West is part and parcel of the current downward spiral of the North Atlantic Arena, and that far from being a *“Christian Club”,* which Europe long seized to be in contrast to the famous, often-quoted opinion of the former Turkish Prime Minister Mesut Yilmaz, the current European value landscape is characterized above all by a more and more generalized disrespect of the Ten Commandments. We debate the positions by Inglehart and Baker (2000) and Mehmet Goermez (2012), the current President of the religious affairs institute (DIYANET) of the NATO member country Turkey on secularization, and factor-analyze the self-declared observance of the Western publics of the Ten Commandments. This item was still contained in the first wave of the *World Values Survey*, and unfortunately was dropped in later surveys, while it reveals many insights, which are important to this day.

In statistical terms, we can observe that factor 1 and factor 2 – the two types of non-observance of the Ten Commandments in the mid-1980s - are highly intercorrelated with each other, but that factor 1 has the highest factor loading with the erosion of property rights, and factor 2 has the highest loading with the non-observance of the Command regarding the Sainthood of God's name. Table 9.1 in its different components reveals the shattering reality that *"Europe is a club of the erosion of property rights"* and a *"club of the erosion the Sainthood of God's name"* above anything else. This phenomenon is strongest in France, Sweden and the Netherlands, and weakest in Ireland, the United States and Malta. Religious observance in the United States of America is indeed different from the rest of Western democracies.

In his rightly famous 1998 book on global economic history, the world system scholar Andre Gunder Frank came to the conclusion that the days of the Euro-Atlantic arena are counted, and that countries like the United Kingdom and the North Atlantic arena in general become *"underdeveloping" countries,* while the dynamics of the world economy return to where they were always centered until the mid-18th Century: Asia. This global shift has many important conclusions also for the Arab world and the Muslim world/the Indian Ocean arena. To underline the point that Muslim societies are recovering from a long trough of relative decline in world society, and that they face a long-term U-shaped recovery path, we re-analyze data from the Angus Maddison database. This database is is the most complete database on real income convergence or divergence in the world over the long term.

Although very long-term very long-term data series about Muslim country GDP per capita in real purchasing power from the times before the First World War are available only for Java/Indonesia and the Ottoman Empire/Turkey, there are enough available post 1918 or post 1945 data for several other Muslim and Arab countries in particular so that we can draw at least some cautious general conclusions. Comparing country GDPs per capita to global averages or the leading Western powers on a very long-term basis reveals astonishing results.

All the countries of the region with available data currently recover from their relatively weak position into which they were sliding in the 1950s and 1960s and beyond. From the trough of the 50 years, long-term Kondratiev/Kondratieff cycle in the 1980s and early 1990s, the following majority Muslim semi-periphery countries managed to recover well: Albania; Algeria; Egypt; Iran; Jordan; Malaysia; Morocco; Tunisia. Several of the Muslim countries are showing quite a considerable recent relative ascent. Turkey's case, which is the biggest Muslim economy in the world today is a special case. It experienced already an upsurge during the heyday of the Kemalist reforms in the 1920s. Also, Tunisia and Malaysia experienced a strong convergence process over the last four decades relative to global GDP per capita levels or GDP per capita levels of leading Western powers. Iran and Iraq experienced their peaks in their relative world system fluctuations during the world oil-price hikes of the 1970s, and experienced a strong divergence process afterwards. Syria's case is an interesting confirmation of a more short term cycle (20 years cycle, often called Kuznets-cycle) in the convergence process, which came to a halt due to the recent civil war.

In this Chapter, we also debate what some Muslim social scientists described as the *"worshipful nature of economic activity,"* in Islamic societies as an application of a general principle considering any action a Muslim performs – whether related to economy or not – eventually as a worship deserving God's reward, if the intention was to please God.

10. The Implosion of "Self-expression" in the West and Further Doubts about the Correctness of Inglehart's Theory in the Light of the New Direct Measurements of Self-expression Values from the *World Values Survey*, 2010-2014

Data emerging from the *World Values Survey* in the first decade of the 2000s seemed to suggest that the precariousness, which more and more characterizes the economies of leading Western countries, leads toward an implosion of what Inglehart and his sociological school of thought interpreted as "self-expression values".

Our analysis of the time series element in the *World Values Survey* data shows that indeed, global value change seems to correspond to various ups and downs. To this end, we calculated which countries – in descending order – had very high increases or decreases in non-traditional values over preceding *World Values Survey* surveys from the original WVS website Inglehart's own data (http://www.worldvaluessurvey.org/wvs/ articles/folder_ published/article_base_54). The very idea that self-expression values in the West are imploding, while in other regions of the world they are rising, is a challenge to existing value theories.

The world, described by Inglehart and Baker, 2000, where in advanced industrial societies people pay large sums of money and travel long distances to experience exotic cultures no longer seems to exist for the "1.000 Euro" generation born after 1975, which experiences more and more job insecurity and hardly finds full-time tenured work opportunities, let alone the financial means to travel to long-distant countries. No wonder then that "self-expression" is dramatically declining in the West.

We also highlight the fact that the latest wave of *World Values Survey* data, wave 6, from 2010 - 2014, released in May 2014 contains an item which directly asked 74,044 respondents in 52 countries whether they think that self-expression is an important value for child education. The correlation between these data and Inglehart's self-expression index is negative and the R^2 between the two variables is almost 20%.

Among the twenty countries of our globe with a strong resilience of the self-expression tendencies, there is a greater number of Muslim countries (i.e., members of the Organization of Islamic Cooperation) among them. Let us think for an instance Inglehart's theory to its end: according to the *World Values Survey* data, among the twenty superstars of a resilient trend towards self-expression we find Jordan; Pakistan; Bangladesh; Nigeria; Turkey; Algeria; Egypt; and Uganda!

The most notable implosions or slow developments of self-expression – independent from the secularization process – had to be noted by contrast in western democracies. The resilience of self-expression is explaining more than 1/5 of economic growth in the world system. Muslim countries are among the trend leaders in both directions, i.e.,the resilience of self expression, and economic growth during the crisis years. Our Choropleth maps in this Chapter underline our contentions. Even a pure Inglehartian world values analysis would have to come to the conclusion that the value basis of Western society is eroding.

So while the methodology of the two approaches – Inglehart's and our own – is different, the same conclusions can be drawn from it. Map 10.5 is the final verdict then in the debate about these issues: it just shows how powerful the Tsunami of the crisis of 2008 hit the North Atlantic arena, and how the geography of global development is changing to the detriment of the hitherto dominant West.

We then invite our readers to have a renewed look at Map 4.15 in Chapter 4 and at the left, first column of Table 10.1. This would be the world view, which Inglehart's theory would suggest. This world view is very much in the spirit of his modernization theory. The Western countries, especially the Protestant countries, are seen as islands of modernity. Their self-expression culture would be a robust precondition of *"effective democracy"*. The data, emerging from the first four waves of the *World Values Survey*, provided some large-scale evidence to maintain such a theory. However, with all the extensions of the *World Values Survey* project over the last decades, both in terms of geography as well as the completeness of the data, the world of Map 4.15 recedes into the memory about a world order, which no longer exists and which was severely shattered in its foundations by the *tsunami* of the global economic crisis of 2008. As we tried to show in this book, it was also shattered by the long shadows of the internal corrosion, which social decay and the loss of values brought about long before the 2008 crisis hit the North Atlantic arena.

11. What the Importance Given to God by Humans Does to Society

If religions survived throughout human history, it is not by chance. Religion actually plays a more a positive role in society than acknowledged by some social scientists like Comte, Spencer, and Marx. We find religion not only at the basis of human values, but also at the end of the human strive for happiness. That is why religion, arts, science, laws, and good government are more complementary in society than contradictory. It is noteworthy that in classical Islamic thought, medieval scholars (like al-Kindi, al-Farabi, Ibn Sina, Ibn Rushd, *et cetera)* have always attempted to find common points between philosophy (science, reason) and religion. As most of them also were deeply involved with scientific experiments and theories, their attempts at reconciling Reason and Religion, Science and God, were probably the first of its kind in human history. They were precursors even of the Enlightenment that occurred in the eighteenth-century Europe. Most of the points later debated by Enlightenment in Europe have been discussed at length in classic Islamic thought, by the Mu'tazilites, Ash'arites, Ikhwan al-Safa, and several renowned philosophers.

We then debate the meaning of law: If we don't have laws, we don't have society; if we don't have society, we don't have civilization. The law is what distinguishes humans from animals (Ibn Khaldun). It lays the foundations of human civilization. Ibn Khaldun thought that *"social laws can only be discovered by gathering a large number of facts and observing comparable cases,"* - a task that is presently performed by our modern social sciences. Ibn Khaldun showed that *"Both past and present events can supply such facts. Social laws operate the same way in the same kinds of structure."* (Martindale, 1960: 132)

Taking the broad sociological and philosophical reflections developed in this Chapter into consideration, we also look into the discernable patterns of connection between the importance of God to one's life and social behavior. Instead of a clear-cut linear connection between the expressed depth of a respondent's attachment to God in one's life and patterns of social behavior or opinions, there is a U-shaped or inverted U-shaped curve emerging in most cases. This is reminiscent of so many phenomena encountered in the social sciences ever since the discovery by the Nobel laureate in economics, Simon Kuznets, 1955 that the extent of inequality in a society corresponds to an inverted U-shaped curve (y-axis) of the level of a society's development (x-axis). Correspondingly, we can discern for both global humanity as well as the global *Umma* (i.e., the community of global Muslims) the following very interesting and perhaps also very surprising patterns to hold:

- The importance of God to one's life and the acceptancy of taking a bribe resemble an inverted U-shape, with highest levels of a disposition towards illegal behavior at middle levels of self-declared religiosity
- The importance of God to one's life and the acceptancy of cheating on taxes again resembles an inverted U-shape, with highest levels of a disposition towards illegal behavior at middle levels of self-declared religiosity
- The importance of God to one's life and the rejection of the idea that democracy is better again resemble an inverted U-shape at the world level, with highest levels of a disposition towards rejecting the idea that democracy is better to be found at middle levels of self-declared religiosity. For Muslim respondents, the pattern resembles rather a "W", with highest levels of the rejection of the idea that democracy is better at self-declared very low, middle and very high levels of religiosity
- The importance of God to one's life and the opinion that competition is harmful resembles an inverted U-shape, with highest levels of a disposition against competition at middle levels of self-declared religiosity
- The importance of God to one's life and life satisfaction is weakly resembling a W-shape, with highest levels of life satisfaction for the global sample at very low, moderately high and very high levels of self-declared religiosity. The highest levels of life satisfaction for the global *Umma* are to be found at moderately low, medium level and very high levels of self-declared religiosity.

Table 11.1 contains still more data on what faith does to society. Faith is understood here as the assigned importance to God in one's life. We distinguish here between the effects on the level of the global sample and on the level of global Orthodoxy, Protestantism, Hinduism, Muslims, and Catholicism. The size of the statistical linear effects is rather small and rather seems to suggest that again, U-shaped patterns seem to be at work.

The two leading "Hayek" question items in the *World Values Survey*,

- is competition good or harmful? and
- is private or government ownership of the means of production to be preferred?

are not related at all with the importance of God to one's personal life, and the correlations are below +-.100.

Other notable tendencies are:

- religious Protestants tend to distrust the police more than secular Protestants, but else, the variables of trust in institutions and social capital exhibit little surprising results
- religious Roman Catholics tend to have a higher trust in the Armed Forces, and religious Roman Catholics and Muslims tend to have a higher trust in NATO, reflecting especially the situation in the Southern Mediterranean countries (i.e., Portugal, Spain, France, Italy, Greece, Turkey, moving from West to East)
- the correlation between the importance assigned to God in one's personal life and the self-classification on a left-wing to right-wing scale, is below +.100 for Protestantism and Islam, and is below +-.200 for Orthodoxy, Hinduism and Catholicism, and is also below +-.200 on a global level. The political left, in a way, is still a prisoner of the anti-religious philosophy of the Germany of the 19th Century, which was the cornerstone of historical Marxism.

Our debate also includes the discussion of sociological concepts in comparison to classical Islamic thought. We can never stress enough the depth, religions reach in societies, regardless of the kind of religion (Islam, Judaism, Buddhism, Christianity, etc...) and the kind of society (European, Asian, Middle-Eastern, etc...). Nor can we underestimate the influence religion may exert in shaping the mind and attitudes of millions of people, or in creating group solidarity *(Asabiyya,* according to Ibn Khaldun), or in developing militant movements and political blocks, or in determining the outcome of elections (in democratic societies, the USA is a good example) and other social processes, or in the outcome of power conflicts.

We also show in this Chapter the effects of the importance an individual assigns to God for the individual social and ethical convictions as documented in the *World Values Survey* for the different major global denominations with sufficient numbers of respondents to be included in Table 11.1.

For most major denominations and for the global publics as a whole (believers and non-believers alike, people with and without denominations), the effects of the importance assigned to God in one's life seem to concern mainly attitudes about religious affairs in general. But it's important to remember that on a global level, 82% of the variance of the variable: *Politicians who don't believe in God are fit for public office* is NOT determined by the importance an individual assigns to God in one's life. People "distant" or "near" to God in their personal life will have a preference for either secular or religious politicians. Our materials confirm the hypothesis that individual faith has its most direct consequences for individual morality and family values. In quantitative terms, the strongest effects of belief systems from all denominations are on the acceptancy or rejection of abortion, while even for the acceptancy of homosexuality and prostitution, the percentages of explained variances are already rather low.

Finally, we also look into the much-debated issue of "Islamophobia", measured here simply by the rejection of Muslim neighbors. It is not determined at all by the extent of religiosity of their neighbors. Whether an individual with Orthodox, Protestant, Hindu, Muslim, or Catholic faith rejects a Muslim neighbor is not determined at all by by the importance the individual assigns to God in one's life. The same happens with the rejection of neighbors who are homosexuals, immigrants or foreign workers, neighbors of a different race, or people who suffer from aids. The only exception, interestingly enough, is Protestantism, where believers seem to be rather more conservative concerning attitudes towards sexual minorities than the rest of society. Religious Protestants also tend to be more sexist in their attitude concerning University studies of female children, and they also favor income inequality.

12. Hofstede, Schwartz, Inglehart and the Future of World Development

This Chapter is intended to provide new and additional insights into our rethinking of the *World Values Survey* data and analysis. For one, we would like to present a still more conclusive proof of the interrelationship between the different types of permissiveness and the weight these factors have in relationship to the other variables, contained in the *World Values Survey* data. Based on our analysis of the complete available data based on 28 items from the *World Values Survey* from 70 countries of the world, including the OIC (Organization of Islamic Cooperation) member countries Albania; Azerbaijan; Bangladesh; Bosnia and Herzegovina; Burkina Faso; Indonesia; Jordan; Kyrgyzstan; Mali; Nigeria; Turkey; and Uganda we attempt to show the interrelationships between permissiveness, the shadow economy, educational values, and other socio-political variables, like fundamental positions on the market economy and democracy.

Tables 12.2 and the following Tables show the percentages of the total variance explained by our investigation for each variable of our model.

The nine factors to be extracted from the data for more than 90.000 representative respondents in 70 countries are the following:

- moral (sexual) permissiveness ('Permissiveness 1')
- acceptancy of the shadow economy('Permissiveness 2')
- distance from religion ('Permissiveness 3')
- educational values: independence and imagination

- distance to market economy values
- education values: responsibility and tolerance
- educational values: determination and perseverance and being against saving
- right wing acceptance of inequality
- educational values: favoring unselfishness, rejecting hard work

Contrary to Inglehart's expectations about a positive role of the low importance given to religion in society, and divorce and abortion being fully accepted, it emerges that the two factors of permissiveness (permissive family values and the loss of hard-core Max Weberian economic values) are closely interrelated with one another and with the loss of religious values.

Table 12.3 shows the factor loadings for each of the variables analyzed here. The variables with a high importance for "effective democracy", i. e., tolerance and respect for other people, rejection or acceptance of corruption, and the assessment of democracy as such and vis-à-vis military rule, are highlighted in our Table 12.3. Nowhere there is a notable negative or positive factor analytical loading of beyond 0. 333 (>10% of variance explained) confirming that religious people are antidemocratic, right-wing, and pro or anti-market.

In addition, the structure of the factor loadings even suggests the following:

a) distance from religion is even a motive to reject a democratic political system
b) moral/sexual permissiveness goes hand in hand with economic and social decay

Table 12.4 shows the correlations between the promax factors, extracted from the correlation matrix between the variables of our model. Table 12.5 and Maps 12.1 to 12.9 show the country values for our analysis ("factor scores") as well as the cascades of moral and social decay in the Western countries and also the evidence for the Muslim countries with available data. Graph 12.5 finally summarizes the pessimistic research findings, which rather support the views of Barro and Schumpeter against the logic, proposed by Inglehart.

In Table 12.6 we provide our readers with clear-cut Pearson-Bravais correlation coefficients between the data presented by Hofstede and Inglehart and the factor scores from our own analytical dimensions, presented in this work on the bases of promax factor analysis with individual data from up to more than 80 countries. Table 12.7 shows the Pearson-Bravais correlations between the Schwartz dimensions and our results. In many ways, we can show that Hofstede's Power Distance, Individualism versus Collectivism, Long-Term Orientation, and Indulgence versus Restraint very well correspond to our own factor analyses. The same happens with Inglehart's main dimensions, traditional versus secular, and survival versus self-expression, which we can well interpret in our own system. In all cases, however, we could avoid some of the problematic assumptions, still inherent in the research by Hofstede and Inglehart.

Schwartz's factors Affective Autonomy and Harmony do not achieve any correlations which have more than 25% of variance in common with our own factors, and in addition, the following dimensions from our own research are untapped, it seems, by the Schwartz' factors (to judge from the less than 25% of variance they have in common with the Schwartz factors):

Chapter 5: racism
Chapter 5: higher education of the younger generation (education gap between the generations)
Chapter 5: distrust of the army and the press
Chapter 5: female rejection of the market economy and democracy
Chapter 12 re-analysis: shadow economy mentality ('Permissiveness 2')
Chapter 12 re-analysis: educational values: independence and imagination
Chapter 12 re-analysis: education values: unselfishness and being against hard work
Chapter 6: feminism analysis: Generation gap in educational level

In addition, the following factors from Hofstede and Inglehart are untapped; it seems, by Schwartz's theories (again to judge from the less than 25% of variance they have in common with the Schwartz factors)

Hofstede: Masculinity versus Femininity
Hofstede: Uncertainty Avoidance Index
Inglehart: Self-Expression Values (WVS 1-4, 2006)

Table 12.8 shows the correlations of the country scores from Schwartz' work with standard socio-economic indicators. Interestingly enough, Muslim population shares and OIC membership present high correlations with the Schwartz factors "Embeddedness", "Hierarchy" and "Mastery".

We then debate current contentious political cleavages, especially in Europe, in the light of the empirics, as suggested by the *World Values Survey*. These days, in the leading world newspapers we read stories which tell us a lot about the conflicts about global values in countries like Europe today. Is prostitution justifiable? Is homosexuality justifiable? The French socialists, it seems, for example seem to think that one is not, and the other is. President Hollande and his administration put considerable political energy into legalizing homosexual marriages and prohibiting prostitution. But global citizens hold another view, and there is a high positive correlation of 0.632 between the two items in the *World Values Survey*, based on 218877 individuals from around the globe, i.e., people in favor of the complete acceptability of homosexuality will also be in favor of the complete acceptability of prostitution and vice versa.

Graph 12.1 highlights the politically, socially and ethically robust and globally applicable message of our book on the drivers of "effective democracy": a sound gender political agenda, ending the political discrimination of women, and economic freedom will be conducive to "effective democracy". Nevertheless the path towards "effective democracy" will be one of ups and downs, and especially in developing countries, there will be also certain limits for a too rapid economic liberalization in terms of "effective democracy".

As the manuscript to this book was about to be finished, the new data of the *World Values Survey,* 2010-2014 were released, containing yet another enormous wealth of new data, including on the Muslim world. We have chosen to concentrate on two phenomena, which received a large attention on the pages of this book – tolerance and democracy.

In Table 12.12 we calculate a simple UNDP Human Development Index type of Index of Tolerance, minimizing the rejection of neighbors with the following characteristics among the publics of the above mentioned countries of wave 6 of the *World Values Survey*:

- People who speak a different language
- People of a different religion
- Immigrants/foreign workers
- People of a different race

According to the *World Values Survey* data, the most tolerant nation on earth today is Uruguay, followed by Sweden; New Zealand; Spain; Trinidad and Tobago; Poland; Rwanda; Colombia; Chile and Australia.

Uzbekistan, Morocco and Kazakhstan are nowadays ahead of Germany; and Pakistan, Qatar and Tunisia are more tolerant than the EU-member country Romania. Some Muslim countries such as Turkey (which is still ahead of the OECD-member country South Korea), have still a poor performance.

Table 12.13 and Maps 12.9 to 12.12 list the *World Values Survey* results for the average importance given by the global publics to democracy and the standard deviation of this indicator. Where the standard deviation is low, opinions on democracy – either way – are undivided, while high standard deviations indicate that the publics are – often bitterly – divided on the issue of democracy.

Countries with an above than average importance assigned to democracy, and very high internal divisions on this issue are Tunisia; Mexico; Romania; Armenia and Yemen. While there is a general consensus that democracy is important, there are important dissenting voices. Nostalgia for past more authoritarian patterns of government can go hand in hand with economic discontent with present conditions. Countries with an above than average importance assigned to democracy, and very low internal divisions on this issue are the Netherlands; Egypt; Sweden; Turkey; and Cyprus. For anyone, attempting to turn back the clocks of history in such countries could result to be a very costly error. The recent introduction of internet censorship in Turkey would be just one example showing the relevance of this hypothesis.

Countries with still a below than average importance assigned to democracy, but already very high internal divisions on the issue are Libya; Philippines; Qatar; the Occupied Palestinian Territories; and Russia. In these countries and territories, debates on the issue of democracy will surge, one way or the other. While the average importance assigned to democracy is still lower than the world average, the divisions on the issue are already very high, and unforeseen events could trigger a popular movement for more participation and democracy. Finally, countries with a below than average importance assigned to democracy, and very low internal divisions on this issue are Singapore; Rwanda; South Korea; Estonia; and Lebanon. One might expect that the current stagnation in the democratic development of the country will continue: publics don't assign a great importance to democracy, and they are hardly divided on this issue.

Table 12.13 and our maps also have another, more immediate and direct implication: the dire state of the support of democracy in many Western countries, currently hit by the economic crisis and austerity packages, and the surge of democracy in the Muslim world and the Arab world in particular. That Egypt is ahead of Germany, Uzbekistan ahead of the EU-members Poland and Spain, and a number of other Arab and Muslim countries in general ahead of the United States; and Qatar ahead of the EU-member Estonia with justification should be celebrated by the Arab and Muslim readership of this book. Such comparison could be continued endlessly, and it is time for our readers to lay down this book and start to click: http://www.worldvaluessurvey.org/WVSOnline.jsp

Chapter 1

BACKGROUND

1.1. THE CRISIS OF THE WEST

This book is directed to global audiences and not just Western audiences alone. People in the Arab and Islamic world, currently (i.e., July 2014) being confronted with a rising polarization in many Arab countries, such as Egypt and Tunisia, between what is generally being referred to as *"secularists"* and *"Islamists",* should be reminded that Western societies themselves have not escaped a very deep economic crisis.

Box 1.1. The overall message of this Chapter

This chapter highlights the development of global value research and emphasizes the necessity to gain hard-core, quantitative and valuable insights from comparative knowledge about such phenomena as generalized trust and social capital, individualism, family ties, morality, attitudes toward work, and religious practice for the explanation of economic processes in the tradition of Alesina and Guiliano, 2013. Hitherto existing attempts in sociology to draw maps of human values across nations (i.e., Hofstede and Inglehart) are still unsatisfactory in terms of country coverage, issue coverage and also methodologies used. We introduce our readers to the contributions by Harvard economist Professor Robert Barro, who thinks that religion does positively affect economic outcomes mainly by fostering religious beliefs that influence individual traits such as thrift, work ethic, honesty, and openness to strangers. Economic theory ever since Joseph Alois Schumpeter is interested in these relationships between family values and economic well-being. Also classical Arab sociology (Ibn Khaldūn) paid attention to such phenomena, when it tried to explain the rise and decline of civilizations.

We show the relevance of these approaches in the framework of the shifting centers of gravity in the world economy. We introduce to our readers the freely available *World Values Survey* data base and show some of its applications.

The contribution of science – in our case social science – to the betterment of the social order can only consist of the presentation of rigorous, quantitative, well-founded analysis of underlying major trends, based on solid data and well-recognized procedures to analyze them. Our diagnosis that the *Western crisis* is also moral, and not only economic, might also help audiences beyond the perimeters of the developed Western countries to find new and humane ways to balance religious doctrines, value systems and the prerequisites of a democratic and enlightened society.

The systematic use of large-scale comparative international opinion data above all from the *World Values Survey,* which we undertake here, is not new and is not anymore restricted to the sociological discipline.[3] More and more, the economics discipline becomes aware of the enormous possibilities to address the contentious issues of culture in the economics of a global society, characterized more and more by migration and the globalization of world religions. Global value research, which originally grew out of the desire of large transnational corporations (i.e., the IBM Corporation was the first) to adapt to ever more complex and pluralistic cultural patterns of their clients and staff (Minkov, 2009; Minkov and Hofstede, 2011, 2013) by now is a long-established social science discipline, and it is a pity how little the overall global public and especially global decision makers take notice about its results so far.

Sociology, for sure, did its "homework" and there were solid attempts to arrive at realistic assessments of global Muslim and Arab Muslim opinion in a comparative perspective. Sociology, looking already back on a very long established tradition of the empirical sociology of religions, which dates back to the 19th Century (Morel, 2003), contributed greatly over the last two decades towards understanding global Muslim opinion (Diez-Nicolas, 2007, 2010; Moaddel, 1996, 1998, 2001, 2002a, 2002b, 2002c, 2003a, 2003b, 2003c, 2004a, 2004b, 2007a, 2007b, 2008, 2009, 2010; Moaddel and Abdel-Latif, 2006; Moaddel and Azardamaki, 2002; Moaddel and Karabenick, 2008, Moaddel, de Jong and Dagher, 2011; Moaddel, Tessler and Inglehart, 2008a, 2008b; Tessler, 2002, 2003, 2008; Tessler and Gao, 2009; Tessler and Jamal, 2006; Tessler and Robbins, 2007; Tessler, Moaddel and Inglehart, 2006a, 2006b; just to mention a few).

At this stage, the comparative social scientist and economist might wonder perhaps whether all the public debate that rages on the issues under scrutiny here in many countries around the globe, i.e., global values, multiculturalism *et cetera* - is really well-informed by the hard-core evidence from quantitative social science on the subject.[4]

It is necessary to recall at the outset that the question about the relationship between economics and religion is not new for advanced research in the economics discipline. The essay by Barro and McCleary, 2003 is a good example of how today economic research uses data from the *World Values Survey Project (WVS)* to study the relationship between religion, denominations and economic growth. Alesina, 2013; Alesina and Angeletos, 2005; Alesina and Fuchs-Schündeln, 2007; Alesina and Guiliano, 2010, 2011, 2013; Alesina, Cozzi and Mantovan, 2012; and Alesina, di Tella, and MacCulloch, 2004 all show how the economic discipline can indeed gain hard-core, quantitative and valuable insights from comparative knowledge about such phenomena as generalized trust and social capital, individualism, family ties, morality, attitudes toward work and perception of poverty, and

[3] In general, we use the data from waves 1-5 of the *World Values Survey*. The data of wave 6, 2010-2014, which were released in May 2014, will be the subject of further research. Some of these data were used in the concluding Chapters 10-12. Our *World Values Survey* data **"wvs1981_2008_v20090914.sav"** were freely available via the Website of the WVS project as an SPSS file.

[4] http://www.theguardian.com/world/2010/oct/17/angela-merkel-german-multiculturalism-failed

religious practice for economic processes. In our book, we will attempt without hesitation to define "cultures" on a global scale largely following Alesina and Guiliano, 2013.

We refute the often unspoken, but politically very relevant contention by many members of the global Western political class and the global Western public at large that at the end of the day there is an unbridgeable gap between the West and the Muslim countries.[5] For this purpose, we use advanced international data and advanced quantitative methods, like Promax oblique factor analysis, based on up to 30 indicators from the *World Values Survey*. Although some of our indicators are different from those in Alesina and Guiliano, 2013, there is lots of resemblance between Alesina's and our own approach. There is also a high correspondence between the choropleth geographical maps of global values in Alesina and Guiliano, 2013 and our own.

Erich Weede, a neo-liberal German sociologist, observed in 2006 that Islam contains a chance of promoting capitalism and economic freedom which other religions lack:

> "Think of Christ who [i.e., according to the Christian Gospel] threw the traders out of the temple, or of Buddha who became a hermit and delivered his first sermon to deer. But Mohammed was a full-time trader before he became a prophet, a ruler, and a military leader. This is as close to a capitalist background as was possible in the 7th Century. [...] Here is an unexploited resource which might permit a unique religious legitimation of capitalism or economic freedom and thereby of limited government. [...] There is another [...] support for the establishment of economic freedom in Islam. In principle, God's law is believed to be complete. The ultimate legislator in Islam is God, not rulers or elected politicians [....] If one thinks of taxation, it is obvious that legislation by God has some advantages over legislation by greedy secular authorities who tend to overtax their subjects or citizens. Just imagine how much better entrepreneurs and taxpayers could live in Western democracies, if God were believed to have forbidden tax or deficit increases to our politicians. In principle, Islam provides its own justification for constraining rulers and for limited government." (Weede, 2006)

In designing this research project, we made the discovery that hitherto existing attempts in sociology to draw maps of human values across nations (i.e., Hofstede and Inglehart) are really still unsatisfactory in terms of country coverage, issue coverage and also in terms of the methodologies used.

Besides background variables, like age, gender, education level, life satisfaction, educational values, values of responsibility, general political attitudes on the left-right scale, identification with democracy, general social capital variables (trust in people, in the armed forces, in the Press) and religious attitudes, our study in particular looks at hard-core economic values in all countries of the world with available data, namely,

1.	Competition good or harmful (competition policy, admission of a free market)
2.	Migration policy (prevent people from coming) (migration policy)
3.	Important child qualities: hard work (attitudes towards work)
4.	Justifiable: avoiding a fare on public transport (attitudes towards public services; limited morality)
5.	Justifiable: cheating on taxes (tax evasion and shadow economy)
6.	Justifiable: claiming government benefits even if one is not entitled to them (attitudes towards

[5] http://www.dw.de/turkey-not-fit-for-eu-accession-sarkozy/a-14875593

	public services, work attitudes)
7.	Justifiable: someone accepting a bribe (corruption)
8.	Rejecting Neighbors: Immigrants/foreign workers (economic and social discrimination)
9.	Neighbors: People of a different race (economic and social discrimination)
10.	University is more important for a boy than for a girl (economic and social discrimination)

Fascinating relationships between the underlying factors, statistically explaining our variables, emerge. Our approach allows us to determine the exact position of each country on those factors.

One basic message of our analyses, based on hundreds of thousands of interviews recorded in the *World Values Survey* data base, calculated by advanced statistical multivariate analysis, is that democratic and liberal values can also correspond to a civilization, characterized by an enlightened religion, in the West and in the Muslim world alike.

In our re-analysis of the *World Values Survey* data we show that Inglehart's strategically important self-expression values, associated by him with high levels of subjective well-being, good health, and high interpersonal trust, as well as tolerance of outgroups, support for gender equality, postmaterialist values, and environmental activism, are nowadays dramatically declining in the leading Western countries, especially in Europe. We also show that the world economic rise of the global South, among them the BRICS countries, is no coincidence. Economic growth in the post-crisis period from 2008 onwards is highly and positively correlated with basic human values.

Dostoevsky better underlined the dilemma, when he wondered about the value of life without God. His answer is known: *"If God doesn't exist, everything is permitted."* Indeed, no other author better reflected in his works this idea than Marquis de Sade, pushing it to the extreme. Without the belief in God, all crimes are permitted. There are no morals; it is just a question of point of view.

This book analyses the weight of such factors as the loss of religion and the rise of the shadow economy, including in leading Western countries, and starts out from the theoretical frame of reference provided by Robert Barro. Religion affects economic outcomes mainly by fostering religious beliefs that influence individual traits such as thrift, work ethic, honesty, and openness to strangers. For Barro, beliefs in heaven and hell might affect these traits by creating perceived rewards and punishments that relate to "good" and "bad" lifetime behavior. Barro's perspective is largely confirmed in this study. The networks and interactions fostered by churches and religious denominations are important elements of social capital.

For Inglehart, phenomena such as bribery, corruption, tax evasion, cheating the state to get government benefits for which one wouldn't be entitled, hardly exist, while the rich data base of the *World Values Survey* provides ample evidence about these phenomena. Our multivariate statistics and choropleth maps of global value adherence and global value change underline the correctness of the preoccupation about these phenomena. A growing number of members of the economics profession, starting with Joseph Alois Schumpeter and his hypothesis about the waning of family values and the capitalist family enterprise as the basis of the crisis of capitalism, share these views.

Policy pointer:

We show that Inglehart's strategically important self-expression values, associated with high levels of subjective well-being, good health, and high interpersonal trust, as well as tolerance of out-groups, support for gender equality, postmaterialist values, and environmental activism, are dramatically declining in the leading Western countries, especially in Europe. We also show that the world economic rise of the global South, among them the BRICS countries, is no coincidence: economic growth in the post-crisis period from 2008 onwards is highly and positively correlated with basic human values.

Dostoevsky better underlined the dilemma, when he wondered about the value of life without God. His answer is known: *"If God doesn't exist, everything is permitted."*

We analyse the weight of such factors as the loss of religion and the rise of the shadow economy, including in leading Western countries, and start out from the theoretical frame of reference provided by Robert Barro. Barro thinks that religion affects economic outcomes by fostering religious beliefs that influence individual traits such as thrift, work ethic, honesty, and openness to strangers. For Barro, beliefs in heaven and hell might affect these traits by creating perceived rewards and punishments that relate to "good" and "bad" lifetime behavior. Barro's perspective is largely confirmed in this study. The networks and interactions fostered by churches and religious denominations are important elements of social capital.

But our present research is not the work of theologians, but the work of economists, political scientists, and sociologists, attempting to analyze the existing value orientations and the interrelationships between value orientations *"on the ground"*. Is the decay of religious values in industrialized societies concomitant with a rise or a decline of democratic orientations; or is secularization part and parcel of the decay of family values and economic values, which threaten to affect negatively the very existence of the capitalist market economy? Joseph Alois Schumpeter, the great non-Marxist pessimist of capitalism, already predicted so in 1950.

If overall values decline, then, also tax morale and the human work ethic, considered to be so vital for economic progress since the days of the sociologist Max Weber, will decline as well.

At this stage, our readers might ask themselves whether or not classical Islamic thought already foresaw this. We thus attempt to remind our readers of Ibn Khaldūn (1332-1406), who was actually the first sociologist ever in the world. The contemporary world can learn a lot, actually, from this particular classical Arab writer and his description of the devastating effects of the loss of social values on a society. The issues being dealt with here in this book are of a fundamental importance not only to Western countries, confronted by a concomitant decline of their economic role in the world economy and their religious values, they are also of a fundamental importance to the Muslim, especially the Arab countries in search of democracy. From the very start, Muslim scientists had also a great affinity towards mathematics, which are so necessary in the development of empirical sociology. We remind our readers of such figures as Muḫammad ibn Mūsā al-Khawārizmī (780-850), the Muslim mathematician and astronomer, whose works introduced Hindu-Arabic numerals and the concepts of algebra and algorithm to European mathematics.

There is another level of contention as well with the Inglehart paradigm: the relationship between feminism and religion. Since more and more cultures and religions of the world are being integrated into the *World Values Survey* data base, the emerging results more and more contradict some of the interpretations, which were based on generalizations, gained from earlier and smaller *World Values Survey* samples with less religious and cultural diversity. Not only democracy, but also feminism can be compatible with moderate expressions of religiosity in all civilizations. Analyzing data on feminism and religion not only on a national, but also on a regional level, we show this exactly to be the case. Analyzed regions include many Arab countries, including Saudi-Arabia.

The reality of religious feminism is as important a subject for further research, as are tendencies for a new marginalization processes along gender lines in the new democracies which emerged after the democratic changes in Latin America and Eastern Europe in the 1980s. What can the Arab world learn from this?

In our work, we discover that value change is by far from linear. Most factors, representing core items of the *World Values Survey,* present a curve-linear relationship with the *UNDP-Human Development Index,* which is a valid indicator of *"human security".* Nobel Laureate in economics, Simon Kuznets, was correct in predicting that inequality is a curve-linear function of development levels. Today we are presented with several *"Kuznets"* curves of human values.

Thus, many of the still existing linear hypotheses and expectations in current global value research have to be submitted to a fundamental rethinking. We will discover as well that the trade-off of religiosity with social behavior is not linear. U-shaped and inverted U-shaped curves of the relationship between religiosity and human behavior hold for both the global population and for the global Muslim population.

For one, the representatives of organized Western religion struggle with secularization tendencies for decades now, if not for centuries. In the words of the former Roman Catholic Pope Benedict XVI:

> 'Ever since the Enlightenment the criticism of religion has been gathering momentum; history has also come to be marked by the presence of atheistic systems in which God was seen as a mere projection of the human mind, an illusion and the product of a society already misled by so many alienating factors. Moreover the past century experienced a strong process of secularization under the banner of the absolute autonomy of the human being, considered as the measure and architect of reality, but impoverished by being created "in the image and likeness of God". A particularly dangerous phenomenon for faith has arisen in our times: indeed a form of atheism exists which we define, precisely, as "practical", in which the truths of faith or religious rites are not denied but are merely deemed irrelevant to daily life, detached from life, pointless. So it is that people often believe in God in a superficial manner, and live "as though God did not exist" (et si Deus non daretur). In the end, however, this way of life proves even more destructive because it leads to indifference to faith and to the question of God.
>
> In fact human beings, separated from God, are reduced to a single dimension — the horizontal — and this reductionism itself is one of the fundamental causes of the various forms of totalitarianism that have had tragic consequences in the past century, as well as of the crisis of values that we see in the current situation. By obscuring the reference to God the ethical horizon has also been obscured, to leave room for relativism and for an ambiguous conception of freedom which, instead of being liberating, ends by binding human beings to idols. The temptations that Jesus faced in the wilderness before his public

ministry vividly symbolize which "idols" entice human beings when they do not go beyond themselves. Were God to lose his centrality man would lose his rightful place, he would no longer fit into creation, into relations with others. What ancient wisdom evokes with the myth of Prometheus has not faded: man thinks he himself can become a "god", master of life and death.' BENEDICT XVI, GENERAL AUDIENCE, Paul VI Audience Hall, Wednesday, 14 November 2012, available at: http: //www. vatican. va/holy_father/benedict _xvi/audiences/2012/documents/hf_ben-xvi_aud_20121114_en. html)

Several Islamic thinkers also contributed to this debate, which may be described as a debate about *"Islamic modernism"*. As John L. Esposito explained,

> "Islamic modernists asserted the need to revive the Muslim community through a process of a reinterpretation or reformulation of their Islamic heritage in light of the contemporary world. It is a respond to the political, cultural, and scientific challenge of the West and modern life. The modernist attempted to show the compatibility (and thus acceptability) of Islam with modern ideas and institutions, whether they be reason, science and technology, or democracy, constitutionalism and representative government" (Esposito, 1984).

In his view, Islamic modernism was a response both to the continued weaknesses and to the external political and religious-cultural threat of colonialism. (Ahmad N. Amir, Abdi O. Shuriye, Ahmad F. Ismail: 2012) Let's take two examples:

> Muhammad Abduh, hailed as the "first leader of modern thought in Egypt," "the father of the 20th century Muslim thought" and "pioneer of the modernist movement in the Arab world", favorably embraced the ideas of modernity which principally geared to the advancement of Muslim society. The basic foundation of modern ideal proposed by him was largely grounded in rationalism, liberalism, nationalism and universalism of Islam." Abduh believed the rigid structures of Islamic culture were holding back what was essentially a logical and fluid religion." (Sedgwick 2009: 1)
>
> Abduh advocated the necessity to accept modernity founded in Islamic construct and civilization. For "it is only as Islam is reformed to agree with modern conditions, it is believed, that its true character as a world religion will be apparent." (C. C. Adams: 204) Abduh "believed that it was not only possible but imperative to show how 'modern' values and institutions could be reconciled with authentically Islamic ideals." His inclusivity in taking "more conciliatory and evolutionary approach" characterized his worldview and principle outlook of modernism, with strong conviction that adopting the western ideals will liberate the Muslim from stagnation, decadence and backwardness and lead the ways to reap the benefit of science and technology. (Ahmad N. Amir, Abdi O. Shuriye, Ahmad F. Ismail, 2012)
>
> Another great Islamic thinker, the Algerian Malik Bennabi, "was distinguished from other Arab thinkers by his use of the functional approach in interpreting the possible role of religion in social change. He therefore did not look to Islam as a metaphysical creed, but rather as a motivating ideology, a major part in a complex social system." (Bariun F. 1991) He argued that the problem "was not in how to prove God's existence to the Muslim, but to make him/her sense that His existence fills up his/her soul as a source of energy." (Bennabi M. 1981)
>
> Bennabi accorded much importance to the term "civilization" which was used as subtitle for all his books. However, unlike ibn Khaldūn, Bennabi did not adopt the concept of group cohesion (*Asabiyya*)

"as the means of material and immaterial progress. To Bennabi, religion was a prerequisite to the rise of any civilization, for he viewed it as the 'compound' which gives man the time, the soil, and the spark to start a cycle of civilization. He considered the principles of current Western civilization to be based on Christian ethics and morals. Despite the fact that Christianity developed long before Islam, he pointed out that the civilization of Islam flourished long before that of Christianity. Civilization, Bennabi concluded, is born twice: 'first when the religious idea is born, and second when the idea becomes recorded in souls and is entered into the events of history.' (Bennabi M. 1986) Islamic civilization, on the other hand, had 'both births' at once, Bennabi suggested." (Bariun F. 1991)

1.3. The Sociology of Religions, the Crisis in the West, and Decay of Religious Values and the Rise of the Shadow Economy and Social Anomaly

Generations of Western sociologists have told the world that secularization is the fate of Western society. French sociology from the 19th Century onwards in particular (Auguste Comte, Emile Durkheim…) has emphasized this point with great detail, and considering the background of French society after the Revolution of 1789, this message was all too logical.

International sociology has been studying global values and global value change *empirically* for a number of years now.[6] This research tradition is connected, as we already mentioned, with the *'World Values Survey'* project at the University of Michigan, headed by Professor Ronald F. Inglehart. The *World Values Survey* (WVS), in collaboration with the EVS (*European Values Study),* carried out representative national surveys of hitherto 97 societies containing almost 90 percent of the world's population. The five waves of surveys started in 1981, and the fifth survey ended in 2007.

But increasingly, the *economics profession,* that is mathematical, quantitative economics, already makes large-scale use of these data, integrating the *World Values Survey* country level results into international economic growth accounting. Alesina and Guiliano, 2013 define culture to comprise the following *measurable* tendencies in the following variables:

- Generalized trust
- Individualism versus collectivism
- Family ties
- Generalized vs. limited morality
- Attitudes toward work and perception of poverty
- Religion

The most studied cultural trait, according to Alesina and Guiliano, 2013 is the measure of *generalized trust* toward others. The importance of this trait cannot, Alesina and Guiliano, 2013 argue, be overemphasized. Every commercial transaction has within itself

[6] http: //www. worldvaluessurvey. org/

an element of trust; and economic backwardness in the world can be explained, Alesina and Guiliano, 2013 argue following economist Kenneth Arrow, precisely by the lack of mutual confidence. The *World Values Survey* asks respondents around the globe: *"Generally speaking, would you say that most people can be trusted or that you can't be too careful when dealing with others?"* Nowadays, there is, Alesina and Guiliano highlight, a vast literature showing a close connection between trust and economic development.

Generalized vs. limited morality are also, Alesina and Guiliano, 2013 argue, relevant in fostering economic development. Limited morality exists where cooperative behavior is extended only towards immediate family members, whereas *generalized morality* exists where cooperative behavior is extended toward everyone in society.

We will discover that the trade-off of religiosity with social behavior is not linear: Such U-shaped and inverted U-shaped curves of the relationship between religiosity and human behavior are analyzed in this study for the global population and for the global Muslim population.

Our aim is thus to continue the empirical research traditions in the *scientific* discourse about the sociology of immigration, integration, and religion, which is now so common today in the global scientific debate. Here we refer interested readers to the current most important studies, studying immigration, integration, religion and economic growth from a rigorous, quantitative perspective. The worst which could happen in any society is that it detaches itself from the social realities in the name of whatever ideology. Empirical social science and economics can contribute towards a more humane society by presenting data on contentious social issues [Alexander, Inglehart and Welzel, 2011; Alexander and Welzel, 2011; Barro and Hwang, 2007; Barro, 2004; Barro and McCleary, 2003, 2004; de Soysa and Nordas, 2006; Diez-Nicolas, 2003, 2007, 2009, 2010; Hofstede, 2001; Hofstede and Minkov, 2010; Hofstede, Hofstede and Minkov,, 2010; Inglehart, 1970, 1971, 1977, 1990, 1997, 2000, 2002, 2003, 2003a, 2003b, 2003c, 2007; Inglehart and Baker, 2000; Inglehart and Norris, 2003a, 2003b, 2003c; Inglehart and Welzel, 2003, 2005, 2009, 2010a, 2010b; Inglehart et al. 2004; McCleary and Barro, 2006a; Moaddel, 1994, 1996, 1998, 2001, 2002a, 2002b, 2002c, 2003a, 2003b, 2003c, 2004a, 2004b, 2006, 2007a, 2007b, 2008, 2009, 2010; Moaddel and Abdel-Latif, 2006; Moaddel and Azadarmaki, 2002; Moaddel and Karabenick, 2008; Moaddel, de Jong J and Dagher, 2011; Moaddel, Tessler and Inglehart, 2008a, 2008b; Noland, 2004; Noland, 2005; Noland and Pack, 2004; Norris, Inglehart, and Welzel, 2003; Norris and Inglehart, 2003, 2004; Tessler, 2002; Tessler, 2003; Tessler and Jamal, 2006; Tessler and Robbins, 2007; Tessler, Moaddel and Inglehart, 2006a, 2006b; Welzel, 2007; Welzel and Alexander, 2008; Welzel and Klingemann, 2008; Yesilada and Noordijk, 2010]. We aready highlighted that for Schumpeter the decline of family values is intimately linked to the decline of the capitalist order. No public opinion survey item better highlights the decline of family values than the opinion that marriage is an outdated institution. In Map 1.1., we attempt to provide our readers with the basic facts about this: except for the United States, Poland, the Czech Republic, Croatia, Romania and Japan, most Western democracies are already above or far above a threshold of 15% of adult people saying that marriage is an outdated institution.

Starting with the usual IMF World Economic Outlook data (as of October 2012), we immediately see that the share of the Western "triade" (USA, the European Union, and Japan) in global purchasing power has been reduced dramatically since the 1990s.

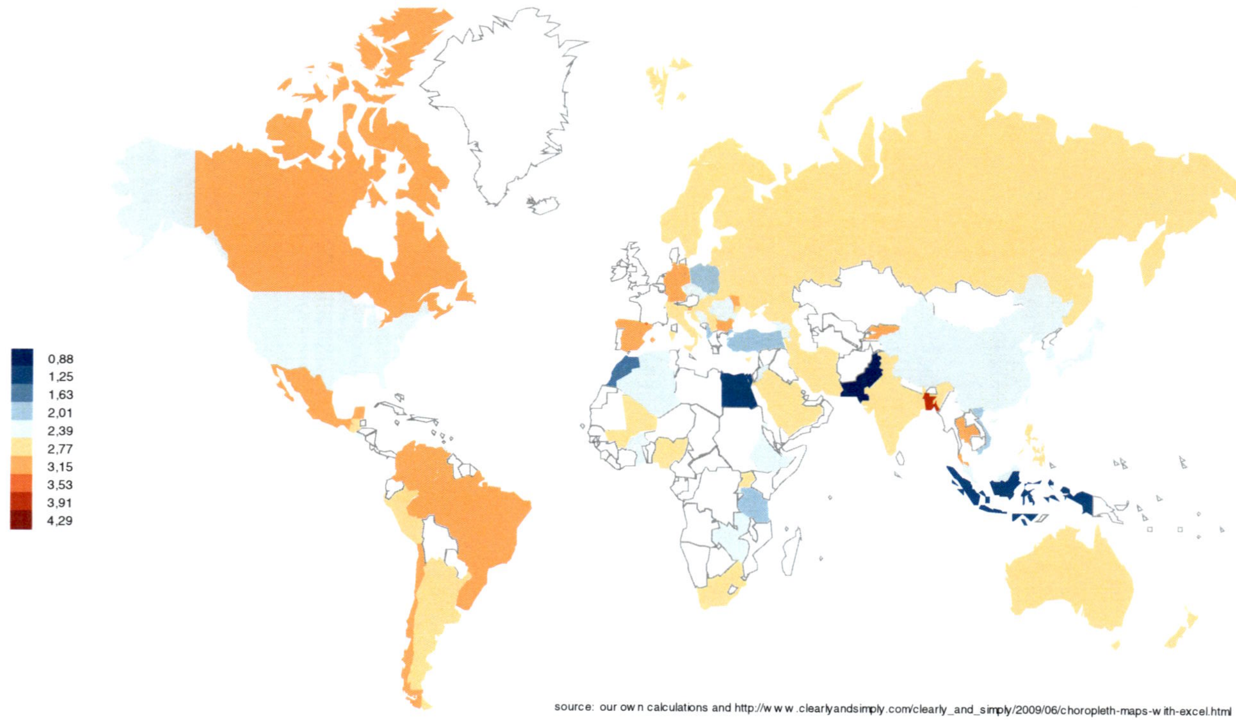

Map 1.1. Was Schumpeter right after all? The loss of capitalist family values as measured by the *World Values Survey* response *"marriage is an outdated institution"* [natural logarithms (+1) of the percentages consenting with that opinion. The threshold from blue to yellow and red colors is around 15%.]

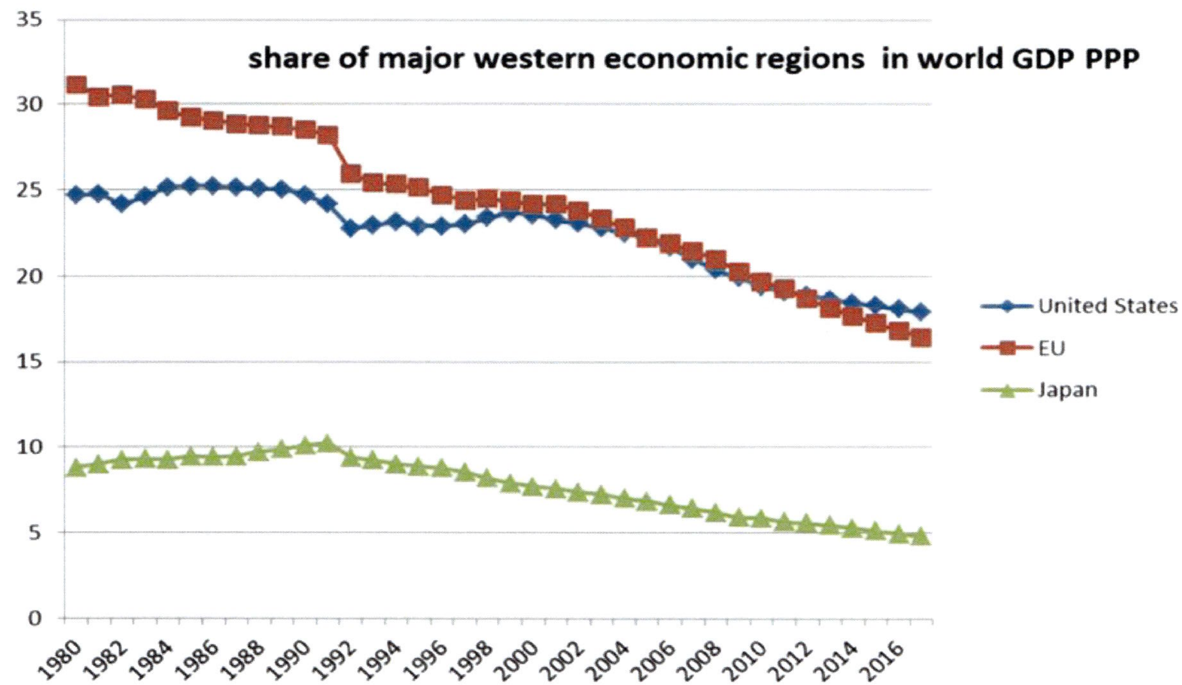

Graph 1.1. the declining share of major western regions in global GDP.

Even in their wildest anti-Western dreams, the opponents of the West would not have been able to imagine what has come true today – the tremendous reduction of Western economic power within only three decades. What radical Islamist movements could not achieve – the undermining of Western military power – the economic crisis and the rise of new economic global players accomplished within the timespan of a few years. Not the bombs of Mr. Osama Ben Laden, but the relentless economic crisis starting in 2007 achieved this decline.

Today, we see a substantial rise in the economic power not only of the so-called BRICS (Brazil, Russia, India, China, South Africa), especially China and India, but also of the Arab states and other emerging Muslim economic powers, among them Turkey, the single most important market in the Muslim world today.

1.3. Ibn Khaldūn (1332-1406) and the Role of Values in Civilization

The first global scholar ever to predict to happen what Schumpeter foresaw in 1950, i.e.,the loss of family values leading towards the decline of the social order, was the classical Arab sociologist and historian Ibn Khaldūn (1332-1406), who explained in his most important book *Al-Muqaddimah* (*"Introduction to history")* that historical change and the succession of dynasties are a function of the interactions between nomadic culture and urban civilization. Ibn Khaldūn is well known for his explanation of the nature of state and society and for being "the founder of the new discipline of sociology". He has *"was brought to the attention of the modern world by Gumplowics and treated with great respect by Franz Oppenheimer."* (Martindale, 1960: 132)

> "Ibn Khaldūn fully realized that he had created a new discipline, 'ilm al-'umran, the science of culture, and regarded it as surprising that no one had done so before and demarcated it from other disciplines. This science can be of great help to the historian by creating a standard by which to judge accounts of past events. Through the study of human society, one can distinguish between the possible and the impossible, and so distinguish between those of its phenomena which are essential and those which are merely accidental, and also those which cannot occur at all." (Issawi and Leaman, 1998: 4: 623 - 627).

The core of his sociology is found in the concept of *"Asabiyya"* (social solidarity, or group feeling). The state and society emerge together out of necessity as form and substance of the same thing. As an individual cannot survive in nature without cooperating with others, no sooner do men band together and solve the problems of nature than

> "There arises the need for restraining force to keep men off each other in view of their animal propensities for aggressiveness and oppression of others. Now the weapons with which they defend themselves against wild beasts cannot serve as a restraint, seeing that each man can make equal use of them. Nor can the restraint come from other than men, seeing that animals fall far short of men in their mental capacity. The restraint must therefore be constituted by one man, who wields power and authority with a firm hand and thus prevents anyone from attacking anyone else, i. e., by a sovereign. Sovereignty is therefore peculiar to man, suited to his nature and indispensable to his existence…The state is… to society as form is to matter, for the form by its nature preserves the matter and, as

> the philosophers have shown, the two are inseparable." (Ibn Khaldūn, I: 1; Martindale, 1960: 132-133)

Ibn Khaldūn considers both nomads and sedentary people as natural groups, based on *"the different ways in which they make their living. Social organization enables them to cooperate toward that end and to start with the simple necessities of life, before they get to conveniences and luxuries."* (Ibn Khaldūn, Tr. F. Rosenthal, 1958, II: 1; available online: http://www.muslimphilosophy.com/ik/Muqaddimah/index.htm).

Ibn Khaldūn was the first to propose a model of value change within the time span of four generations. He says:

> "As a rule no dynasty lasts beyond the life (span) of three generations. A generation is identical with the average duration of the life of a single individual, namely, forty years, (the time) required for growth to be completed and maturity reached (...)"
>
> The first generation retains the desert qualities, desert toughness, and desert savagery. (Its members are used to) privation and to sharing their glory (with each other); they are brave and rapacious. Therefore, the strength of group feeling continues to be preserved among them. They are sharp and greatly feared. People submit to them.
>
> Under the influence of royal authority and a life of ease, the second generation changes from the desert attitude to sedentary culture, from privation to luxury and plenty, from a state in which everybody shared in the glory to one in which one man claims all the glory for himself while the others are too lazy to strive for (glory), and from proud superiority to humble subservience. Thus, the vigor of group feeling is broken to some extent. People become used to lowliness and obedience. But many of (the old virtues) remain in them, because they had had direct personal contact with the first generation and its conditions, and had observed with their own eyes its prowess and striving for glory and its intention to protect and defend (itself). They cannot give all of it up at once, although a good deal of it may go. They live in hope that the conditions that existed in the first generation may come back, or they live under the illusion that those conditions still exist.
>
> The third generation, then, has (completely) forgotten the period of desert life and toughness, as if it had never existed. They have lost (the taste for) the sweetness of fame and (for) group feeling, because they are dominated by force. Luxury reaches its peak among them, because they are so much given to a life of prosperity and ease. They become dependent on the dynasty and are like women and children who need to be defended (by someone else). Group feeling disappears completely. People forget to protect and defend themselves and to press their claims. With their emblems, apparel, horseback riding, and (fighting) skill, they deceive people and give them the wrong impression. [...] When someone comes and demands something from them, they cannot repel him. The ruler, then, has need of other, brave people for his support. He takes many clients and followers. They help the dynasty to some degree, until God permits it to be destroyed, and it goes with everything it stands for (...) In the course of these three generations, the dynasty grows senile and is worn out. Therefore, it is in the fourth generation that (ancestral) prestige is destroyed." (Ibn Khaldūn, F. Rosenthal, 1958, III: 12).

Furthermore, he was the first author to comprehend the strategic importance of human values for the growth and development trajectory of a society. His major contribution in this context then consisted in the analysis of the correlation between *Asabiyya* (social solidarity), the social cohesion of a society, and political power.

As he put it:

> " (...) Aggressive and defensive strength is obtained only through group feeling which means (mutual) affection and willingness to fight and die for each other.
>
> Now, royal authority is a noble and enjoyable position. It comprises all the good things of the world, the pleasures of the body, and the joys of the soul. Therefore, there is, as a rule, great competition for it. It rarely is handed over (voluntarily), but it may be taken away. Thus, discord ensues. It leads to war and fighting, and to attempts to gain superiority. Nothing of all this comes about except through Asabiyya, as we have also mentioned." (*Ibn Khaldūn, F. Rosenthal, 1958, III: 1).*

He argued that the strong social cohesion of tribal peoples enabled them to conquer urbanized regions and build regimes and civilizations, but that these conquests were undone by the tribes' gradual loss of *Asabiyya* in the urban setting, leading to new conquests by tribal peoples still strong in desert cohesiveness. Although power was the basis of rule and royal authority was established through military might, the *'glue'* that held societies together (Tibi, 1996) was *Asabiyya,* based on kinship and religion and stronger in tribal than in urban society. Conquerors, whose social cohesion was weak, were soon overcome by the civilization of the conquered and gave way to a new conquering group. Rulers would be successful only so long as they remained just; as the rulers' level of luxury increased, so did their level of exploitation, and injustice soon produced division and *'the ruin of civilization'* (Tibi, 1996). In Ibn Khaldūn's own words:

> "It should be known that attacks on people's property remove the incentive to acquire and gain property. People, then, become of the opinion that the purpose and ultimate destiny of (acquiring property) is to have it taken away from them. When the incentive to acquire and obtain property is gone, people no longer make efforts to acquire any. The extent and degree to which property rights are infringed upon determines the extent and degree to which the efforts of the subjects to acquire property slacken. When attacks (on property) are extensive and general, extending to all means of making a livelihood, business inactivity, too, becomes (general), because the general extent of (such attacks upon property) means a general destruction of the incentive (to do business). If the attacks upon property are but light, the stoppage of gainful activity is correspondingly slight. Civilization and its well-being as well as business prosperity depend on productivity and people's efforts in all directions in their own interest and profit. When people no longer do business in order to make a living, and when they cease all gainful activity, the business of civilization slumps and everything decays. People scatter everywhere in search of sustenance, to places outside the jurisdiction of their present government. The population of the particular region becomes light. The settlements there become empty. The cities lie in ruins. The disintegration of (civilization) causes the disintegration of the status of dynasty and ruler, because (their peculiar status) constitutes the form of civilization and the form necessarily decays when its matter (in this case, civilization) decays." (*Ibn Khaldūn, Rosenthal, 1958, III: 41).*

Policy pointer:

Today, we see a substantial rise in the economic power not only of the so-called BRICS (Brazil, Russia, India, China, South Africa), especially China and India, but also of Arab states and other emerging Muslim economic powers, among them Turkey. Our book is conscious about these important tectonic shifts in the world economy, but it tries to relate them also to the shifting global values and the loss of many values, important for the good functioning of democracies and market economies in the West. Many centuries ago, the classical Arab sociologist and historian Ibn Khaldūn (1332-1406), explained in his most important book *Al-Muqaddimah ("Introduction to history")* that historical change and the succession of dynasties are a function of the interactions between nomadic culture and urban civilization. The core of his sociology is found in the concept of *"Asabiyya"* (social solidarity, or group feeling). Ibn Khaldūn was the first to propose a model of value change within the time span of four generations (i.e., for Khaldūn, 160 years). The first generation retains the desert qualities, desert toughness, and desert savagery. Under the influence of royal authority and a life of ease, the second generation changes from the desert attitude to sedentary culture, from privation to luxury and plenty. Thus, the vigor of group feeling is broken to some extent. People become used to lowliness and obedience. The third generation, then, has (completely) forgotten the period of desert life and toughness, as if it had never existed. Luxury reaches its peak among them, because they are so much given to a life of prosperity and ease. Group feeling disappears completely. When someone comes and demands something from them, they cannot repel him. The ruler, then, has need of other, brave people for his support. He takes many clients and followers. They help the dynasty to some degree, until God permits it to be destroyed, and it goes with everything it stands for (…) In the course of these three generations, the dynasty grows senile and is worn out. Therefore, it is in the fourth generation that (ancestral) prestige is destroyed.

In the field of economics, "Ibn Khaldūn understands very clearly the supply and demand factors which affect price, the interdependence of prices and the ripple effects on successive stages of production of a fall in prices, and the nature and function of money and its tendency to circulate from country to country according to demand and the level of activity." (Issawi and Leaman, 1998: 4: 623 - 627).

Ibn Khaldūn also foresaw the tendency towards economic cycles, which he linked, quite in a modern fashion, to 'value change' (Inglehart and Norris, 2003a, 2003b, 2003c; 2010, 2012; Norris and Inglehart, 2003, 2004). In the Muqaddimah, he even writes the following:

> "We have stated before that a city with a large civilization (population) is characterized by high prices in business and high prices for its needs. (The prices) are then raised still higher through customs duties; for sedentary culture reaches perfection at the time when the dynasty has reached its greatest flourishing, and that is the time when the dynasty levies customs duties because then it has large expenditures, as has been stated before. […] Customs duties raise the sales (prices), because small businessmen and merchants include all their expenses, even their personal requirements, in the price of their stock and merchandise. Thus, customs duties enter into the sales price. The expenditures of sedentary people, therefore, grow and are no longer reasonable but extravagant. The people cannot escape this (development) because they are dominated by and subservient to their customs. All their profits go into (their) expenditures. One person after another becomes reduced in circumstances and indigent. Poverty takes hold of them. Few persons bid for the available goods. Business decreases, and the situation of the town deteriorates. All this is caused by excessive sedentary culture and luxury. They corrupt the city generally in respect to business and civilization." (*Ibn Khaldūn, Rosenthal, 1958, IV: 18).*

1.4. TOWARDS STUDYING THE KHALDŪN/BUDDENBROOKS CYCLE

Contemporary quantitative long economic cycle (Kondratiev-cycle) researchers, looking into the question of long economic cycles (Kondratiev-cycles), like Rennstich, 2007, stumbled onto the issue again, without duly taking into account the theoretical forerunner Ibn Khaldūn. But Rennstich linked the four generation cycle to the Nobel-laureate winning four generation family saga *'Buddenbrooks. The Decline of a Family'* written by the German novelist Thomas Mann (Mann, 1901/1924), calling the four-generation cycle of social systems the *Buddenbrooks cycle.*

In that family saga *Buddenbrooks* Thomas Mann describes the decline of a wealthy north German merchant family over the course of four generations in the years from 1835 to 1877. The work led to Mann's Nobel Prize in Literature in 1929. Our readers in such countries as in the Arab world, in Iran and in Turkey easily can get access to the translations of that novel in Arabic, Farsi and Turkish.

According to the *Buddenbrooks economic cycle school,* described by Rennstich, 2007 the mentioned novel by Thomas Mann contains a dire and general message for all societal systems and is a model of analysis, also to be applied to the current contemporary, post-World-War II world order. Independently of Rennstich, Devezas, 2006, Devezas/Corredine, 2001, developed a similar and mathematical model, also based on a four-generation model. Schumpeter, in his value-conservative and pessimistic interpretation, also touched upon this issue in his 1950 *Capitalism, Socialism and Democracy,* when he stated that capitalism will not disappear from the face of the earth by a socialist revolution, but by the gradual erosion of capitalist individual family values and the rise of a socialist-oriented intellectual elite and bureaucracy.

Norris and Inglehart, in their widely-received work, 2004/2011, provide a lot of empirical evidence from surveys in 76 countries from 1981 to 2001, which include the North Atlantic arena, other developed Western capitalist nations, and also Russia, China, Eastern Europe, most of Latin America, India, Vietnam, several states of Sub-Saharan Africa and several Muslim nations, including Morocco, Algeria, Egypt, Jordan, Turkey, Iran, Pakistan, Bangladesh and Indonesia, and came to the conclusion that *the driving force behind what they and many others call "the secularization process" is growing human security.*

The basic contradiction with Inglehart's sociology arises because we perceive the problem of the *shadow economy* and the *loss of economic and social values* as a big threat to the fabric of a free society. The *'secular values'* of society combine according to Inglehart a low importance given to religion, a low importance given to obedience in child education, a weak sense of national pride, a distance from the attitude that the main goal of one's life is directed towards making one's parents proud, full acceptance of divorce and abortion, no limits on selling foreign goods, and little respect for authority in society.

One of the aims of our book is to provide an empirical link between this important dimension of economics and the more traditional question writing of the hitherto existing uses of the *World Values Survey* data in cross-national comparative research, such as the emphasis on religion and the secularization process. Possible negative impacts of the secularization process on economic morality have never been really probed in cross-national value research, and yet, economic theory places an ever-increasing emphasis on such processes as the driving force of economic decline (see Chapter 3, below). In fact, the shadow economy data contained in the *World Values Survey* provide a rich, as yet untapped resource on research about the shadow economy.

The data 2005/2008 contain the following relevant items for shadow economy research:

- Justification of social behaviors
- (V198) Justifiable: claiming government benefits even if one is not entitled to them
- (V199) Justifiable: avoiding a fare on public transport
- (V200) Justifiable: cheating on taxes
- (V201) Justifiable: someone accepting a bribe

The earlier surveys contained also additional data on

- Justifiable: paying cash (F131)
- Justifiable: manipulation of food (F138)
- Justifiable: buy stolen goods (F139)
- Justifiable: keeping money that you have found (F140)

Economic theory today offers a clear verdict on the destructiveness of the shadow economy for the processes of economic growth (see Chapter 3, below). But what are the relationships between the process of secularization and the shadow economy, especially at middle income levels? The Inglehart school goes as far as to say that what they term *'self-expression'* i.e., the priority of the quality-of-life, a feeling of happiness, the preparedness to be politically active, the acceptancy of homosexuality, and social capital are the most robust pre-conditions of what Inglehart calls *'effective democracy'*. This is an indicator, which combines performance in terms of *'human rights'* and the rule of law (concept of the World Bank).

In Inglehart and Welzel, 2005, we learn that two dimensions are said to dominate the picture of the *World Values Survey*: *(1) Traditional vs. Secular-rational values and (2) Survival vs. Self-expression values.* These two dimensions explain, according to Inglehart, more than 70 percent of the cross-cultural variance on scores of more specific values. According to Inglehart and Welzel, 2005, the Traditional/Secular-rational values dimension reflects the contrast between societies in which religion is very important and those in which it is not. A wide range of other orientations are closely linked with this dimension. Societies near the traditional pole emphasize the importance of parent-child ties and deference to authority, along with absolute standards and traditional family values, and reject divorce, abortion, euthanasia, and suicide. These societies have high levels of national pride, and a nationalistic outlook, measured by the rejection of the import of foreign goods. Societies with secular-rational values have the opposite preferences on all of these topics.

The second major dimension of cross-cultural variation in Inglehart and Welzel, 2005 is linked to the transition from industrial society to post-industrial societies, where priorities have shifted from an emphasis on economic and physical security toward increasing emphasis on subjective well-being, self-expression and the quality of life.

As we show in Chapter 4 of this book, there was frequent criticism in the social science literature of Inglehart's approach, but there has never been an all-encompassing attempt to arrive at a totally new factor analytical re-interpretation of the totality of the *World Values Survey* data. Survival values/self-expression values (and not traditional values versus secular values) include items on divorce and abortion and an item on limits on selling foreign goods, while the economic "bread and butter" item on incomes and jobs, currently listed in the traditional versus secular values dimensions, very plausibly might be listed under the survival dimension. Also in mathematical-methodological terms, criticism directed at the dominant Inglehart approach could be manifold, as we show in our methodological chapter: Inglehart and his associates use standard factor analysis with orthogonal varimax rotation, which does not allow for correlations between the factors. More contemporary approaches allow for such correlations, and especially prefer, among others, the promax factor rotation model, which we use in our analysis.

Policy pointer:

In Inglehart and Welzel, 2005, we read that two dimensions dominate the picture of the *World Values Survey*: (1) Traditional vs. Secular-rational values and (2) Survival vs. Self-expression values. These two dimensions explain, according to Inglehart, more than 70 percent of the cross-cultural variance on scores of more specific values. According to Inglehart and Welzel, 2005, the Traditional/Secular-rational values dimension reflects the contrast between societies in which religion is very important and those in which it is not. A wide range of other orientations are closely linked with this dimension. Societies near the traditional pole emphasize the importance of parent-child ties and deference to authority, along with absolute standards and traditional family values, and reject divorce, abortion, euthanasia, and suicide. These societies have high levels of national pride, and a nationalistic outlook, measured by the rejection of the import of foreign goods. Societies with secular-rational values have the opposite preferences on all of these topics.

The second major dimension of cross-cultural variation in Inglehart's system is linked to the transition from industrial society to post-industrial societies, where priorities have shifted from an emphasis on economic and physical security toward increasing emphasis on subjective well-being, self-expression and the quality of life.

The theoretical foundations and empirical definitions, offered by Inglehart/Baker, 2000, on the factor 'traditional values' are nothing else than the combination of a low importance given to religion in society, a low importance given to obedience in child education, a weak sense of national pride, distance to the goal of one's life to make one's parents proud, full acceptancy of divorce and abortion, no limits on selling foreign goods, and little respect for authority. In purely conceptual terms, there is indeed no better definition of what Christian [liberation] theologians call capitalist "idolatry" and the "ideological weapons of death" (Hinkelammert, 1983, 1986).

Many readers could remark at this point that the theoretical foundations and empirical definitions, offered by Inglehart/Baker, 2000, on the factor *'traditional values'* are nothing else than the combination of a low importance given to religion in society, a low importance given to obedience in child education, a weak sense of national pride, distance to the goal of one's life to make one's parents proud, full acceptancy of divorce and abortion, no limits on selling foreign goods, and little respect for authority. At this point, many readers will concede that in a way, and in purely conceptual terms, there is indeed no better definition of what Christian [liberation] theologians [in the traditions of the theologian Franz Hinkelammert from Honduras] call capitalist "idolatry" and the "ideological weapons of death" (Hinkelammert, 1983, 1986).

We offer a counter-model: that a moderately interpreted religion is a necessary contribution to human civilization. With the increase of permissiveness in one sphere of life, there is the danger of permissiveness also spreading to other fields, especially economic morality, like the tax morale, the avoidance of the shadow economy, the avoidance of corruption *et cetera.* Religion in the framework of the philosophy of the Enlightenment is a welcome contribution to global civilization. This does not imply that non-believers do not contribute positively to civilization; but there will be limits on future economic growth if the combination of a low importance given to religion, a weak sense of national pride, and a low respect of parents and little respect for authority in society *et cetera* become the generalized norms of society.

1.5. Robert Barro and the Study of Human Values

We already remarked that Barro, Harvard Professor of economics, in many ways has to be regarded as an antagonist of Inglehart's cherished traditional value decline. Barro speaks decidedly in favor of the importance of religion for sound economic growth and long-run economic well-being. In Barro/McCleary, 2003 the authors are well aware that there are large positive tradeoffs for the Muslim religion with the belief in heaven and, even more so, with the belief in hell. They base their analysis on the mathematical statistical analysis of world-wide patterns of economic growth, driven or impeded by determining variables, among them variables pertaining to religious beliefs. Barro and McCleary, 2003, instead of viewing *'Islam'* or *'religious beliefs'* in general as an *'impediment'* of economic growth, tend to see them as requirements of a resilient society today:

> 'Our central perspective is that religion affects economic outcomes mainly by fostering religious beliefs that influence individual traits such as thrift, work ethic, honesty, and openness to strangers. For example, beliefs in heaven and hell might affect these traits by creating perceived rewards and punishments that relate to "good" and "bad" lifetime behavior. In this perspective, organized religion—and, more specifically, attendance at religious services—would affect economic performance mostly indirectly, that is, through influences on the religious beliefs. Hence, we envision a chain whereby church attendance [i.e., religious service attendance] affects religious beliefs, which affect individual traits, which affect economic outcomes. [...] Although religiosity tends to decline overall with economic development, the partial relations depend on the specific dimensions of development. For example, the measures of religiosity are positively related to education, negatively related to urbanization, and positively related to the presence of children. Increased life expectancy tends to be negatively related with church attendance [i.e., religious service attendance] but positively related to religious beliefs.'
>
> [...] For given church attendance [i. e., religious service attendance], *increases in some religious beliefs—notably in hell, heaven, and an after-life—tend to increase economic*

> *growth. There is also some indication that the stick represented by the fear of hell is more potent for growth than the carrot from the prospect of heaven.* We should stress that these patterns of growth effects apply when we control for reverse causation by using the instrumental variables suggested by our analysis of the determinants of religiosity. '
>
> 'The instruments are the existence of a state religion, the presence of government regulation of religion, the extent of religious pluralism, and the composition of adherence among the main religions. The results remain intact when we enter the composition of religions directly into the growth equations. Based on the arguable exogeneity of the instrumental variables, we think that our estimates reflect causal influences from religion to economic growth, rather than the reverse. *Our conjecture is that higher religious beliefs stimulate growth because they help to sustain aspects of individual behavior that enhance productivity.* '

In the following, we will highlight the urgency of the matters under debate here. Not to answer these questions already in the present *"background chapter"*, but just to give our readers an indication as to how relevant these issues might be for the politics and economics of our globe. Combining the economic growth figures for the period 2008-2011 in a single indicator via the technique of principal components analysis (see Chapter 3, below), based on the data and projections of the IMF (see variable 200 in our freely available data base, created for the purpose of our present studies http: //www. hichemkaroui. com/?p=3709), we arrive at the following global map of a combined factor of economic rates,[7] 2008-2011. This map shows how the economic crisis of 2008 hit the countries, whose value systems are especially characterized by low importance given to religion, a weak sense of national pride, and a low respect of parents and little respect for authority in society. It is very tempting to see this map as the realization of the Khaldūn/Buddenbrooks cycle:

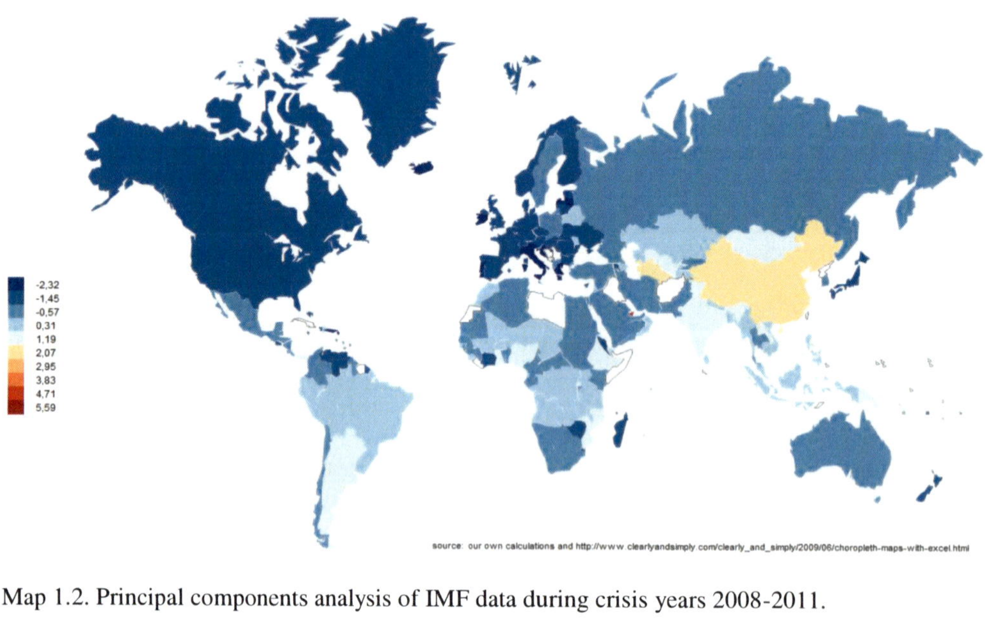

Map 1.2. Principal components analysis of IMF data during crisis years 2008-2011.

[7] It is important to emphasize that the numbers are factor scores and not economic growth rates

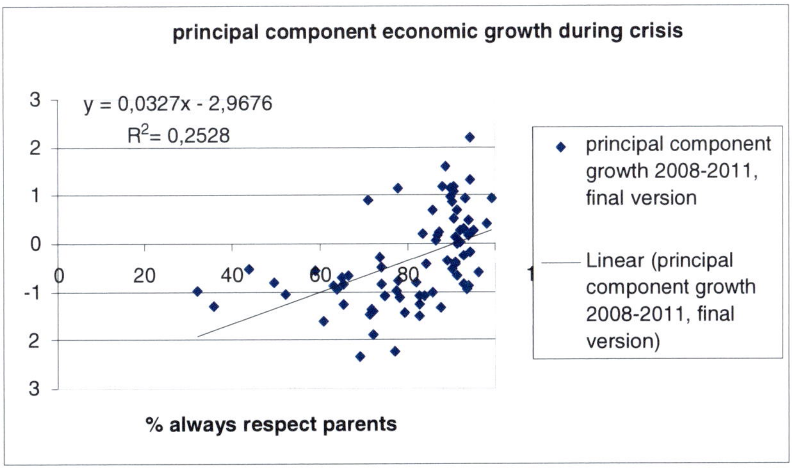

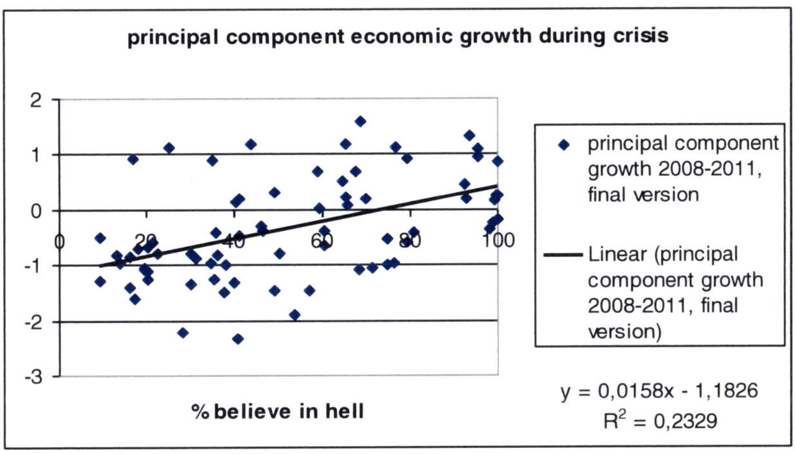

Graph 1.2. Is the resilience of economic growth during the current global economic crisis culture-driven? *World Values Survey* data and current economic growth in the world system.

A first inspection of the facts now seems to suggest that *value change is becoming destructive indeed and undermines the very basis of economic growth in the West.* Respect of parents is related in a clear-cut positive manner to economic growth, as well as the belief in hell, as correctly predicted by Barro:

> *Policy pointer:*
> Robert Barro, Harvard Professor of economics is decidedly speaking in favor of the importance of religion for sound economic growth and long-run economic well-being. Instead of viewing 'Islam' or 'religious beliefs' in general as an 'impediment' of economic growth, he tends to see them as requirements of a resilient society today. Barro thinks that there are causal influences from religion to economic growth, rather than the reverse. His conjecture is that higher religious beliefs stimulate growth because they help to sustain aspects of individual behavior that enhance productivity. Respect of parents is related in a clear-cut positive manner to economic growth, as well as the belief in hell, as correctly predicted by Barro.

1.6. Barro and the Cycles of Convergence and Divergence

The time has come to present our readers with the facts of long-term economic convergence and divergence. The very idea of a Khaldūn/Buddenbrooks cycle would suggest that there are dramatic ups and downs in the fate of nations across time. For a number of years now, the economic discipline debated these issues rationally and without emotions (Barro and Sala-i-Martin, 1992; Barro and Ursúa, 2008; Barro, Sala-i-Martin et al., 1991; Jaeger and Springler, 2012; and Mankiw, Romer and Weil, 1992). The recent upsurge in the economic power of many Muslim countries and the onset of the decline of the West makes such a debate very necessary, and puts all the discourse on alleged Muslim poverty and alleged Muslim radicalism on the one hand and Western secularism as a precondition of modernization, democracy and economic growth on the other hand truly on its head.

For Barro's political economy of growth (Barro, 2012, 2013), the *"iron law of convergence"* would suggest that countries eliminate the initial gaps in levels of real per capita GDP at a rate around 2% per year. Convergence at a 2% rate implies for Barro, 2012 that it takes 35 years for half of an initial gap to vanish and 115 years for 90% to disappear. Now this "Barro" convergence is conditional on differences in varying cross-economy characteristics, such as saving rates or fertility rates or the quantity of human capital or the institutional quality or colonial history or geographical features. Because of the conditioning variables, Barro's results were more pessimistic than the iron-law convergence rate would originally have suggested (Barro, 2012, 2013). Many sub-Saharan African countries or North Korea or Burma or Bolivia or Venezuela might not converge at all if key underlying variables, such as the quality of human capital and institutions, were not improved, Barro said in 2012, 2013.

For Barro, 2013, the convergence property in the neoclassical model derives from diminishing returns to capital. Economies that have less capital per worker (relative to their long-run capital per worker) tend to have higher rates of return and higher growth rates (Barro, 2013). The convergence is described by this kind of political economy of economic growth as being "conditional" because the steady-state levels of capital and output per worker depend in the neoclassical model on the propensity to save, the growth rate of population, and the position of the production function characteristics that may vary across economies (Barro, 2013). For neoclassical growth economics, additional sources of cross-country variation in growth rates might be especially government policies with respect to levels of consumption spending, protection of property rights, and distortions of domestic and international markets.

For Barro, 2013, it's clear that given the values of the other explanatory variables the neoclassical model predicts a negative coefficient on initial GDP, which enters in the system in logarithmic form. *The coefficient on the log of initial GDP is interpreted as a conditional rate of convergence.* If the other explanatory variables are held constant, then the economy tends to approach its long-run position at the rate indicated by the magnitude of the coefficient. The estimated coefficient has a magnitude of -0. 025 (s. e. = 0. 003) and is highly significant and implies a conditional rate of convergence of 2.5% per year.

In their neoclassical, more pessimistic counter position, referred to more than 1000 times in the literature, Mankiw, Romer and Weil, 1992, started from Solow, 1956, model which assumed a standard neoclassical production function with decreasing returns to capital. Taking the rates of saving and population growth as exogenous, Solow indeed showed that these two variables determine the steady-state level of income per capita. Because saving and population growth rates vary across countries, different countries reach different steady states. Mankiw, Romer and Weil, 1992 started from the assumption that Solow's model wields simple testable predictions about how these variables influence the steady-state level of income.

- The higher the rate of saving, the richer the country
- The higher the rate of population growth, the poorer the country.

The position taken up by Barro and associates in this context is quite different and more optimistic in terms of the possibility to reach the goal of convergence for the periphery (Barro and Sala-i-Martin, 1992; Barro and Ursúa, 2008; Barro, Sala-i-Martin et al., 1991): a country's per capita growth rate tends to be inversely related to its starting level of income per person. Therefore, in the absence of shocks, poor and rich countries would tend to converge in terms of levels of per capita income. *Beta-convergence* relates to poor economies growing faster than rich ones, and *Sigma-convergence*, involves a decline over time in the cross-sectional dispersion of per capita income or product. The convergence coefficient Beta depends on the productivity of capital and the willingness to save. There is a similarity in the estimated rates of Beta-convergence in different contexts. This applies to U.S. States and to 73 regions of seven European countries (Germany, the United Kingdom, Italy, France, the Netherlands, Belgium, and Denmark) from 1950 to 1985. The estimated rates of Beta-convergence are similar to those found for the United States; and there would be no evidence to suggest that poor regions, such as those in southern Italy, are being systematically left behind in the growth process.

However, as we aim to show in our Graph 1.3 and 1.4, there are severe and very large upswings and downswings of real GDP per capita along the paths of history. This holds for both the very long haul for the last two millennia as well as the time period for the last two centuries, all documented in Maddison's work (Bolt and van Zanden, 2013). The *historical* unprecedented relative well-being of the European Mediterranean and the Muslim world emerges from these long-term data:

The center seems to be characterized at first sight during the last 150 years by tendencies towards a successful convergence over time, while the periphery of the world system suffers recurring structural breaks and even reversals in their convergence path. We show here the trajectories of the real GDP in purchasing power parities in constant 1990 $ in Argentina, Chile, and the BRIC countries, as well as in Japan and the United States, all in relationship to the real GDP in constant 1990 $ in the United Kingdom, the world's leading industrial country since the age of the industrial revolution. The two countries for which we have very long data series and a large Muslim population, India and Indonesia, seem to suggest that after a longer process of divergence, there is now a stronger trend towards convergence setting in again. The same holds true for the other two countries with very large Muslim populations, covered in the Maddison data set: China and Russia.

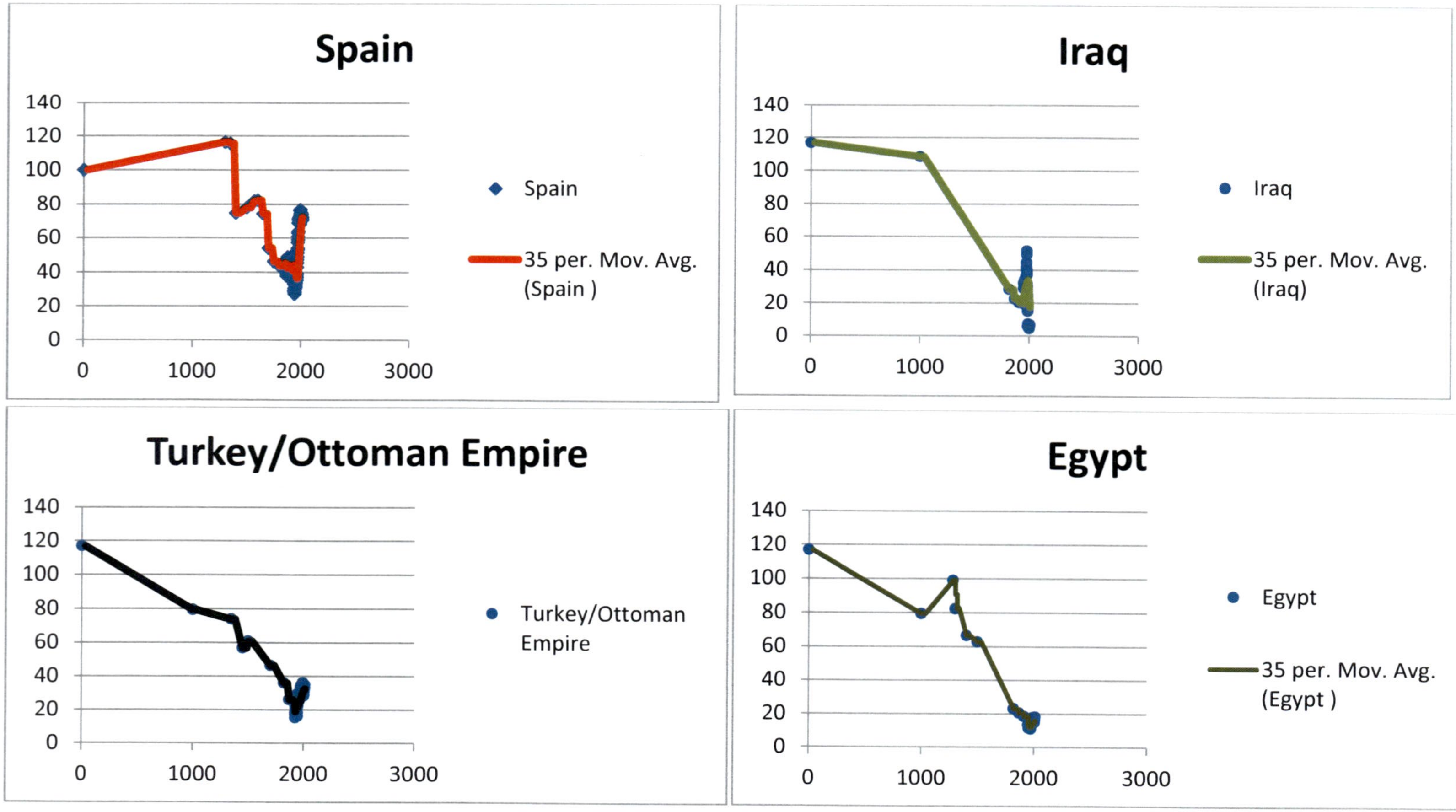

Graph 1.3. (Continued)

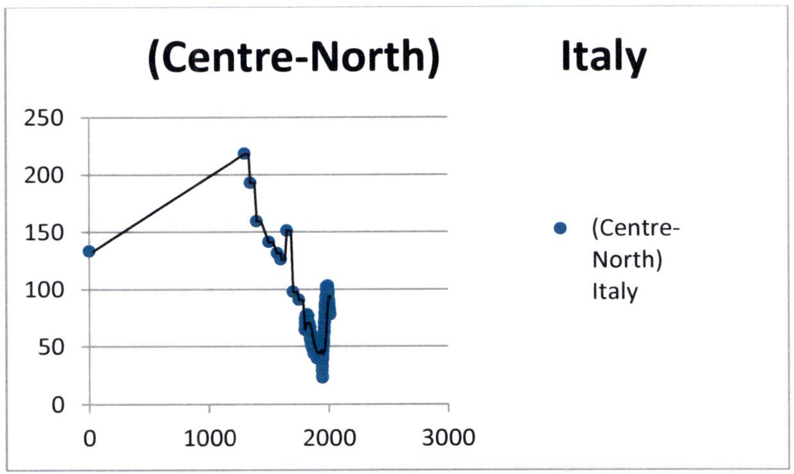

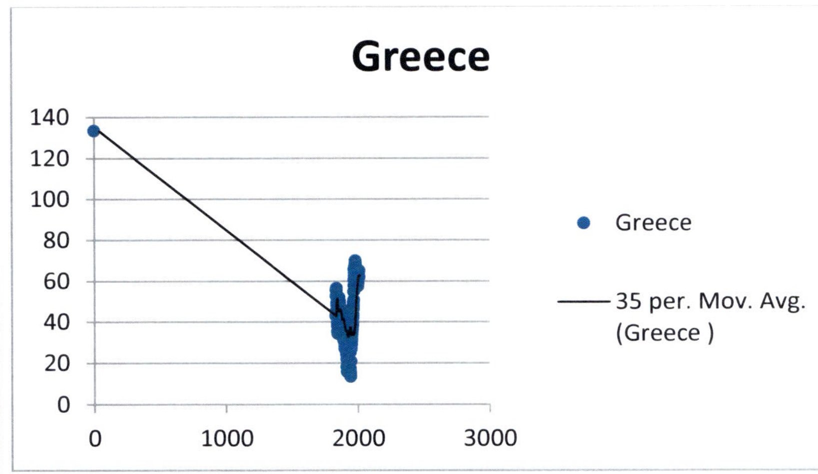

Graph 1.3. The very long two millennia haul and the Maddison data: countries of the world system, constant real GDP per capita in relationship to the England/UK (England = 100). Results for Spain, Iraq, Turkey/Ottoman Empire, Egypt, Italy and Greece and projected onto the Western calendar. 35 year period sliding averages.

Legend: our own compilations, based on the Maddison data sets, as documented in Bolt and van Zanden, 2013.

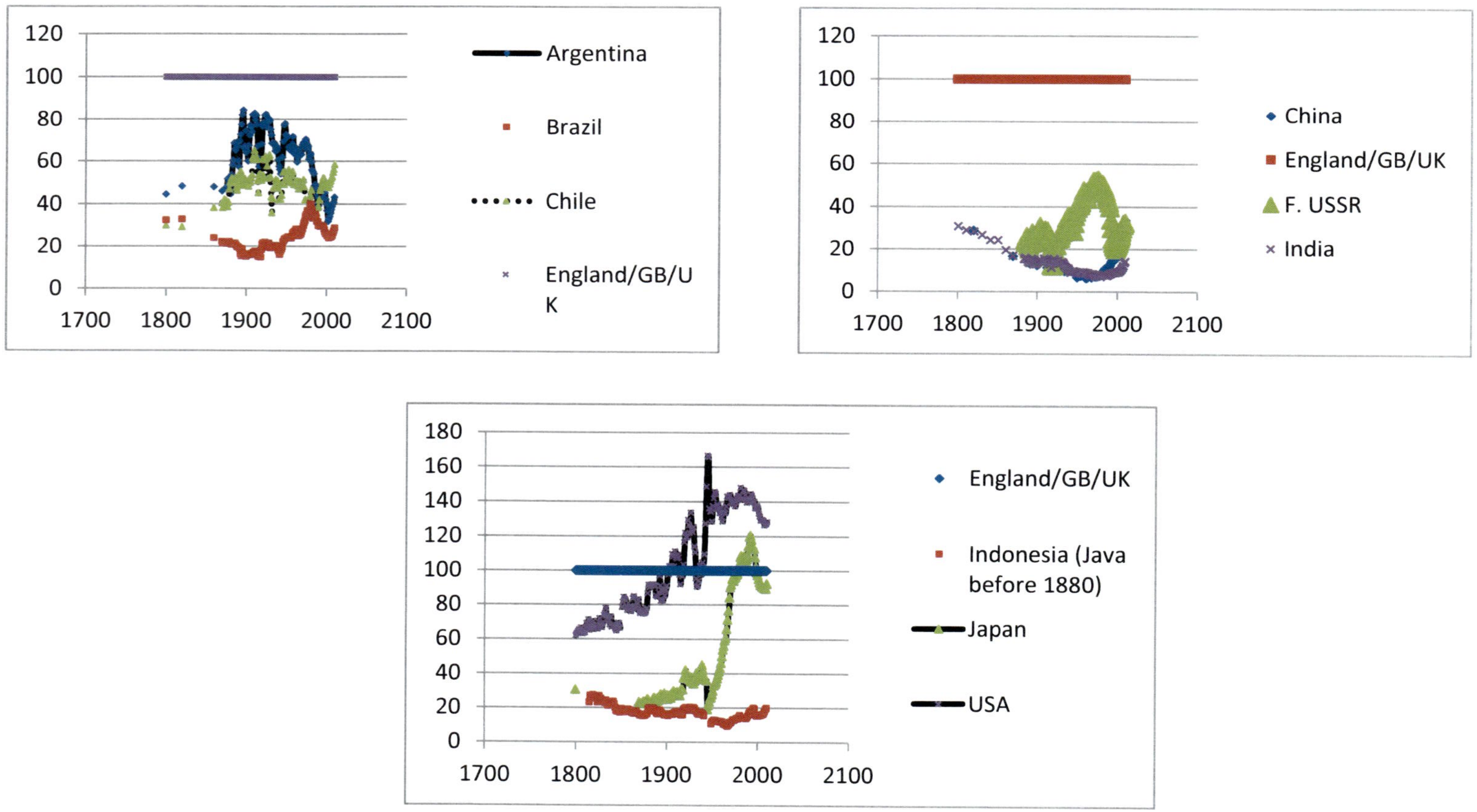

Graph 1.4. The two century long haul since industrialization and the Maddison data: countries of the world system, constant real GDP per capita in relationship to the England/UK (England = 100) and projected onto the Western calendar.
Legend: our own compilations, based on the Maddison data sets, as documented in Bolt and van Zanden, 2013.

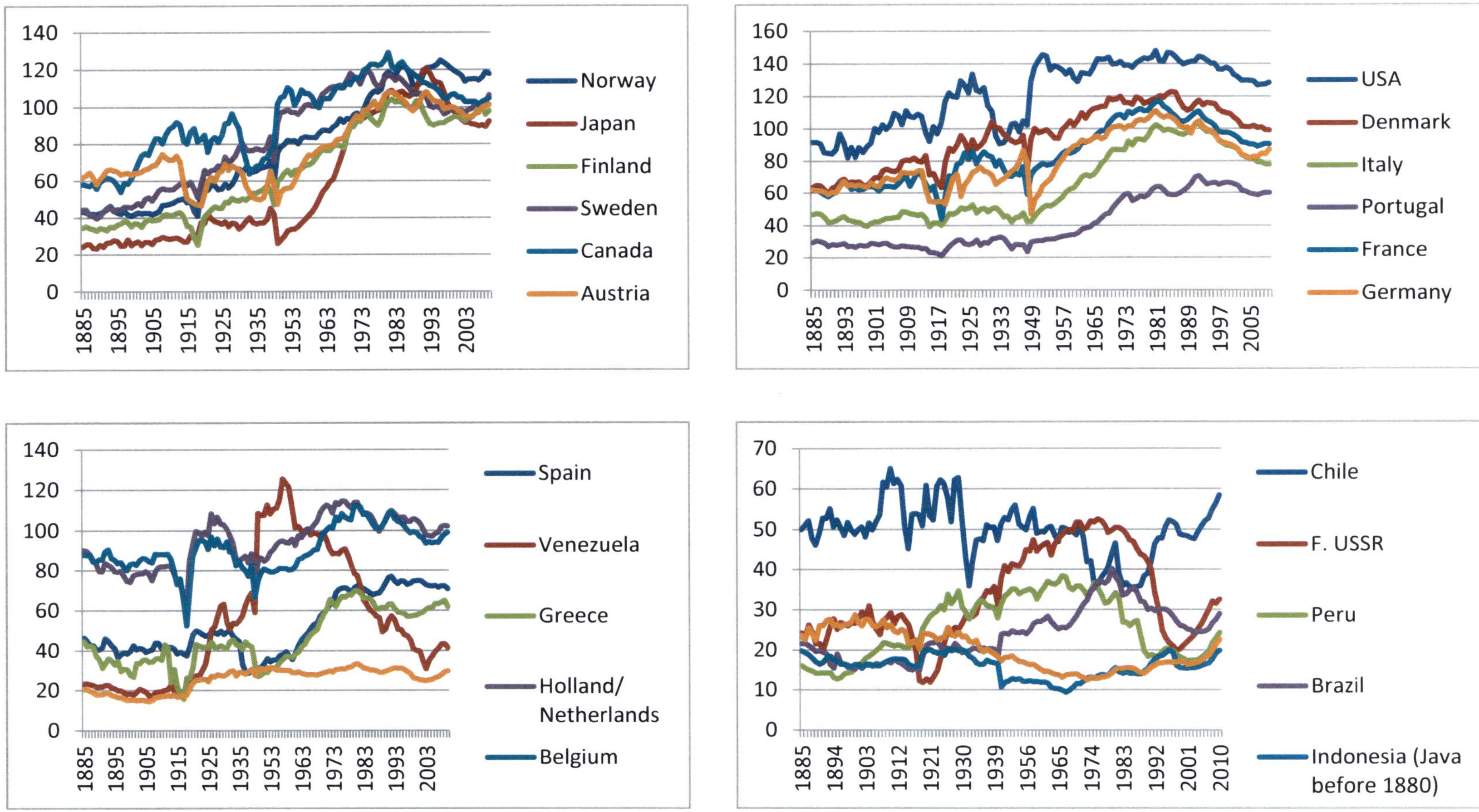

Graph 1.5. (Continued)

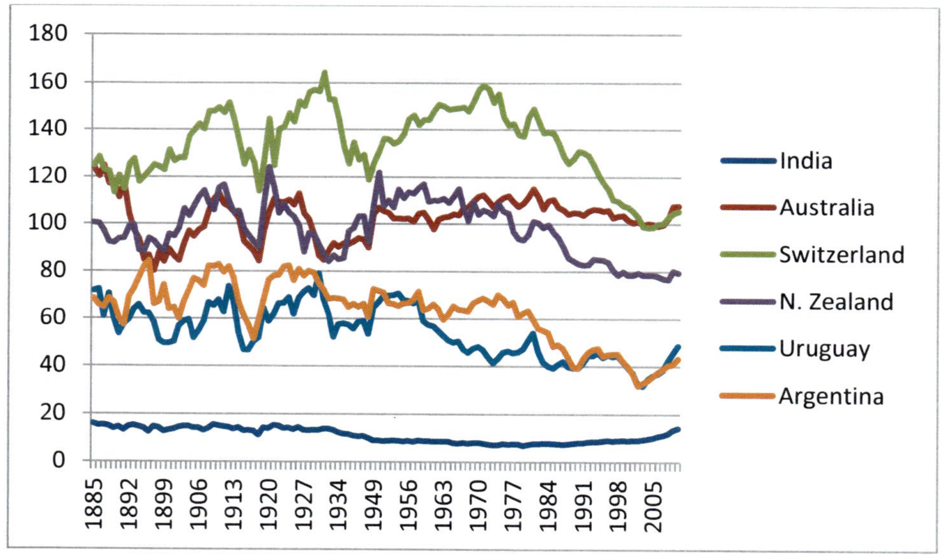

Graph 1.5. The paths of convergence, 1885-2010. Country GDP per capita in relationship to the UK GDP per capita. Legend: our own compilations, based on the Maddison data sets, as documented in Bolt and van Zanden, 2013.

Graph 1.5 further illustrates the skeptical position, taken up by Mankiw, Romer and Weil, 1992. Even the countries that raced ahead over a very long period, like Japan, Finland, Sweden, Canada and Austria fell back against the UK from the times of Thatcherism onwards, and the same can be said about the US, Denmark, Italy, Portugal, France and Germany. As is evident from Graph 1.5, we have ordered the sequence of the mentioning of the different countries by their percentage size of relative income convergence with the United Kingdom, rightly considered by Karl Marx as the "typical case" of capitalism.

Spain, Venezuela, Greece, the Netherlands, Belgium and Columbia are not very stable in their *"convergence cycles"*, rather than to speak about *"convergence paths"*. They and Chile, Russia/the USSR, Peru, Brazil, Indonesia, Sri Lanka, India, Australia, Switzerland, New Zealand, Uruguay and Argentina often present very sharp and cyclical ups and downs of their convergence behavior over time.

Policy pointer:

For Barro, 2013, it's clear that given the values of the other explanatory variables, the neoclassical model predicts a negative coefficient on initial GDP, when economic growth should be explained in the long run. The coefficient on the log of initial GDP is interpreted as a conditional rate of convergence. If the other explanatory variables are held constant, then the economy tends to approach its long-run position at the rate of convergence of 2.5% per year.

A country's per capita growth rate tends to be inversely related to its starting level of income per person. Therefore, in the absence of shocks, poor and rich countries would tend to converge in terms of levels of per capita income. Beta-convergence relates to poor economies growing faster than rich ones, and Sigma-convergence involves a decline over time in the cross-sectional dispersion of per capita income or product. The convergence coefficient Beta depends on the productivity of capital and the willingness to save.

The Japanese economist Akamatsu, 1961, puts the "differentiation" of the world economy into the center of his theoretical developments of economics. The differentiation of the world economy leads to the rapid diffusion of new techniques to rising industrial nations, which starts with the import of new commodities by these nations. In time, techniques and capital goods are imported as well, and homogenous industries are being established. The uniformization of both industry and agriculture gave rise to the fierce and conflictive competition between Europe, the United States and Japan in the last quarter of the 19th Century. When an innovation occurs in some industry in an advanced nation, investment is concentrated there, causing a rise in the trade cycle. Innovation leads to an increase in exports, and the nation's prosperity creates and increases the import of raw materials and foodstuffs. Akamatsu sees a counter-movement in other parts of the world, centered on the rising production of gold, which, according to him, leads to an increase in effective demand and further stimulates exports of the innovating nation. In that way, world production and trade expand, prices increase and a world-wide rise in the long-term trade cycle results.

However, innovations spread from the innovating nations to other nations, leading to the development of industries in those countries, with the result of a conflictive relationship with the industries of the innovating nation. Exports of the innovating nation become stagnant, and on the world level, there is a tendency towards overproduction, prices turn downwards, and the rates of growth of production and trade fall. What later research tended to call the first, rising A-phase of the long economic cycle was for Akamatsu a period of differentiation in the world economic structure. The "falling period" (or B-phase of the cycle) Akamatsu argues, coincides with a process of uniformization in world economic structure.

Arrighi, Silver and Brewer (2003) further developed these arguments, put forward by Akamatsu, and consciously linked their theoretical advances with the models implied in the works of Raymond Vernon (1966, 1971), which specify the life cycle of a product as defined by introduction, growth, maturity, saturation, and decline.

> "In this representation, profit-oriented innovations (and their impact on competitive pressures) cluster in time, generating swings in the economy as a whole from long phases of predominating "prosperity" to long phases of predominating "depression." *Yet it is plausible to hypothesize that they also cluster in space.* That is to say, we can substitute "where" for "while" [...] and read it (i.e., a mathematical formulation by the authors] as a description of a spatial polarization of zones of predominating "prosperity" and zones of predominating "depression" [...]Moreover, as innovators in wealthy countries reap abnormally high rewards relative to effort, over time the environment for innovations in these countries improves further, *thereby generating a self-reinforcing "virtuous circle" of high incomes and innovations. The obverse side of this virtuous circle is a second tendency--the tendency, that is, for the poorer countries at the receiving end of the process to reap few, if any, of the benefits of the innovations. [...]As a result, even when they do not initiate the innovations, wealthy countries tend to be in an incomparably better position than poor and middle-income countries to reap their benefits and shift their costs and disruptions onto others.* In short, opportunities for economic advance, as they present themselves successively to one country after another, do not constitute equivalent opportunities for all countries." (Arrighi, Silver and Brewer, 2003, our emphasis added)

Policy pointer:

Heterodox economic theories, influenced by the thinking of long-term economic cycles and dependency theory tended to view the problem of convergence in different terms from the Barro-model, described above.

Already the Japanese economist Akamatsu, 1961, puts the "differentiation" of the world economy into the center of the theoretical developments of economics. The differentiation of the world economy leads to the rapid diffusion of new techniques to rising industrial nations, which starts with the import of new commodities by these nations. In time, techniques and capital goods are imported as well, and homogenous industries are being established. The uniformization of both industry and agriculture gave rise to the fierce and conflictive competition between Europe, the United States and Japan in the last quarter of the 19th Century. When an innovation occurs in some industry in an advanced nation, investment is concentrated there, causing a rise in the trade cycle. Innovation leads to an increase in exports, and the nation's prosperity creates and increases the import of raw materials and foodstuffs. Innovations spread from the innovating nations to other nations, leading to the development of industries in those countries, with the result of a conflictive relationship with the industries of the innovating nation. Exports of the innovating nation become stagnant, and on the world level, there is a tendency towards overproduction, prices turn downwards, and the rates of growth of production and trade fall. The first, rising A-phase of the long economic cycle will be a period of differentiation in the world economic structure, while the "falling period" (or B-phase of the cycle) will coincide with a process of uniformization in the world economic structure.

Arrighi, Silver and Brewer (2003) further developed these arguments, put forward by Akamatsu, and consciously linked their theoretical advances with the models implied in the works of Raymond Vernon (1966, 1971), which specify the life cycle of a product as defined by introduction, growth, maturity, saturation, and decline. Profit-oriented innovations (and their impact on competitive pressures) cluster in time, generating swings in the economy as a whole from long phases of predominating "prosperity" to long phases of predominating "depression." Yet it is plausible also to hypothesize that they also cluster in space.

The kind of wealth that First World countries had attained through industrialization appears to have been an instance of "oligarchic wealth" in contrast to "democratic wealth." The intense competition that ensued from generalized industrialization efforts did not just prevent Third World countries from attaining their objective; it also tended to undermine the industrial foundations of the oligarchic wealth of First World countries. But today, the major emerging other new Muslim economic powers are: Bangladesh; Indonesia; the Islamic Republic of Iran; Malaysia; Pakistan; and Turkey.

Our long-term optimistic assessment of global changes to the benefit of non-Western regions is also supported by a rigorous UNDP-type-infant mortality reduction index on the basis of World Bank data since 1960.

Contrary to the endless talk on poverty and social exclusion in the Muslim world in the Western media, a very different picture emerges: some of the countries with the most rapid social improvements on the globe were Arab countries, and on an overall basis, especially since 1995, Muslim and other ascending Third World nations by far outperform the leading Western countries in their social growth dynamics. Many leading Western nations are confronted with an implosion of their development dynamics.

With such ongoing changes in global power structures, the time has come to evaluate anew also value changes, taking place on the international level.

For Arrighi, Silver and Brewer, 2003 the following further tendencies hold:

- The kind of wealth that First World countries had attained through industrialization appears to have been an instance of *"oligarchic wealth"* in contrast to *"democratic wealth."* The intense competition that ensued from generalized industrialization efforts did not just prevent Third World countries from attaining their objective; it also tended to undermine the industrial foundations of the oligarchic wealth of First World countries. This tendency was especially in evidence in the 1970s, when the worldwide intensification of competitive pressures on industrial producers appeared to be affecting First World countries more negatively than Third World countries.

- The ease with which the United States succeeded in mobilizing resources in global financial markets to defeat the USSR in the 1980s, and then to sustain a long domestic economic expansion and a spectacular boom in the New York Stock Exchange in the 1990s, led to the belief that "America's Back!" Even assuming that U. S. global power was resuscitated as much as this belief implies, it would be a very different kind of power than the one deployed at the height of U. S. hegemony. The new power that the United States came to enjoy in the 1980s and 1990s, rested on the capacity of the United States to outcompete most other states in global financial markets. The sustainability of this negative-sum game for much longer is doubtful.
- The most important tendency for understanding the present and future of the global hierarchy of wealth may be the continuing economic expansion of China. The rise of East Asia seems to be the most hopeful sign that the extreme global inequalities created under European colonial imperialism and consolidated under U. S. hegemony will eventually give way to a more just and equal world.

IMF-data for the Arab countries are available from Algeria; Bahrain; Comoros; Egypt; Jordan; Kuwait; Lebanon; Libya; Mauritania; Morocco; Oman; Qatar; Saudi Arabia; Sudan; Tunisia; and the United Arab Emirates to further analyze these claims. The major emerging other new Muslim economic powers are: Bangladesh; Indonesia; the Islamic Republic of Iran; Malaysia; Pakistan; and Turkey. For analytical purposes, we mention the data for Iran and Turkey separately.

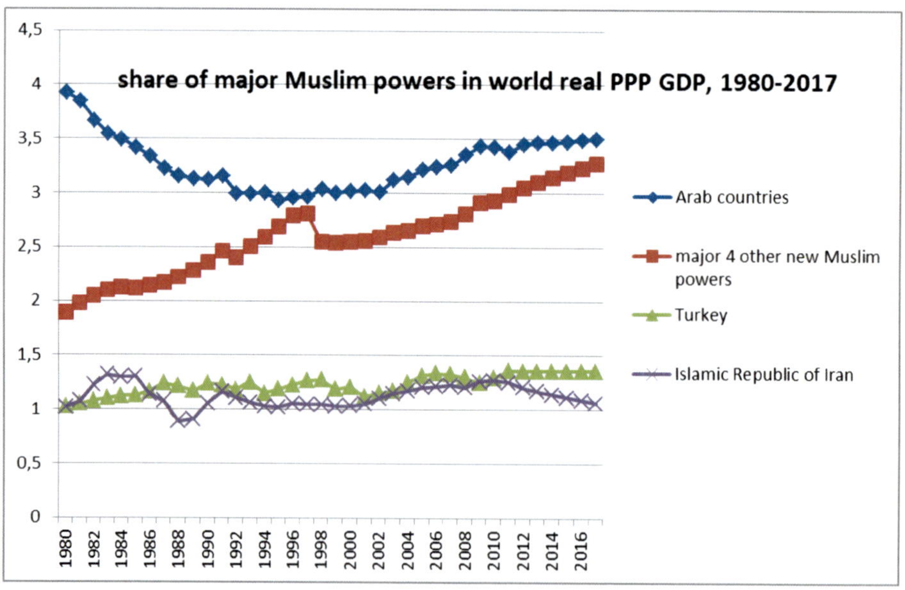

Major 4 other rising Muslim powers: Bangladesh; Indonesia; Malaysia; Pakistan.

Graph 1.6. The rising share of major western regions in global GDP - IMF World Economic Outlook data.

Equally astonishing is the rise of the importance of the BRICS countries in the international system, evidenced by their share in the world purchasing power today:

Our long-term optimistic assessment of global changes to the benefit of non-Western regions is also supported by a rigorous UNDP-type-infant mortality reduction index on the basis of World Bank data since 1960, which could be regarded as a good proxy indicator for Inglehart's "human security", available for many countries in annual time series data. Infant mortality is one of the most telling and powerful indicators of poverty in the world. Projecting the data onto an index, which ranges from 0.0 (worst performance) to 1.0 (best performance) is a powerful and also easy to handle tool of the social sciences. As it is amply known from the UNDP Human Development Reports, such an index weights the value of each country against the worst performance *for each year since 1960,* as compared to the difference between the best and the worst performance in the group of countries with full data and under consideration. The very up-to-date World Bank infant mortality rate data permit such a continuous, country-to-country and year-to-year assessment for none the less than 111 nations of the world, shedding a light on the social dimension of development with a precision thought unthinkable two or three decades ago. We will present these comparisons on the following pages.

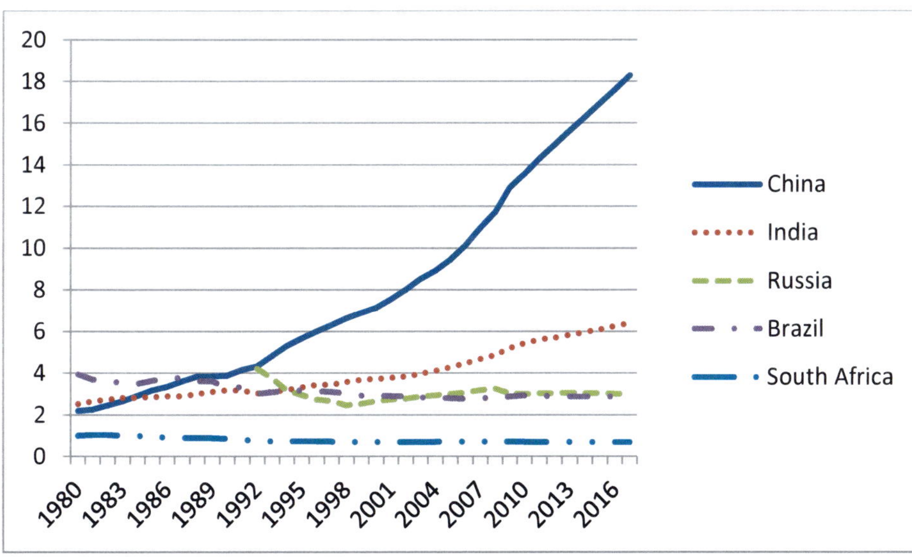

Graph 1.7. The rising power of the BRICS – IMF World Economic Outlook data.

1.7. The Decline of Western Countries, the Analysis of Global Values and the World System Position of the Arab and the Muslim World

Contrary to the endless talk on poverty and social exclusion in the Muslim world in the Western media, a very different truth emerges: some of the countries with the most rapid social improvements on the globe were Arab countries, and on an overall basis, especially

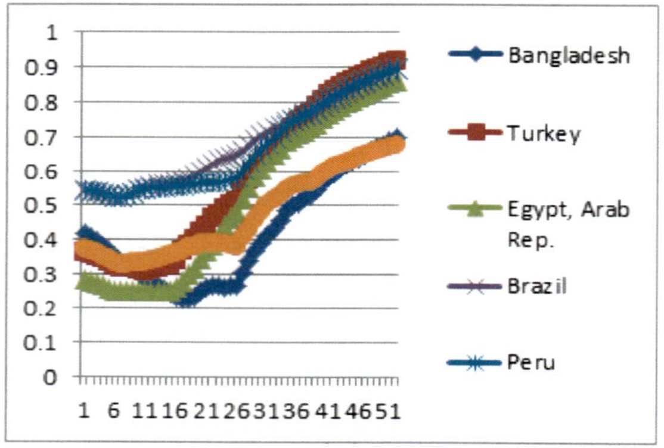

a

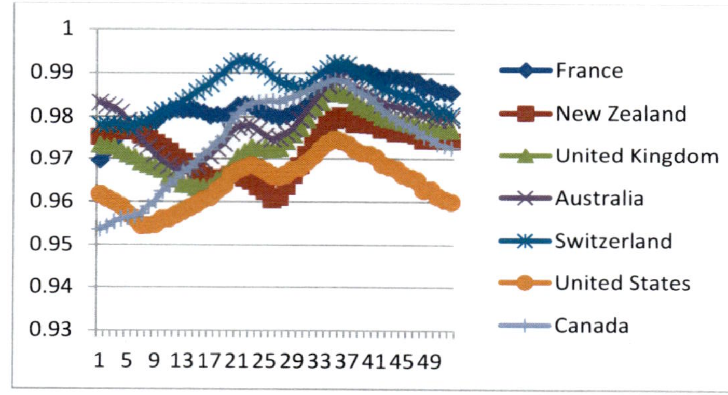

b

Graph 1.8a and 1.8b. The waning of the West: a UNDP-type index of overcoming infant mortality in 111 nations of the world since 1960 – some of the best and some of the worst performers, especially since 1995.

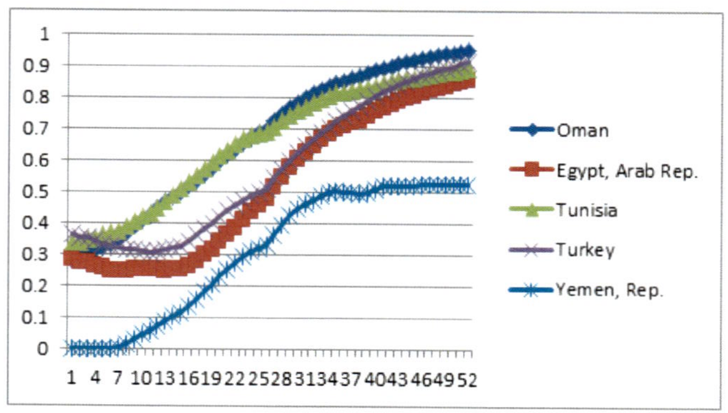

c

Graph 1.8c. The myth of Arab and Muslim poverty - UNDP-type index of overcoming infant mortality in 111 nations of the world since 1960 – results for the Arab nations.

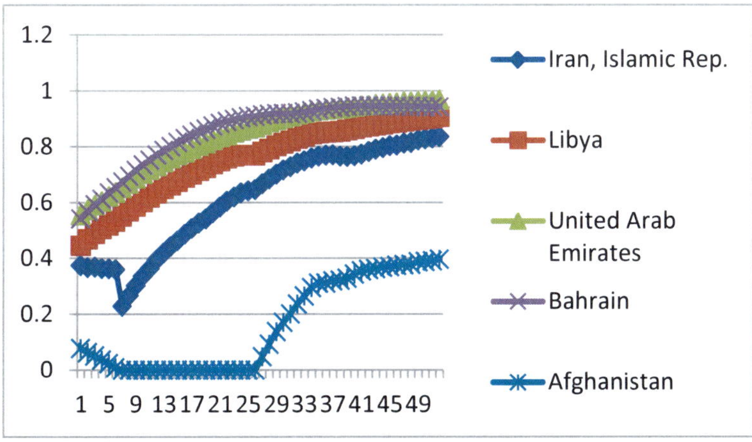
1.2
1
0.8
0.6
0.4
0.2
0
1 5 9 13 17 21 25 29 33 37 41 45 49
Iran, Islamic Rep.
Libya
United Arab Emirates
Bahrain
Afghanistan

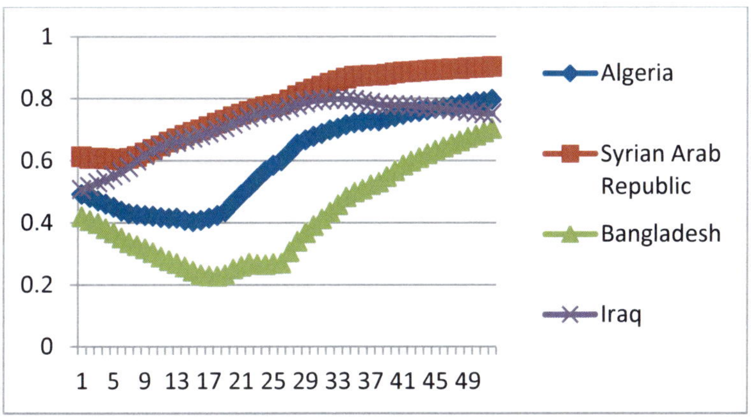
1
0.8
0.6
0.4
0.2
0
1 5 9 13 17 21 25 29 33 37 41 45 49
Algeria
Syrian Arab Republic
Bangladesh
Iraq

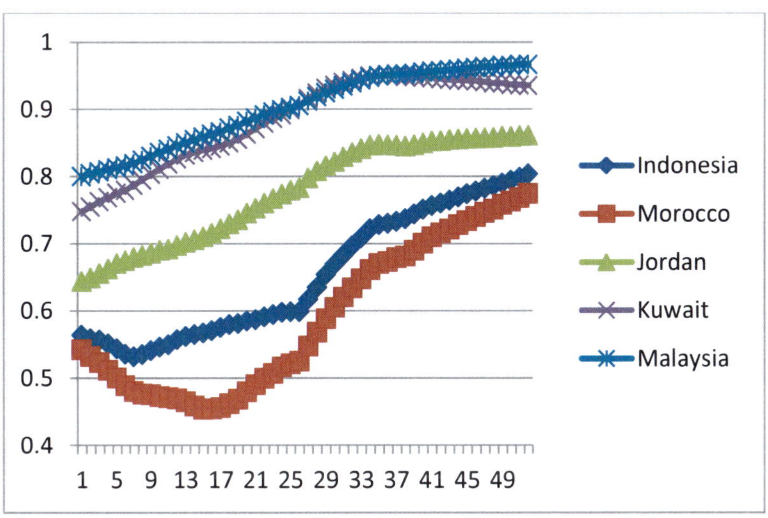
1
0.9
0.8
0.7
0.6
0.5
0.4
1 5 9 13 17 21 25 29 33 37 41 45 49
Indonesia
Morocco
Jordan
Kuwait
Malaysia

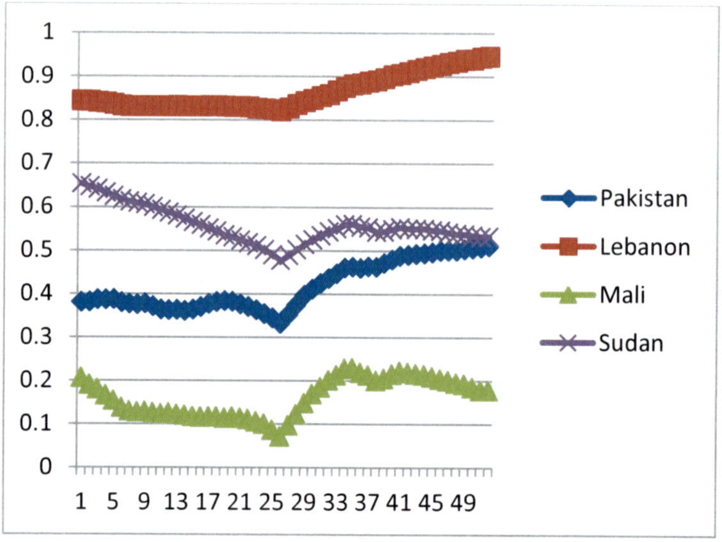

since 1995, Muslim and other ascending Third World nations by far outperform the leading Western countries in the social growth dynamics. Many leading Western nations are confronted with an implosion of their development dynamics. To assess empirically these tendencies, we constructed *for each year since 1960 to 2012* a UNDP type *overcoming of infant mortality index* for all 111 countries of the world with complete World Bank data (http: //data. worldbank. org/indicator/all) – the end result being that *most of the Arab and Muslim countries had a spectacular progress while major Western countries showed a declining position:*

Only a few countries (like Iraq, Sudan, Mali…) do not correspond to our overall picture of the often-spectacular social progress which we are witnessing in so many ascending nations around the globe.

With such ongoing changes in global power structures, the time has come to evaluate anew also value changes, taking place on the international level. Are the cultural codes of the West in crisis still the cultural codes of all the emerging socio-economic forces, which will shape the destiny of the 21st Century? Even a naïve and first impressionistic look at the dominant cultures in the leading industrial countries of the West (Canada, France, Germany, Italy, Japan, the United Kingdom, and the United States) will suggest to us that perhaps in many ways the values in Brazil, India, and South Africa, let alone the emerging new Muslim powers, are characterized by a still greater role of religion in public life and less emphasis on secularism, while in China and Russia, the long years in power of a ruling communist party left their often very devastating hallmarks on society.

1.8. The *World Values Survey* Project – A Necessary Tool to Assess Global Value Change

To make our point in a simple way, and to draw also the attention of skeptics of Western civilization to the utility of the *"World Values Survey"* project, here a few statistics, which further highlight our argument: in the surveys of the last wave of the *World*

Values Survey in the Western countries, comprising Andorra [2005], Australia [2005], Canada [2006], Finland [2005], France [2006], Germany [2006], Great Britain [2006], Italy [2005], Japan [2005], Netherlands [2006], New Zealand [2004], Norway [2007], Spain [2007], Sweden [2006], Switzerland [2007], United States [2006], only *57. 1% of respondents* were of the opinion, that it is *never justifiable to avoid a fare on public transport.* In *highly developed Sweden,* this percentage is *already a minority (45.8%), and in secularist France and in the United States of America, half the population (50.0%) would abide by the law here,* while the other half – to differing degrees, already would be prepared to break the law and take a free ride, from time to time at least, or worse.

What we perceive to be a real crisis of economic and social morality in Western countries is also reflected in the fact that among the 15-29 years old population, only 36.9% openly stated that taking a free ride on public transport is never justifiable, while among the population aged 50 years or more still 70% thought that taking a free ride on public transport is never justifiable.

Our following table, designed to allow our readers a quick glance at the potentials of the *World Values Survey* data, now runs what statisticians call a 'cross-table' between the *acceptancy of prostitution* and the *acceptancy of taking a free ride on public transport* in the West:

Table 1.1. Typical empirical data, available from the *World Values Survey*

	Avoiding a fare on public transport never justifiable
Prostitution never justifiable	75.30%
2	44.50%
3	44.90%
4	42.60%
5	52.40%
6	47.50%
7	45.60%
8	43.00%
9	42.70%
Prostitution always justifiable	56.20%

Source: our own compilations from *World Values Survey*, online data analysis, March 2014, based on the data for: Andorra [2005], Australia [2005], Canada [2006], Finland [2005], France [2006], Germany [2006], Great Britain [2006], Italy [2005], Japan [2005], Netherlands [2006], New Zealand [2004], Norway [2007], Spain [2007], Sweden [2006], Switzerland [2007], United States [2006].

While 70.8 % of the inhabitants of the core-Western countries, who attended religious acts of their respective denomination more than once a week, said that taking a free ride on public transport is never justifiable, only 53.2 % of the secularists, who never attend a religious act, stated that taking a free ride on public transport is never justifiable.

The *World Values Survey* project even allows free internet access to such data from a growing number of Arab countries and Muslim countries in general. Table 1.2 shows the data for the Saudi regions, but of course the low number of representative interview partners in these Saudi regions causes a higher error margin at the 95% confidence interval.

Table 1.2. Typical *World Values Survey* data: the acceptance of a free ride in Saudi society

	Justifiable: avoiding a fare on public transport (BASE=1480)	
Region where the interview was conducted	Never justifiable	Base for mean
SA: Taif	5.30%	75
SA: Al Madinah	13.00%	100
SA: Tabuk	18.70%	75
SA: Dammam \ Khobar	44.40%	225
SA: Makkah	49.00%	100
Total	57.40%	1480
SA: Hail	58.70%	75
SA: Riyadh	68.50%	330
SA: Jeddah	73.00%	300
SA: Qassim	90.00%	100
SA: Abha \ Khamis	90.00%	100

If we now focus our attention on the most important issue of the Arab world today, the issue of democracy and transformation from former dictatorships and semi-dictatorships, we all of a sudden realize how important empirical value research is for the future trajectory of democratic transformation.

1.9. THE *WORLD VALUES SURVEY* PROJECT AND DATA FROM THE MUSLIM WORLD

One of the most robust trends in international opinion surveys during the last two decades has been the desire for democracy and for religion in the Muslim world at the same time. Yes, religion and democracy, that is the wish of the people.

If we look at the results from the Muslim countries in the 2005-2008 wave of *World Values Survey* investigations, 76.8% of Muslims in Egypt [2008], Indonesia [2006], Iraq [2006], Iran [2005], Jordan [2007], Malaysia [2006], Mali [2007], Morocco [2007], Turkey [2007] say that *"God is very important in their life";* and only 5.7% of Muslim respondents in these countries say that democracy is bad or very bad, while an overwhelming majority of them – 94.3% - say that *democracy is very good or fairly good.*

If one looks at the results from earlier wave of surveys in the late 1990s and the beginning of the current millennium (according to the Western calendar), 89.0% of Muslims, and also the religious minorities in Algeria [2002], Egypt [2000], Iran [2000], Iraq [2004], Jordan [2001], Morocco [2001], Morocco [2001], Pakistan [2001], and Saudi Arabia [2003] say that *"God is very important in their life";* and again only 6.6% of

Muslim respondents[8] in these countries say that *democracy is bad or very bad,* while an overwhelming majority of them again say (93.4%) that democracy is very good or fairly good.

In Table 1.3, we highlight the global trends of values and opinions according to the *World Values Survey* and distinguish between the Muslim and the non-Muslim respondents on a global scale. Thus we are presented with the following global mainstream majority Muslim opinions, worldwide, which hardly can be distinguished from family-oriented mainstream pro-democracy global opinion, worldwide. The following 38 main opinions characterize Muslim conservative, family oriented but democratic majority opinion around the globe:

1. Believe in: God
   agree/strongly agree: One of main goals in life has been to make my parents proud
2. Get comfort and strength from religion
3. Believe in: people have a soul
4. Believe in: heaven
5. Moments of prayer, meditation ...
6. How important is God in your life: very important (8, 9, 10)
7. Believe in: hell
   A religious person
8. Believe in: life after death
9. Important child qualities: good manners
10. never justifiable: prostitution
11. Willingness to fight for country
12. Believe in: sin
13. never justifiable: someone accepting a bribe
14. agree/strongly agree: Husband and wife should both contribute to income
15. Traits in a woman: Woman religious very important
16. never justifiable: buy stolen goods
17. never justifiable: cheating on taxes
18. never justifiable: euthanasia
19. never justifiable: avoiding a fare on public transport
20. Human beings should coexist with nature
21. Believe in: devil
22. agree + strongly agree - politicians who don´t believe in God are unfit for public office
23. never justifiable: For a man to beat his wife
    agree/strongly agree: Would give part of my income for the environment
24. Important child qualities: tolerance and respect for other people
25. never justifiable: claiming government benefits even if one is not entitled to them
26. Important child qualities: religious faith
27. agree/strongly agree: religious leaders should not influence government
28. Traits in a woman: Woman educated very important
29. Important child qualities: feeling of responsibility

[8] And similar low percentages from the other religious groups living in these countries.

30. Society must be gradually improved by reforms
31. Child needs a home with father and mother
32. Competition is good, scale 1-3
33. Important child qualities: hard work
34. An essential characteristic of democracy: Governments tax the rich and subsidize the poor (8,9,10)
35. agree/strongly agree: Job best way for women to be independent
36. Hard work brings success, 1-3
37. Importance of democracy: absolutely important (10)
38. Protecting environment vs. Economic growth

Only a *minority of global Muslims* holds the following 36 opinions:

1. Equality above freedom
2. An essential characteristic of democracy: Religious authorities interpret the laws (8,9,10)
3. Political system: Having the army rule (very good+fairly good)
4. An essential characteristic of democracy: The army takes over when government is incompetent (8,9,10)
5. If only one child allowed: boy
6. Agree/strongly agree: University is more important for a boy than for a girl
7. Political system: Having a strong leader (very good/fairly good)
8. Rejecting neighbors: Immigrants/foreign workers
9. Rejecting neighbors: People of a different religion
   No confidence: The United Nations
10. Extreme right (8,9,10: Self positioning in political scale (1 - left ... 10 right))
11. Rejecting neighbors: People of a different race
12. No confidence: The Political Parties
13. Agree/strongly agree: More than one wife
14. Extreme left (1-3: Self positioning in political scale (1 - left ... 10 right))
15. No confidence: The European Union
16. People can only get rich at the expense of others, scale 1-3
17. No confidence: Major Companies
18. No confidence: Labor Unions
19. No confidence: Parliament
20. Rejecting neighbors: People not from country of origin
21. No confidence: The Government
22. Society must be radically changed
23. No confidence: The Press
24. No confidence: The Civil Services
25. No confidence: The Police
26. Competition is harmful, scale 8-10
27. No confidence: Television
28. Disagree + strongly disagree: religious leaders should not influence how people vote
29. Disagree or strongly disagree: Democracy may have problems but is better

30. Ideal number of children 5 or more
31. No confidence: Justice System
32. Satisfaction with the way democracy develops: not at all satisfied
33. No confidence: Armed Forces
34. Political system: Having a democratic political system (very bad, fairly bad)
35. Always justifiable for a man to beat his wife
36. No confidence: Education System

Table 1.3. Results from the latest combined version of the *World Values Survey* – global values and Muslim values (percentages from the global WVS surveys, unweighted by population sizes of the Muslims and non-Muslims in the different countries)

Opinions regarded **as well** compatible with the future construction or present functioning of democracy **(marked in green)**

	global values and conditions of living	global Muslim values and Muslim conditions of living	global values =100, Muslims = x
Believe in: God	88.9	98.9	111.2
agree/strongly agree: One of main goals in life has been to make my parents proud	82.8	94.5	114.1
Get comfort and strength from religion	77.7	94.5	121.6
Believe in: people have a soul	82.8	93.0	112.3
Believe in: heaven	71.4	92.1	129.0
Moments of prayer, meditation …	76.6	91.4	119.3
How important is God in your life: very important (8, 9, 10)	66.4	90.4	136.1
Believe in: hell	59.1	89.9	152.1
A religious person	71.7	86.4	120.5
Believe in: life after death	67.3	84.5	125.6
Important child qualities: good manners	75.9	83.9	110.5
never justifiable: prostitution	61.3	83.0	135.4
Willingness to fight for country	74.4	82.0	110.2
Believe in: sin	72.7	81.8	112.5
never justifiable: someone accepting a bribe	75.3	81.4	108.1
agree/strongly agree: Husband and wife should both contribute to income	84.3	81.2	96.3
Traits in a woman: Woman religious very important	77.6	79.6	102.6
never justifiable: buy stolen goods	73.7	75.2	102.0
never justifiable: cheating on taxes	63.7	74.2	116.5

Table 1.3. (Continued)

	global values and conditions of living	global Muslim values and Muslim conditions of living	global values =100, Muslims = x
never justifiable: euthanasia	45.5	72.7	159.8
never justifiable: avoiding a fare on public transport	58.3	69.9	119.9
Human beings should coexist with nature	81.8	69.7	85.2
Believe in: devil	47.9	69.5	145.1
agree + strongly agree - politicians who don´t believe in God are unfit for public office	47.0	69.0	146.8
never justifiable: For a man to beat his wife	75.5	68.4	90.6
agree/strongly agree: Would give part of my income for the environment	69.1	68.3	98.8
Important child qualities: tolerance and respect for other people	67.0	66.1	98.7
never justifiable: claiming government benefits even if one is not entitled to them	59.6	66.1	110.9
Important child qualities: religious faith	39.6	65.5	165.4
agree/strongly agree: religious leaders should not influence government	63.2	65.1	103.0
Traits in a woman: Woman educated very important	64.4	63.4	98.4
Important child qualities: feeling of responsibility	68.8	63.3	92.0
Society must be gradually improved by reforms	70.0	62.1	88.7
Child needs a home with father and mother	89.6	61.7	68.9
Competition is good, scale 1-3	56.0	61.7	110.2
Important child qualities: hard work	54.7	60.4	110.4
An essential characteristic of democracy: Governments tax the rich and subsidize the poor (8,9,10)	46.7	59.2	126.8
agree/strongly agree: Job best way for women to be independent	63.5	57.4	90.4

Opinions regarded as **well compatible** with the future construction or present functioning of democracy **(marked in green)**

	global values and conditions of living	global Muslim values and Muslim conditions of living	global values =100, Muslims = x
Hard work brings success, 1-3	46.6	56.7	121.7
Importance of democracy: absolutely important (10)	49.6	55.2	111.3
Protecting environment vs. Economic growth	52.0	50.0	96.2
Equality above freedom	41.6	40.8	98.1
An essential characteristic of democracy: Religious authorities interpret the laws (8.9.10)	20.2	40.2	199.0
Political system: Having the army rule (very good+fairly good)	22.2	40.0	180.2
An essential characteristic of democracy: The army takes over when government is incompetent (8.9.10)	23.4	37.1	158.5
If only one child allowed: boy	31.6	35.6	112.7
agree/strongly agree: University is more important for a boy than for a girl	24.4	34.4	141.0
Political system: Having a strong leader (very good/fairly good)	37.9	33.1	87.3
Rejecting neighbors: Immigrants/foreign workers	20.5	31.4	153.2
Rejecting neighbors: People of a different religion	20.5	31.3	152.7
No confidence: The United Nations	18.6	29.4	158.1
extreme right (8.9.10: Self positioning in political scale (1 - left ... 10 right))	23.5	26.7	113.6
Rejecting neighbors: People of a different race	16.8	26.6	158.3
No confidence: The Political Parties	27.9	25.4	91.0
agree/strongly agree: More than one wife		24.8	
extreme left (1-3: Self positioning in political scale (1 - left ... 10 right))	17.5	21.4	122.3
No confidence: The European Union	18.6	20.3	109.1
People can only get rich at the expense of others. scale 1-3	17.8	20.3	114.0

Opinions regarded as **not very well compatible** with the future construction or present functioning of democracy **(marked in yellow)**

Opinions regarded as **not very well compatible** with the future construction or present functioning of democracy **(marked in yellow)**

	global values and conditions of living	**global Muslim values and Muslim conditions of living**	**global values =100, Muslims = x**
No confidence: Major Companies	15.4	18.2	118.2
No confidence: Labor Unions	20.9	17.6	84.2
No confidence: Parliament	20.9	16.8	80.4
Rejecting neighbors: People not from country of origin	12.6	16.5	131.0
No confidence: The Government	18.0	16.0	88.9
Society must be radically changed	12.3	15.9	129.3
No confidence: The Press	13.1	14.2	108.4
No confidence: The Civil Services	15.6	13.7	87.8
No confidence: The Police	15.8	13.6	86.1
Competition is harmful. scale 8-10	9.6	13.0	135.4
No confidence: Television	11.9	12.4	104.2
disagree + strongly disagree: religious leaders should not influence how people vote	16.9	12.2	72.2
disagree or strongly disagree: Democracy may have problems but is better	14.0	12.2	87.1
Ideal number of children 5 or more	6.9	11.6	168.1
No confidence: Justice System	13.7	9.7	70.8
Satisfaction with the way democracy develops: not at all satisfied	14.2	9.2	64.8
No confidence: Armed Forces	10.4	8.2	78.8
Political system: Having a democratic political system (very bad. fairly bad)	9.6	6.2	64.6
always justifiable for a man to beat his wife	2.6	4.7	180.8
No confidence: Education System	6.6	3.5	53.0

Legend: our own calculations based on SPSS 20, from the *World Values Survey* data, freely available at http: //www. worldvaluessurvey. org/

Policy pointer:

To draw the attention of skeptics of Western civilization to the utility of the *"World Values Survey"* project, here a few statistics, which highlight our argument: only 57.1% of respondents in leading Western countries were of the opinion, that it is never justifiable to avoid a fare on public transport. In highly developed Sweden, this percentage is already a minority (45. 8%), and in secularist France and in the United States of America, half the population (50.0%) would abide by the law here, while the other already would be prepared to break the law and take a free ride, from time to time at least, or worse.

While 70.8 % of the inhabitants of the core-Western countries, who attended religious acts of their respective denomination more than once a week, said that taking a free ride on public transport is never justifiable, only 53.2 % of the secularists, who never attend a religious act, stated that taking a free ride on public transport is never justifiable.

The *World Values Survey* project even allows free internet access to such data from a growing number of Arab countries and Muslim countries in general, even at the regional level.

One of the most robust trends in international opinion surveys during the last two decades has been the desire for democracy and for religion in the Muslim world at the same time. Yes, religion and democracy, that is the wish of the people. If we look at the results from the Muslim countries in the 2005-2008 wave of *World Values Survey* investigations, 76.8% of Muslims say that *"God is very important in their life";* and only 5.7% of Muslim respondents in these countries say that democracy is bad or very bad, while an overwhelming majority of them – 94.3% - say that democracy is very good or fairly good.

Chapter 2

THE METHODOLOGY OF THIS PROJECT

2.1. INTRODUCTION

In this chapter we present an overview of the various research methodologies used in this book and discuss their combined use, strengths and limitations. The methodologies are grouped into: sampling surveys and error margins, correlation, regression and factor analysis. We are well aware that this Chapter might be of limited use for the general public readers of this book. The Chapter is primarily intended for use by graduate students of the social sciences, academic research and teaching personnel, statisticians at national governments or international organizations, analytical research officers, working at transnational corporations or NGOs and other practitioners. Despite this chapter is aimed at the abovementioned professionals, it is yet easily accessible to the general public.

Box 2.1. The overall message of this Chapter

This Chapter informs our readers about the methodologies used in this book.

We analyze the original *World Values Survey* data with more than 200,000 global representative respondents and the aggregate country data at our disposal, using composite index methodology, bivariate relationships, correlation coefficients, partial correlation analysis, simple linear regression, and different multiple regression procedures.

We debate the different factor analytical procedures, necessary to analyze the *World Values Survey* data in detail, and allow interested readers – especially social scientists, graduate students, and employees of international bureaucracies and think tanks to look at the simple mathematical models used in this analysis, and the various methodological assumptions for our IBM-SPSS 20/21 statistical analyses.

We also attempt to make our readers familiar with the principles of survey analysis, which are at the heart of the *World Values Survey* project.

In recent years, the interest in and the amount of research on the relationships between sub-processes of development, such as human development, global competitiveness, innovativeness, infrastructure for development, *et cetera,* and their predictors have been increasing due to increased availability of secondary data and improvements of the methodology (see e.g., UNDP, 2013 and WEF, 2013). The growing interest is mainly attributed to the rapid development of international relations, openness, regional shifts in concentration of production, and opportunities provided by information and communication technology and its rapid diffusion. The theoretical literature on each aspect of development is vast, but the empirical evidence on the relationship between the different factors of interest remains poorly investigated. Thus, there is an urgent need for further research and methodological development at different levels of and on different aspects of development. This chapter, by contributing along these lines, is aimed to improve our understanding of the problems, policies and evaluation of their outcomes.

The rest of the chapter is organizes as follows. In Section 2.2 the composite index methodology, including non-parametric, principal component and factor analysis, Varimax and Promax rotation procedures, other parametric indices and their extension are presented. Section 2.3 is about measurement of correlation relationship. The items discussed include bivariate relations, scatter plots, correlation coefficients and partial correlation analysis. Section 2.4 covers simple, multiple and stepwise linear regression analysis. Survey analysis, aggregation of data and sampling are discussed in Section 2.5. In this section, also sampling error and margin and sample size are discussed.

2.2. The Composite Index Methodology

Following the introduction of the meanwhile world-famous Human Development Index and its annual updates in the Human Development Report in recent years (see UNDP, 2013) a more rich literature on the quantitative measurement of development outcomes has been developed. These outcomes are often multidimensional and each of the dimensions is represented by several indicators with both positive and negative effects on the development outcome. In this study the objective is not only to evaluate the effects of certain policy programs, but also to quantify the state of the outcome. The multidimensionality of the outcome requires the creation of composite indices to have a single measure of performance and also to aggregate the indicators to rank the (country) units in one unique way.

Examples of such indices are studies of globalization and its impacts on inequality, poverty and economic growth. Concerning the measurement of globalization, Heshmati (2006a and 2006b) introduces two composite indices of globalization. The first index is based on the Kearney/Foreign Policy magazine (2002) and the second is obtained from a principal component analysis. They indicate the level of globalization and show how it has developed over time for different countries. The indices are composed of four main components: economic integration, personal contact, technology and political engagement, each generated from a number of indicators. The indices were also used in a regression analysis framework to study the causal relationships between income inequality, poverty, economic growth and globalization. The results show evidence of a weak and negative relationship between globalization and income inequality and poverty. An important index of globalization based on similar methodology, but comprehensive data is the KOF index[9] (see also Dreher, 2006; and Dreher, Gaston and Martens, 2008).

In this study, the focus is on the construction of indices of the development process that are multidimensional and decomposable into different dimensions. Such indices will be a

[9] The 2013 version of the index and underlying data are available at http://globalization.kof.ethz.ch/

useful tool in the quantification of the state of development and the evaluation of predictors' impacts on development. In this section, we introduce the two main approaches of non-parametric and parametric indices and their extensions frequently used in the construction of such indices.

2.2.1. Non-parametric Indices

The non-parametric index is a composite index constructed to aggregate indicators of a certain process or outcome. Such indices are used for the measurement of many phenomena such as globalization[10], the state of the environment[11], human development[12], development strategy, technology and research[13], Child well-being[14] and other areas of application. For instance, the globalization index is a simple combination of forces driving the integration of ideas, people, and economies, worldwide. In existing studies it is often composed of 3-5 major components: economic integration, personal contact, internet technology, political engagement, and social development, where each component in turn is being generated from a number of indicators. This index serves as a model for computation of our sub-indices of development.

The composite indices are constructed by transformation of each of the indicators to an index on a scale of one to a hundred, where one hundred is the maximum value and one is the minimum value. Higher values denote a greater level of the index. In the aggregation of indicators, a weight must be attached to each indicator. The composite index is computed non-parametrically, based on the normalization of the different indicators and their subsequent aggregation, using an ad hoc weighting system as follows:

$$INDEX_i = \sum_{j=1}^{J}\sum_{m=1}^{M}\omega_{jm}\{(X_{jmi} - X_{jm}^{\min})/(X_{jm}^{\max} - X_{jm}^{\min})\} \quad (1)$$

where i indicate country; m and j are within and between component variables; ω_{jm} are the weights attached to each contributing X-variable within a component and weights attached to each of the major components; and *min* and *max* are minimum and maximum values of respective variables across countries in a given year. The index is similar to the commonly-used index, the United Nations Human Development Program (UNDP) Human Development Index (HDI), which is based on aggregation of 3 indicators of educational attainment, life expectancy and real GDP or GNI per capita. The index has been generalized accounting for poverty, inequality and empowerment of women (UNDP, 2013).

---

[10] Heshmati, 2006a; Heshmati and Tausch, 2007; Andersen and Herbertsson, 2003; Dreher, 2006; Kearney, 2002 and 2003; Lockwood, 2004; Lockwood and Redoano, 2005.

[11] Kang, 2002.

[12] Noorbakhsh, 1998.

[13] Heshmati and Oh, 2007; Archibugi and Coco, 2004; Archibugi et al., 2009; Grupp and Mogee, 2004; Al-Mutavakkil et al., 2009.

[14] Heshmati et al., 2008.

Policy pointer:

Modern social science research increasingly uses the kind of indices, first designed by the UNDP Human Development Programme. While their use in socio-economic development accounting is already quite common, their use in cross-national value research has been rather limited, and yet there are great potentials for such a type of research. Future research could concentrate for example on such dimensions as avoiding economic permissiveness, avoiding racism and the authoritarian character, and support for the market economy and democracy.

It is important to note that the index in (1) is suitable for indicators with an expected positive effect on the index outcome, like GDP per capita, where a higher value is desirable from a development perspective. In cases where the indicators are expected to have a negative impact on the outcome, like CO2 emission, with a lower value being more desirable, the index in (1) is rewritten as:

$$INDEX_i = \sum_{j=1}^{J}\sum_{m=1}^{M}\omega_{jm}\{(X_{jm}^{\max} - X_{jmi})/(X_{jm}^{\max} - X_{jm}^{\min})\}. \qquad (2)$$

Alternatively, prior to the normalization in (1) and its aggregation, the negative indicators are transformed to inverses, (*1/X*), reversing their expected impact from negative to positive. The index component's weights in equations (1) and (2) are constant across countries and chosen on an ad hoc basis. In this study, following the tradition of human development index, all indicators are given equal weight (see Noorbakhsh, 1998).[15]

Ideally, the weights attached to each indicator should differ by countries and over time. In certain cases, the researchers assign some indictors a double weight. A heterogeneous weighted system is important as countries are differently endowed by resources and their dependency on resources changes over time. A simple weighting system is to use the square of the normalized indices as a base for the aggregation to obtain the composite index (Altmann et al., 2013). This weighted system implies that the higher normalized values receive a higher weight than the lower ones. It is a reasonable assumption as higher normalized values are considered to be result of specialization and higher performance.

2.2.2. Parametric Indices

There are at least two parametric indices employed for computing an index of a development process: the principal component (PC) or factor analysis (FA). Since the two methods in normalized form give PC scores with unit variance, the PC is more frequently used in the analysis of a development process.[16]

[15] This simple non-parametric index can be used as a benchmark index. Lockwood (2004), in computation of a globalization index, finds the ranking of countries to be sensitive to the way the indicators are measured, normalized and weighted prior to their aggregation (see also Decancq and Lugo, 2009).

[16] Principal Component analysis was originally developed by Pearson (1901) and further improved by Hotelling (1933). The method has been employed in many areas including in computation of a globalization index (Heshmati, 2006a; Andersen and Herbertsson, 2003), an environmental index (Kang, 2002) and a simple globalization index using trade and financial openness by Agénor (2003). Heshmati and Oh (2007) used the

PC analysis is a multivariate technique used for examining relationships within a set of interrelated quantitative variables. Given a dataset with J indicators, at most P principal components can be computed; each is a linear combination of the original indicators with coefficients equal to the *Eigen vectors* of the correlation of the covariance matrix. The principal components are sorted according to the descending order of the *Eigen values*, which are equal to the variance of the components. PC analysis is a way to uncover approximate linear dependencies among the indicators. This method gives a least square type solution to the following model:

$$Y = XB + E \tag{3}$$

where Y is a $n \times p$ matrix of the centered observed indicators, X is the $n \times j$ matrix of scores of the first j principal components, B is a $j \times p$ matrix of *Eigen vectors* or factor patterns, E is a $n \times p$ matrix of residuals, n is the number of observations, p is the number of partial indicators, and j the number of indicators of development process. Unlike in a traditional least squares estimation method case, where the vertical distance between the observed and the fitted line is minimized, here the sum of the squared residuals is measured as distances from the point to the first principal axis.

As part of the analysis, the *Eigen values* and *Eigen vectors* are investigated. The *Eigen values* bigger than 1.0 are used in the computation of the development process index. The *Eigen values* are declining from the first component to other components. By looking at the *Eigen vectors,* it becomes evident which indicators form a specific component and the nature of their effects. In each sub-component, an indicator with an *Eigen vector* exceeding 0.30 is considered statistically as a significant contributor to the principal component.

In practice the researchers use only the first principal component in the computation of a parametric index and in the ranking the countries studied. This method has the disadvantage in that it ignores the information embodied in the remaining indicators. One alternative to account for the information embodied in all principal components with an *Eigen value* bigger than one is to use a weighted average PC index. In the aggregation of the principal components, one can use their explained share of the total variance as weights. This method of aggregation will allow the utilization of information from all indicators of an outcome.

Each of the parametric indices has its own advantages and disadvantages. They can be used to measure the state of development among countries and attribute it to the possible underlying causes. A breakdown of the index into major components provides possibilities to identify positive and negative factors contributing to the development. The parametric approach does not provide decomposition of sub-components. In the non-parametric case, the structure of the components is determined by the researcher. The advantage of the parametric approach is that the components' distribution is determined by the indicators' statistical relationship rather than based on an ad hoc selection of indicators. The results can be used in the design and the evaluation of economic policy measures.

---

method for computation of Lisbon Development Strategy Index while Heshmati et al. (2008) used it to study child well-being in the high and middle income countries.

Policy pointer:

In contrast to the research, which still characterizes the reflexion on *World Values Survey* data in the social sciences, we think that the time has come to use techniques which properly allow for stronger relations between the "factors" underlying the correlations between the variables. We think that the Promax Factor Analysis is such an analytical technique, and we use it throughout this book. Global value research would also be unthinkable today without factor analytical index construction. In our work, we present for example such a combined global value development index (avoiding permissiveness, racism, distrust of the army and the press, authoritarian character, [predominantly] female rejection of the market economy and democracy; and practicing the values of tolerance and respect (weighted by the Eigen values of the promax factor analytical model)). Our research results and the statistical foundations presented here are an invitation for decision makers and researchers to start for themselves developing indices based on the freely available *World Values Survey* data. All they need are competent statisticians, modern statistical software and the firm intention and will to develop projects facilitating the path towards a more mature and encompassing democracy – everywhere around the globe.

It should be emphasized that the PC method is generally a very useful method to reduce the complexity of the data with multi-dimensions. However, the linear combinations of the different dimensions of interest may not be always easy to interpret. In sum, what is statistically reasonable may be neither economically nor normatively reasonable in the evaluation of development. The non-parametric and parametric indices can be computed at each point of the data. In the context of globalization, they can show how globalization has developed for different countries and regions over time. A breakdown of the index into major components provides possibilities to identify sources of globalization. The indices can be used to study the causal relationship between globalization, inequality, poverty, growth and a number of other variables.

2.2.3. A New Parametric Decomposable Index

In previous studies, the causal links between globalization, inequality, poverty and growth have been investigated in a two-step procedure. In the first step, the globalization index is computed as described above non-parametrically or parametrically, and in the second step, the poverty, inequality and growth variables are regressed on globalization to investigate the impact of globalization on the dependent variables. A two-step procedure has the disadvantage in that it often ignores the directions of causality and simultaneity issues. Heshmati and Lee (2010) introduced a new approach in which both steps are estimated parametrically and in a single step. The model is based on a multidimensional analysis of globalization. In the new index, the contribution from each indicator, the shares of individual components and the aggregate composite index are configured by their relationship with economic development. For the matter of sensitivity analysis, two different methods, decomposed and composite models that are estimated iteratively, are used to compute the new globalization index and explain its variations. The indices are used to investigate their impacts on the economic growth and income inequality of the sample countries. The regression analyses show that globalization is positively linked to economic growth, while it has a negative relationship with inequality, but the effects are confounded with unobserved regional effects.

The new model is using two approaches: a composite single index and a decomposed four component globalization index. In the empirical part, Heshmati and Lee adopt the same index structure and components of the non-parametric and parametric PC indices described above. In the decomposed index approach, the impact of each of the globalization component on different economic growth (EG) measures is written as follows:

$$\log EG_{it} = a_o + a_1 \log ECON_{it} + a_2 \log PERS_{it} + a_3 \log TECH_{it} + a_4 \log POLI_{it} + u_{it} \tag{4}$$

where each of the main components is specified as:

$$\log Comp_{it} = \sum_{j=1}^{J-1} b_j \log X_{jit} + (1 - \sum_j b_j) \log X_{Jit} \tag{5}$$

In (4), the log EG is represented by the logarithm of different measures of GDP, specified as a function of the logarithm of the four globalization components: ECON, PERS, TECH and POLI. Each of the globalization components in (5) is estimated as a function of J normalized indicators (X) specific to that component. The component ECON includes economic variables, PERS includes personal variables, TECH includes technology variables, POLI includes political variables, u is an error term and subscripts j, i and t refer to an indicator, a country and time periods.

Each indicator variable has its own parameters (b), reflecting its share contribution to the globalization component as in the non-parametric index, although here they are estimated parametrically and not assigned equal weights on an ad hoc basis. For example, the economic factor, ECON, has four sub-indicators such as trade, foreign direct investment, portfolio investment, and income payments and receipts. Because the sum of $b_{j:}$ s is assumed to be unity, we only need three coefficients to calculate the ECON components value. The b_4 can be derived by $1-b_1-b_2-b_3$. The parameters, a_0-a_4 in model (4), are obtained through regression analysis of economic growth variables and globalization sub-components, and these represent the weights of each component or the effects of globalization on economic growth. Here no restrictions are imposed on the total sum of the effects. The model is non-linear in parameters and the estimation is conducted in an iterative procedure where the parameters of the globalization components (their effects) on economic growth are estimated.

Alternatively, one can estimate the relationship based on the principal component analysis, which does not distinguish between the four globalization components and lumps all indicators into one single index. The model corresponding to (4) and (5) is written as:

$$\log EG_{it} = a_o + a_1 \log Index_{it} + u_{it} \tag{6}$$

where log EG is again representing the different GDP measures, and INDEX is specified as a function of all globalization indicators without distinguishing the four components and u is an error term. The log *Index* is:

$$\log Index_{it} = \sum_{j=1}^{J-1} b_j \log X_{jit} + (1 - \sum_{j=1}^{J-1} b_j) \log X_{Jit} \tag{7}$$

Parameters b1-b3 (and c1-c2, d1-d2, e1-e2 not shown explicitly) are the coefficients of globalization indicators associated with different components of the index. This model is also estimated in an iterative procedure where the whole indicators are employed to estimate the composite globalization index (7) and then estimate the index impacts on economic growth (6). From the estimation of (4) through (7), we could obtain estimates of the contributions of each indicator on the globalization index and the effects of globalization on the economic growth of countries.

With the help of the adjusted R^2 values and root mean square errors (RMSE), one can distinguish evidence of more adequate specification of the relationships and to explain variations in GDP levels and its growth rates. The new globalization indices of countries are calculated based on the coefficients obtained from (5) and (7) as shown below in (8):

$$\begin{aligned} \hat{Gindex}1_i &= \hat{a}_o + \hat{a}_1 ECON_i + \hat{a}_2 PERS_i + \hat{a}_3 TECH_i + \hat{a}_4 POLI_i \\ \hat{Gindex}2_i &= \hat{a}_o + \hat{a}_1 INDEX \end{aligned} \tag{8}$$

where *Gindex*1 and *Gindex*2 represents the decomposed and composite estimated measures of globalization.

Policy pointer:

The methodology developed here describes how data gained in index research are being used to explain dependent – process variables, like economic growth, poverty and inequality. Future social science research will increasingly combine data from aggregate statistics and opinion surveys to arrive at a far more encompassing picture of realities. To put it in simple words: former research on such phenomena as economic growth or income distribution only considered what people had in their pockets, not on their minds. Today, increasingly, the mind and the pocket both matter.

The distribution of different indices can be studied in different dimensions such as countries, regions, over time and by country's level of development. Different components and the overall index as well as the indices measured in metric efficiency terms where countries are compared with the best country in a given year can be analyzed. The efficiency indices, where the individual country observations are compared to the most globalized country in a given year, is better than the standard index, as they provide better metric information than the simple ranking. The method allows the observation of trends showing yearly shifts in the frontier and also a catching up of the sample countries to the frontier consisting of most developed and most globalized countries.

2.2.4. Varimax and Promax Rotation Procedures

The factor analysis examined above and various methods of factor rotation have been conducted in the context of linear factor analysis of continuous variables. The standard linear factor-analytic methodologies do not work well for dichotomous items or indicator variables responses. This limitation has led to the development of nonlinear methods. In both cases, the determination of the association between items and factors is the same and is made using factor loading which is considered by Tabachnick and Fidell (2001) as the correlation between the factor and items. Strongest loadings above a threshold value 0.3 or 0.4 are preferred. Nonlinear factor analysis is often used to identify both the presence and nature of multidimensionality in a set of test items. The loaded matrix is rotated in order to amplify the presence of simple unidirectional latent structures. A simple structure from a set of items was defined in Goruch (1983) as when each factor has a few items with high loadings and the rest with loadings near zero.

Finch (2006), using an item response function (see also Hambleton et al., 1991) conducted a simulation study to compare the performance of two commonly used methods of rotation, namely orthogonal (Varimax) and oblique (Promax) to identify the presence of a simple structure. Factor rotation involves a transformation of the initial factor loadings to obtain a greater simple structure without changing the underlying mathematical relationships in the data. Finch suggests the nonlinear factor analysis rotation method as the preferred method. Orthogonal rotations assume the factors are uncorrelated, while the oblique rotations assume the factors are correlated. The former contains the correlation between the factors, while the later measures the relationship between the individual factors and items. Promax takes the rotated matrix provided by Varimax and raises the loadings to powers where the transformed loading values reflect the simple structure better than in the case of Varimax (McLeod et al., 2001). Each method provides advantages and disadvantages in their application in some circumstances. McDonald (1997), Kieffer (1998) and DeVellis (2003) provide guidelines on decision regarding which rotation procedure to use. The results from the Flinch (2001) simulation study suggest that the two approaches are equally able to recover the underlying factor structure, though the Promax method is better in the identification of the simple structure.

Given conflicting recommendations in the literature, Dien et al. (2005) present a standard protocol for applying PCA to event-related potential datasets. The focus is on optimizing PCA with emphasis on matrix type, factor loading weighing, extraction and rotations. The effects of a covariance versus a correlation matrix, Kaiser normalization versus covariance loadings, truncated versus unrestricted solutions, and Varimax versus Promax rotations are tested on 100 simulation datasets. The parameters were evaluated according to time course reconstruction, source localization results, and misallocation of ANOVA effects. The results show that the correlation matrices resulted in a dramatic misallocation of variance. The Promax rotation yielded much more accurate results than the Varimax rotation. Covariance loadings were inferior to Kaiser Normalization and unweighted loadings. Thus, the evidence supports the use of a covariance matrix, Kaiser normalization, and Promax rotation. When these parameters are used, the unrestricted solutions did not materially improve the results. We argue against their use. Results also suggest that optimized PCA procedures can measurably improve source localization results.

Policy pointer:

The choice of the factor analytical method used to reduce the number of variables of the *World Values Survey* project to its unerlying dimensions is not just a matter for the specialist but it also has many different consequences. Inglehart and most other researchers rely on standard linear factor analysis, which is basically a statistical methdology already developed before the Second World War.

The standard linear factor-analytic methodologies do not work well for dichotomous items or categorical variables responses. This limitation has led to development of nonlinear methods. The loaded matrix is rotated in order to amplify the presence of simple unidirectional latent structure. A simple structure from a set of items was defined in Goruch (1983) as when each factor has a few items with high loadings and the rest with loadings near zero.

Orthogonal rotations assume the factors are uncorrelated, while the oblique rotations assume the factors are correlated. The results from Flinch (2001) suggest that our chosen Promax method is best suited for the identification of a simple structure. The Promax rotation yielded much more accurate results than Varimax rotation.

2.3. Measurement of Correlation Relationships

2.3.1. Bivariate Relations

The economics literature contains extensive discussion of relationships between pairs of variables such as: quantity and prices; consumption and income; saving and interest rate; income and education; unemployment and inflation rates; emission and production; and many more (see e.g., Davidson and MacKinnon, 1993; Johnston and DiNardo, 1997; Gujarati, 1999; Greene, 2008, and Wooldridge, 2009). Such relationships between indicators are called bivariate relations which are a simple and not necessarily adequate analysis of the real situations.

2.3.2. Scatter Plots

The information between two indicators can be shown in form of a scatter plot. Scatter diagrams have three main characteristics: a sign of a positive or negative association, the strength or steepness of the association, and the general shape or linearity/non-linearity of the association between the indicators. The plots are important as building blocks for the analysis of more complicated multivariate situations. A typical example of bivariate relations is a time series plot of key indicators like income, saving, emissions, employment, showing their development and variations over time. When the sample size of pairs of observations is large, the data is often printed as a bivariate frequency distribution where the ranges of the two indicators like height and weight are split into intervals. Each cell of the table shows the number of observations in the corresponding pairs of intervals. The cell frequencies and means and variances can suggest a positive association between the two indicators.

2.3.3. Correlation Coefficient

The correlation coefficient measures the first two characteristics namely the direction and closeness or strength of the linear relations between two indicators, X and Y. The data can be shown in terms of deviations from the sample mean, resulting in a scatter diagram with four quadrants. The diagonal quadrants show the positive association between the two indicators, while the off-diagonals quadrants indicate the negative association. The covariance of the standardized deviations is the correlation coefficient, r, computed as:

$$r = \sum x_i y_i / n s_x s_y = \frac{n\sum XY - (\sum X)(\sum Y)}{\sqrt{n\sum X^2 - (\sum X)^2}\sqrt{n\sum Y^2 - (\sum Y)^2}} \tag{9}$$

where the first part is measured in terms of standard deviations, while the second in terms of the raw data. The correlation coefficient is invariant to the units of measurement. It lies in the interval -1 to +1 and its square $r^2 \le 1$. A low value of r may indicate the possibility of a strong non-linear, positive or negative correlation if the observations are located in a particular segment of the non-linear relation. Correlation coefficients should be interpreted with care. Many large and statistically significant coefficients may contain no meaningful information. The sign and size of a correlation coefficient may lend support to a theory about joint variations of X and Y, but in the absence of a theory it may be classified as a spurious correlation. If two variables are correlated both in level and first difference forms it is an evidence of a causal relationship between the two indicators.

2.3.4. Partial Correlation Analysis

In cases with two or more regressors, partial correlation analysis becomes relevant. Assume three variables Y (income), X (sunspots) and T (trend) numbered as 1, 2 and 3. Define r, b and e as the correlation, the slope of the regression and the residual between 2 variables. The partial correlation coefficient between Y and X, with the influence of T removed or held constant, is defined as the correlation coefficient between two sets of residuals (e) denoted as:

$$r_{12.3} = \frac{\sum e_{1.3} e_{2.3}}{\sqrt{\sum e_{1.3}^2}\sqrt{\sum e_{2.3}^2}} = \frac{r_{12} - r_{13} r_{23}}{\sqrt{1 - r_{13}^2}\sqrt{1 - r_{23}^2}} \tag{10}$$

where e's are LS residuals with mean zero used to compute the correlation coefficient between them. It is simpler to express $r_{12.3}$ in terms of three simple correlation coefficients. It measures the net association between Y and X when any common influence from T has been allowed for. In a similar way, partial correlations $r_{13.2}$ and $r_{23.1}$ can be computed. The simple correlation coefficients are called zero order coefficients while the partial correlation is called first order coefficient. In general, the first order coefficients can be larger or smaller than the corresponding zero order coefficients, and may even have the same or opposite signs.

Starting with X, the relation can be expressed in terms of the increment in explained sum of squares residual (ESS) due to the addition of trend (T) expressed as:

$$R_{1.23}^2 = r_{1.2}^2 + r_{13.2}^2(1 - r_{1.2}^2) \tag{11}$$

In a similar way, starting with T, the increment due to X can be computed. The method allows determining the relative importance of each variable in explaining the variations in Y. In a case where there is a zero correlation between two variables like X and T, $r_{23}=0$, the variables are orthogonal. The partial correlation coefficient is reduced to the sum of simple zero order coefficients, $R_{1.23}^2 = r_{12}^2 + r_{13}^2$.

2.4. Simple, Multiple and Stepwise Regression Analysis

2.4.1. Simple Linear Regression

Classical statistical inference is based on the presumption that there exists some population distribution of all possible observations on the variables of interest. The distribution is characterized by crucial parameter values. The sample statistics computed serve as a basis for inference about the population parameters. The probability approach used extensively in econometrics adapts classical inference procedure to deal with problems raised and due to unavailability of controlled economic experiments. The development is in the area of the discrete bivariate probability distribution and bivariate normal distribution. Assuming a population and sample of sizes N and n, the conditional distribution of Y given X is:[17]

$$E(Y \mid X) = g(X) = \alpha + \beta X \tag{12}$$

Given time series, cross section or panel data and the appended discrepancy (U) between the conditional expectation E(Y| X) and observed Y are available, the cross sectional relation is written as:

$$Y_i = \alpha + \beta X_i + U_i \tag{13}$$

where the subscript i indicates units of observation, the α and β are unknown parameters to be estimated, and U is assumed to be identically and independently distributed with mean 0 and constant variance, U_i are iid $(0, \sigma^2)$. The disturbance U represents the net influence of

[17] For detailed explanations about regression analysis see Davidson and MacKinnon (1993), Johnston and DiNardo (1997), Gujarati (1999), Greene (2008) and Wooldridge (2009).

everything other than the X. Least squares estimator is the common method of estimating the unknown parameters α, β and σ^2. The intercept and slope parameters are chosen to minimize the total deviations in the Y variable (TSS) into those explained (ESS) and unexplained (RSS) parts by the X variable, respectively. The proportion of the variations in Y explained by X (R^2 =ESS/TSS) is called the coefficient of determination.

Having estimated the relation, one may ask what are the properties of these estimators and their usefulness in making inferences about α and β? The answers depend on the sampling distribution of the least squares estimators in repeated applications to infinite set of samples drawn from the population. X are treated as fixed or non-stochastic in repeated sampling. Given certain conditions are fulfilled, the LS estimator has minimum variance in the class of unbiased estimators, and is said to be a best linear unbiased estimator, or BLUE. The significance of X in explaining variations in Y may be based on a test of the correlation coefficient on the regression slope (t-test) or decomposition of the sum of squares (F-test). The point or interval values of Y can be predicted for given changes in X for outside the sample observations accompanied by an estimate of the prediction error.

2.4.2. Multiple Linear Regressions

The multiple linear regressions model is used to study the relationship between a dependent variable like development outcome (Y) and several independent variables like indicators of development, or, in the case of our book, structures of human values (X). If the set of independent variable consists of only one, then the model is reduced to a simple regression case, else a multiple regression case with K explanatory variables is expressed as:

$$Y_i = \alpha + \beta_1 X_{1i} + \beta_1 X_{1i} + \beta_2 X_{2i} + ... + \beta_k X_{ki} + U_i \qquad (14)$$

Here the movement in the dependent development process is defined as a response to changes in some independent or exogenous stimulus factors identified as determinants of development. The added disturbance term (U) is expected to capture the random factors disturbing the stable relationship, omitted variables and measurement errors in the Y variable. The models are specified with the aim to minimize the residuals - the vertical distance between the observed and fitted values of Y - or their sum of squares.

In the context of development, the linear regression model is the single most useful method in the analysis of the relationships between indicators of development and the outcome of a development process. In empirical research, the linear regression model is used as a departure point for deeper analysis, where the linear models are extended to non-linear formulation and testing of their causal relationships. The non-linearity is often introduced in form of adding second order or squares of the explanatory variables or their interactions, in addition to their direct and non-linear effects, capturing substitution and complementarity among the explanatory variables as well.

In the empirical analysis of the development relationship, the primary focus is on the functional form of the relationship between development outcome and its determinants and the relationship hypothesis testing. These are crucial to improve the specification of the model to enable drawing useful inferences about the population of countries. Broader techniques are used by researchers in their search for the set of alternative competing models and for a final choice of a specific selected model. This will help the model specification and model building to avoid omission of relevant variables and inclusion of superfluous variables. It is known that the former leads to the so called omitted variable bias, while over-fit of the model in the latter case leads to a reduced precision of the estimates. For more information see Greene (2008).

There are two common strategies in model building: a general-to-simple and its opposite, namely a simple-to-general formulation. In maintaining simplicity, model builders begin with a small model specification and gradually build up the model by adding more identified explanatory variables. However, this approach, due to the omitted variable bias is less recommended. The increased computational power in recent decades and development of varieties of statistical packages have led to the fact that the alternative general-to-simple model strategy involving long and complex lag structures and functional forms but downward reducing the model to a preferred specification is increasingly attractive. This approach has been automated in some statistical packages to ease their implementation. For instance in the 'kitchen sink' regression case, every relevant variable is included in the model in which some variables turn out to become significant by accident. The same problem arises in stepwise model building, where elements of unexplainable effects may result.

In the specification and selection of models a number of criteria are frequently used. One is the theoretical relationship between the explanatory variables and the dependent variable which justifies their incorporation. Another is their empirical relevance not necessarily supported (as yet) by theory. A third is the nature of the relationship in form of flexible functional forms. The adjusted R^2, the mean-squared error criterion, the pre-test estimator, the Akaike Information Criterion, the Schwartz or the Bayesian Information Criterion, the encompassing principle, the non-nestedness of models are all used as model specification, estimation and selection criteria in the literature. In addition, if necessary, one accounts for issues of endogeneity, dynamics, heteroskedasticity, sample selection, *et cetera*. In our book, we use the abovementioned approaches to investigate aspects of the pivotal rule of various value-related indicators for economic growth and effective democracy.

2.4.3. Stepwise Regression Analysis

As mentioned previously, multiple regression is an extension of the simple regression case. The goal is to enable the assessment of the relationship between a dependent variable and several independent variables. There are several types of multiple regression analyses. These include standard, hierarchical, setwise and stepwise. The choice of the type depends on the question of interest and the researcher preference. Standard multiple regressions are used to address to questions: (i) what is the size of the overall relationship between the dependent and independent variables, and (ii) how much does each independent variable contribute to that relationship? The above section elaborated the standard multiple regressions case. Below we discuss the stepwise case.

Stepwise regression is an approach to selecting a subset of explanatory variables for a regression model. It is used when there is little theory to guide the predictor selection, and the model builder wants to use all available relevant information to provide a good fit in explaining variations in the dependent variable. Stepwise multiple regressions are used to answer a different question: what is the best combination of independent variables to predict the dependent variable of interest? The predictor variables enter into the regression equation one at a time, based upon some statistical criteria. The predictor variables which contribute the most to the prediction equation in terms of increasing the multiple correlations, R, enter first. This process stops when there are no additional variables that add statistically to the regression equation. Thus, not all predictor variables may enter the equation in stepwise regression. The approach is somewhat controversial because the model has been selected rather than tested within a fixed model. On the positive side, the approach is effective in reducing the number of terms. Miller (2002), originally published in 1990, discusses the issue of variable subset selection in regression. Sala-i-Martin (1997) runs four million regressions to find that a large number of variables can be found which are strongly related to economic growth.

Policy pointer:

Our methodological survey presented the wide variety of multivariate techniques used in our research. Apart from bi-variate and partial correlations, our analysis also uses different multiple regression approaches. Stepwise multiple regressions are used to answer a different question: what is the best combination of independent variables to predict the dependent variable of interest? The predictor variables enter into the regression equation one at a time, based upon some statistical criteria. The predictor variables which contribute the most to the prediction equation in terms of increasing the multiple correlations, R, enter first.

2.5. Survey Analysis, Aggregation of Data and Sampling

2.5.1. Survey Analysis

Analysis of surveys is about using collected survey data like World Value Survey (WVS) or different national household consumption expenditure, income or labor market surveys to reveal information that is important to policy making decisions. The analysis is often included in a report which is an assessment of what the survey results suggest. The survey results help to answer questions that we want to know. The results can be presented in different ways and aimed to different audiences. Survey analysis may also lead to new questions raised and the need of additional rounds of analysis. Once a research question is formulated an analysis plan is created. Here the sample analysis plan shows how the survey question is used to answer the stated research question. For details about analysis of survey data the readers are referred to Chambers and Skinner (2003).

There are two approaches to survey analysis and the presentation of survey results: basic and advanced analysis. The former often utilizes built-in analysis and online survey software graphing capabilities, while the later utilizes more sophisticated analysis, creating customized charts, graphs and advanced regression analysis. Often a combination of the two approaches is employed. Excel, IBM-SPSS and SAS are among the software programs that are frequently used in analysis of survey data. Two basic types of analysis that can provide useful information about a sample survey are descriptive statistics and cross tabulations. Descriptive statistics include the quantitative information about the distribution of the sample in form of the mean, the median, the standard deviation, the minimum, the maximum, the ranges, the frequencies, *et cetera.* The cross-tabulations allow comparison of different groups of respondents. A more advanced analysis of the data includes regression analysis described above.

Surveys, based on cohorts or an observational design, generate samples of data. Samples for cohort studies have been convenience or simple random samples. It is desirable to use stratified designs to improve the correspondence between the population and the sample. These issues are discussed, among others, by Dargaz and Hill (1996) in the context of epidemiological risk-factor-analysis generalized to a wider population to estimate the association between risk factors and outcomes. The authors suggest that the data from surveys should be weighted using expansion weights to avoid biased estimates. The design procedure for the calculation of expansion weights has an impact on the variance of the estimates. An accounting for the design in estimating the variance will require specialized analysis procedures and software.

Policy pointer:

Analysis of surveys is about using collected survey data like World Value Survey (WVS) or different national household consumption expenditure, income or labor market surveys to reveal information that is important to policy making decisions. There are two approaches to survey analysis and the presentation of survey results: basic and advanced analysis. The former often utilizes built-in analysis and online survey software graphing capabilities, while the later utilizes more sophisticated analysis, creating customized charts, graphs and advanced regression analysis. Often a combination of the two approaches is employed. Excel, IBM-SPSS and SAS are among the software programs that are frequently used in survey analysis. Surveys, based on cohorts or an observational design, generate samples of data. Samples for cohort studies have been of convenience or simple random sample types. It is desirable to use stratified designs to improve the correspondence between the population and the sample. In contemporary economic theory, for example, Jen et al. (2009) investigated the global variation in health by evaluating the well-known global public health research hypothesis (the so-called Wilkinson's income inequality hypothesis) using individual level data derived from the WVS. Johnson and Mislin (2012), in a study of trust game replication across societies aim to validate the commonly used "trust" question from the WVS. They find that trust as measured by the WVS is positively correlated with experimentally measured trust or trustworthiness across countries. In another study, Morselli and Passini (2012) investigated disobedience and support for democracy using the WVS data.

Jen et al. (2009) investigate the global variation in health by evaluating the well-known global public health research hypothesis (the so-called Wilkinson's income inequality hypothesis) using individual level data derived from the WVS. The global health research hypothesis of a strong relationship between health and income inequality, Jen contends, not supported when individual factors are taken into account. However, the substantial differences between countries remain even after taking account of micro- and macro-variables. This is but one recent application of the fruitful interaction between cross-national research and survey data research. Johnson and Mislin (2012), in a study of trust game replication across societies aim to validate the commonly used "trust" question from the World Values Survey. They find that trust as measured by the WVS is positively correlated with experimentally measured trust or trustworthiness across countries. In another study, Morselli and Passini (2012) investigate disobedience and support for democracy using the WVS data. They suggest that different psychosocial processes may distinguish between different types of movements and protests with focus on the psychological processes triggered in obedience/disobedience. The results show that when disobedience is associated with attitudes of inclusiveness, it is also positively linked to prodemocratic individual attitudes and to the enhancement of democracy at institutional levels.

2.5.2. Aggregate Data

Aggregate data refers to a specific form of data. In statistics, it describes data combined from several measurements. In many instances, instead of disaggregated data summary statistics based on observation are used. In economics, it describes a high level of data that is composed from a multitude of combination of individual or firm level datasets. Examples of aggregate macroeconomic data are levels of general consumer and producer prices, inflation and unemployment rates, while in microeconomics, data of an entire sector of an economy composed of many firms, or of all households in a city or region are aggregated. The most known examples of aggregate data at the country level is the World Development Indicators (WDI) database, available at http://data.worldbank.org/indicator/all which is based on secondary data sources supplied by national statistical agencies of countries around the world and international organizations. Aggregated data may contain a bias due to the aggregation of various components measured in different units of measurement. In addition, it is also subject to measurement error, missing observation and their imputations or projection, error in data, et cetera. Recently, the World Bank expanded its freely downloadable time-series data base for all the countries of the world and hundreds of different economic and social indicators at the website: http://data. worldbank.org/ indicator/all. This website, offering data for each variable on a cross-national and time series perspective in EXCEL formats, also allows users to produce choropleth maps of global development. Readers of our book, for example, may compare our World Values Survey data with income inequality, measured by the share of the lowest 20% in total incomes, by simply employing the offered routine: http://data.worldbank.org/ indicator/ SI.DST.FRST.20/countries?display=map

2.5.3. Sampling, Sampling Error and Margins

In statistics, samles are considered when facing difficulties in the collection of data from the entire population due to high time and cost requirements. Sampling is concerned with the selection of a subset of individuals from the population. The most important aspects of sampling include: the definition of the population, sampling frame, probability sampling, sampling methods, and replacement of selected units, sample size, data collection, error in sample surveys, survey weights and methods of producing random samples. Population is defined as including all households, firm or items with the characteristics we wish to understand and the sample is drawn from. A sampling frame has the property that we can identify every single element and include any in our sample. In the event of using probability sampling, every unit in the population has the same chance of being selected in the sample. This allows unbiased estimates of the population to be obtained. A combination of different sampling methods including simple, systematic, stratified, proportional to size, cluster, sampling, quota, accidental, line intercept and panel samplings is used. In sampling without replacement no element can be selected more than once, while in the case of replacement an element may reappear multiple of times in the same sample.

Sample size is chosen to determine the number of observations. The sample size is an important feature of any empirical study. The aim is to make inferences about a population from a sample. In practice, the sample size is determined by the expense of data collection, and the need to have sufficient statistical accuracy or power. Sample sizes may be chosen in several different ways such as: expedience and reducing the risk of error, using a target variance for an estimate, and using a target for the power of a statistical test. Larger sample sizes generally lead to increased precision when estimating unknown parameters. In statistics, this phenomenon is described as the law of large numbers and the central limit theorem. Procedures for determining minimum sample size in different cases like estimating proportions and means, hypothesis testing and stratification are provided in Bartlett et al. (2001). Recording non-response in data collection is important. Selection bias and random sampling error arise due to differences in selection and calculating probabilities, random variation in the results and non-sampling errors from over- and under-coverage, measurement error, processing error and non-response. In many situations, the sample fraction has to be weighted to correctly represent the population.

For the correspondence between the sample and the population, the sample is randomly selected in a stratified fashion from the household or firm register to estimate the characteristics of the whole population. The advantages of representative sampling are the lower cost and the faster data collection than measuring the entire population. In sampling, each observation measures one or more properties of the observable objects or individuals. In survey sampling, weights are applied to the data to adjust for the sample design to recover the population structure. In business, medical research, agricultural research, firm and industry studies, household income, consumption and poverty studies, sampling is widely used for gathering information about a population and its analysis to generate results useful for policy analysis. For more information on sampling, survey data analysis and analysis methods see Wooldridge (2009), Greene (2008), Bartlett et al. (2001) and Chambers and Skinner (2003).

Sample margin error is a statistic which expresses the amount of random sampling error in a survey result. The larger the margin of the error, the less confidence, that the reported results are close to the true numbers for the population. Thus, margin of error occurs whenever a population is incompletely sampled, e.g., not accounting for all of the characteristics discussed above. The margin of error is defined as the radius of a confidence interval for a statistic from a survey. It can be expressed in percent or absolute quantity. In similarity to the confidence interval, the margin of error can be defined for any different confidence levels. The most common levels chosen are 90%, 95% or 99%. Appendix Graph 2.1 provides statistical margins due to the sampling process at the 95% level of confidence. It is interpreted as the probability that a margin of error around the reported percentage would include the true value. The confidence level, sample design and sample size jointly determine the magnitude of the margin of error. For instance, a larger sample size produces a smaller margin of error.

2.5.4. Sample Size

Statistics is the study of collecting, organizing, analyzing, summarizing data and drawing inferences from the data for decision making. A population consists of all elements that are studied for this purpose. Since due to financial and temporal constraints, the study of an entire population is not feasible, a representative sample is randomly chosen from the population. These sample statistics are then generalized to describe the population parameters, and inferences are drawn from this. The number of elements in such a sample is called the sample size. Sample size is denoted by the lower case letter n while the population size is denoted by upper case letter N. Sample size is the number of observations used for calculating estimates of a given population. Sample size can be smaller or larger but the larger the sample size, the more accurate the results will be. When the sample size is small ($n<30$), labeled often as the magical number, we use the t distribution in place of the normal distribution. If the population variance is unknown and the sample size is small, then we use the t statistic to test the null hypothesis. A large sample size generates more accurate estimates but a too large sample size might cause difficulties in interpreting the tests of significance. Thus, neither too large nor too small sample sizes help research projects. Sample size Tables showing minimum sample size required for given population sizes and confidence intervals are also used to determine the optimal sample size thus avoiding costly data collection and tedious calculations. See Lehmann and Romano (2005).

2.6. Brief Summary

This chapter presented an overview of popular research methodologies used to compute multidimensional indices to rank countries by their performance. We have discussed the different methodologies advantages, strengths and limitations. The methodologies were grouped into sampling surveys and error margins, correlation relationships, factor analysis and regression analysis. The chapter was written such as to be accessible to both professionals and the general public readers of this book.

In recent years, international organizations and national states have shown much interest in research on the relationships between processes of development, such as human development, global competitiveness, innovativeness, environmental degradation, investment flows, infrastructure for development, changes in health conditions, attendance rate and quality of education, economic growth, energy and water supply, *et cetera.* and their predictors have been increasing. The change is attributed to the rapid development of international relations, openness, regional shifts in concentration of production, new opportunities provided by information and communication technology and its rapid diffusion, and not least to the increased availability of secondary and surveys data sources and improvements of the methodology to compare performance of countries.

Statistical Margins due to the sampling process
(at the 95% level of confidence)

various sample sizes are in rows — *various observed results are in columns*

	5%	10%	15%	20%	25%	30%	35%	40%	45%	50%	
	95%	90%	85%	80%	75%	70%	65%	60%	55%	50%	
N=50	6,0	8,3	9,9	11,1	12,0	12,7	13,2	13,6	13,8	13,9	N=50
N=500	1,9	2,6	3,1	3,5	3,8	4,0	4,2	4,3	4,4	4,4	N=500
N=1000	1,4	1,9	2,2	2,5	2,7	2,8	3,0	3,0	3,1	3,1	N=1000
N=1500	1,1	1,5	1,8	2,0	2,2	2,3	2,4	2,5	2,5	2,5	N=1500
N=2000	1,0	1,3	1,6	1,8	1,9	2,0	2,1	2,1	2,2	2,2	N=2000
N=3000	0,8	1,1	1,3	1,4	1,5	1,6	1,7	1,8	1,8	1,8	N=3000
N=4000	0,7	0,9	1,1	1,2	1,3	1,4	1,5	1,5	1,5	1,5	N=4000
N=5000	0,6	0,8	1,0	1,1	1,2	1,3	1,3	1,4	1,4	1,4	N=5000
N=6000	0,6	0,8	0,9	1,0	1,1	1,2	1,2	1,2	1,3	1,3	N=6000
N=7000	0,5	0,7	0,8	0,9	1,0	1,1	1,1	1,1	1,2	1,2	N=7000
N=7500	0,5	0,7	0,8	0,9	1,0	1,0	1,1	1,1	1,1	1,1	N=7500
N=8000	0,5	0,7	0,8	0,9	0,9	1,0	1,0	1,1	1,1	1,1	N=8000
N=9000	0,5	0,6	0,7	0,8	0,9	0,9	1,0	1,0	1,0	1,0	N=9000
N=10000	0,4	0,6	0,7	0,8	0,8	0,9	0,9	1,0	1,0	1,0	N=10000
N=11000	0,4	0,6	0,7	0,7	0,8	0,9	0,9	0,9	0,9	0,9	N=11000
N=12000	0,4	0,5	0,6	0,7	0,8	0,8	0,9	0,9	0,9	0,9	N=12000
N=13000	0,4	0,5	0,6	0,7	0,7	0,8	0,8	0,8	0,9	0,9	N=13000
N=14000	0,4	0,5	0,6	0,7	0,7	0,8	0,8	0,8	0,8	0,8	N=14000
N=15000	0,3	0,5	0,6	0,6	0,7	0,7	0,8	0,8	0,8	0,8	N=15000
	5%	10%	15%	20%	25%	30%	35%	40%	45%	50%	
	95%	90%	85%	80%	75%	70%	65%	60%	55%	50%	

Readers are reminded that survey results are estimations, the accuracy of which, everything being equal, rests upon the sample size and upon the observed percentages (i.e., the survey result data). The real percentages vary within the following confidence limits:

Source: Eurobarometer, Standard Eurobarometer 80, available at http://ec.europa.eu/public_opinion/archives/eb/eb80/eb80_en.htm

Graph 2.1. Statistical margins due to the sampling process at the 95% level of confidence.

This chapter, by combining different methodologies and their integrated use, has contributed to an improved understanding of the problems associated with data collection, data management and analysis, measurement tools and evaluation of their outcomes. Policy implications of results and recommendations to practitioners and decision makers are also presented. Therefore, the present review of the research methodology, in addition to the enhancing of the understanding of the readers of the present book, hopefully may influence the practice of methodologies elsewhere.

Chapter 3

The Dark Side of Economics: Why It Is Good to Analyze World Values Survey Data on the Shadow Economy and Why Corruption Is a Development Impediment

3.1. Introduction

Shadow economic activity has been on rise and causing violations of laws and regulations, lowering tax revenue collections, statistical discrepancies, inequality, corruption, and public budget deficit and public debt problems for the state and its organizations. The shadow economy captures all the activities beyond measurement by official activity. In all countries, there is evidence that the shadow economy has a significant share of the overall economy. It is also labelled as hidden, black, underground, unobserved, unofficial, unrecorded, and parallel economies. The rise of the shadow economy around the world is attributed to the stronger presence of government activity, increase in tax rates, and the desire to escape taxes and regulatory restrictions. Tanzi and Schunecht (1997), Tanzi (1999), Schneider (2005), Eilat and Zinnes (2002), Ahumada et al. (2008) and Chaudhuri et al. (2006) shed light on the shadow economic activities, its measurement and development across developed, developing and transition economies.

In reviewing the recent literature, this chapter focuses on a number of areas related to the shadow economy, in particular economic freedom, corruption, effective democracy and their relationship with economic growth. There are several factors identified by researchers that are expected to have a negative association with the size of the shadow economy. These include trust (D'Hernoncourt and Meon, 2012), and tax morale and quality of institutions (Torgler and Schneider, 2009). Three key factors that their extent has impacted the size of shadow economy are: debt, default risk, corruption and financial development (Elgin and Uras, 2012; Blackburn et al., 2012), Information Communication Technologies (ICT) (Indjikian and Siegel, 2005) and environmental violations (Biswas et al., 2012).

Considering economic freedom and corruption (Apergis et al., 2012), we particularly looked at economic development and its variations (Pieroni and d'Agostino, 2013), income and corruption (Saha and Gouder, 2013), well-being (Belasen and Hafer, 2012), benefits from economic freedom (Gehring, 2013), and industrial policy and biased redistribution

(Holcombe, 2013). In some studies, there is evidence of a causality between economic freedom and economic growth (Narayan et al., 2011), competition law and health policy (Mossialos and Lear, 2012), economic policies (Compton et al., 2011), capitalism as a necessary condition for economic freedom (Pryor, 2010), transition from planned to market economies (Pääkkönen, 2010), and inflows of FDI (Azman-Saini et al., 2010).

Considering effective democracy, the focus in the literature hitherto has been on the impacts of democracy on FDI (Asiedu and Lien, 2011), the effectiveness of aid (Bjørnskov, 2010) and benefits to political elites in the recipient countries. Political culture (Feld and Kirchgässner, 2000) and press freedom (Kalenborn and Lessmann, 2013) are found to affect the outcome of economic policy. There is also the argument for degrowth policies (Johanisova and Wolf, 2012) to face of multiple crises with environmental, social and economic dimensions and different questions raised (Romano, 2012). State capacity, democracy and rapid growth in industries (Rock, 2009), and the importance of good institutional structures are important for economic performance (Knutsen, 2013) and the effects of economic and political inequality on institutions (Savoia et al., 2010).

The rest of this chapter is organized into three major parts. The shadow economy and its measurement and magnitude across country and country groups, factors determining its rate of increases and its linkages with environment and Information Communication Technologies (ICT) are discussed. The second section is about economic freedom, corruption and their relationship with economic growth. The third section is about effective democracy and how it impacts economic growth of nations under different resource endowments. The final section concludes.

Box 3.1. The overall message of this Chapter

It is surprising that the items available from the *World Values Survey* data base have not yet been hitherto used to a sufficient degree in the dominant theory construction of empirical social science.

The shadow economic activity has been on the rise and causing violations of laws and regulations, lower tax revenue collections, statistical discrepancies, inequality, corruption and public budget deficits and public debt problems for the state and its organizations. The rise of the shadow economy around the world is attributed to the stronger presence of government activity, the increase in tax rates, and the desire to escape taxes and regulatory restrictions.

We have reviewed a large number of recent studies investigating the theoretical and empirical aspects of the shadow economy, in particular the measurement of the shadow economy and its causality with economic growth. In the process, emphasis is made on issues such as corruption, economic freedom, state and institutions capacity and quality, as well as morale and trust. Many of these studies use large data sets allowing for cross country and country groups comparisons.

A negative relationship is found between shadow economic activity and the level of development, democracy and press freedom. Countries with abundant natural resources for export, consistent with the theory of curse of natural resources, are facing a larger degree of shadow economic activity, corruption, inefficiency, inequality and lower economic growth.

3.2. Shadow Economy

3.2.1. Shadow Economic Activities and Their Measurement

The shadow economy is a key source of the gap between observable and actual economic measures. It captures all the activities beyond measurement by official activity and consists of both legal and illegal activities outside the reach of government. It makes up a significant share of the overall economy around the word. Other synonyms for the shadow economy are the hidden, black, underground, unobserved, unofficial, unrecorded, and parallel economies. There is evidence that underground activities have been on the rise since the 1970s. The rise is attributed to the stronger presence of government activity in the economies; the increase in tax rates to finance larger public spending programs, and in parallel, the desire to escape taxes and regulatory restrictions has gained prominence (Tanzi and Schunecht, 1997). Tanzi (1999) suggests that the shadow economy reaps because of the presence of activities that are difficult to measure and tax.

Schneider (2005) considers shadow economic activities a fact of life. Most societies attempt to reduce its magnitude by controlling activities through legal measures such as punishment and persecution or by preventive measures with investment in welfare and education. Despite significant investment in the collection of data on shadow economic activities, it is rather difficult to obtain accurate information about its nature and magnitude. Schneider mentions the existence of a comprehensive literature on particular aspects of the shadow economy, but the subject remains controversial. Furthermore, there is disagreement among researchers about the definition, estimation procedures and their use in economic analysis and policy making.

Policy pointer:

In contrast to current theories about global values, advanced by sociology, Schneider (2005) considers shadow economic activities a fact of life. Most societies attempt to reduce its magnitude by controlling activities through legal measures such as punishment and persecution or by preventive measures with investment in welfare and education. Despite significant investment in the collection of data on shadow economic activities, it is rather difficult to obtain accurate information about its nature and magnitude. Schneider mentions the existence of a comprehensive literature on particular aspects of the shadow economy, but the subject remains controversial. Furthermore, there is disagreement among researchers about the definition, estimation procedures and their use in economic analysis and policy making.

In a common approach, Schneider (2005) defines the shadow economy to include all market-based legal production of goods and services that are deliberately concealed from public authorities for the following reasons: (i) to avoid payment of income, value added or other taxes, (ii) to avoid payment of social security contributions, (iii) to avoid having to meet certain legal labor market standards, and (iv) to avoid complying with certain administrative procedures. However, this definition does not include economic activities that are illegal and fit the characteristics of classical crime, as well as the informal household economy or tax evasion.

In a common approach, Schneider (2005) defines the shadow economy to include all market-based legal production of goods and services that are deliberately concealed from public authorities for the following reasons: (i) to avoid payment of income, value added or other taxes, (ii) to avoid payment of social security contributions, (iii) to avoid having to meet certain legal labor market standards, and (iv) to avoid complying with certain administrative procedures. However, this definition does not include economic activities that are illegal and fit the characteristics of classical crime, as well as the informal household economy or tax evasion.

In another approach Eilat and Zinnes (2002) treat the shadow economy as a distinct entity, instead of seeing it just as a symptom of policy failures of the transition countries. They examine its short-term and dynamic consequences for development. The shadow economy is measured in two different ways: first, the electricity method which attributes growth in total electricity consumption in excess of growth in GDP to the shadow economy, and the second measure is a modified electricity approach correcting for limitations in the first approach.

Yet in a third approach, Ahumada et al. (2008) look at the monetary measure of the shadow economy where the money demand function, observed cash balances, and its variation which is explained by variables which induce agents to make hidden transactions is used to estimate the size of the shadow economy. However, on econometric grounds, researchers have criticized the quantitative accuracy of this method. The critique is attributed to time series properties, structural breaks and sensitivity to units of measurement to lag the dependent variable and its initial condition.

3.2.2. Shadow Economic Activities in Industrialized and Transition Economies

Limited statistics from high income countries point to a positive trend in shadow economic activities development, but yet little is known about its magnitude in transition, low-income and emerging economies. Schneider (2005) estimates the shadow economy for 110 countries (66 developing, 23 transition and 21 industrialized OECD) observed for 1990/1991, 1994/1995 and 1999/2000. The results provide some insights about the main causes and studies the dynamic effects of the shadow economy. The main causes of the shadow economy are found to be the high tax and social security contribution burdens, the intensity of regulations and the low quality of public sector services.

The transition economies have undergone major changes. Increased unemployment, decline in GDP, a paralyzed bureaucracy and government corruption during this period saw a surge in the growth of shadow economic activities. Eilat and Zinnes (2002) conducted research on the shadow economy in transition countries. The objective was to use a policy perspective to find out whether shadow economy is a “friend” or a “foe”. The research was conducted in three parts: it lays out the theoretical and empirical backgrounds, it estimates the size of the shadow economy, and it examines its effects and discusses issues of policy implementation.

For the empirical part, Eilat and Zinnes (2002) measures the relative size of the shadow economy to official GDP in 25 transition countries for the period 1990 to 1997. The patterns show that, once established, the shadow economy is hard to remove. Estimation results show that a dollar decline (rise) in official GDP is attenuated by a shadow economic expansion (contraction) of 31 (25) cents. Finally the authors examine whether the shadow economy prevents, slows down, or promotes economic growth and competitiveness, and through what mechanisms. In addition, they consider implications for policymaking that address the key questions. The policy recommendations include: actions with multiple benefits (foreign exchange management, better regulation, institutional strengthening, oversight, transparency and public participation, bank privatization and capital market development, decentralization and better local public finance, rule of law, further liberalization, macro-stabilization), actions that directly target the shadow economy (pay your bills, taxation, stricter, more strategic enforcement), actions whose effectiveness is changed by the shadow economy, and implementation considerations (market exit, policy complementarity and coordination, selective targeting, dynamic considerations).

Schneider (2005), in the context of industrialized and transition economies, mentions that the shadow economy is expected to influence the tax system and its structure, the efficiency of resource allocation between sectors, and the official economy in a dynamic sense. Therefore, several studies have integrated underground economies into macroeconomic models to facilitate investigation of the effects of monetary and fiscal policies on the formal and informal economies and economic growth. In the neoclassical view, the underground economy is assumed to provide the economy with dynamic entrepreneurial spirit. It can lead to greater competition and higher efficiency, help to create markets, increase financial resources, enhance entrepreneurship, and transform the legal, social, and economic institutions necessary for assets accumulation providing a higher potential for economic growth.

Policy pointer:

Schneider (2005), in the context of industrialized and transition economies, mentions that the shadow economy is expected to influence the tax system and its structure, the efficiency of resource allocation between sectors, and the official economy in a dynamic sense.

Schneider (2005) concludes that for all countries investigated, the shadow economy as share of GDP has reached a remarkably large size (Africa 33.9-41.2; Americas 34.2-41.5; Asia 20.9-26.3; Transition countries 31.5-37.9 and highly developed OECD countries 13.2-16.8). The average percentage shares of GDP in all cases are increasing over time. The author demonstrates empirically a strong interaction of the shadow economy with government policies and with the official economy. He draws three further conclusions. First, an increasing burden of taxation and social security payments, combined with rising state regulatory activities, are the major driving forces underlying the size and growth of the shadow economy. Second, the shadow economy has a statistically significant and quantitatively important influence on the growth of the official economy. Increases in the shadow economy have a negative effect on the official growth in a developing country, but a positive effect in the developed industrialized and transition countries. People engage in shadow economic activity because of government actions, most notably high levels of taxation and regulation.

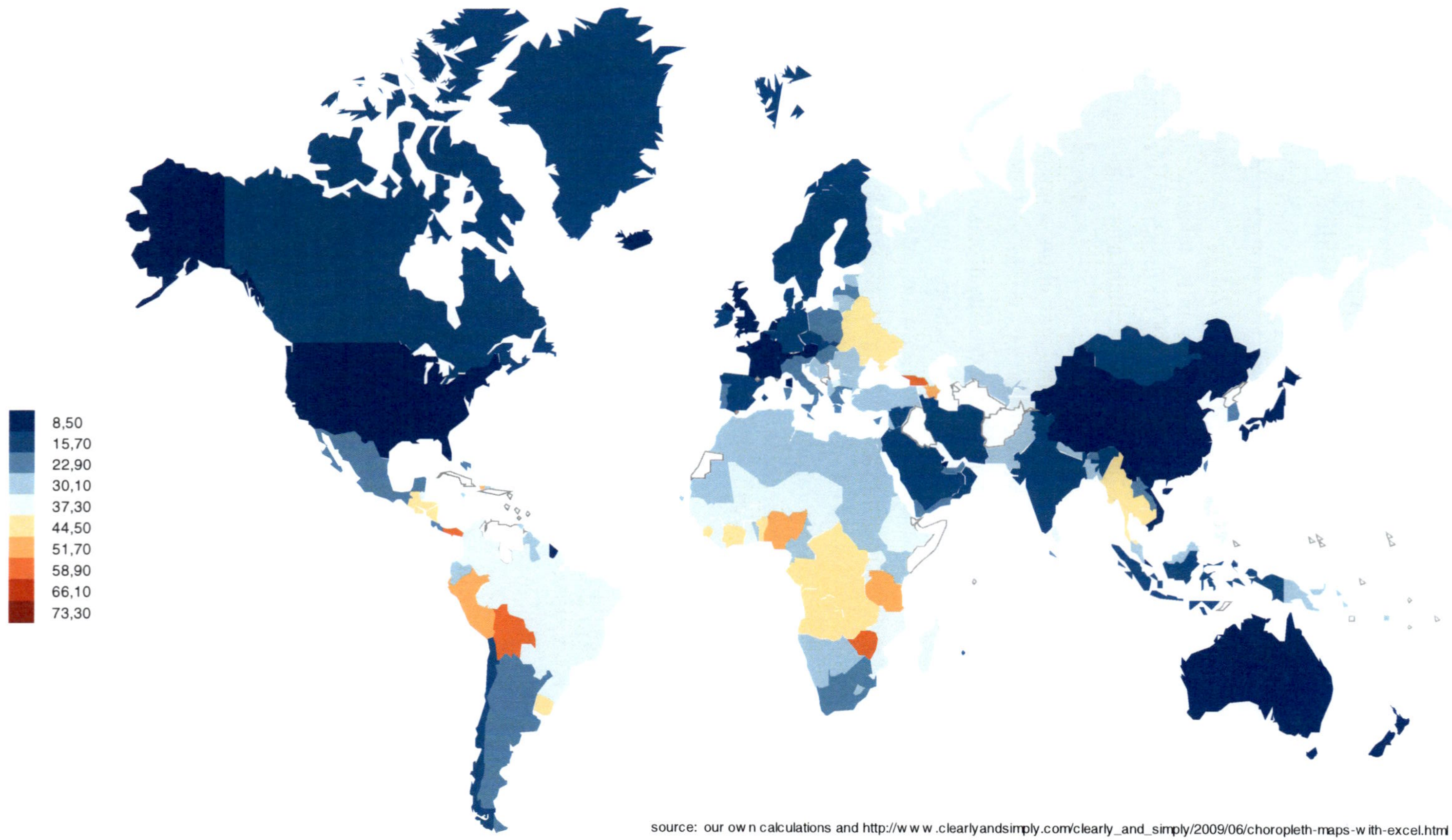

Map 3.1. The shadow economy in the world per GDP, according to Schneider, 2012.

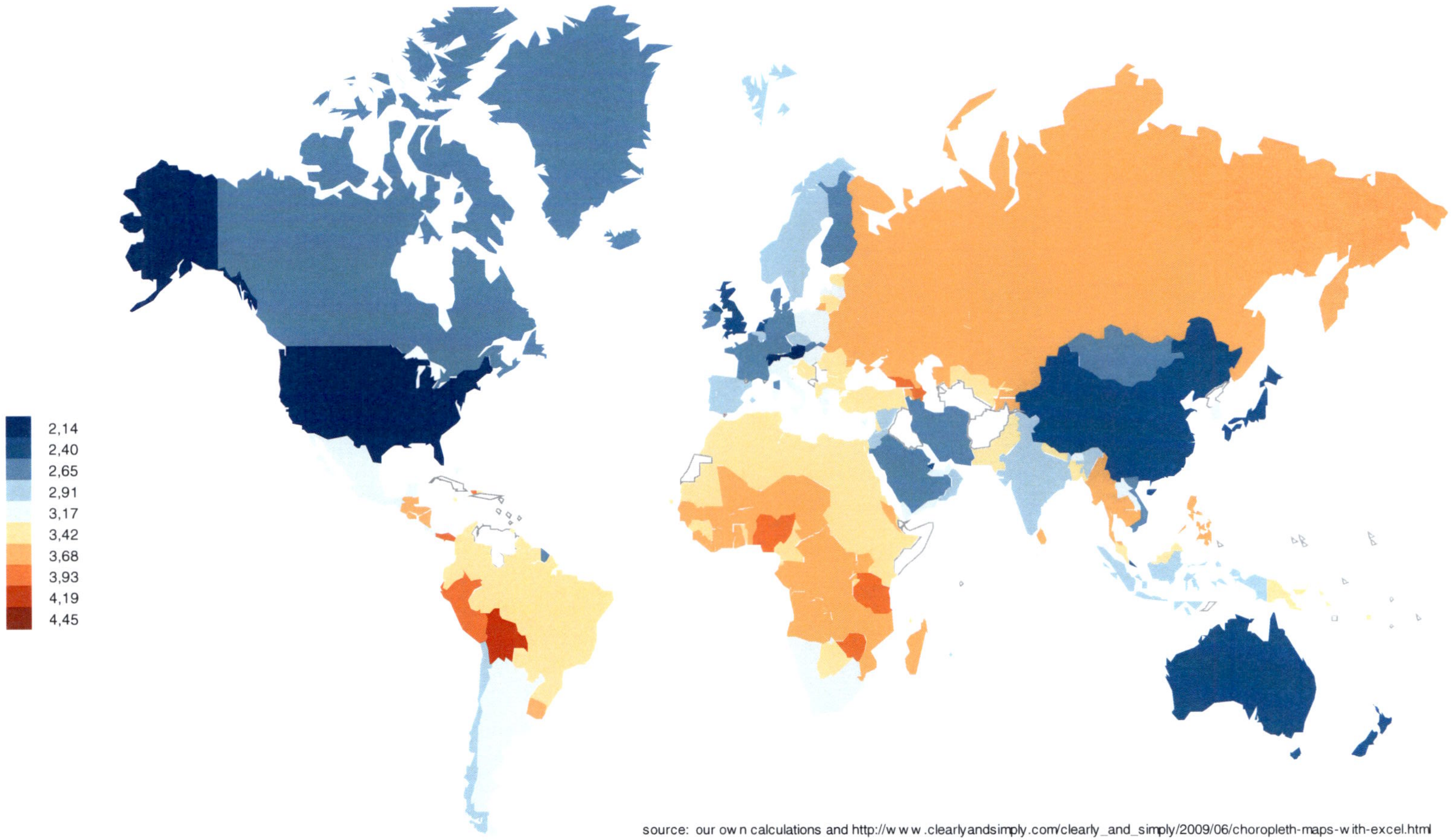

Map 3.2. Log shadow economy per GDP, calculated from Schneider.

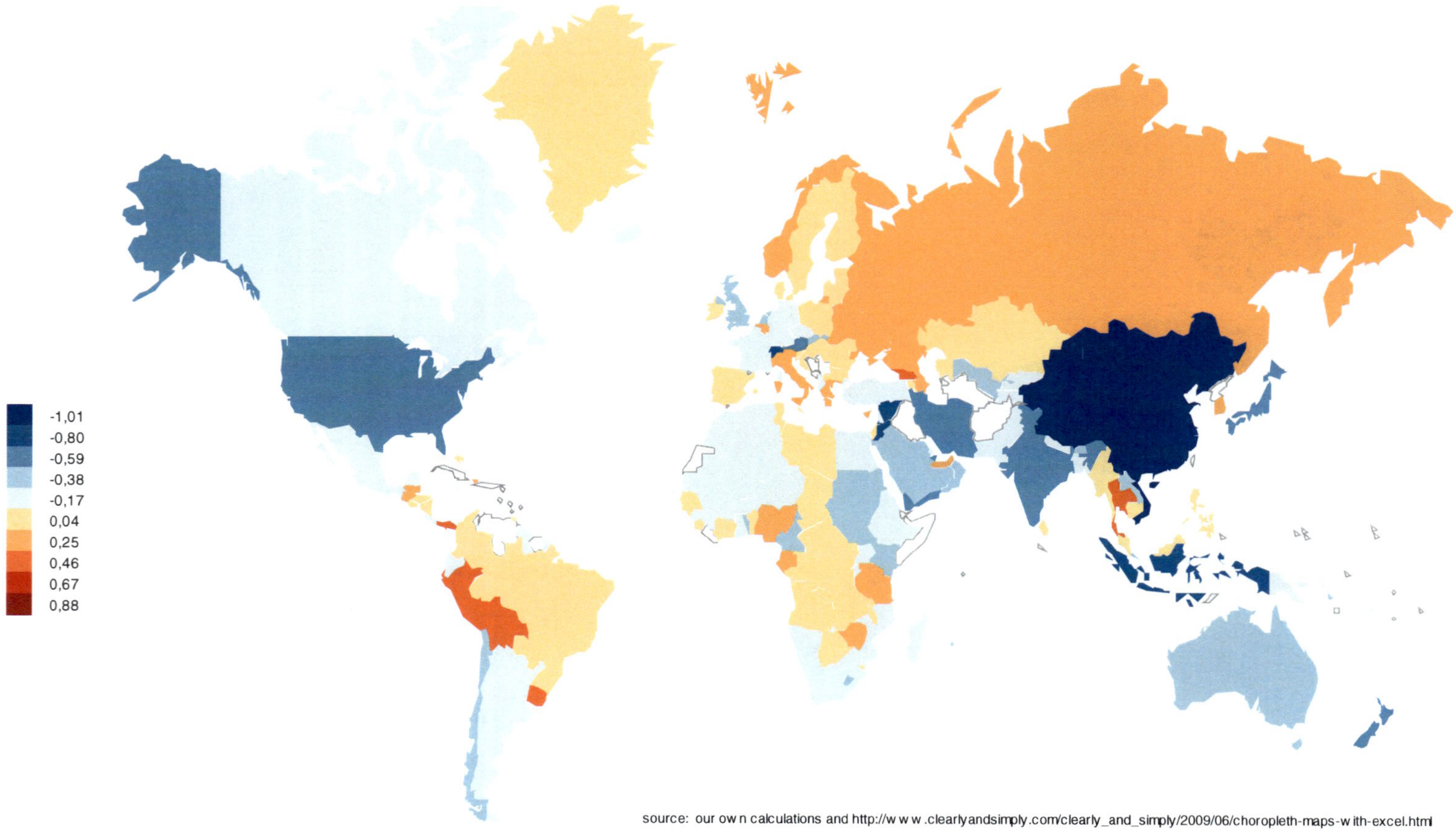

Map 3.3. Log shadow economy per GDP per capita in the world according to Schneider, 2012, irrespective of log GDP per capita and log GDP per capita^2.

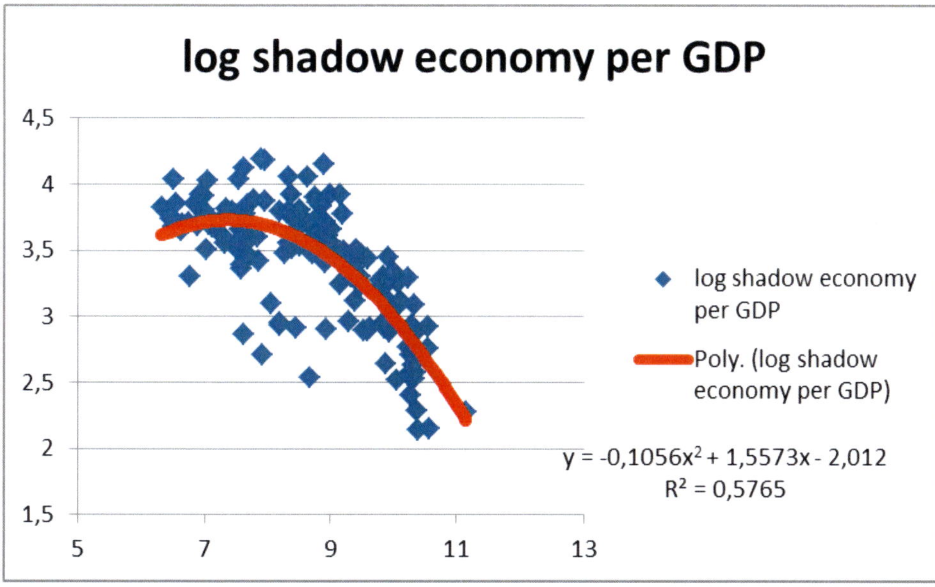

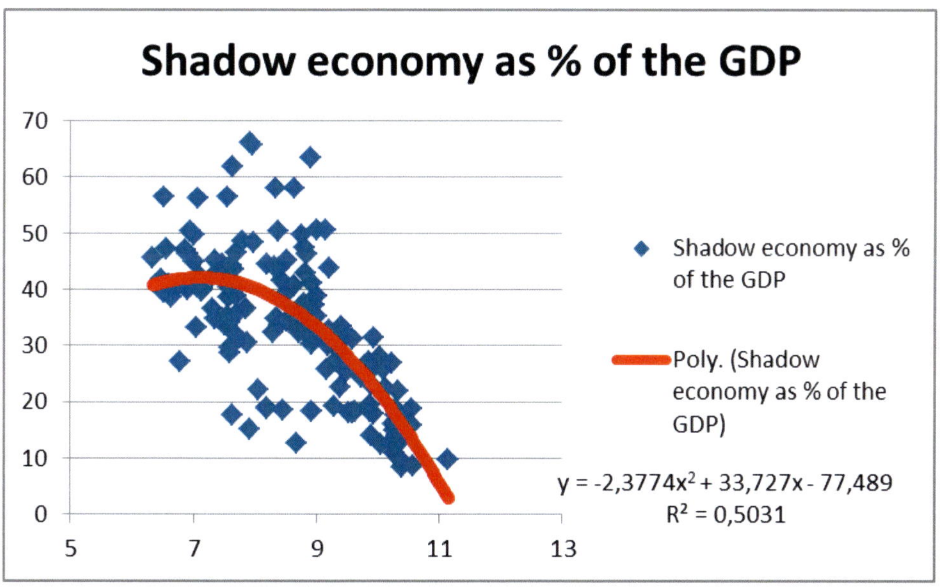

Graph 3.1. The trade-off between log GDP per capita (x-axis) and (log) shadow economy per GDP (y-axis).

Schneider (2005) concludes that for all countries investigated, the shadow economy as share of GDP has reached a remarkably large size (Africa 33.9-41.2; Americas 34.2-41.5; Asia 20.9-26.3; Transition countries 31.5-37.9 and highly developed OECD countries 13.2-16.8). The average percentage shares of GDP in all cases are increasing over time. The author demonstrates empirically a strong interaction of the shadow economy with government policies and with the official economy. He draws three further conclusions. First, an increasing burden of taxation and social security payments, combined with rising state regulatory activities, are the major driving forces underlying the size and growth of the shadow economy. Second, the shadow economy has a statistically significant and quantitatively important influence on the growth of the official economy. Increases in the shadow economy have a negative effect on the official growth in a developing country, but a positive effect in the developed industrialized and transition countries. Finally, shadow economies are a complex phenomenon, and are present in all types of economies. People engage in shadow economic activity because of government actions, most notably high levels of taxation and regulation.

Chaudhuri et al. (2006), using state level data from India, investigate the size of the hidden economy in Indian states over the period 1974/75 to 1995/96. The results show evidence of a varying size of the hidden economy in Indian states (10.2 till 48.3 percent of GDP), and it provides evidence in favor of liberalization of the Indian economy in 1991/92, which reduced the growth in the size of the hidden economy. In addition, the results show that the growth in the size of the shadow economy is less in election years, lower in states governed by coalition of political parties, and increased growth of newspapers and the literacy rates translate into cleaner governance and helps to lower the amounts of shadow economy activities.

In Choropleth Map 3.1, we portray the latest data series by Schneider (2012) on the extent of the shadow economy (averages for 1999-2007, given by Schneider, 2012), and also run an analysis of residuals, explaining the shadow economy by log GDP per capita and its square. Countries above the trend line are prone to shadow economic activity irrespective of the size of the GDP per capita. Choropleth Map 3.2 portrays these unstandardized residuals, while Graph 3.1 shows the trade-off between log GDP per capita and the shadow economy, based on Schneider's data and https://www.academia.edu/5258245/ Choropleth_maps_from_EXCEL_data_designed_by_Robert_Mundigl_FRG_on_Smart_Global_Development_.

3.2.3. Tax Morale, Trust, Institutional Quality and Shadow Economy

There are several factors that are expected to have a negative association with the size of shadow economy. These include trust, tax morale, and quality of institutions. This section briefly discusses these factors rule considering the shadow economic activities magnitude and development.

Some researchers, explaining the shadow economy, go beyond the objective variables such as the tax burden, rate of public expenditure, or the density of regulation, and use subjective variables such as perceptions, expectations, attitudes and motivations such as tax morale or institutional quality. The relationship between tax morale and institutional quality and the shadow economy is investigated by Torgler and Schneider (2009). The shadow economy is measured as a percentage of the official GDP using the DYMIMIC-method where one estimates the parameters for determining the size of the shadow economy. European Values Survey data on cheating on taxes and Latinobarometer data on the justifiability of avoiding paying taxes are measures of tax morale, and the quality of governance index is used as a proxy for institutional quality. These allow then an investigation of the impact of such factors at the macro level. They use a multivariate analysis to examine the quantitative impact of these factors on the level of and changes in the shadow economy. After controlling for a variety of potential factors, they find strong support for the hypothesis that higher tax morale and a higher institutional quality lead to a smaller shadow economy. The results show that improving social institutions (enhancing tax morale, voice and accountability, the rule of law, government effectiveness and its regulatory quality and by reducing corruption) helps to reduce the incentive to "go underground". In addition, the legal structure and security of property rights are among important factors that influence the size of the shadow economy.

Another factor with significant potential impact on the size of shadow economy is trust. Trust can be a substitute to formal and legal contracts in a situation when the agents involved in shadow transactions cannot rely upon the formal legal system to enforce agreements or settle disputes. This view suggests that trust increases the size of the formal sector by negatively impacting the size of the informal sector. In this regard, D'Hernoncourt and Meon (2012) investigate the relationship between trust and the size of shadow economy. They report a negative relationship between the size of the shadow economy and generalized trust. The data include 145 countries from both developed and developing countries observed over the period 1999-2003. Trust defined as a view that most people can be trusted is measured as the trust index provided in the WVS data. Comprehensive sensitivity analyses conducted confirm that the relationship is robust to controlling for various sets of economic, policy, and institutional variables, to changing the estimate of the shadow economy and the estimation period, as well as to controlling for endogeneity. Trust and the shadow economy are negatively related and trust matters more for developing countries. The impact runs through agents' propensity to shy away from paying taxes. However, the shadow economy is found to be independent from trust in institutions and from income inequality. It is mainly present in developing countries. The findings suggest that the tax compliance effect of trust dominates its role as a substitute for the legal system.

Policy pointer:

Although *World Values Survey* (WVS) data are used by the economics profession in their attempt to estimate the drivers of the shadow economy, contemporary sociological WVS research has failed hitherto to integrate the shadow economy into general theories of values and value change.

3.2.4. Shadow Economy, Corruption, Information and Communication Technologies (ICT) and Environment

Three key factors that their extent, use and abuse have impacted the size of the shadow economy are corruption, Information and Communication Technologies (ICT) and environmental violations. Water quality and air pollution are serious problems in many developing countries threatening the health of citizens. Many of the environmental problems are fostered by the sizable shadow economy in developing and transition economies. Many countries lack regulations or their governments are corrupt or unable to implement effective environmental policies.

Biswas et al. (2012) are surprised that there is still a lack of research on the shadow economy-environment nexus. They fill the gap in the literature by testing the extent to which the informal sector contributes to pollution and corruption undermines environmental policy.

The empirical study depending on the model and the sample size varies between 107 and 134 countries for the period 1999 to 2005. Two hypotheses are tested: the shadow economy increases pollution, and corruption exacerbates the effect of the shadow economy on environmental degradation. Pollution is measured as per capita environmental pollution. Shadow economy is measured as a share in the GDP and to measure corruption, the corruption index from Political Risk Services is used. They find that the larger the shadow economy, the greater pollution, but the effect declines with controlling corruption. The results hold as well when controlling for economic and demographic covariates. Account is also made for country group heterogeneity by level of income. The findings imply that policy makers may address the effects of corruption and the shadow economy before increasing environmental standards and regulations.

The diffusion and spread of internet technology in recent decades has been having consequences for different sectors of the economy. The effects span from business to banking, e-government, education, information, productivity, foreign direct investment, inflation and political economy issues. In some cases, researchers point to its positive impacts on economic outcomes such as inflation reduction, higher volume of trade, increased productivity and higher economic growth. Indjikian and Siegel (2005) provide a survey of the economic effects of the spread of ICT. In this regard, Elgin (2013) asks whether the spread of the internet aided or abetted the shadow economy. The theoretical framework argues that internet usage creates two effects on the size of the informal sector: through increasing productivity and reducing the size of the shadow economy and tax evasion. Using a panel data of 152 countries over nine years from 1999 to 2007, Elgin examines the empirical relationship between the degree of internet usage and the size of the shadow economy. Informal sector size, TFP (total factor productivity) growth and taxes are estimated as a system where internet usage explains TFP growth and taxes which in turn determine the size of the informal economy accounting for various control variables. The estimation results indicate that the association between internet usage and the shadow economy size strongly interacts with GDP per-capita and level of development. In designing an economic policy toward the shadow economy, it is recommended that policymakers should take into account for the two opposite effects on productivity and tax aversion. The policy measure should be taken to put the productivity enhancing effect forward relative to the tax-evasion effect. Subsidizing ICT investment and better infrastructure, improving institutions and the bureaucratic quality are among steps needed to be taken by governments.

3.2.5. Debt, Default Risk, Financial Development and Size of Informal Economy

The data on sovereign debt and the size of the shadow economy vary substantially among countries. The macroeconomic implications include high interest rate payments which constraints economic well-being, risk of debt default, limited enforcement of tax collection and presence of the informal sector, which influences fiscal instruments, tax revenues and government's ability to repay public debt. Elgin and Uras (2012) address the interactions between government's indebtedness, sovereign default risk and the size of the informal sector. They test a theory suggesting that in societies with limited tax enforcement, the presence of informality constrains the set of fiscal policy alternatives. It increases public debt and the implied probability of sovereign debt restructuring. In the empirical part a number of hypotheses are tested: a larger size of the informal sector is associated with higher public indebtedness, higher interest rates paid on sovereign debt, a higher level of financial instability and a higher probability of sovereign default. These are represented by the ratio of public debt to GDP, the interest rate spread, and the financial stress index (FSI), and default probability obtained from the World Development Indicators (WDI) database.

The empirical results in Elgin and Uras (2012) from cross-country panel regressions of 152 countries over the period of 1999-2007 show that after controlling for previously highlighted variables identified in the literature that could explain the variation in financial instability, sovereign default risk and public indebtedness, the size of informality in the economy remains as a significant determinant of these variables. In order to show that a larger shadow economy size is associated with the four possible undesirable outcomes, elasticities are computed. A 1% increase in the informal sector size led to a 6% increase in the ratio of public debt to GDP, a 127% increase in the financial stress index, a 14% increase in the interest rate spread and a 4% increase in the probability of default. It is concluded that improving enforcement through enhancing law and order and implementing policies for lower taxes and unemployment can have significant effects on sovereign default risk. The policy implications suggest that public policy should focus more on taking measures to reduce the size of the shadow economy. Increasing law enforcement, enhancing institutional development, policies towards lowering income taxes and unemployment are among the prescribed measures recommended to policy-makers in societies with a large informal sector size.

In a recent study, the theoretical relationship between the underground economy and financial development in a model of tax evasion and bank intermediation is studied by Blackburn et al. (2012). Here the presence of an underground sector is a reflection of individuals' incentives to conceal their economic activities or circumstances for two reasons. First, these activities would be less rewarding if practiced in the formal sector. A second reason is because the activities are illegal. The shadow economy has potentially serious implications for economic performance and public policy. The primary objective of this study is to shed lights on the determinants of the underground activity and their influence on individual's incentives. The results show that the lower the financial development, the higher the incidence of tax evasion, and the greater the size of the underground economy. This negative relationship implies that the marginal net benefit of income disclosure increases with the level of financial development.

3.3. Economic Freedom, Corruption and Economic Growth

3.3.1. Corruption, Economic Freedom and Development

There are few studies on the substitutability of corruption and economic freedom. The aim is to analyze how economic freedom affects corruption. However, the analysis of their relation with economic growth is of great interest. It is expected that the more the government intervenes in markets, the more level of economic freedom declines. Apergis et al. (2012) investigate the relationship between economic freedom and corruption using data from 50 U.S. states covering 1981-2004. This study contributes to the literature in a number of respects. First, it uses a more objective measure of corruption, namely, the number of government officials convicted in a state for crimes related to corruption. Secondly, it exploits both time series and cross-sectional variations in the data. Three different measures of economic freedom are used. The results show that in the long-run, economic freedom, per capita income, and education have a negative and statistically significant impact on corruption whereas income inequality has a positive impact. The causality tests reveal bidirectional causality between economic freedom and corruption in both the short-run and long-run.

It is expected that economic freedom reduces the levels of corruption. Empirical studies support this expectation, but with some inconsistency in the results. Therefore, the prediction that economic freedom is beneficial in reducing corruption is not found to be universally robust in empirical studies. Pieroni and d'Agostino (2013) reviews this relationship by using firms' data in a cross-country survey from 1999. They argue that approaches using aggregated macro data have not been able to explain it appropriately. They model cross-country variations of the economic freedom-corruption relationship using multilevel models. They also disentangle the determinants for several components of economic freedom with unequal effects. The results show that the extent of the macro-effects on the measures of microeconomic freedom for corruption can explain why a lack of competition policies and government regulations may yield more corruption but with large variability across countries. Estimations for Africa and transition economy subsamples confirm these conjectures.

Policy pointer:

The data base of the *World Values Survey* contains very precise items in connection with a neo-liberal interpretation of economic processes, like the acceptancy or rejection of free competition, or various items on the acceptability or desirability of state intervention. In contrast to hitherto published research, we try to integrate these elements into the core-model of global value change. For this good reason, we present here an extensive survey how contemporary economic theory already integrates economic freedom in its empirical and theoretical research. Our survey is intended to be an overview especially for readers who are more familiar with other disciplines of the social sciences and not economics. In the presentation, we use a non-mathematical language.

Maps 3.4 to 3.11 portray the extent of economic freedom indicators in the world in 2014, showing the rather large degree of economic, business, fiscal and financial freedom in several Arab and Muslim countries according to the well-known Heritage Foundation data series (see http://www.heritage.org/index/explore). Surely, the state chancelleries and finance and economics ministries of countries like Jordan, Malaysia or the Gulf countries are better disciples of Nobel laureate Hayek than the political class in France ...

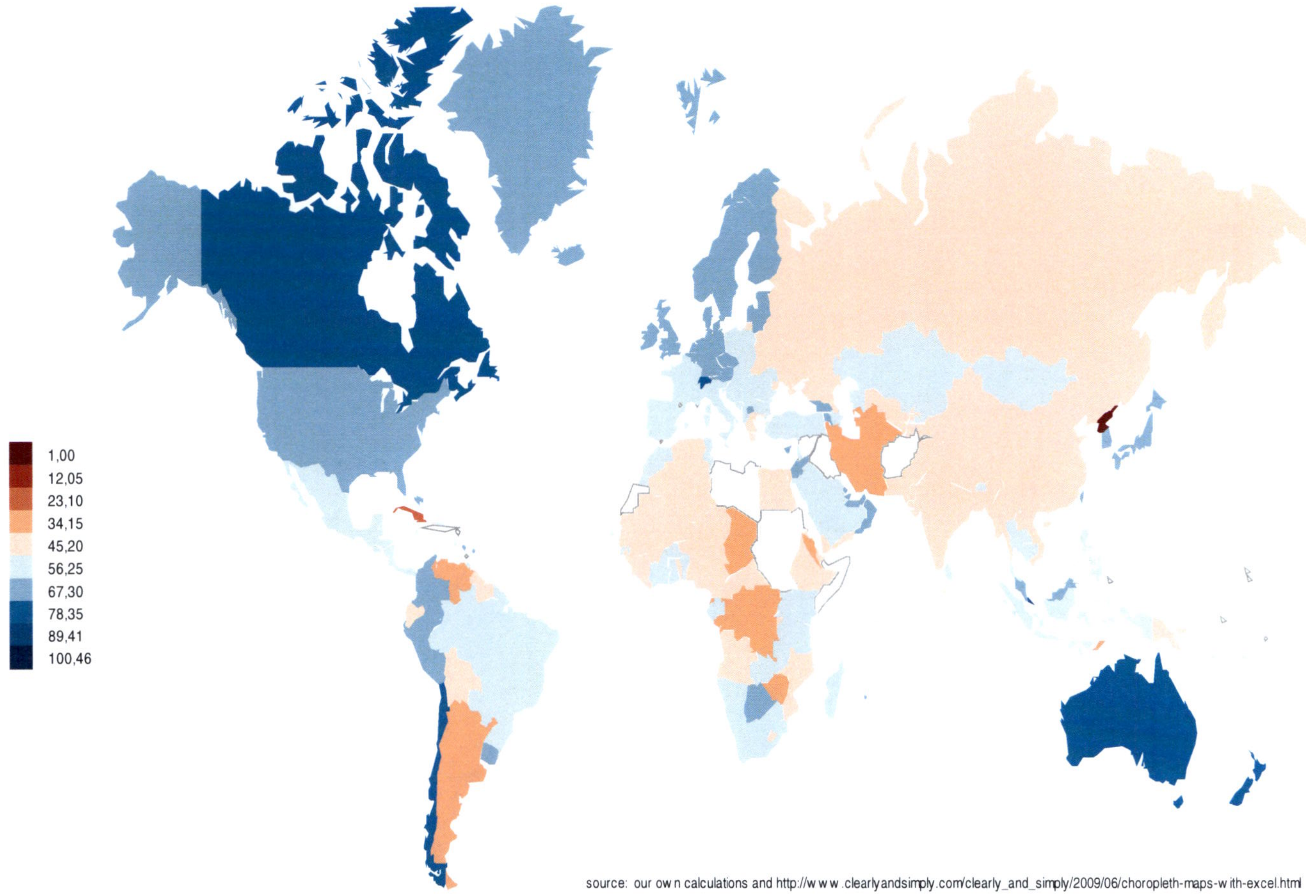

source: our own calculations and http://www.clearlyandsimply.com/clearly_and_simply/2009/06/choropleth-maps-with-excel.html

Map 3.4. Economic Freedom in the world – The Muslim countries on the global maps of the Heritage Foundation data.

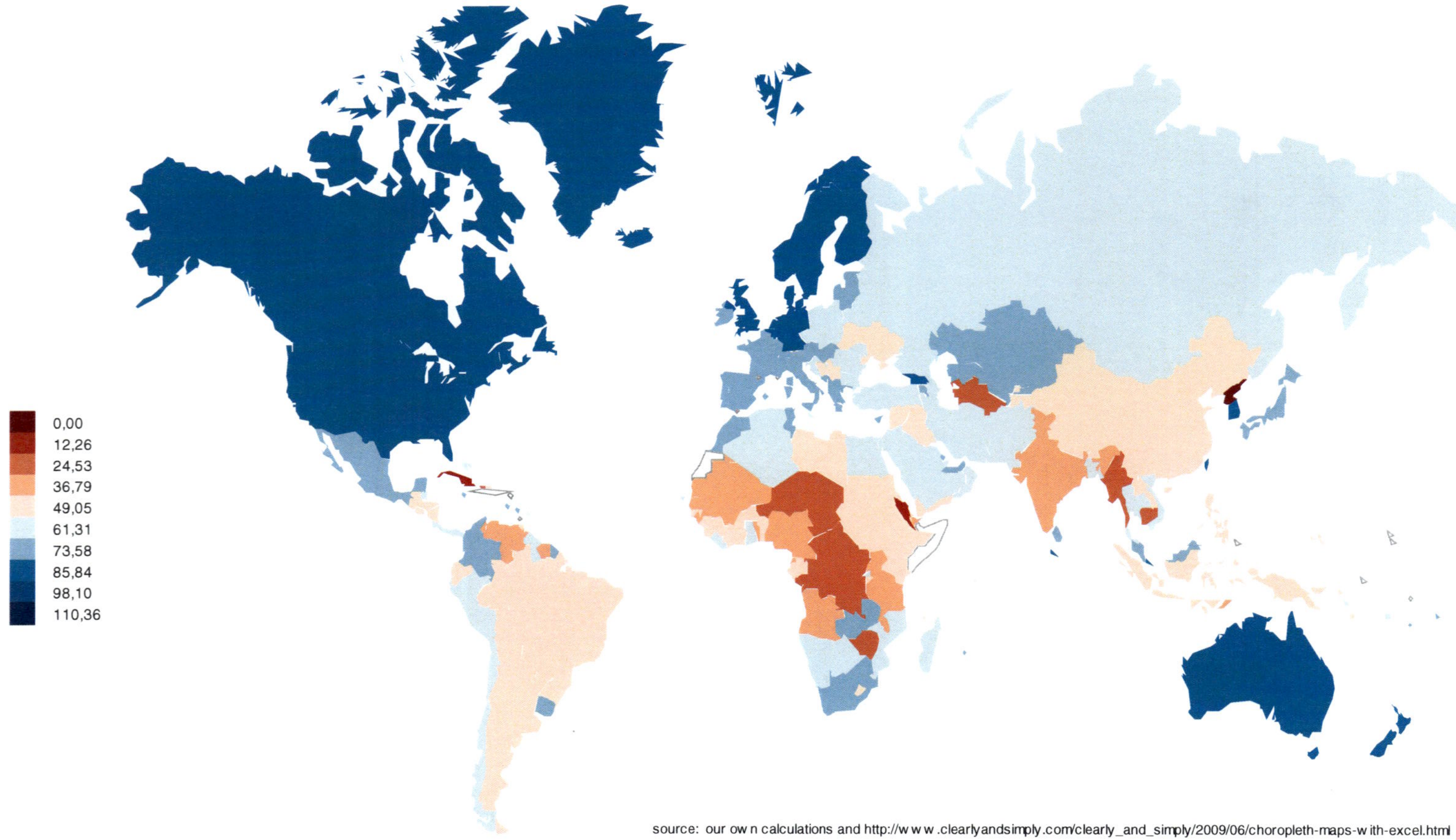

Map 3.5. Business Freedom in the world– The Muslim countries on the global maps of the Heritage Foundation data.

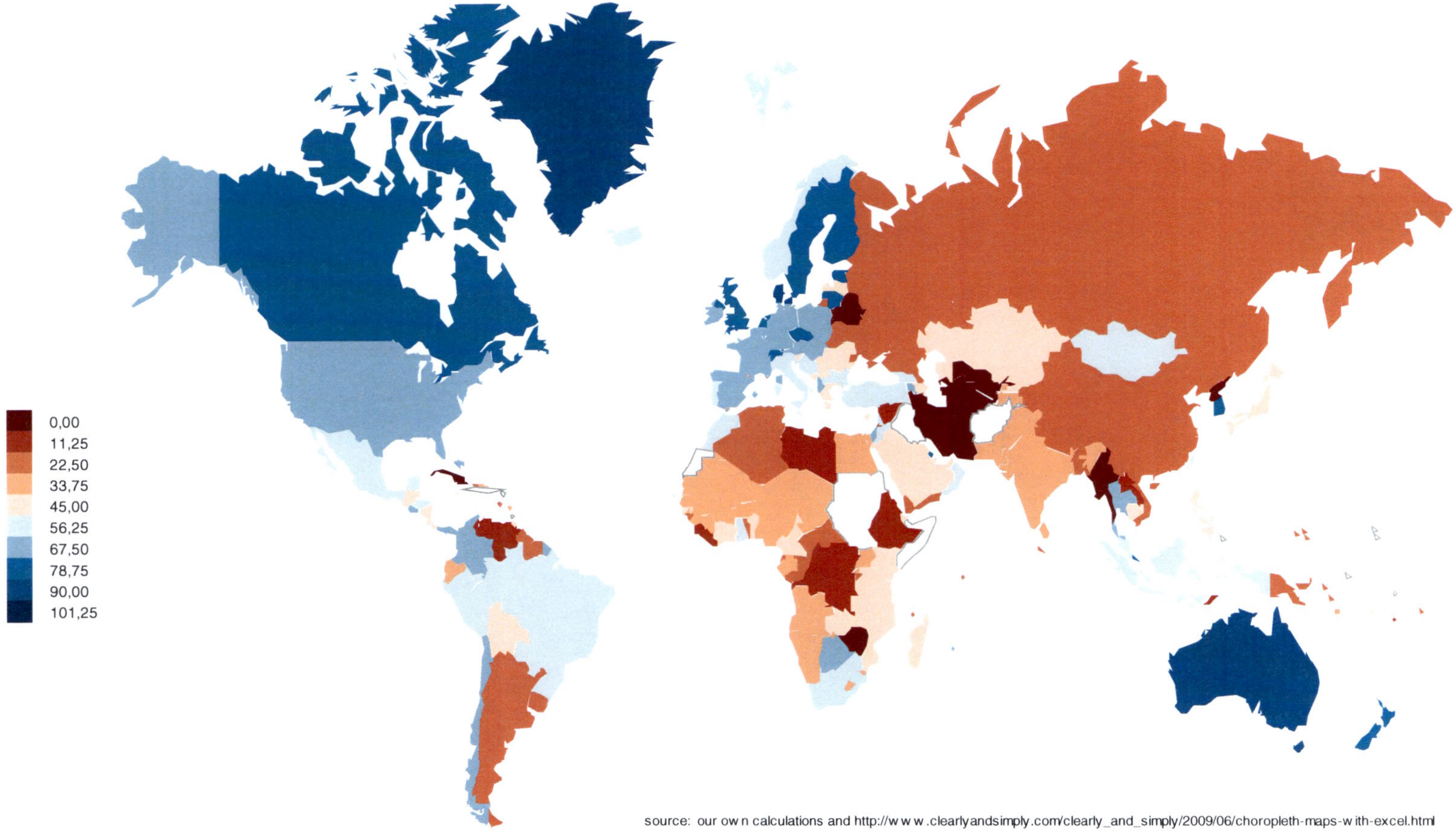

Map 3.6. Financial Freedom in the world – The Muslim countries on the global maps of the Heritage Foundation data.

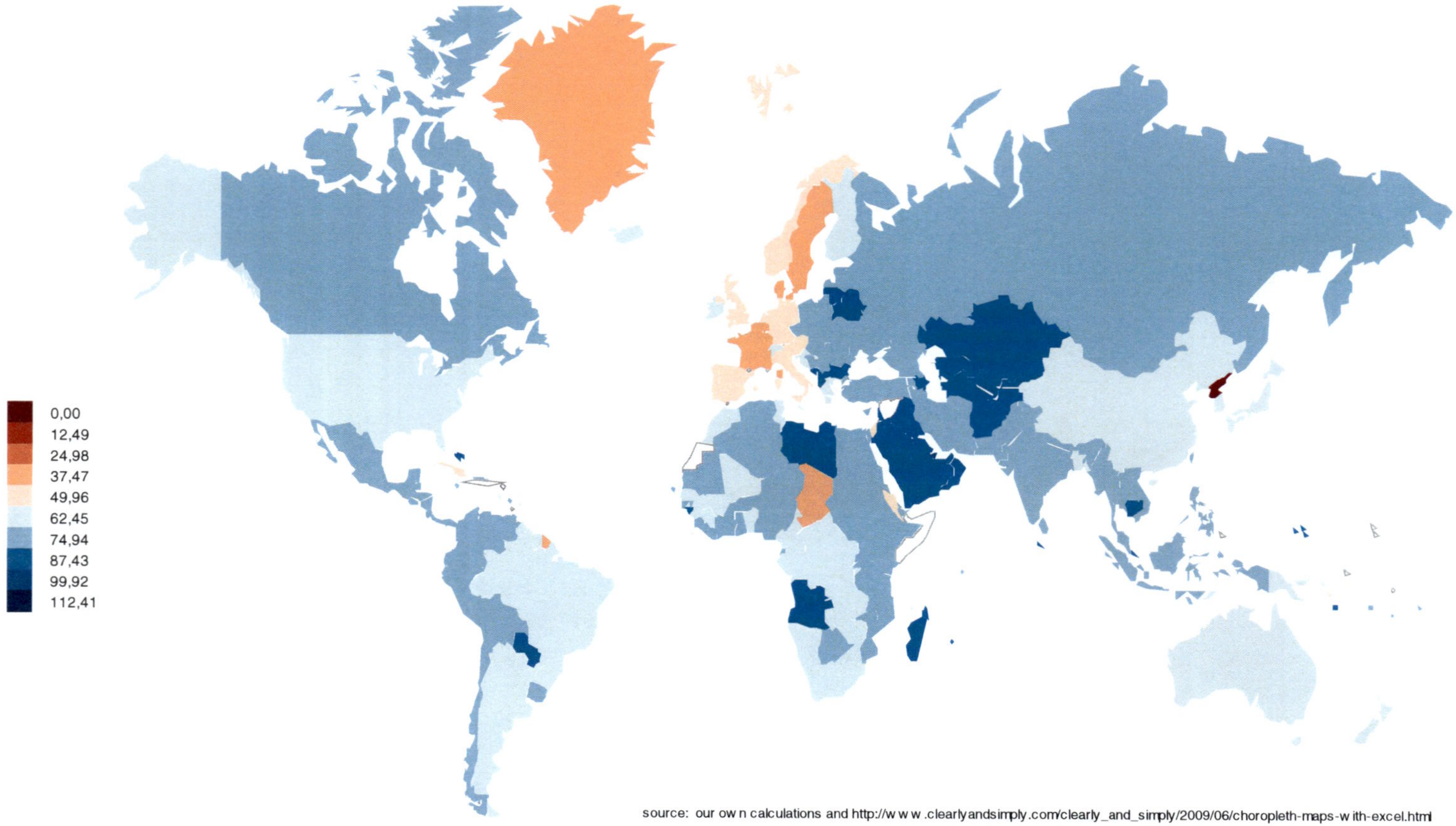

Map 3.7 Fiscal Freedom in the world – The Muslim countries on the global maps of the Heritage Foundation data.

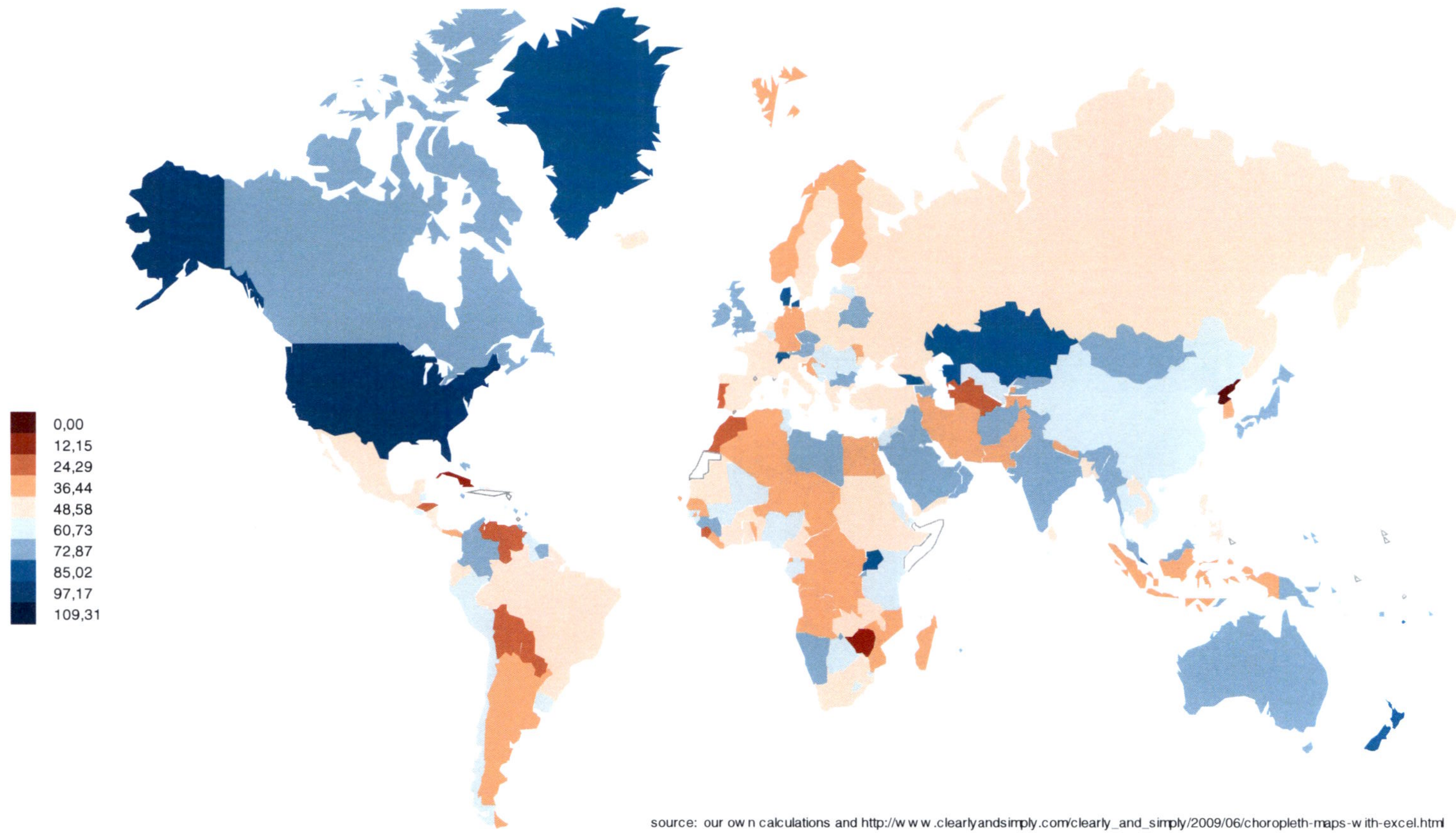

Map 3.8. Labor Freedom – The Muslim countries on the global maps of the Heritage Foundation data.

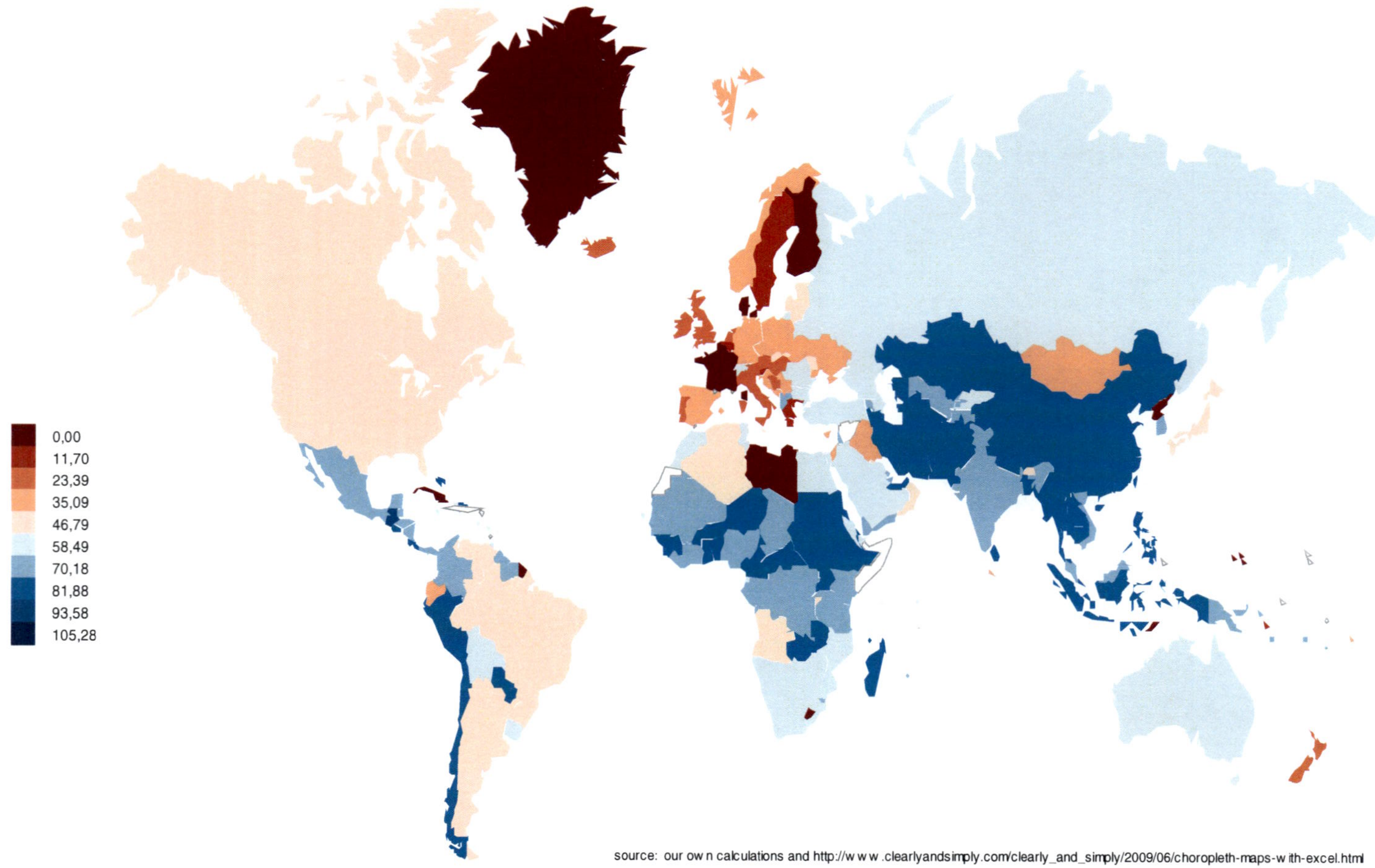

Map 3.9. Low Government Spending – The Muslim countries on the global maps of the Heritage Foundation data.

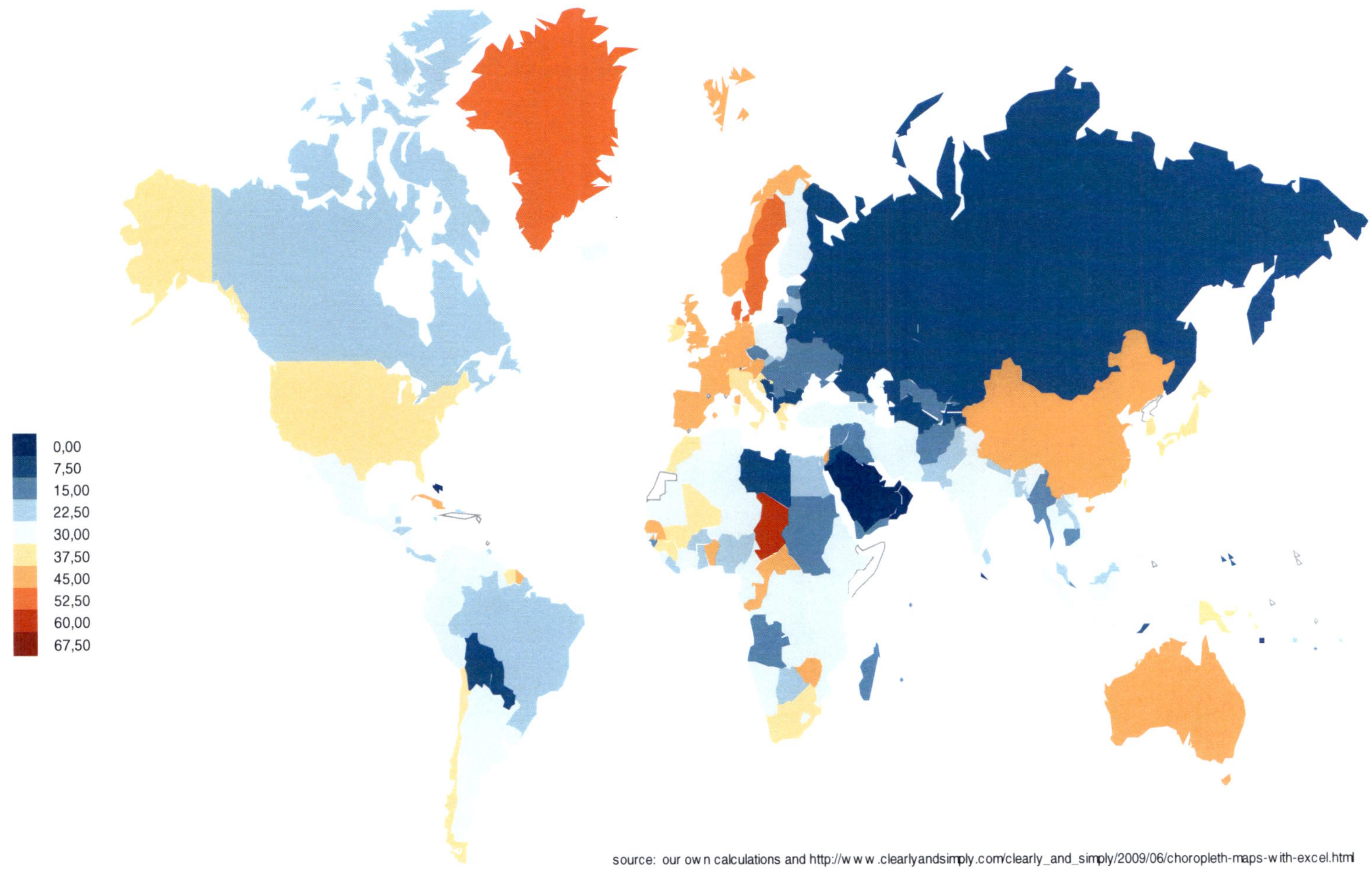

Map 3.10. Income Tax rate – The Muslim countries on the global maps of the Heritage Foundation data.

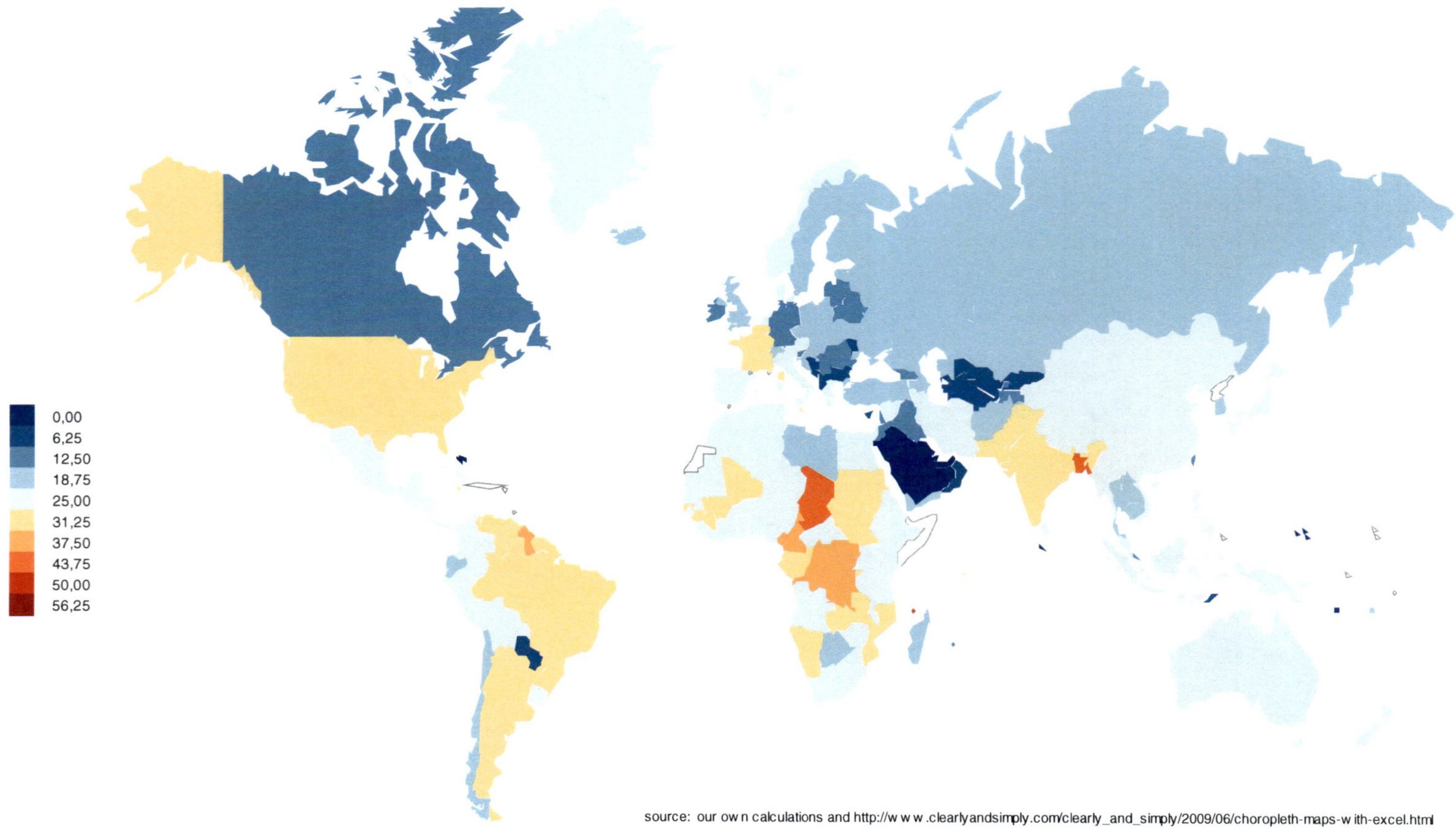

Map 3.11. Corporate tax rates – The Muslim countries on the global maps of the Heritage Foundation data.

Policy pointer:

Muslim societies sometimes perform very well on indices of economic freedom, irrespective of their development level. As defined by the well-known think tank *"Heritage Foundation"* (http://www.heritage.org/index/book/chapter-5) economic freedom is the condition in which individuals can act with maximum autonomy and minimum obstruction in the pursuit of their economic livelihood and prosperity. As Hayek observed, to be controlled in our economic pursuits would mean to be controlled in everything.

Erich Weede, the great neo-liberal German sociologist, observed in 2006 that there is a real chance that Islam contains a chance of promoting capitalism and economic freedom which other religions lack.

"Think of Christ who [i.e., according to the Christian Gospel] threw the traders out of the temple, or of Buddha who became a hermit and delivered his first sermon to deer. But Mohammed was a full-time trader before he became a prophet, a ruler, and a military leader. This is as close to a capitalist background as was possible in the 7th Century. [...] Here is an unexploited resource which might permit a unique religious legitimation of capitalism or economic freedom and thereby of limited government. [...] There is another [...] support for the establishment of economic freedom in Islam. In principle, God's law is believed to be complete. The ultimate legislator in Islam is God, not rulers or elected politicians [....] If one thinks of taxation, it is obvious that legislation by God has some advantages over legislation by greedy secular authorities who tend to overtax their subjects or citizens. Just imagine how much better entrepreneurs and taxpayers could live in Western democracies, if God were believed to have forbidden tax or deficit increases to our politicians. In principle, Islam provides its own justification for constraining rulers and for limited government." (Weede, 2006)

For the Heritage Foundation, a comprehensive view of economic freedom encompasses all liberties and rights of production, distribution, or consumption of goods and services.

The Heritage Foundation Index of Economic Freedom takes a broad and comprehensive view of country performance, measuring 10 separate areas of economic freedom. Each economic freedom is individually scored on a scale of 0 to 100. A country's overall economic freedom score is a simple average of its scores on the 10 individual freedoms. The 10 economic freedoms are grouped into four broad categories:

Rule of law (property rights, freedom from corruption);
Government size (fiscal freedom, government spending);
Regulatory efficiency (business freedom, labor freedom, monetary freedom); and
Market openness (trade freedom, investment freedom, financial freedom).

Business freedom is about an individual's right to establish and run an enterprise without undue interference from the state. Burdensome and redundant regulations are the most common barriers to the free conduct of entrepreneurial activity.

Financial Freedom. An accessible and well-functioning formal financial system ensures the availability of diversified savings, credit, payment, and investment services to individuals. By expanding financing opportunities and promoting entrepreneurship, an open banking environment encourages competition in order to provide the most efficient financial intermediation between households and firms as well as between investors and entrepreneurs.

Fiscal freedom is a direct measure of the extent to which government permits individuals and businesses to keep and manage their income and wealth for their own benefit and use. A government can impose fiscal burdens on economic activity through taxation, but it also does so when it incurs public debt that ultimately must be paid off through taxation.

It is well-known that corruption is detrimental to economic performance, especially in developing countries. However, it is unclear whether an increase in income consistently reduces corruption. Saha and Gouder (2013) investigate the relationship between income and corruption to provide insight into the changes in the level of corruption and economic development. For the empirical part data for 100 countries classified by region and income for the period 1995 to 2008 is used in this study. Unlike most previous studies, they use hierarchical polynomial regression to evaluate the existence of non-linear relationships after controlling for socio-economic and institutional factors. A quadratic function is found to best fit the data. The results show that corruption is explained by various socio-economic, political and institutional factors. The results also challenge some of the findings of a negative income–corruption association. Despite an upsurge of corruption among the low-to-medium income countries, advanced stages of development eventually reduce corruption levels. Policy implications suggest that a combination of economic, institutional and social policies can effectively reduce and lower the effects of corruption. Achieving a high level of educational attainment, employment opportunities and equal income distribution are necessary to discourage corrupt activities.

3.3.2. Gains and Well-being from Economic Freedom

Interest for factors that increase well-being, happiness and life satisfaction has increased over the past decades. Belasen and Hafer (2012) mentions that there is plenty of evidence that well-being is positively related to the level of general intelligence and economic freedom across countries. Belasen and Hafer aim to determine whether economic freedom and well-being are related at the state level. Well-being is measured as a multidimensional index and economic freedom as an index of freedom. Regression analysis of the link between well-being and economic freedom accounting for different control variables indicates that, across the 50 U.S. states, improvements in economic freedom lead to higher levels of well-being. The relationship between well-being and economic freedom is found to differ significantly across regions in the U.S. It is suggested that the government plays a major role in influencing individual well-being.

In relation with the link between economic freedom and well-being Gehring (2013) recently asked the question: who benefits from economic freedom? The focus is on the effects of economic freedom on subjective well-being. Well-being is defined as happiness obtained from the World Value and the European Value Surveys. Results from a panel of 86 countries over the 1990 to 2005 period suggest that economic freedom has a significant positive effect on subjective well-being. Economic freedom dimensions including legal security and property rights, sound money, and regulation are strong predictors of higher well-being. Different sensitivity analysis tests show that the positive effect is not affected by socio-demographic variables. Developing countries profit more from higher economic freedom from reducing the regulatory burden which strengthen the effect. Culture seems to moderate the effect. It is concluded that societies that are more tolerant and have a positive attitude toward the market economy profit the most in the form of well-being.

The issue of inequality in the distribution of freedom, growth and well-being is an important issue. For instance, many South Koreans believe that the country's remarkable economic successes are the product of an industrial policy that made Korean industry competitive. Holcombe (2013) mentions that, while there is widespread sentiment in favor of maintaining the emphasis on industrial policy, others believe that industrial policy has generated unequal benefits at the expense of the welfare of working-class. In recent years, there are tendencies toward economic democracy to share more equally the gains from increased productivity. Neither industrial policy nor economic democracy is in the best interest of Koreans as each reward particular interest groups. A laissez faire policy of minimal government interference is suggested to provide the best environment to foster South Korea's continued economic progress.

3.3.3. Causality between Economic Freedom and Economic Growth

The causality and its direction between economic freedom and economic growth has been another focus of attention of the kind of cross-national comparative literature under review here. Narayan et al. (2011) examines the relationship between democracy and economic growth in 30 Sub-Saharan African countries for the period 1972 to 2001. Democracy is defined as democracy index constructed by Freedom House and the Legislative Index of Electoral Competitiveness. The results show support for the Lipset hypothesis indicating that in the long run, real GDP Granger causes[18] democracy and an increase in GDP results in an improvement in democracy for a number of countries. Support for the compatibility hypothesis that in the long run democracy Granger causes real income and an increase in democracy has a positive effect on real income is also found for a number of countries. Support for the conflicting hypothesis that in the long run democracy Granger causes real income and an increase in democracy has a negative effect on real income is found for two countries. For most countries there is long run Granger neutrality between democracy and real GDP supporting the skeptical hypothesis for majority of the countries.

European community (EC) welfare states developed compulsory public social insurance to promote social cohesion among citizens. The European health systems are based on principles of solidarity, equity, and universality. In recent years the EC health policy shows signs of tension between EC economic freedoms and social policy. The EC competition law restricts national health policy options of member states and limits the ability to adequately accounting for non-economic gains in consumer welfare. Mossialos and Lear (2012) define the policy and legal parameters of the debate between competition law and health policy in EC. A sample of cases is used to analyze how the European Court of Justice, national courts, and National Competition Authorities have applied competition laws to the health services

[18] A time series X is said to Granger-cause Y if it can be shown through a series of statistical tests on lagged values of X that the X values provide significant information about future values of Y.

sector. Health policy does not enjoy a special exclusion from the competition law which emphasizes above all trade. It is concluded that implications of the convergence of recent trends in competition law enforcement and health system market reforms do not undermine the social model and its goals of equity and social cohesion.

Compton et al. (2011) analyze which economic policies are most favorable to economic growth. In the literature, policies that promote economic freedom have been suggested as a viable path towards sustained economic growth. Using the multidimensional measures of economic freedom index the authors investigate the nature of the relationship between economic freedom, the size of government, taking taxes, labor market and economic growth for the fifty U.S. states spanning the period 1981 to 2004. The results from comprehensive sensitivity analysis, in respect with variable definitions and model specification as well as accounting for endogeneity and selection bias and after controlling for various control variables, show evidence of a significant positive relationship between economic freedom and economic growth at the state level. However, not all selected components of economic freedom affect growth equally. As policy recommendation, constraining excessive government expenditure, minimizing tax burden and maintaining an open labor market are suggested.

3.3.4. Capitalism, Economic Transition, Freedom and Growth

Over the last two centuries the spread of capitalism and the level of political freedom have risen closely. The connection between economics and politics has made economic freedom indispensable for the achievement of political freedom. Pryor (2010) tests Milton Friedman's conjecture that capitalism is a necessary condition for political freedom. Analysis of data shows that for the decade around 2000 indices of the degree of capitalism and the degree of political freedom with few exceptions are highly correlated. This provides plausibility for Friedman's conjecture. However, results based on data for the nineteenth century refute the Friedman's conjecture. These contradictory results are related to per capita GDP and the educational level of the population. Despite their correlations, capitalism is neither a necessary nor sufficient condition for political freedom and no confirmation is found for the reverse proposition.

Policy pointer:

Given the *World Values Survey* data presence in many of the transition countries, it has to be recalled here that economic freedom is considered by many economic researchers as the central issue of the transition process.

The 1991 collapse of the Soviet Union lead to the creation of 15 new states and a transition from centrally planned economy to market economy for 25 states for the period 1998 to 2005. These countries experience heterogeneous growth and development. Pääkkönen (2010) reviews the political economy of economic growth in post-communist economies transition to free markets. The focus is on the role of economic policy and institutions in the transition process. The author tests the hypothesis that better institutions, measured in terms of economic freedom, contribute to growth. The empirical results confirm this hypothesis. There is an interactive effect between economic freedom and investment, where institutions affect growth indirectly by affecting the marginal effects of investment. It is concluded that non-linearities are present in the specified growth model, where the marginal effect turns to become negative. Increased government consumption also has a negative effect on growth suggesting the presence of wasteful spending and hindrance to growth.

The relationship between foreign direct investment (FDI) and its technology, management and skill spillovers and improved productivity and economic growth is intensively analyzed with conflicting conclusions. Efforts have been made to identify determinants of inflows of FDI to design suitable policies and to remove restrictions imposed on foreign capital flows. Azman-Saini et al. (2010) investigate the systemic link between economic freedom as absorptive capacity, FDI and economic growth in a panel of 85 countries for the period 1976 to 2004. The empirical results based on the system estimator, reveal that FDI has no direct positive effect on output growth. Instead, the effect of FDI is contingent on the level of economic freedom in the host countries. The countries promoting greater freedom of economic activities gain from the presence of multinational corporations. The results suggest that economic freedom is important to the host countries absorptive capacity. Policy makers should promote better freedom of economic activities.

Policy pointer:

One particularly politically relevant research tradition from the discipline of economics at the crossroads with the research traditions based on the *World Values Survey* project have been studies on economic freedom.

Over the last two centuries the spread of capitalism and the level of political freedom have risen closely. The connection between economics and politics has made economic freedom indispensable for the achievement of political freedom. Pryor (2010) tested Milton Friedman's conjecture that capitalism is a necessary condition for political freedom. The analysis of data shows that for the decade around 2000 the indices of the degree of capitalism and the degree of political freedom, with few exceptions, are highly correlated. This provides plausibility for Friedman's conjecture. However, results based on data for the nineteenth century refuted the Friedman's conjecture. These contradictory results are related to per capita GDP and the educational level of the population. Despite their correlations, capitalism is neither a necessary nor a sufficient condition for political freedom and no confirmation is found for the reverse proposition.

The 1991 collapse of the Soviet Union lead to creation of 15 new states and a transition from centrally planned economies to market economies for 25 states for the period from 1998 to 2005. These countries experienced heterogeneous growth and development. Pääkkönen (2010) reviewed the political economy of economic growth in the post-communist economies transition to free markets. The focus is on the role of economic policy and institutions in the transition process. The author tested the hypothesis that better institutions, measured in terms of economic freedom, contributed to growth. The empirical results confirm this hypothesis. However, the finding points to an interactive effect between economic freedom and investment, where institutions affect growth indirectly by affecting the marginal effects of investment. It is concluded that non-linearities are present in the specified growth model, where the marginal effect turns to become negative.

Increased government consumption also has a negative effect on growth suggesting the presence of wasteful spending and hindrance to growth.

3.4. Effective Democracy and Economic Growth

3.4.1. Democracy, Natural Resources and Aid

In recent decades, the attitude towards foreign direct investment (FDI) has changed. Researchers and international organizations encourage developing countries to attract FDI as a tool to fight poverty. However, most developing countries have a weak democracy. Asiedu and Lien (2011) examine the impact of democracy on FDI assuming that the relationship between democracy and FDI is the same for resource exporting and non-resource exporting countries. The objective is to examine whether natural resources in host countries alter the relationship between democracy and FDI. Results based on data from 112 developing countries over the period 1982 to 2007 show that democracy promotes FDI. However, the effect is observed only if the share of natural resources in total exports is less than some critical value. Of the sample, 90 countries are identified, where an expansion of democracy may enhance FDI inflows, while in 22 countries it may reduce FDI. The effect of democracy on FDI is found to depend on the size and not the type of natural resources exported. It is also well-known that in resource rich countries democracy is weak and the curse of natural resources leads to lower economic growth.

The literature on the measurement of effectiveness of aid is voluminous. Foreign aid is found to be ineffective in creating economic growth. Bjørnskov (2010) refers to a popular argument for the absence of beneficial effects is that it is skimmed by political elites in the recipient countries. Some studies also suggest that aid may be more effective in relatively democratic developing countries. Data for 88 developing countries for the period 1960 to 2000 with a five-year average is explored to test the relationship between democracy and effectiveness of aid. The results derived indicate that foreign aid and democracy in conjunction are associated with a higher share of income held by the upper quintile as a result of rent seeking during the transition to democracy. The results show that foreign aid leads to a more skewed income distribution in democratic developing countries while the effects are negligible in autocratic countries. The paper closes with a discussion of potential mechanisms generating this perverse effect and question whether equitable income distribution and democratization can be achieved through foreign aid.

3.4.2. Political Culture, Press Freedom and the Outcome of Economic Policy

Feld and Kirchgässner (2000) in their research show that the political culture in Switzerland is influenced by the country's system of direct democracy. In order to support their statement the outcomes of two political decisions of Swiss citizens from 1989 vote on an initiative that proposed to abolish the Swiss army and 1979 vote on a possible exit from nuclear energy system is studied. The outcome in both cases has been unexpected. The two cases showed that compared to purely representative systems, direct democracy leads to a different type of communication among citizens and also between citizens and their representatives. The results show that the opportunity of deciding for themselves on political issues provides citizens with greater incentives to collect more information. Better informed citizens and focus on common interest leave less room for politicians to pursue their personal interests. As a consequence of better informed citizens who enjoy and apply direct democratic rights, public expenditure and public debt are lower. The increased participation in decision making and feeling of more responsible for their community has resulted in lower tax evasion compared to representative democratic systems with less participation.

Policy pointer:

Economic research provides very important insights into the conditions of "effective democracy", which cannot depend – as contemporary World Values Research often contends – on "self-expression" values alone. Also, all the dimensions, which were shown to be relevant in our own empirical value research (see Chapter 5) already surfaced in prior economic research, like economic permissiveness, which most strongly affects large parts of Latin America (except Venezuela, Peru, and the countries of the Southern Cone), some but not all nations of Africa, most of the former USSR and some other countries of East Central and Southeastern Europe, and Thailand and the Philippines.

Corruption is widespread in particular among developing countries. A precondition for achieving growth and reducing poverty is to deal with corruption in an effective way. This view is supported by the growing literature. Potentially effective instruments in rooting corruption are democratic elections and press freedom. Since their effects on corruption interdepend on each other a simultaneous interactive model is used to estimate the conditional effects. Kalenborn and Lessmann (2013) empirically analyze the joint impact of democracy and press freedom on corruption. They argue that both institutional features are complements rather than substitutes in controlling corruption. Corruption is proxied by the corruption perception index of Transparency International, democracy is represented by Vanhanen's democratization index and press freedom is provided by Freedom House. Estimation results based on a cross section of 170 countries covering the period from 2005 to 2010 as well as based on a panel evidence for 175 countries from 1996 to 2010 show that democratic elections only work in controlling corruption, if there is a certain degree of press freedom in a country, and vice versa. The policy implication of the result is that democratic reforms are more effective, if they are accompanied by institutional reforms strengthening the monitoring of rent seeking representatives.

3.4.3. Political Democracy, Economic Democracy and Degrowth Policies

Johanisova and Wolf (2012) mentions that we are living in times of quickly shifting paradigms. Ideals as economic growth and political democracy are falling and losing their appeal in the face of multiple crises with environmental, social and economic dimensions. In contrast to political democracy with power control in the public sector, economic democracy has received much less attention. The authors define economic democracy as "a system of checks and balances on economic power and support for the right of citizens to actively participate in the economy regardless of social status, race, gender, *et cetera.*" They suggest six possible faces of economic democracy and look at their implications for the vision of a sustainable, equitable and non-growing society including: regulation of market mechanisms and corporate activities to curb economic power; support for social enterprises which are better suited to a degrowth economy; democratic money creation processes; and pluralist community currencies. Reclaiming the commons is seen by the authors as an important aspect of enhancing economic democracy, redistribution of income and capital assets is another approach to achieving economic democracy; and finally, a broader view of economic democracy would involve a diversity of production scales and modes. Implementation of these visions will mitigate growth pressure from profit-oriented enterprises and facilitate easier implementation of degrowth policies.

In relation to the link between democracy and degrowth, Romano (2012) elaborates the following four questions: Can the degrowth project contribute to rediscovering the meaning of democracy? Can the establishment of a real democracy lead to building a degrowth society? And, has the project of a "democratic degrowth" a real chance to succeed, from a political point of view? If not, how should degrowth be re-thought to foster democracy? Romano's thesis is that the currently dominant formulation of degrowth and its relationship to democracy are highly problematic. A degrowth proposal reproduces a central paradox where degrowth does not aspire to restore collective linkages. It only works like an intervention offering techniques that will allow the human species merely to stay alive. Moreover, there is a political problem: while the need for degrowth is presented as very urgent, the tactic for spreading the idea is one of an elitist strategy of voluntary simplicity, which can only work very slowly. As a result, degrowth cannot connect to real social processes covering a large part of the population. To make degrowth feasible and to restore democracy, Romano proposes a foundation of degrowth on a human subject of "de-thinking" and discusses the political pathways for such degrowth project.

Policy pointer:

Corruption is widespread in particular among developing countries. A precondition for achieving growth and reducing poverty is to deal with corruption in an effective way. This view is supported by a growing literature. Potentially effective instruments in rooting out corruption are democratic elections and press freedom. Since their effects on corruption interdepend on each other, a simultaneous interactive model was used to estimate the conditional effects in the literature. Kalenborn and Lessmann (2013) empirically analyzed the joint impact of democracy and press freedom on corruption. They argue that both institutional features are complements rather than substitutes in controlling corruption. Corruption is proxied by the corruption perception index of Transparency International, democracy is represented by Vanhanen's democratization index and press freedom is provided by Freedom House. Estimation results based on a cross section of 170 countries covering the period from 2005 to 2010 as well as based on panel evidence for 175 countries from 1996 to 2010 show that democratic elections only work in controlling corruption, if there is a certain degree of press freedom in a country, and vice versa. The policy implication of the result is that democratic reforms are more effective, if they are accompanied by institutional reforms strengthening the monitoring of rent seeking representatives.

3.4.4. Democracy, State Capacity and Economic Growth

In general growth in Asia's authoritarian regimes has been very high. This gives rise to the question: has democracy and/or democratization slowed growth in Asia? Rock (2009) tests these hypotheses linking Asia's democracies and autocracies to growth. The tests are conducted in a within panel regression framework that controls for fixed country and time effects, the other major variables affecting growth, and for endogeneity between explanatory variables resulting from direct effect of democracy and its indirect effect through investment. The *"democracy slows growth"* hypothesis is tested against the bureaucratically capable authoritarian regimes of East Asia's developmentally minded governments. The findings from various model specifications and sensitivity analysis reject the *"democracy slows growth"* hypothesis. It shows that unlike autocracy, democracy by itself causes growth and investment to rise in Asia. The implication of these findings is that concern over the economic growth performance of new democracies in Asia is misplaced.

It is well-known that good institutions are important for economic performance. However, little is known about the effects of specific institutional structures. Knutsen (2013) discusses how regime type and state capacity may interact in jointly affecting economic growth. A number of hypotheses are drawn from the trade-off between state capacity and democratization in enhancing economic growth. State capacity is measured as bureaucracy quality index and other control variables include regime duration, the ethnic fractionalization index, plurality religion and data on the colonizer (if any). The empirical analysis finds a positive and robust effect of democracy on growth in Sub-Saharan Africa, a continent characterized by weak-capacity states. The authors identify a robust interaction effect between democracy and state capacity on growth, both in Africa and globally. The effect of democracy on growth increases when state capacity is reduced. Democracy has a positive effect on growth in weak-capacity states, but not in high-capacity states. The results indicate also that state capacity enhances growth only in dictatorships confirming a trade-off effect between state capacity and democracy in affecting growth. The robustness of the result was confirmed through sensitivity analysis.

Institutions play a major rule as determinants of growth and development. Political economy factors such as inequality and political democracy influence the quality of institutions. Savoia et al. (2010) surveys the theoretical and empirical literature on the effect of economic and political inequality on institutions. The current understanding suggests that unequal societies develop exploitative and inefficient institutions. Most used indicators in empirical research include: property rights, regulatory quality, government effectiveness, political stability, executive constraints, democracy index, income inequality index, and land inequality. Empirical research has mainly concentrated at cross-national level and supports, to some extent, on the existence of an inverse relationship between inequality and institutions. The authors suggest further research but based on country, state, and micro-level analysis as complement to aggregate level studies. A separation of the direct and indirect effects of inequality on institutions through democratization is recommended to be investigated.

Finally, we introduce in maps 3.12 and 3.13 the concept of *"effective democracy",* as understood by the Inglehart School of sociology, and further developed by us. Map 3.12 portrays the values for this variable, as documented by Alexander, Inglehart and Welzel, 2011, and Welzel and Alexander, 2008. The concept combines standard western social science data on political and human rights with World Bank data about the rule of law. These data are presented in Map 3.12. Map 3.13 shows the residuals of a multiple standard OLS regression, explaining effective democracy by log GDP per capita and log GDP per capita^2, i.e., this map tells us, how different nations could achieve effective democracy irrespective of their high or low economic development levels.

The ten global superstars in terms of effective democracy are:

Finland
Iceland
Denmark
New Zealand
Switzerland
Norway
Sweden
Luxembourg
Austria
Netherlands

According to the scale, the ten worst performers are:
Burma
Cuba
Libya
Sudan
Turkmenistan
Syria
Uzbekistan
Saudi Arabia
Equatorial Guinea
Laos

In terms of their economic development level, the outstanding ten best performances were registered in:

Dominica
Saint Lucia
Chile
Cape Verde
New Zealand
Uruguay
Barbados
Mongolia
Bahamas, The
Costa Rica

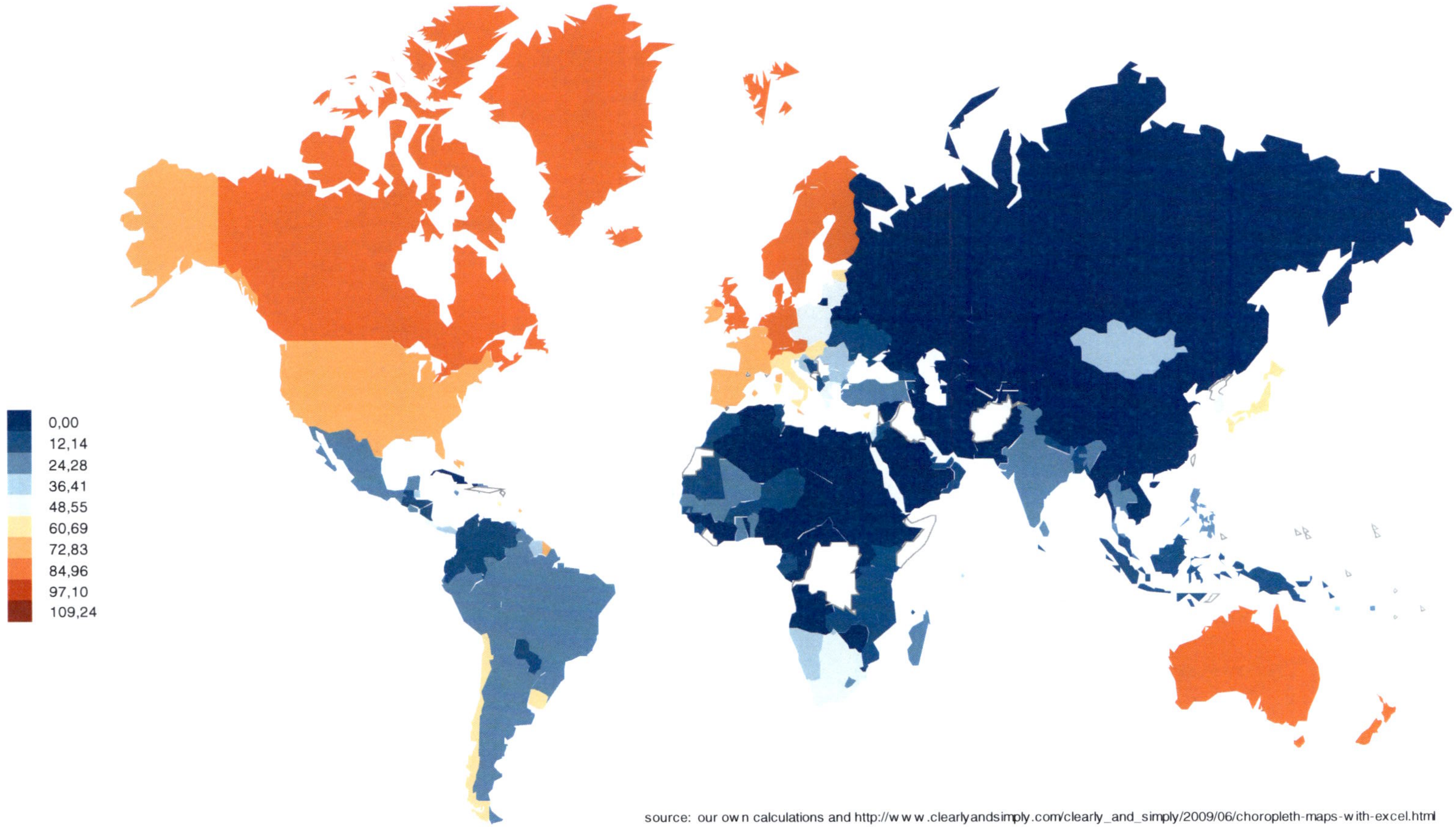

Map 3.12. Effective democracy.

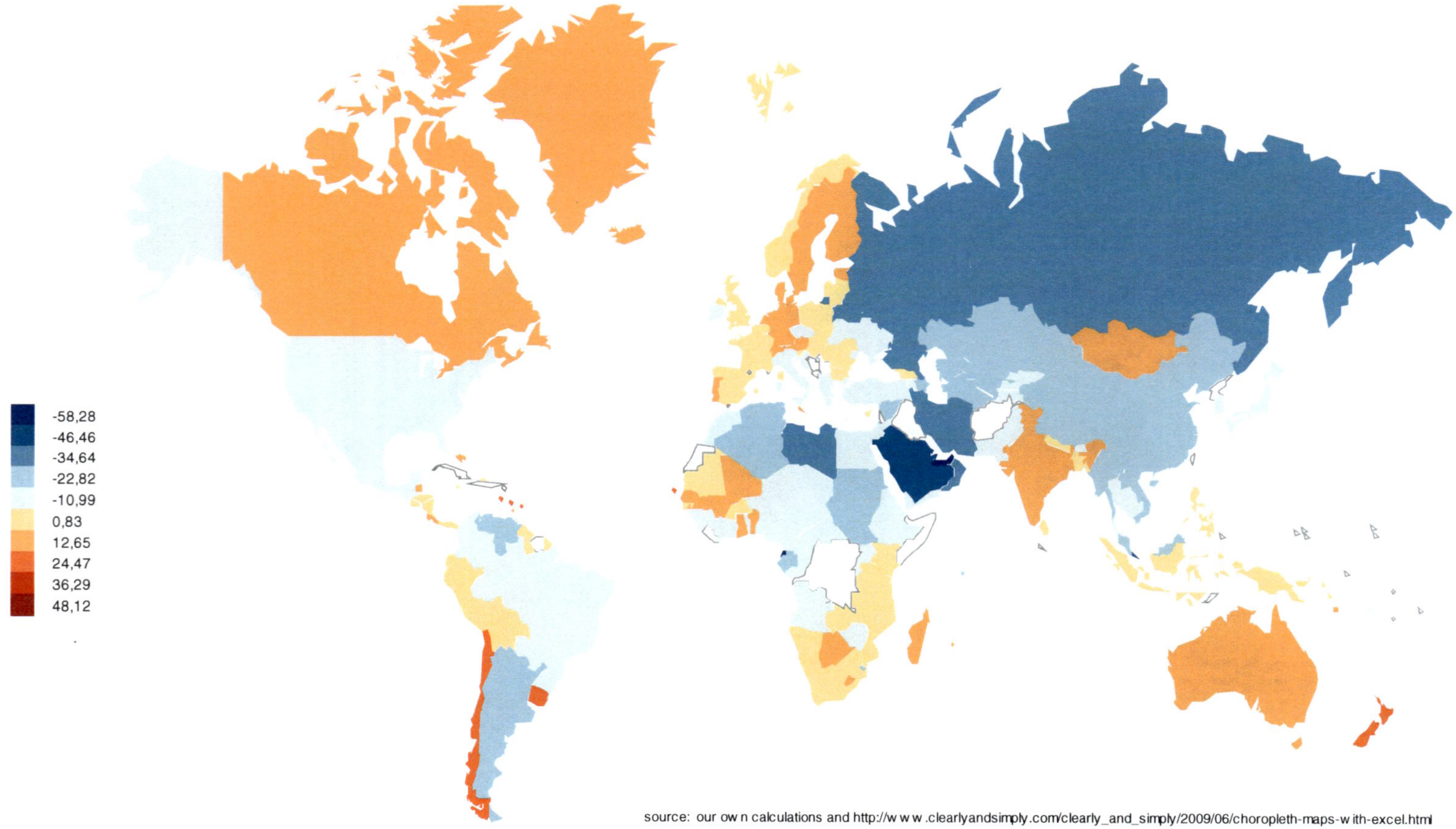

Map 3.13. Effective democracy, irrespective of development level (residuals from a multiple standard non-linear OLS regression, explaining effective democracy by log GDP per capita and log GDP per capita^2).

The ten worst performances in terms of the economic development level of society were registered in:

Equatorial Guinea
United Arab Emirates
Qatar
Saudi Arabia
Bahrain
Singapore
Oman
Libya
Kuwait
Russia

3.5. Brief Summary of the Dark Sides of Economics

Shadow economy and corruption are the two main dark sides of economics. Shadow economic activity has been on rise around the world regardless of level of development, geographic location or type of governance of the countries. It has caused significant violations of laws and regulations, and has reduced tax revenue collections, cause statistical discrepancies, inequality, corruption and increased public budget deficit and subsequent public debt default problems. The literature on the shadow economy and its direct and indirect effects on economic growth are reviewed in this chapter. Several studies attribute the shadow economy to higher taxes, larger government and its interventions, restricted economic freedom, and the desire to escape taxes and regulatory restrictions by individuals and businesses.

We have reviewed a large number of recent studies investigating theoretical and empirical aspects of the shadow economy, in particular the measurement of the shadow economy and its causality with economic growth. In the process, emphasis was made on issues such as corruption, economic freedom, state and institutions capacity and quality, as well as morale and trust. For the empirical parts, these studies use large data sets allowing for cross country and country groups comparisons. A negative relationship is found between shadow economic activity and level of development, democracy and press freedom. Countries with abundant natural resources for export are facing larger degree of shadow economic activity, corruption, inefficiency, inequality and lower economic growth. In the literature this is referred to as the 'curse of natural resources'.

Policy pointer:

It is well-known that good institutions are important for economic performance. However, little is known about the effects of specific institutional structures. Knutsen (2013) discusses how regime type and state capacity may interact in jointly affecting economic growth. A number of hypotheses are drawn from the trade-off between state capacity and democratization in enhancing economic growth. State capacity is measured by a bureaucracy quality index and other control variables include regime duration, an ethnic fractionalization index, the plurality religion and data on the country's colonizers (if any). The empirical analysis finds a positive and robust effect of democracy on growth in Sub-Saharan Africa, a continent characterized by weak-capacity states. The authors identify a robust interaction effect between democracy and state capacity on growth, both in Africa and globally. The effect of democracy on growth increases when state capacity is reduced. Democracy has a positive effect on growth in weak-capacity states, but not in high-capacity states. The results indicate also that state capacity enhances growth only in dictatorships confirming the trade-off effect between state capacity and democracy in affecting growth. Robustness of the results is confirmed through sensitivity analysis.

Institutions play a major rule as determinants of growth and development. Political economy factors such as inequality and political democracy influence the quality of institutions. Savoia et al. (2010) survey the theoretical and empirical literature on the effect of economic and political inequality on institutions. The current understanding suggests that unequal societies develop exploitative and inefficient institutions. Most used indicators used in empirical research include: property rights, regulatory quality, government effectiveness, political stability, executive constraints, democracy index, income inequality index, and land inequality. Empirical research has mainly concentrated at cross-national level and supports, to some extent, the existence of an inverse relationship between inequality and institutions.

Chapter 4

THE SOCIOLOGICAL AND PSYCHOLOGICAL THEORIES OF GLOBAL VALUES

In the preceding Chapter we looked at the relevant contemporary economic theories about the shadow economy. To our great surprise we will now find out that the major existing sociological and psychological theories about global values do not even talk about these phenomena: they as yet do not exist for global value research in the traditions of Hofstede, Schwartz/Davidov and Inglehart, the three major theoretical approaches guiding the discipline.

So, among the most prominent competing international attempts to define and measure the development of human values, we should specify the current three major approaches dominating international social science:

- *Hofstede's* theory of global values
- *the Schwartz/Davidov* approach
- *Inglehart's* and his associates' studies of world values

First we mention *Geert Hofstede.* This Dutch psychologist and his associates really stood at the beginning of comparative international value research. Initially, they based their empirical studies on global culture on the statistical analysis of the staff of the single US transnational enterprise IBM in 40 different countries around the world (see also Hofstede, 2001; Hofstede and Minkov, 2010; Hofstede, Hofstede and Minkov, 2010; Minkov and Hofstede, 2011, 2013). The background to all this was very clear: with IBM production taking place in more and more countries around the globe, IBM needed to know more about the culturally determined mindsets of its employees.

Box 4.1. The overall message of this Chapter

Hofstede, Schwartz/Davidov and Inglehart, the three major existing sociological and psychological theories about global values do not talk about the shadow economy. According to Hofstede there are four to six basic clusters of international value systems, and they are all defined along the scales of how different national societies handle ways of coping with inequality, ways of coping with uncertainty, the relationship of the individual with her or his primary group, and the emotional implications of having been born as a girl or as a boy.

The next theory has been presented by the Israeli psychologist Shalom Schwartz and further developed by Eldad Davidov, a researcher who pays great attention to the statistical-mathematical foundations of comparative value analysis, including the detection of contradictory statements in survey research to reduce the measurement error. In accordance with Centuries of Jewish religious and secular humanism, the values of self-transcendence, culminating in Universalism and Benevolence, are the ultimate aim of human life.

Schwartz, 2006a highlights a map of global values, which arrives at results, which are comparable to those of other researchers in the field. As with Hofstede and the GLOBE Project, influenced by Hofstede, there is a problem of limited country samples. Schwartz began his investigations not on generalized surveys of the total population, but on relatively small samples of global schoolteachers and college students. Most samples only included between 180 and 280 respondents. Schwartz identifies seven basic cultural orientations and the structure of interrelations among them: West European, English-speaking, Latin American, East European, and South Asian, Confucian influenced, and African and Middle Eastern.

The seven dimensions are

1. Embeddedness
2. Hierarchy
3. Mastery
4. Affective Autonomy
5. Intellectual Autonomy
6. Egalitarianism
7. Harmony

A careful reading of the Schwartz categories would suggest that they contradict the well-known interpretation of Muslim civilization by S. Huntington. According to Schwartz, several Muslim countries are at the forefront of the values of affective autonomy, intellectual autonomy, and the quest for harmony.

In his 2008 study on the contradictions of the European integration process, Davidov also highlights the close interrelationship between gender prejudice, homophobia, anti-Semitism, xenophobia, social out-groups (as for example homeless people), Islamophobia and the overall devaluation of newcomers.

Inglehart, in some of his main publications, developed by contrast an interpretation of global value change (Inglehart, 1970, 1977, 1990, 1997, 2000), which rests on a well-known two-dimensional scale of global values and global value change, which is based on the statistical technique of factor analysis.

4.1. Geert Hofstede

According to Hofstede and his school, there are *four to six basic clusters of international value systems,* and they are all defined along the scales of how different national societies handle ways of *coping with inequality, ways of coping with uncertainty, the relationship of the individual with her or his primary group, and the emotional implications of having been born as a girl or as a boy.* Hofstede defines these dimensions of national culture as

- Power Distance
- Individualism vs. Collectivism
- Masculinity versus Femininity
- Uncertainty Avoidance Index
- Long-Term Orientation
- Indulgence versus Restraint

Between 1990 and 2002, Hofstede replicated these dimensions in six other cross-national studies on very different populations from consumers to airline pilots, covering between 14 and 28 countries. In the 2010 third edition of his book *'Cultures and Organizations: Software of the Mind',* scores on the dimensions are listed for 76 countries.[19]

Power distance, according to Hofstede, is the extent to which the less powerful members of organizations and institutions (like the family) *accept and expect that power is distributed unequally*. This represents inequality (more versus less), but it is *defined from below, not from above. 'It suggests that a society's level of inequality is endorsed by the followers as much as by the leaders.'*

Uncertainty avoidance deals with a *society's tolerance for uncertainty and ambiguity.* It indicates to what extent a culture programs its members to feel either uncomfortable or comfortable in unstructured situations. Unstructured situations are novel, unknown, surprising, and are different from the usual.[20] Uncertainty avoiding cultures try to minimize the possibility of such situations by strict laws and rules, safety and security measures, and *on the philosophical and religious level they try to avoid uncertainty by a belief in absolute Truth.* According to Hofstede and his associates,

> "[...] People in uncertainty avoiding countries are also more emotional, and motivated by inner nervous energy. The opposite type, uncertainty accepting cultures, are more tolerant of opinions different from what they are used to; they try to have as few rules as possible, and on the philosophical and religious level they are relativist and allow many currents to flow side by side. People within these cultures are more phlegmatic and contemplative, and not expected by their environment to express emotions."[21]

Individualism on the one side, versus its opposite, *collectivism* is the degree to which individuals are integrated into groups. On the individualist side we find societies in which the

[19] http: //www. geerthofstede. nl/dimensions-of-national-cultures
[20] http: //www. geerthofstede. nl/dimensions-of-national-cultures
[21] http: //www. geerthofstede. nl/dimensions-of-national-cultures

ties between individuals are loose: everyone is expected to look after her/himself and her/his immediate family:

> "On the collectivist side, we find societies in which people from birth onwards are integrated into strong, cohesive in-groups, often extended families (with uncles, aunts and grandparents) which continue protecting them in exchange for unquestioning loyalty. The word collectivism in this sense has no political meaning: it refers to the group, not to the state."[22]

Masculinity versus its opposite, *femininity* is defined by the Hofstede School as referring to the distribution of emotional roles between the genders. According to Hofstede's system this is another fundamental issue for any society to which a range of solutions are found.[23] For Hofstede, female values around the globe differ less than male values from country to country.

With the passage of time, Hofstede and his associates added a *fifth dimension* to their value studies, called the *long-term orientation (LTO).* In part, this dimension is based on data from the *'World Values Survey'* project. According to this interpretation, long-term oriented societies *'foster pragmatic virtues oriented towards future rewards, in particular saving, persistence, and adapting to changing circumstances. Short-term oriented societies foster virtues related to the past and present such as national pride, respect for tradition, preservation of "face", and fulfilling social obligations.'*[24]

In this context, Hofstede and Minkov (2010) also provided a principal components analysis of their own of the different items from the *World Values Survey* on such items as thrift and perseverance, religious faith, efforts to meet expectations of friends, love of parents, dedication of parents to their children's' well-being (rather than having a parallel parental life of their own), the stance on divorce, attitudes on national pride, making parents proud, and the importance of service to others in life.[25]

We re-arranged the original presentation in Hofstede and Minkov (2010) to allow our readers to immediately be able to see the relationship of the long-term orientation scale with the original factor loadings (i. e., principal components), based on the Hofstede/Minkov re-analysis of the *World Values Survey* data:

Table 4.1. Hofstede's LTO score

	Original factor loadings, reported in Hofstede/Minkov, 2010	Multiplied by -100 to arrive at the reformulated LTO-score
thrift	-0,82	82
perseverance	-0,77	77
divorce justifiable	-0,04	4
always love parents	0,10	-10
live up to friends' expectations	0,57	-57
parents do their best	0,62	-62
parental pride	0,63	-63
religious faith	0,65	-65
national pride	0,68	-68
service to others	0,84	-84

[22] http: //www. geerthofstede. nl/dimensions-of-national-cultures
[23] http: //www. geerthofstede. nl/dimensions-of-national-cultures
[24] http: //www. geerthofstede. nl/dimensions-of-national-cultures
[25] http: //www. geerthofstede. nl/dimensions-of-national-cultures

South Korea, Japan and China lead the field, while Egypt, Jordan, but also several Latin American countries, rank lowest among the 38 countries with data.

According to Hofstede and his school, there is also a sixth dimension emerging from their data: *Indulgence versus Restraint.* Indulgence stands *"for a society that allows relatively free gratification of basic and natural human drives related to enjoying life and having fun. Restraint stands for a society that suppresses gratification of needs and regulates it by means of strict social norms."*

Table 4.2 now summarizes the Hofstede scores in the world system. Many Muslim societies find themselves at the opposite ends of the various different scales, i.e., the global dividing lines are different from those portrayed by Huntington.

Table 4.2. The global Hofstede scores (majority Muslim countries are marked in green color)

Country	Power Distance	Individualism vs. Collectivism	Masculinity vs. Femininity	Uncertainty Avoidance Index	Long-Term Orientation	Indulgence vs. Restraint
Argentina	49	46	56	86	20	62
Australia	36	90	61	51	21	71
Austria	11	55	79	70	60	63
Bangladesh	80	20	55	60	47	20
Belgium	65	75	54	94	82	57
Brazil	69	38	49	76	44	59
Bulgaria	70	30	40	85	69	16
Canada	39	80	52	48	36	68
Chile	63	23	28	86	31	68
China	80	20	66	30	87	24
Colombia	67	13	64	80	13	83
Croatia	73	33	40	80	58	33
Czech Rep	57	58	57	74	70	29
Denmark	18	74	16	23	35	70
El Salvador	66	19	40	94	20	89
Estonia	40	60	30	60	82	16
Finland	33	63	26	59	38	57
France	68	71	43	86	63	48
Germany	35	67	66	65	83	40
Greece	60	35	57	112	45	50
Hong Kong	68	25	57	29	61	17
Hungary	46	80	88	82	58	31
India	77	48	56	40	51	26
Indonesia	78	14	46	48	62	38
Iran	58	41	43	59	14	40
Ireland	28	70	68	35	24	65
Italy	50	76	70	75	61	30
Japan	54	46	95	92	88	42
Korea South	60	18	39	85	100	29
Latvia	44	70	9	63	69	13
Lithuania	42	60	19	65	82	16
Luxembourg	40	60	50	70	64	56
Malaysia	104	26	50	36	41	57
Malta	56	59	47	96	47	66
Mexico	81	30	69	82	24	97

Country	Power Distance	Individualism vs. Collectivism	Masculinity vs. Femininity	Uncertainty Avoidance Index	Long-Term Orientation	Indulgence vs. Restraint
Morocco	70	46	53	68	14	25
Netherlands	38	80	14	53	67	68
New Zealand	22	79	58	49	33	75
Norway	31	69	8	50	35	55
Pakistan	55	14	50	70	50	0
Peru	64	16	42	87	25	46
Philippines	94	32	64	44	27	42
Poland	68	60	64	93	38	29
Portugal	63	27	31	104	28	33
Romania	90	30	42	90	52	20
Russia	93	39	36	95	81	20
Serbia	86	25	43	92	52	28
Singapore	74	20	48	8	72	46
Slovak Rep	104	52	110	51	77	28
Slovenia	71	27	19	88	49	48
Spain	57	51	42	86	48	44
Sweden	31	71	5	29	53	78
Switzerland	34	68	70	58	74	66
Taiwan	58	17	45	69	93	49
Thailand	64	20	34	64	32	45
Trinidad and Tobago	47	16	58	55	13	80
Turkey	66	37	45	85	46	49
USA.	40	91	62	46	26	68
United Kingdom	35	89	66	35	51	69
Uruguay	61	36	38	100	26	53
Venezuela	81	12	73	76	16	100
Vietnam	70	20	40	30	57	35

Our own compilation from: http://geerthofstede.com/dimensions-of-national-cultures and http://www.geerthofstede.com/research--vsm. We only considered the countries with complete values for the final analysis.

For Hofstede and his associates, there are very clear *underlying dimensions of these six dimensions* and social and *cultural history*. Some of these conclusions, not necessarily fully shared by us, are nevertheless really breathtaking theoretical perspectives over the last two millennia of global history:

> "*Power distance scores* are high for Latin, Asian and African countries and smaller for Anglo and Germanic countries. *Uncertainty avoidance* scores are higher in Latin countries, in Japan, and in German speaking countries, lower in Anglo, Nordic, and Chinese culture countries. *Individualism* prevails in developed and Western countries, while collectivism prevails in less developed and Eastern countries; Japan takes a middle position on this dimension. *Masculinity* is high in Japan, in some European countries like Germany, Austria and Switzerland, and moderately high in Anglo countries; it is low in Nordic countries and in the Netherlands and moderately low in some [...] countries like France, Spain and Thailand. *Long-term orientation* scores are highest in East Asia, moderate in Eastern and Western Europe, and low in the Anglo world, the Muslim world, Latin America and Africa. *Indulgence* scores are highest in Latin America, parts of Africa, the Anglo world and Nordic Europe; restraint is mostly found in East Asia, Eastern Europe and the Muslim world."

"The grouping of country scores points to some of the *roots of cultural differences*. These should be sought in the common history of similarly scoring countries. All *Latin countries,* for example, *score relatively high on both power distance and uncertainty avoidance*. Latin countries (those today speaking a Romance language, i. e., Spanish, Portuguese, French, Italian or Romanian) have inherited at least part of their civilization from the Roman Empire. The Roman Empire in its days was characterized by the existence of a central authority in Rome, and a system of law applicable to citizens anywhere. This established in its citizens' minds the value complex which we still recognize today: centralization fostered large power distance and a stress on laws fostered strong uncertainty avoidance. The Chinese empire also knew centralization, but it lacked a fixed system of laws: it was governed by men rather than by laws. In the present-day countries once under Chinese rule, the mindset fostered by the empire is reflected in large power distance but medium to weak uncertainty avoidance. The Germanic part of Europe, including Great Britain, never succeeded in establishing an enduring common central authority and countries which inherited its civilizations show smaller power distance. Assumptions about historical roots of cultural differences always remain speculative but in the given examples they are plausible. In other cases they remain hidden in the course of history."[26]

"Power distance is correlated with the use of violence in domestic politics and with income inequality in a country. Uncertainty avoidance is associated with Roman Catholicism and with the legal obligation in developed countries for citizens to carry identity cards. Individualism is correlated with national wealth and with mobility between social classes from one generation to the next. Masculinity is correlated negatively with the percent of women in democratically elected governments. Long-term orientation is correlated with school results in international comparisons. Indulgence is correlated with sexual freedom and a call for human rights like free expression of opinions."[27]

The differences between the Muslim countries in the sample and the global unweighted means reveal no particular pattern:

Table 4.3. The Muslim countries in Hofstede's global sample and their differences from the global, unweighted means

	Power Distance	Individualism vs. Collectivism	Masculinity vs. Femininity	Uncertainty Avoidance Index	Long-Term Orientation	Indulgence vs. Restraint
Bangladesh	21.42	-25.76	5.89	-6.95	-2.76	-27.50
Indonesia	19.42	-31.76	-3.11	-18.95	12.24	-9.50
Iran	-0.58	-4.76	-6.11	-7.95	-35.76	-7.50
Malaysia	45.42	-19.76	0.89	-30.95	-8.76	9.50
Morocco	11.42	0.24	3.89	1.05	-35.76	-22.50
Pakistan	-3.58	-31.76	0.89	3.05	0.24	-47.50
Turkey	7.42	-8.76	-4.11	18.05	-3.76	1.50

Our own compilation from: http: //geerthofstede. com/dimensions-of-national-cultures and http: //www. geerthofstede. com/research--vsm. We only considered the countries with complete values for the final analysis.

[26] http: //www. geerthofstede. nl/dimensions-of-national-cultures

[27] http: //www. geerthofstede. nl/dimensions-of-national-cultures

In the following presentation of Hofstede's theory, we are presenting the global ranks of the countries of the world according to Hofstede's scales. We highlight the results for the majority Muslim countries in green color.

Table 4.4. The global ranks according to Hofstede's scales

Country	Power Distance	Individualism vs. Collectivism	Masculinity vs. Femininity	Uncertainty Avoidance Index	Long-Term Orientation	Indulgence vs. Restraint
Argentina	42	31	24	14	57	6
Australia	52	2	17	46	55	8
Austria	62	25	4	29	22	41
Bangladesh	9	49	25	38	35	28
Belgium	26	9	26	6	6	57
Brazil	19	34	32	26	38	25
Bulgaria	16	42	45	19	14	49
Canada	50	5	28	49	42	21
Chile	29	48	54	17	47	16
China	10	50	12	57	4	59
Colombia	23	61	15	23	61	2
Croatia	14	38	44	24	24	39
Czech Rep	36	24	20	28	13	50
Denmark	61	10	58	61	44	19
El Salvador	24	54	47	7	56	7
Estonia	47	21	53	39	8	55
Finland	56	18	55	41	41	22
France	20	11	38	15	18	45
Germany	53	17	11	34	5	58
Greece	32	37	21	1	37	26
Hong Kong	21	47	22	59	21	42
Hungary	44	4	3	21	23	40
India	12	28	23	53	29	34
Indonesia	11	59	35	50	19	44
Iran	34	32	39	40	60	3
Ireland	59	14	9	55	54	9
Italy	41	8	6	27	20	43
Japan	40	30	2	9	3	60
Korea South	33	55	48	20	1	62
Latvia	45	13	60	37	15	48
Lithuania	46	20	56	35	7	56
Luxembourg	48	22	29	30	17	46
Malaysia	1	45	30	54	39	24
Malta	38	23	34	4	34	29
Mexico	7	41	8	22	53	10
Morocco	17	29	27	33	59	4
Netherlands	51	6	59	44	16	47
New Zealand	60	7	18	48	45	18
Norway	57	15	61	47	43	20

Country	Power Distance	Individualism vs. Collectivism	Masculinity vs. Femininity	Uncertainty Avoidance Index	Long-Term Orientation	Indulgence vs. Restraint
Pakistan	39	60	31	31	31	32
Peru	27	57	43	13	52	11
Philippines	3	39	14	52	49	14
Poland	22	19	13	8	40	23
Portugal	30	44	52	2	48	15
Romania	5	40	42	11	28	35
Russia	4	33	50	5	9	54
Serbia	6	46	40	10	27	36
Singapore	13	51	33	62	12	51
Slovak Rep	2	26	1	45	10	53
Slovenia	15	43	57	12	32	31
Spain	37	27	41	16	33	30
Sweden	58	12	62	60	26	37
Switzerland	55	16	7	42	11	52
Taiwan	35	56	37	32	2	61
Thailand	28	53	51	36	46	17
Trinidad and Tobago	43	58	19	43	62	1
Turkey	25	35	36	18	36	27
USA	49	1	16	51	51	12
United Kingdom	54	3	10	56	30	33
Uruguay	31	36	49	3	50	13
Venezuela	8	62	5	25	58	5
Vietnam	18	52	46	58	25	38

Our own compilation from: http://geerthofstede.com/dimensions-of-national-cultures and http://www.geerthofstede.com/research--vsm. We only considered the countries with complete values for the final analysis.

The following maps show the original Hofstede scores, as they are reproduced in Table 4.2, projected onto world maps. These choropleth maps help us to arrive at a better understanding of the six Hofstede factors.

- Hofstede's *power distance* scale is generally lowest in Scandinavian and in the Anglo-Saxon democracies, and also in the social welfare state Austria, and it is, generally speaking, highest the ex-USSR, in India and South-East Asia and in China, in Mexico and Venezuela and in some countries of Central and Eastern Europe.
- The most *collectivistic* countries are to be found in the Andean region of Latin America, in China and in countries of South and South-East Asia, while the highest *individualism* is to be found in the United States and in the Anglo-Saxon democracies, with high scores also to be found in the Netherlands, Denmark, and also in Italy and in Hungary.

- The highest *masculinity* scores are to be found in some countries of Central and Eastern Europe, in Japan and in Venezuela, while the countries of Scandinavia and the Baltic region, the Netherlands and, interestingly enough, Chile and Portugal are characterized by high *femininity*.
- *Uncertainty avoidance* is highest in Greece, Portugal and Uruguay, and is generally highest in Roman Catholic and Orthodox cultures, and also in Japan and South Korea, while it is lowest in Protestant cultures, but also in China and in Iran, in South Asia and in South-East Asia.
- *Hofstede's long-term orientation (LTO)* is highest in South Korea, Japan, China, in Russia and the Baltic Republics, and in Germany and some other European countries, while it is especially absent in Latin America, in the Anglo-Saxon democracies and Scandinavia, and in Morocco and Iran; with most of the other European countries and the countries of South and South-East Asia classified in the same ranks.
- Hofstede's *Indulgence Factor* is highest in Venezuela, in countries of the Caribbean and in Mexico and, interestingly enough, also in Sweden, New Zealand, Australia, the UK, the US and Canada and in some European countries (Denmark, Netherlands, Switzerland and Austria) and in Chile, while *Restraint* is highest in Pakistan, in most former communist countries, and in China. Most continental European countries seem to be characterized by low indulgence and higher restraint.

Policy pointer:

Hofstede's power distance scale is generally lowest in Scandinavian and in the Anglo-Saxon democracies, and also in the social welfare state Austria, and it is, generally speaking, highest the ex-USSR, in India and South-East Asia and in China, in Mexico and Venezuela and in some countries of Central and Eastern Europe.

The most collectivistic countries are to be found in the Andean region of Latin America, in China and in countries of South and South-East Asia, while the highest individualism is to be found in the United States and in the Anglo-Saxon democracies, with high scores also to be found in the Netherlands, Denmark, and also in Italy and in Hungary.

The highest masculinity scores are to be found in some countries of Central and Eastern Europe, in Japan and in Venezuela, while the countries of Scandinavia and the Baltic region, the Netherlands and, interestingly enough, Chile and Portugal are characterized by high femininity.

Uncertainty avoidance is highest in Greece, Portugal and Uruguay, and is generally highest in Roman Catholic and Orthodox cultures, and also in Japan and South Korea, while it is lowest in Protestant cultures and also in China and in Iran, in South Asia and in South-East Asia.

Hofstede's long-term orientation (LTO) is highest in South Korea, Japan, China, in Russia and the Baltic Republics, and in Germany and some other European countries, while it is especially absent in Latin America, in the Anglo-Saxon democracies and Scandinavia, and in Morocco and Iran; with most of the other European countries and the countries of South and South-East Asia classified in the same ranks.

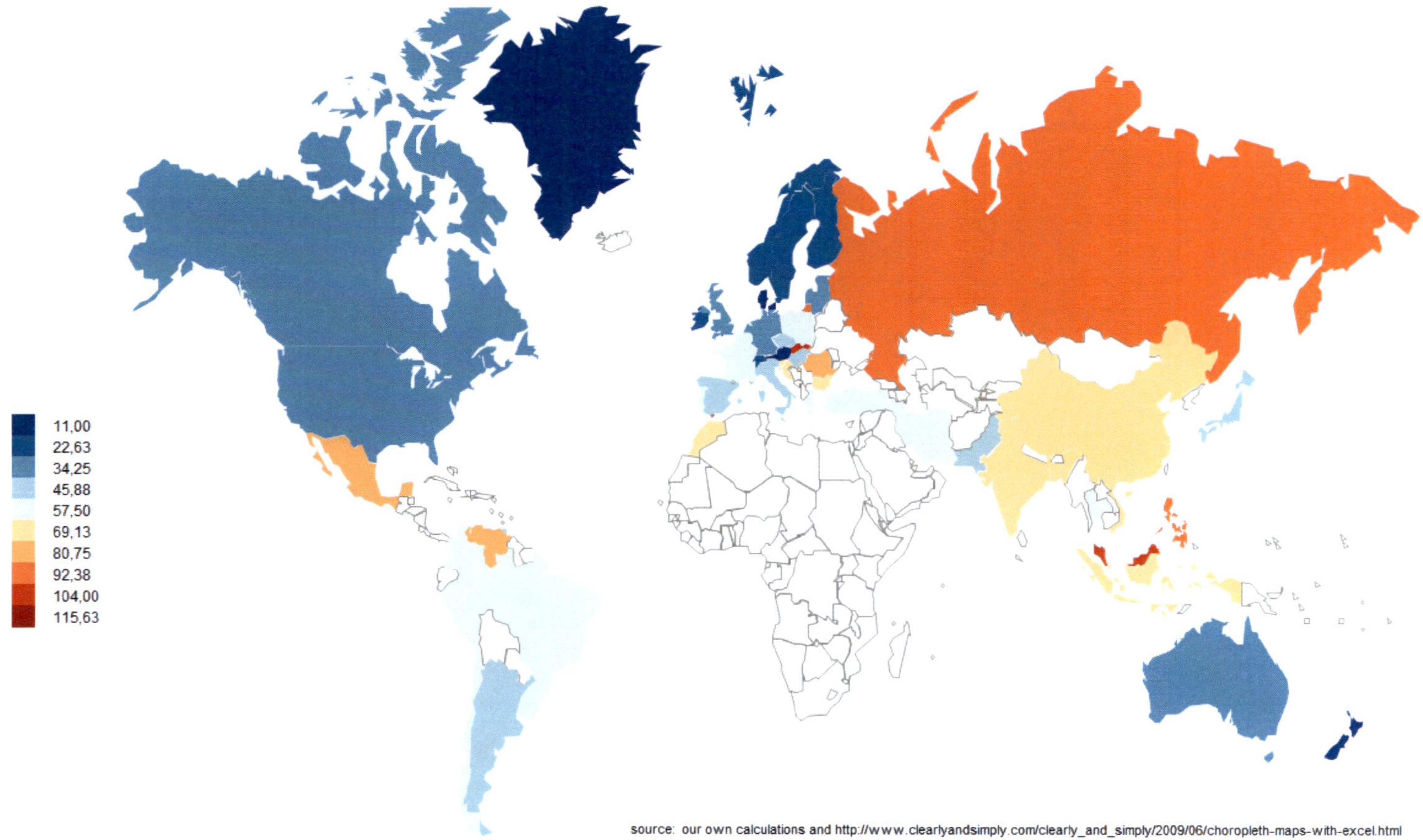

Map 4.1. Hofstede's Power Distance (Maximum: painted in red).

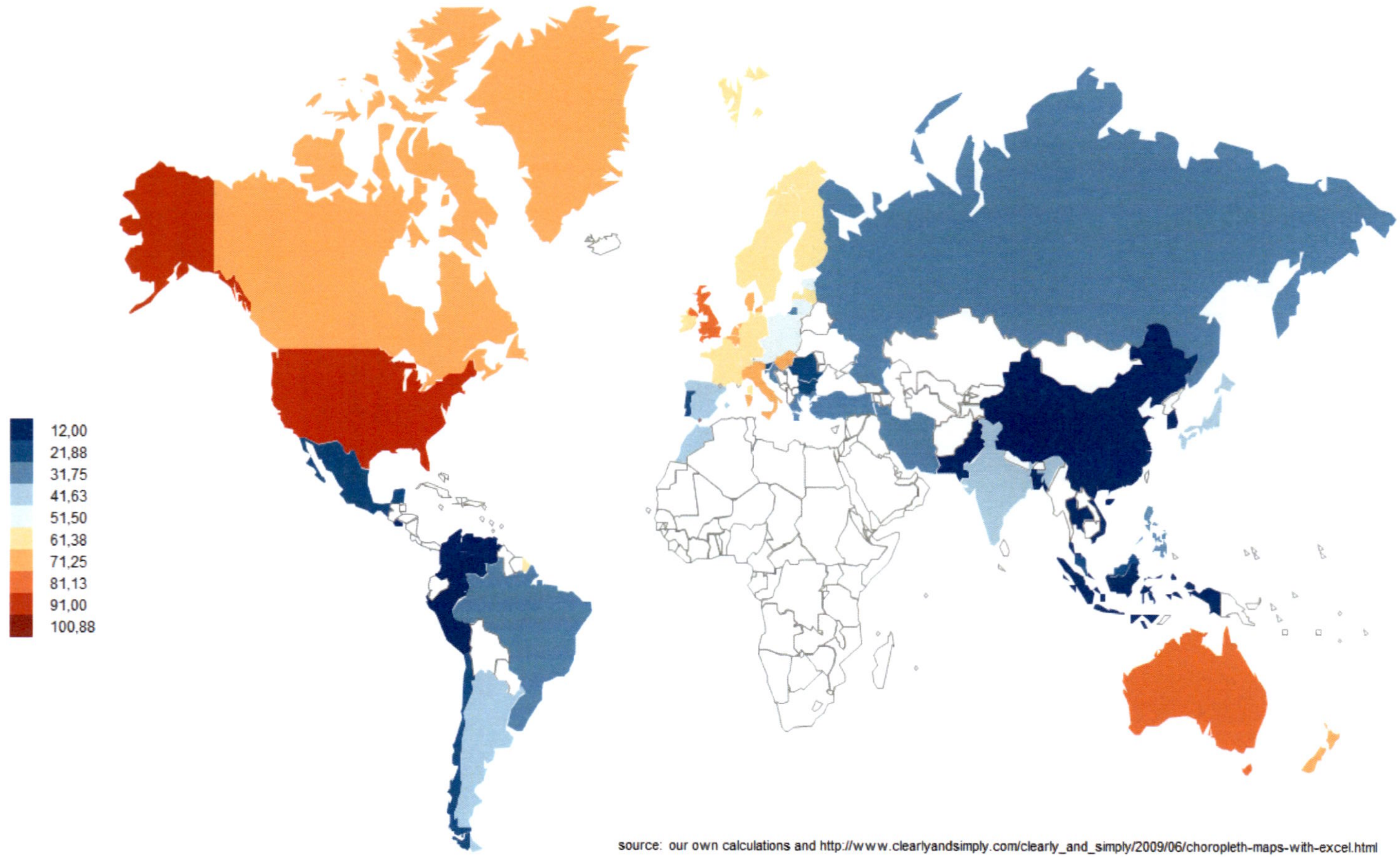

Map 4.2. Hofstede's Individualism (painted in red) vs. Collectivism (Painted in blue).

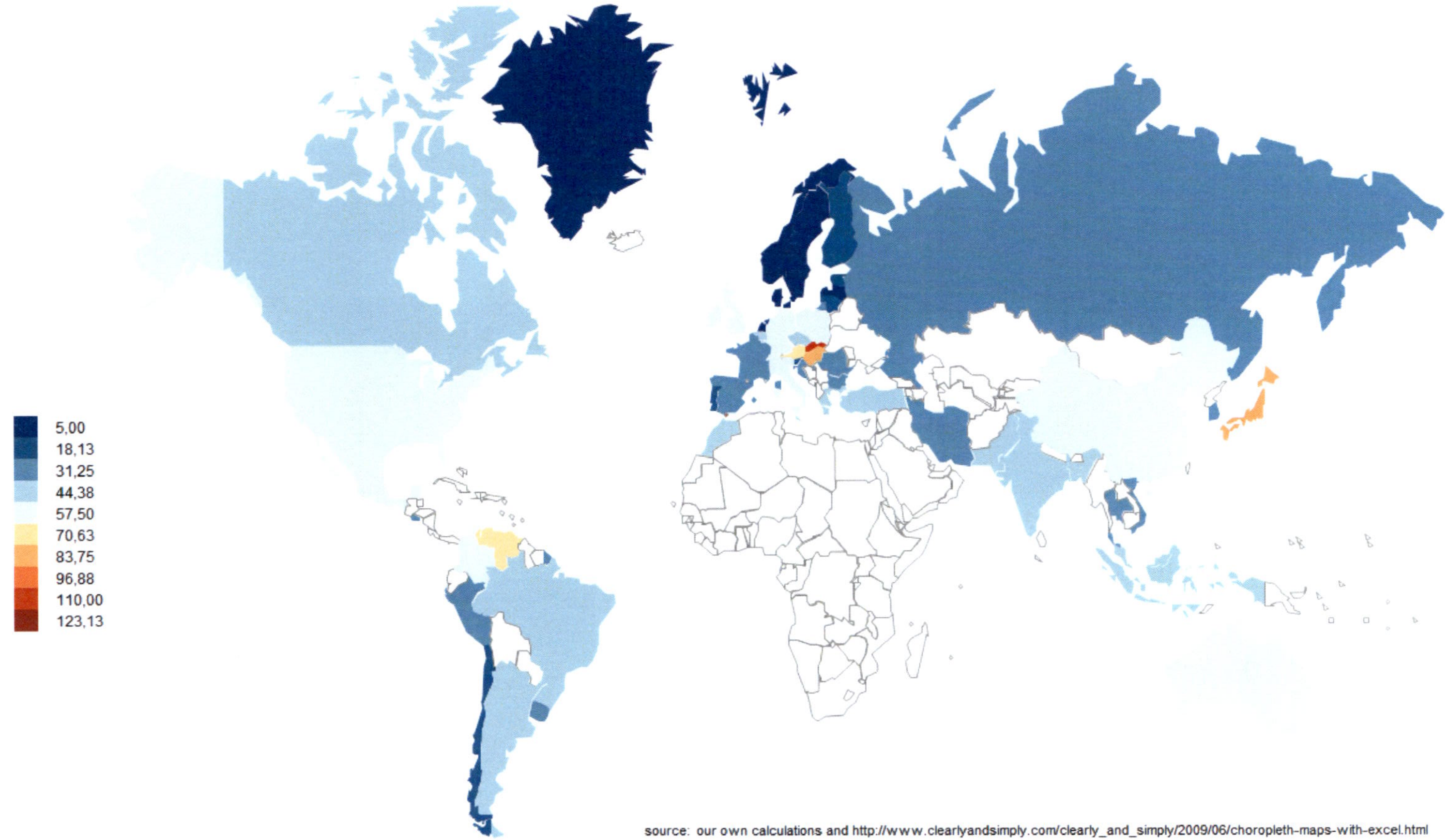

Map 4.3. Hofstede's Masculinity (painted in red) versus Femininity (Painted in blue).

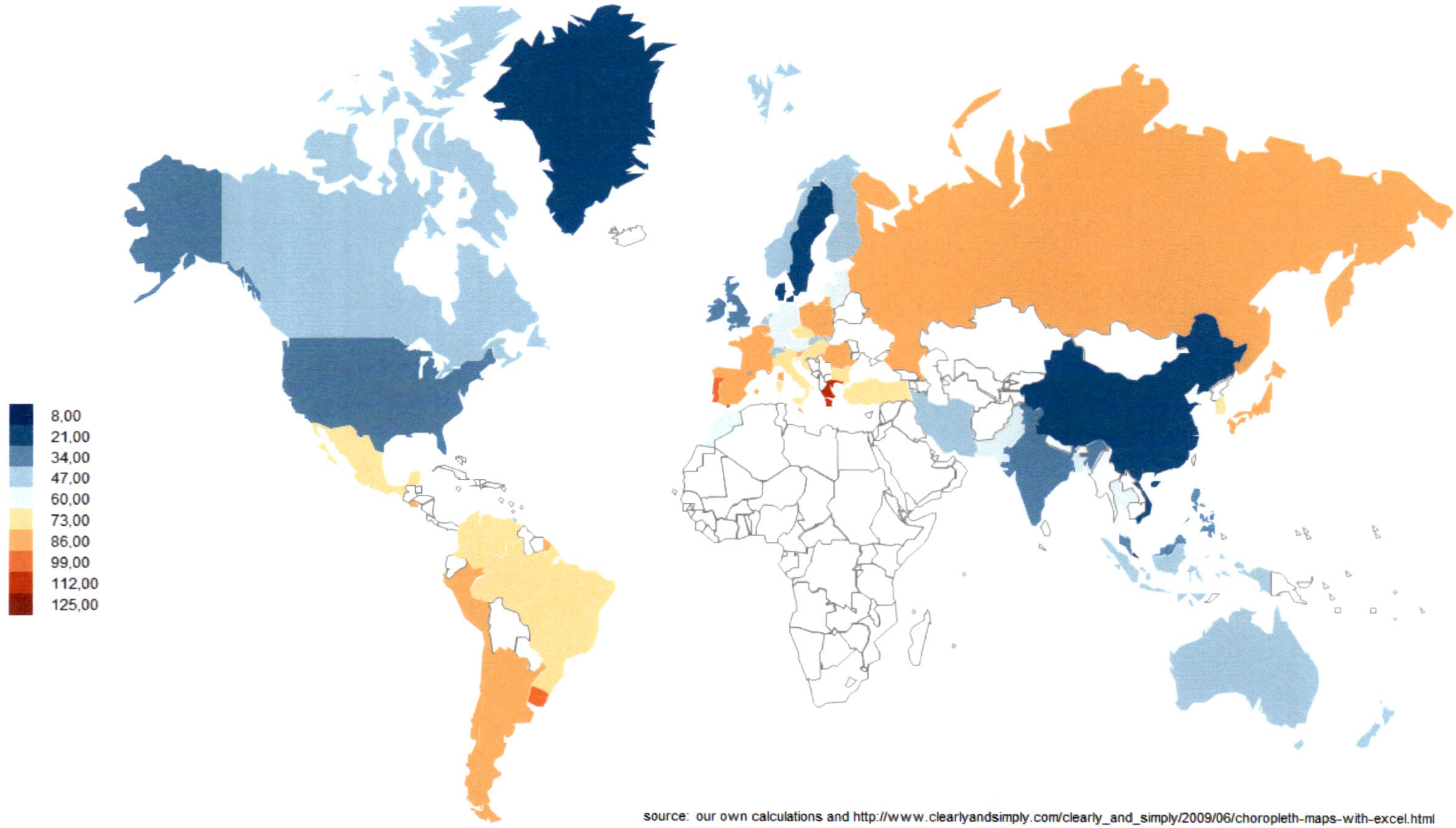

Map 4.4. Hofstede's Uncertainty Avoidance Index (painted in red).

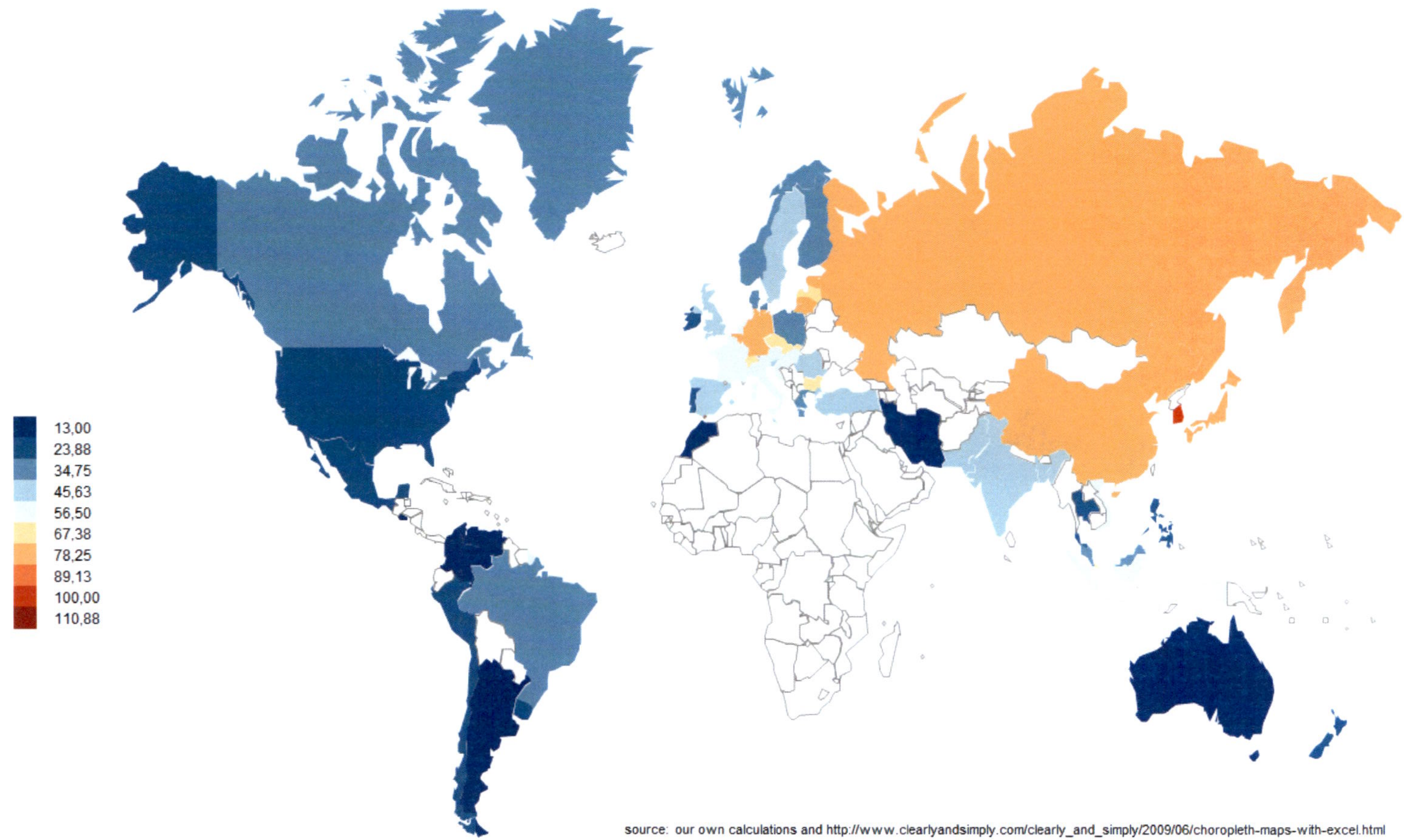

Map 4.5. Hofstede's Long-Term Orientation (painted in red).

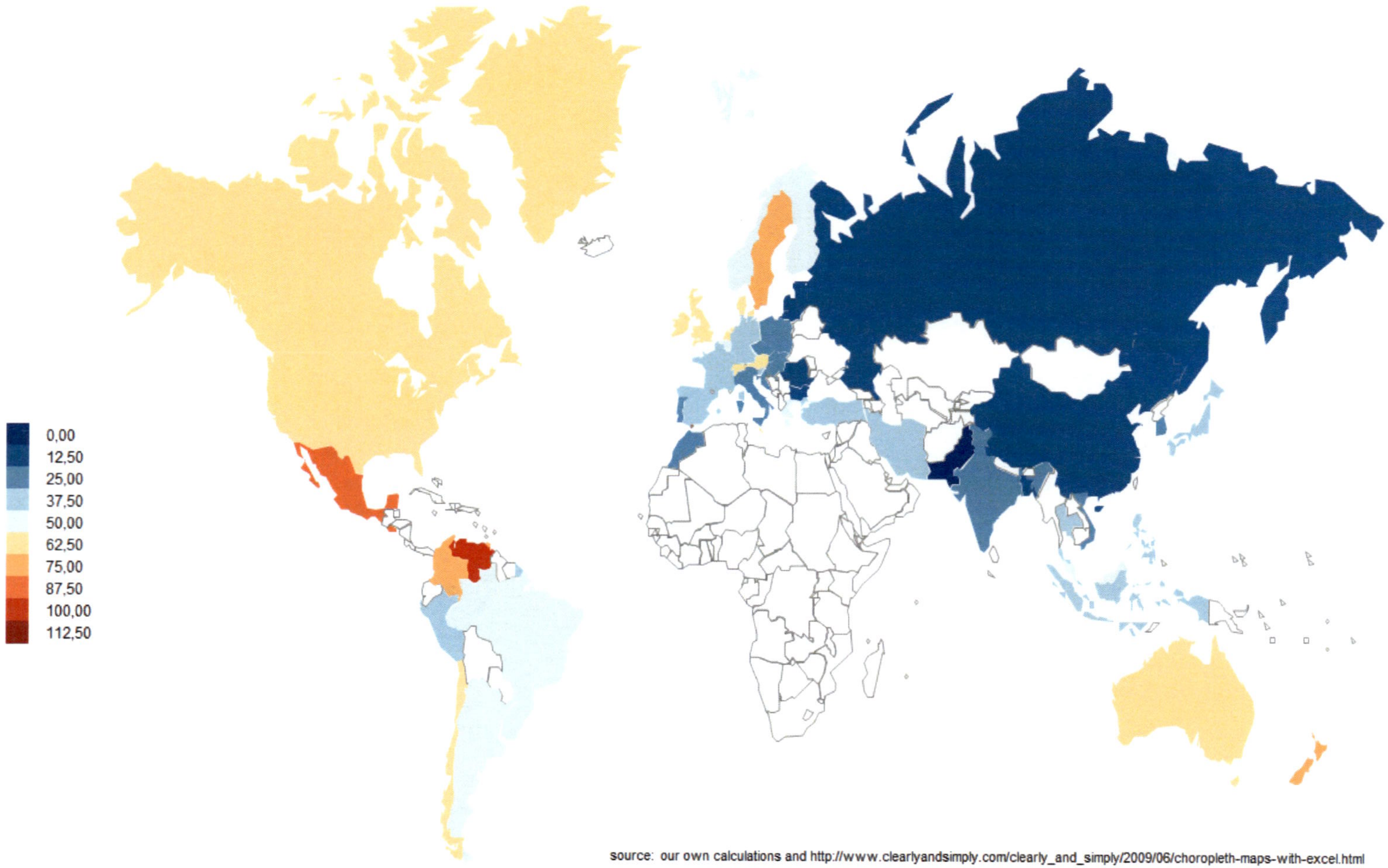

Map 4.6. Hofstede's Indulgence (painted in red) versus Restraint (painted in blue).

Hofstede's challenging categories found a global audience and attracted the attention of hundreds of scholarly articles, among others in the growing field of business studies. More recently, Grovewell International ("GLOBE Project) replicated the Hofstede findings on their own and largely came to the same conclusions.[28]

4.2. SHALOM SCHWARTZ AND ELDAD DAVIDOV

The next theory which we will briefly present here has been developed by Shalom Schwartz, an Israeli psychologist and Professor at the Hebrew University in Jerusalem (Schwartz, 2006a, 2006b, 2007a, 2007b, 2009). It was further developed in mathematical-statistical and theoretical terms by Eldad Davidov. As with Hofstede's studies, there was an enormous international reception, especially in the expanding field of international business studies.

Schwartz, 2006a highlights a map of global values, which arrives at results, which correlate with Inglehart's and Hofstede's scales. In Chapter 12 we re-analyze these correlations and also show the relationships with our own dimensions. As with Hofstede and the GLOBE Project, influenced by Hofstede, there is a problem of limited country samples. Schwartz began his investigations not on generalized surveys of the total population (Schwartz, 2009) but on relatively small global samples of schoolteachers and college students. Participants were 80 samples of schoolteachers (k-12) from 58 national groups and 115 samples of college students from 64 national groups, together constituting 67 nations and 70 different cultural groups. Samples from ethnically heterogeneous nations came from the dominant, majority group. Most samples only included some 180 to 280 respondents. Our readers will recall from Chapter 2 that even with a sample of 500 respondents, error margins will be between +-1.9% and +-4.4%. College students and teachers (Schwartz), just as employees of the multinational corporation IBM (Hofstede), are hardly representative of the global population. At a later stage, Schwartz validated his scales with survey data from the *European Social Survey*.[29]

Schwartz presents analyses of data from up to 73 countries, validating seven basic cultural orientations and the structure of interrelations among them: West European, English-speaking, Latin American, East European, and South Asian, Confucian influenced, and African and Middle Eastern.

The seven dimensions are:

[28] http://www.grovewell.com/ The GLOBE project was initiated by Professor Robert J. House of the Wharton School of the University of Pennsylvania in 1991. Based on interviews with more than 17.000 global managers in 62 cultures, the nine GLOBE cultural competencies are:

1. Performance orientation
2. Assertiveness orientation
3. Future orientation
4. Human orientation
5. Collectivism I: Institutional collectivism
6. Collectivism II: In-group collectivism
7. Gender egalitarianism
8. Power distance
9. Uncertainty avoidance

[29] http://www.europeansocialsurvey.org/

1. Embeddedness
2. Hierarchy
3. Mastery
4. Affective Autonomy
5. Intellectual Autonomy
6. Egalitarianism
7. Harmony

At the basis of Schwartz's theory are ten items from his so-called *Portrait Value Questionnaire,* whose individual dimensions we shall present according to a recent synthesis, provided by Dobewall and Strack, 2013. Items marked like "#a197" or "a193" *et cetera* refer to the original Schwartz *Portrait Value Questionnaire.*

Universalism: #a197 "care for nature": Looking after the environment is important to this person; to care for nature.

Benevolence: #a193 "help others": It is important to this person to help the people nearby; to care for their well-being.

Conformity: #a196 "behave properly": It is important to this person to always behave properly; to avoid doing anything people would say is wrong.

Security: #a191 "secure surroundings": Living in secure surroundings is important to this person; to avoid anything that might be dangerous.

Power: #a190 "wealth": It is important to this person to be rich; to have a lot of money and expensive things.

Achievement: #a194 "successful": Being very successful is important to this person; to have people recognize one's achievements.

Stimulation: #a195 "risk excitement": Adventure and taking risks are important to this person; to have an exciting life.

Hedonism: #a192 "good time spoil self": It is important to this person to have a good time; to "spoil" oneself.

Self-Direction: #a189 "creative original": It is important to this person to think up new ideas and be creative; to do things one's own way.

At one point in his analysis, Schwartz explains:

"I derived value dimensions for comparing cultures by considering *three of the critical issues that confront all societies.* The first issue is the nature of the relation or the boundaries between the person and the group: To what extent are people autonomous vs. embedded in their groups? I label the polar locations on this cultural dimension *autonomy versus embeddedness.* In autonomy cultures, people are viewed as autonomous, bounded entities. They should cultivate and express their own preferences, feelings, ideas, and abilities, and find meaning in their own uniqueness. There are two types of autonomy: Intellectual autonomy encourages individuals to pursue their own ideas and intellectual directions independently. Examples of important values in such cultures include broadmindedness, curiosity, and creativity. Affective autonomy encourages individuals to pursue affectively positive experience for themselves. Important values include pleasure, exciting life, and varied life. In cultures with an emphasis on embeddedness, people are viewed as entities embedded in the collectivity. Meaning in life comes largely through social relationships, through identifying with the group, participating in its shared way of life, and striving toward its shared goals. Embedded cultures emphasize maintaining the status quo and restraining actions that might disrupt in-group solidarity or the traditional order. Important values in such cultures are social order, respect for tradition, security, obedience, and wisdom.

The second societal problem is to guarantee that people behave in a responsible manner that preserves the social fabric. That is, people must engage in the productive work necessary to maintain society rather than compete destructively or withhold their efforts. People must be induced to consider the welfare of others, to coordinate with them, and thereby manage their unavoidable interdependencies. The polar solution labeled *cultural egalitarianism* seeks to induce people to recognize one another as moral equals who share basic interests as human beings. People are socialized to internalize a commitment to cooperate and to feel concern for everyone's welfare. They are expected to act for the benefit of others as a matter of choice. Important values in such cultures include equality, social justice, responsibility, help, and honesty. The polar alternative labeled *cultural hierarchy* relies on hierarchical systems of ascribed roles to insure responsible, productive behavior. It defines the unequal distribution of power, roles, and resources as legitimate. People are socialized to take the hierarchical distribution of roles for granted and to comply with the obligations and rules attached to their roles. Values like social power, authority, humility, and wealth are highly important in hierarchical cultures.

The third societal problem is to regulate how people manage their relations to the natural and social world. The cultural response to this problem labeled **harmony** emphasizes fitting into the world as it is, trying to understand and appreciate rather than to change, direct, or to exploit. Important values in harmony cultures include world at peace, unity with nature, and protecting the environment. *Mastery* is the polar cultural response to this problem. It encourages active self-assertion in order to master, direct, and change the natural and social environment to attain group or personal goals. Values such as ambition, success, daring, and competence are especially important in mastery cultures. *In sum, the theory specifies three bipolar dimensions of culture that represent alternative resolutions to each of three problems that confront all societies: embeddedness versus autonomy, hierarchy versus egalitarianism, and mastery versus harmony.* (Schwartz, 2006a)

Correlations between the sample means were used in a *multidimensional scaling analysis* to assess the presence of seven cultural orientations and the relations among them. The 2-dimensional projection in Graph 4.1, which is the main result of Schwartz's research, portrays the pattern of intercorrelations among values, based on the sample means. Each value item is represented by a point such that the more positive is the correlation between any pair of value items, the closer they are in space, and the less positive their correlation, the more distant they are (Schwartz, 2009). According to Schwartz, this analysis should clearly discriminate the seven orientations: Embeddedness; Hierarchy; Mastery; Affective Autonomy; Intellectual Autonomy; Egalitarianism; and Harmony.

In his 2009 analysis, Schwartz says:

"Equally important, the regions representing each orientation form the integrated cultural system postulated by the theory: They emanate from the center of the circle, follow the expected order around the circle, and form the poles of the three broad cultural dimensions." (Schwartz, 2009)

In Graph 4.1, we refer to Schwartz's summarizing essays, 2006a and 2009 and the overviews of the relationships in his theoretical system.

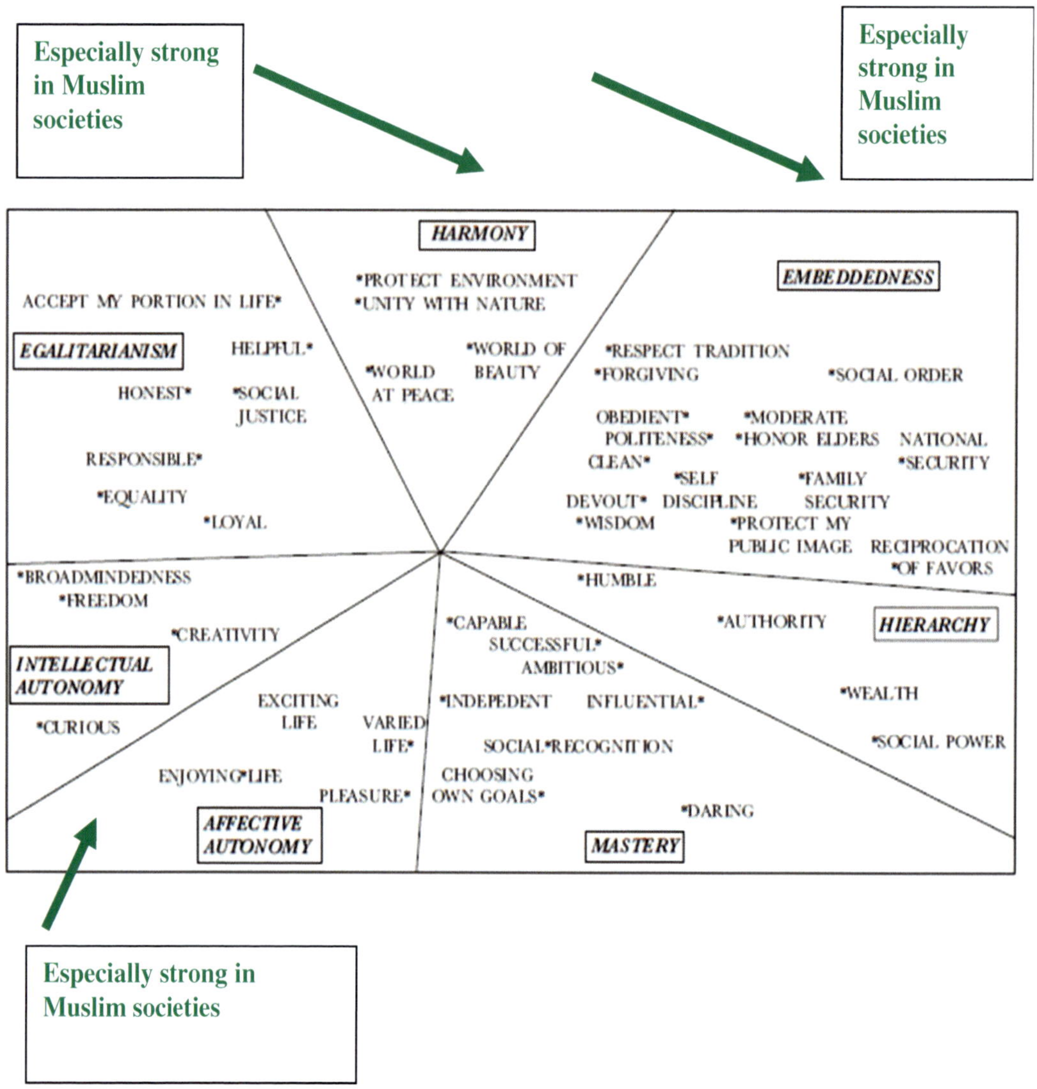

Graph 4.1. The relationship of the seven cultural-level dimensions of societies to underlying values (adapted from Schwartz, 2006a and 2009).

Muslim societies, according to Schwartz, rank very high on values of embeddedness. Embeddedness combines such values as social order, respect tradition, forgiving, moderateness, obedience, politeness, cleanness, national security, devoutness, wisdom, self-discipline, family security, honoring elders, reciprocation of favors, protecting the public image. According to Schwartz, 2006a, in cultures with an emphasis on embeddedness, people are viewed as entities embedded in the collectivity. Meaning in life comes largely through social relationships, through identifying with the group, participating in its shared way of life, and striving toward its shared goals. *"Embedded cultures emphasize maintaining the status quo and restraining actions that might disrupt in-group solidarity or the traditional order. Important values in such cultures are social order, respect for tradition, security, obedience, and wisdom." (Schwartz, 2006a)*

Schwartz attempts to identify seven transnational cultural groupings: West European countries, English-speaking countries, Latin American countries, East European countries, South Asian countries, Confucian influenced countries and African and Middle Eastern countries. Only nine cultures are located outside of their expected region. Four of these are, according to Schwartz, from the culturally diverse Middle East (Turkey, Greek Cyprus, Israeli Arabs, and Israeli Jews). Schwartz maintains that his regions show some parallels with the zones in Huntington (1993) and Inglehart and Baker (2000).

The high profile of Muslim societies on the value scales of Intellectual and Affective Autonomy, Harmony and Embeddedness in the framework of Shalom Schwartz is really noteworthy, because it contradicts many of the popular, sweeping generalizations often drawn from the works of Huntington:

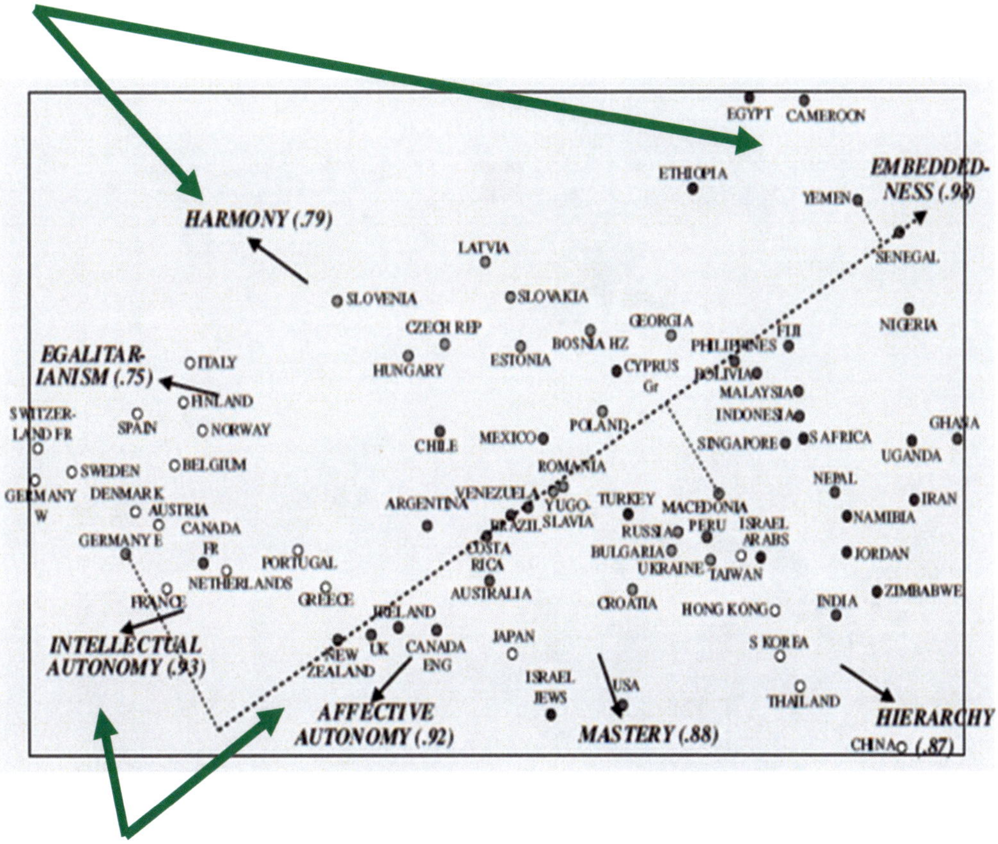

Graph 4.2. The Schwartz global map of human values (Adapted from Schwartz, 2006a and 2009).

Based on Ralston, Egri, Reynaud et al. 2011 and Schwartz, 2006a, we drew the following choropleth maps of the Schwartz value scales:

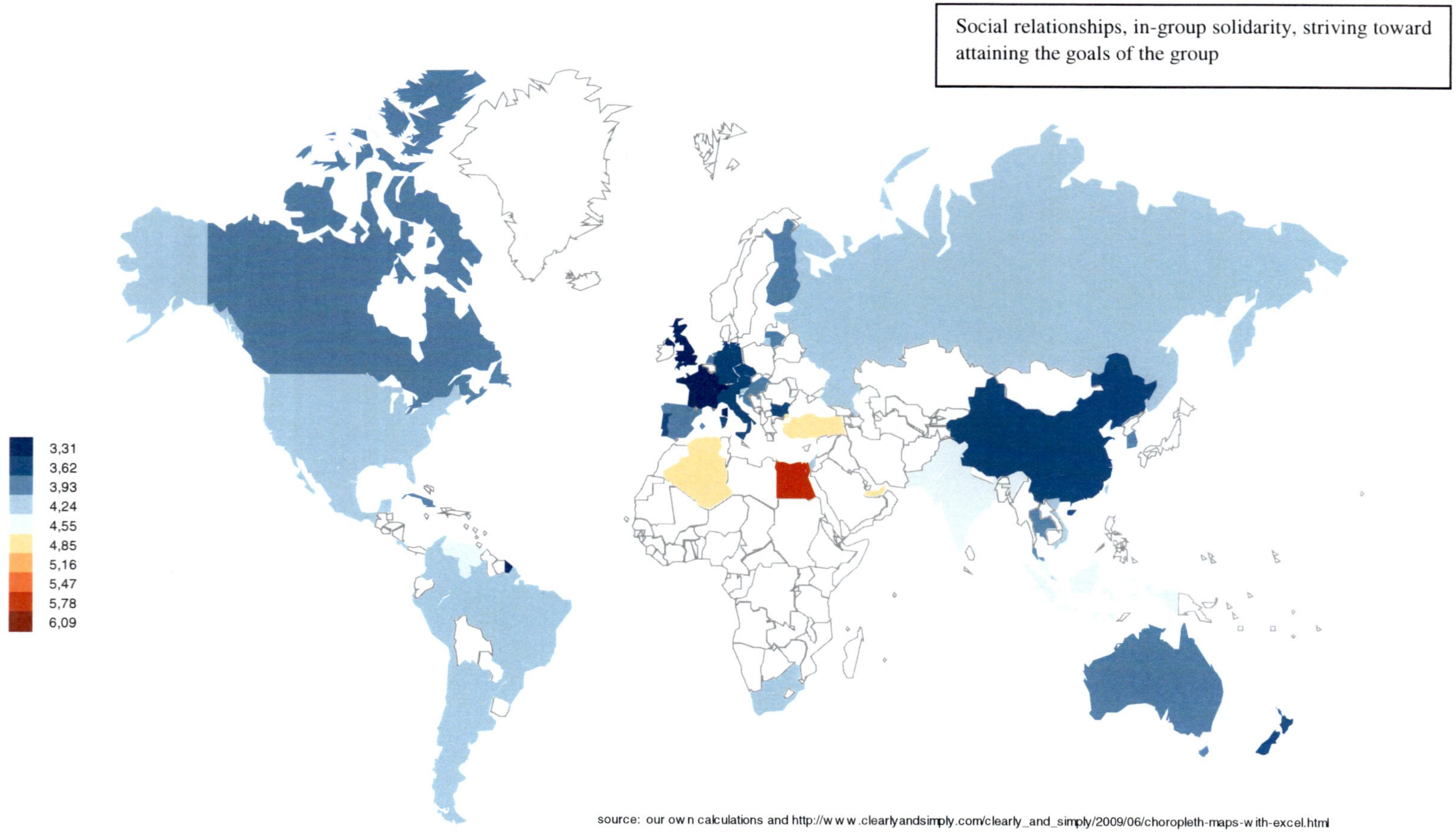

Map 4.7. Schwartz societal dimension: Embeddedness (highest value: painted in red).

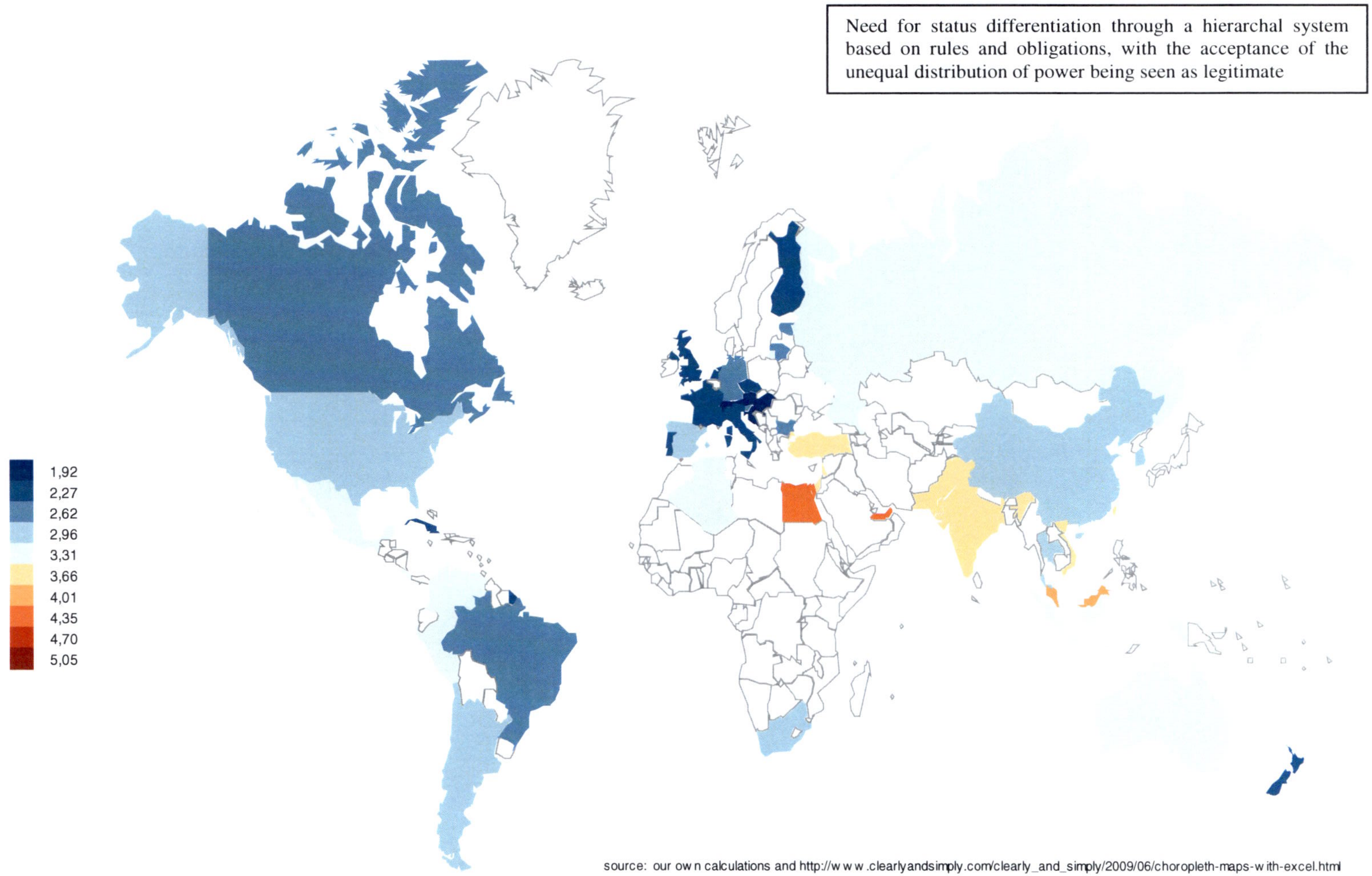

Map 4.8. Schwartz societal dimension: Hierarchy (highest value: painted in red).

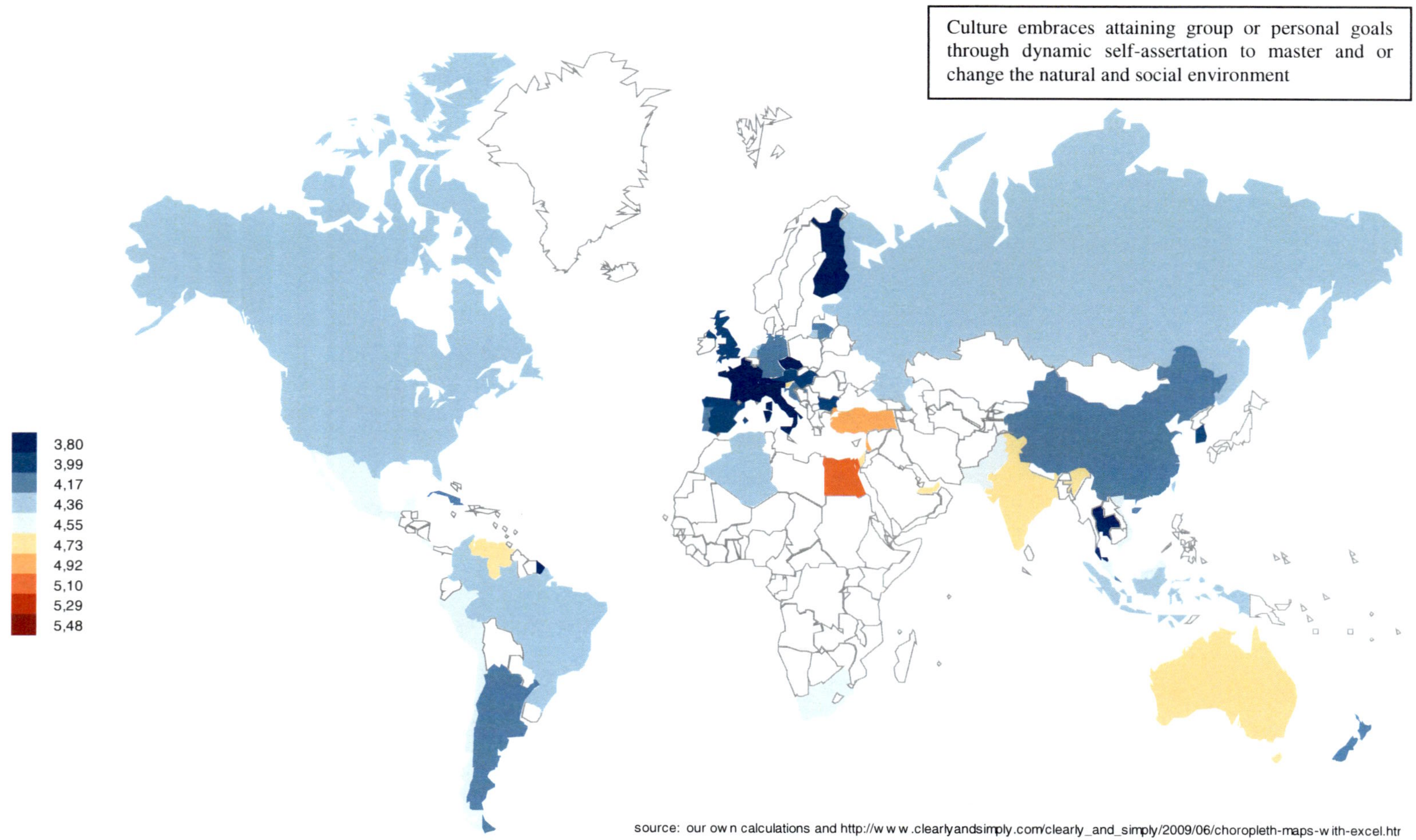

Map 4.9. Schwartz societal dimension: Mastery (highest value: painted in red).

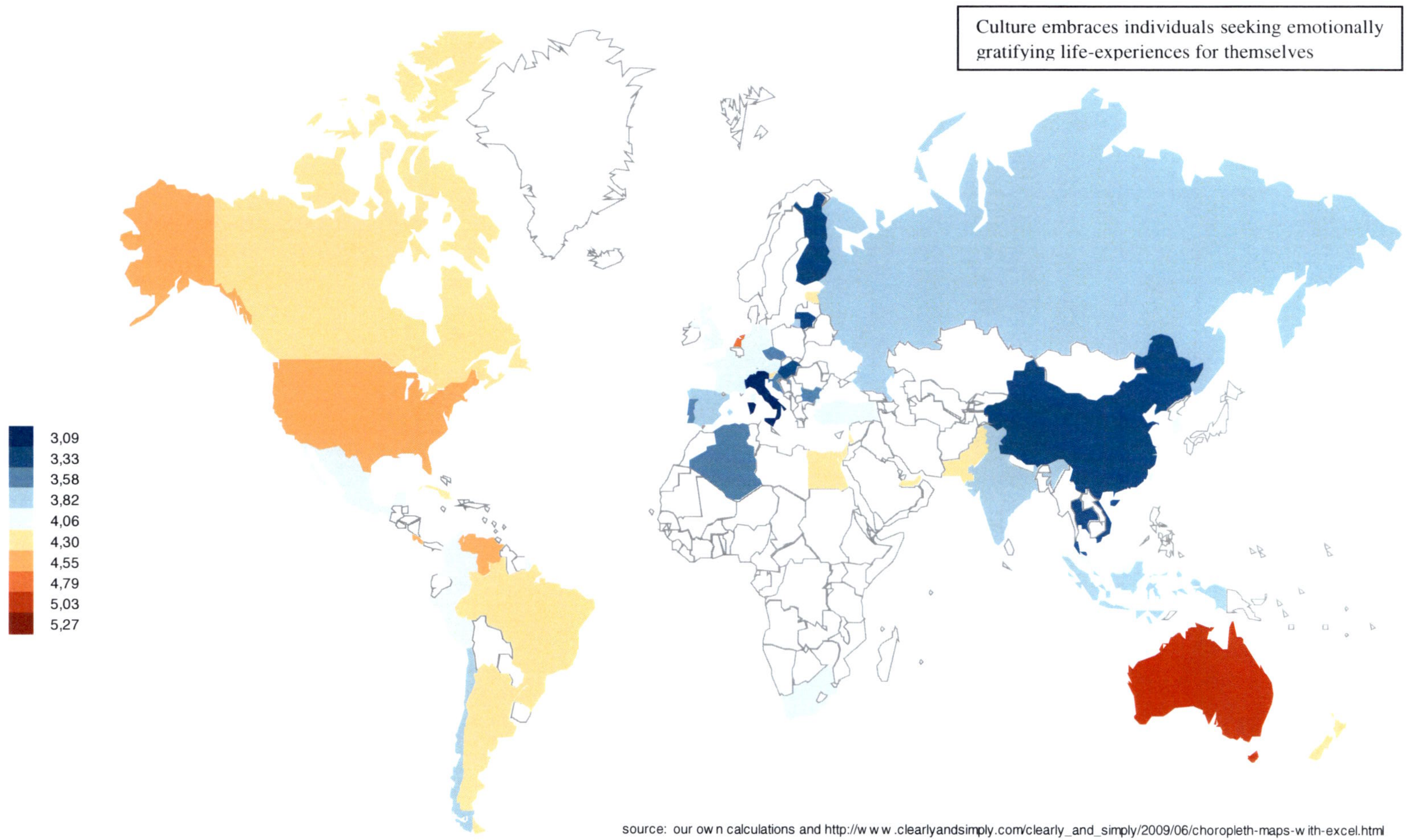

Map 4.10. Schwartz societal dimension: Affective Autonomy (highest value: painted in red).

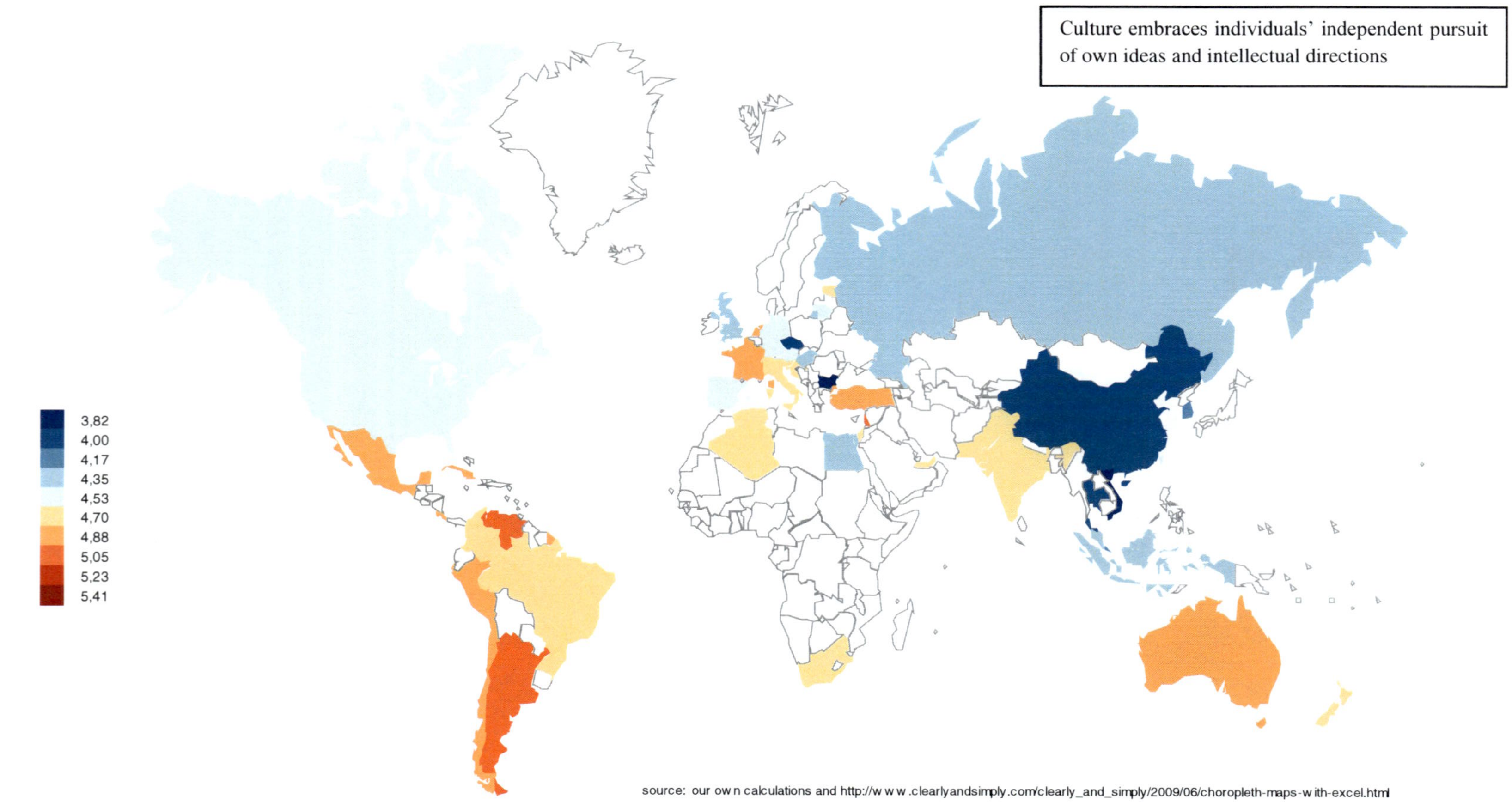

Map 4.11. Schwartz societal dimension: Intellectual Autonomy (highest value: painted in red).

Culture embraces the view that all people are moral equals and that there should be commitment and concern for the welfare of all

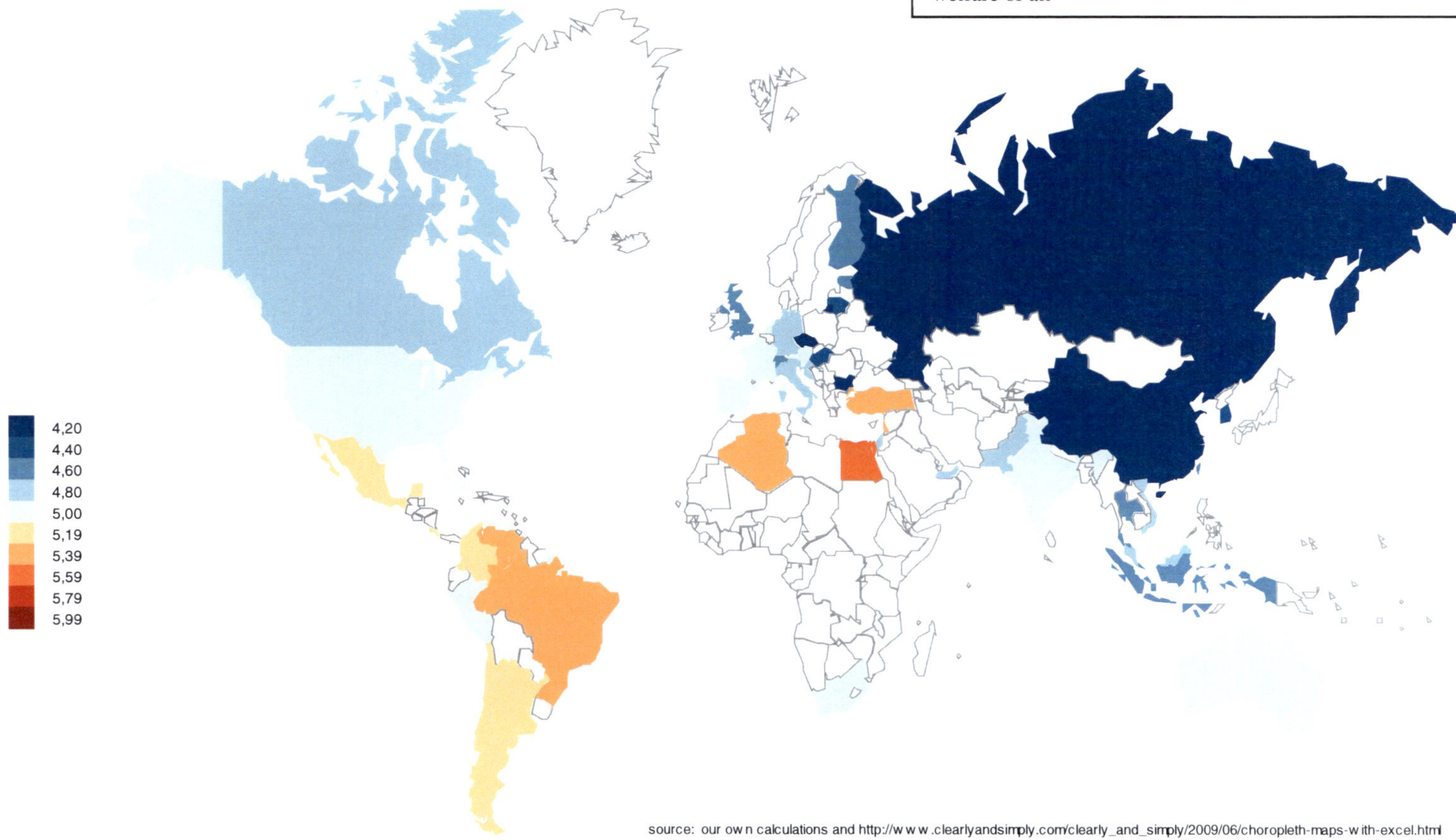

source: our own calculations and http://www.clearlyandsimply.com/clearly_and_simply/2009/06/choropleth-maps-with-excel.html

Map 4.12. Schwartz societal dimension: Egalitarianism (highest value: painted in red).

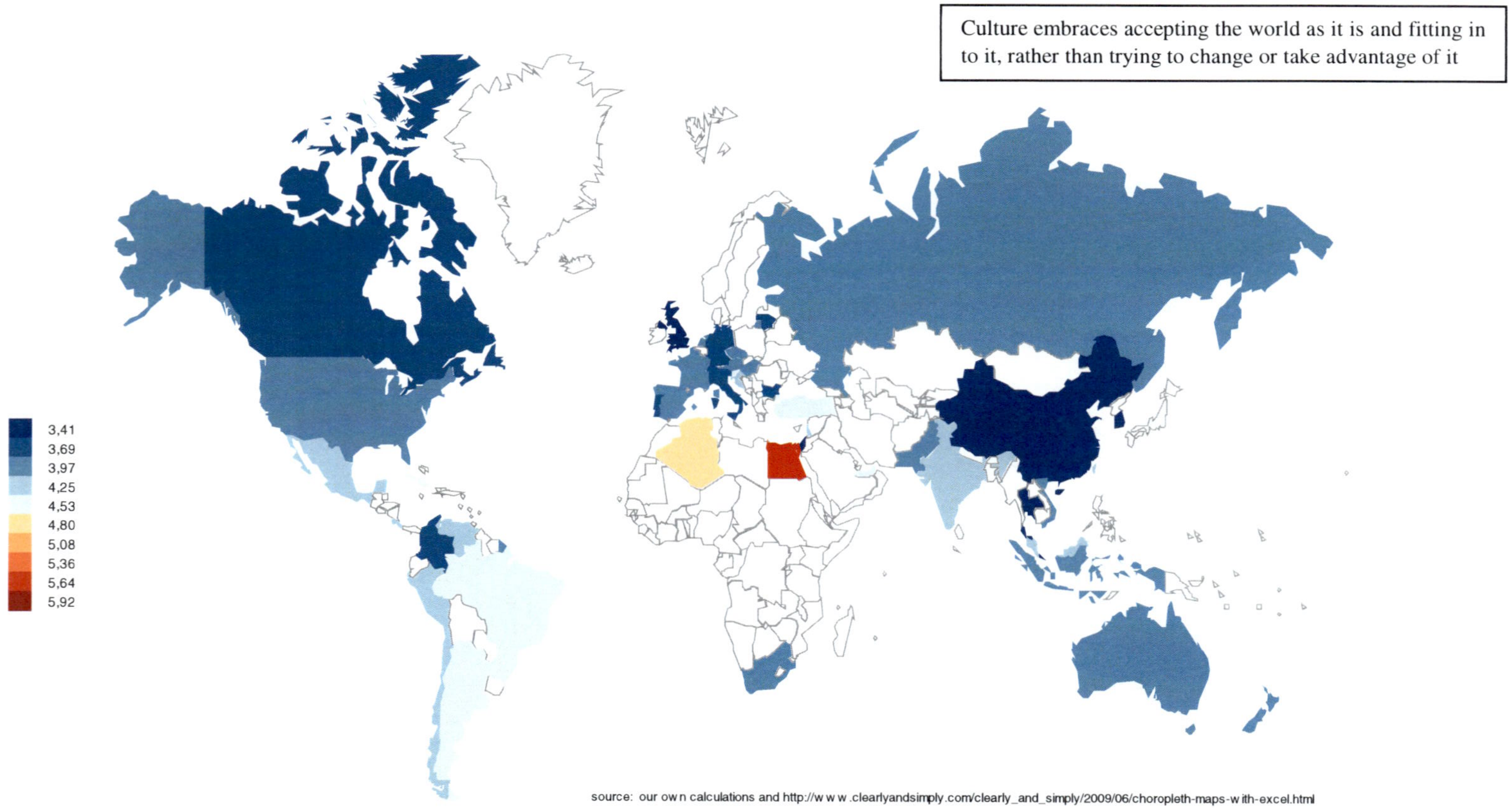

Map 4.13. Schwartz societal dimension: Harmony (highest value: painted in red).

A careful reading of the Schwartz categories would suggest however that there is a considerable discrepancy of his research results with the interpretation of Muslim civilization, inherent in the works of Huntington. We feel that especially the specific country results for mastery, affective autonomy, intellectual autonomy and harmony contradict the generalizations about Muslim countries, inherent in Huntington, 1996.

Eldad Davidov, who is also a researcher of Israeli origin, and who is currently professor of sociology at Zurich University, pays great attention to the statistical-mathematical foundations of comparative value analysis, including the detection of contradictory statements in survey research. Among his major works of mathematical sociology one finds Davidov, 2008, 2010; Davidov, Meuleman, Billiet, and Schmidt, 2008; Davidov, Schmidt and Billiet, 2011; and Davidov, Schmidt, and Schwartz, 2008; where Davidov and associates also base their analysis of human values on the cascade of human values, already developed in the works of Shalom Schwartz.

In his 2008 study on the contradictions of the European integration process, Davidov also highlights the close interrelationship between gender prejudice, homophobia, anti-Semitism, xenophobia, social out-groups (as for example homeless people), Islamophobia and the overall devaluation of newcomers:

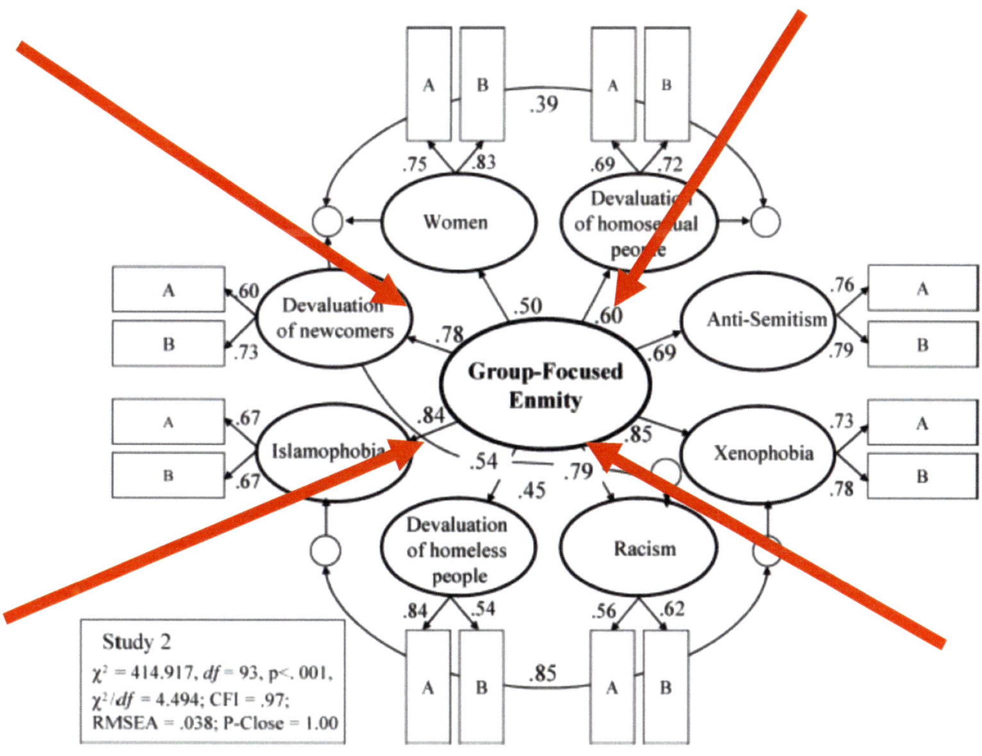

Graph 4.3. Davidov's explanation of group-focused enmity (adapted from Davidov and associates, 2008): group focused enmity explains racism, xenophobia, anti-Semitism, devaluation of homosexuals, of women, newcomers, Muslim people and homeless people at the same time.

Policy pointer:

Just as Hofstede, the works of the Israeli psychologist Shalom Schwartz received a considerable international reception not only in the social sciences, but also in the expanding field of international business studies. Schwartz highlights a map of global values, which arrives at results, which are comparable to those of other researchers in the field. Schwartz identifies seven basic cultural orientations and the structure of interrelations among them: West European, English-speaking, Latin American, East European, and South Asian, Confucian influenced, and African and Middle Eastern.

The seven dimensions are

1. Embeddedness
2. Hierarchy
3. Mastery
4. Affective Autonomy
5. Intellectual Autonomy
6. Egalitarianism
7. Harmony

Muslim societies, according to Schwartz, rank very high on values of embeddedness. Embeddedness combines such values as social order, respect tradition, forgiving, moderateness, obedience, politeness, cleanness, national security, devoutness, wisdom, self-discipline, family security, honoring elders, reciprocation of favors, and protecting the public image.

A careful reading of the Schwartz categories would suggest that there is a considerable contradiction with the interpretation of Muslim civilization, inherent in the works of S. P. Huntington. The specific country results for his dimensions mastery, affective autonomy, intellectual autonomy and longing for harmony suggest that key Muslim countries are at the forefront of these values.

Eldad Davidov pays great attention to the statistical-mathematical foundations of comparative value analysis, including the detection of contradictory statements in survey research to reduce the measurement error.

In his 2008 study on the contradictions of the European integration process, Davidov also highlights the close interrelationship between gender prejudice, homophobia, anti-Semitism, xenophobia, social out-groups (as for example homeless people), Islamophobia and the overall devaluation of newcomers.

4.3. Inglehart's Theory

Inglehart, in some of his main publications, developed by contrast an interpretation of *global value change* (Alexander, Inglehart and Welzel, 2011; Inglehart, 1970, 1971, 1977, 1990, 1997, 2000, 2002, 2003, 2003a, 2003b, 2003c, 2007; Inglehart and Baker, 2000; Inglehart and Norris, 2003a, 2003b, 2003c; Inglehart and Welzel, 2003, 2005, 2009, 2010a, 2010b; Inglehart et al. 2004; Moaddel, Tessler and Inglehart, 2008a, 2008b; Norris, Inglehart, and Welzel, 2003; Norris and Inglehart, 2003, 2004; Tessler, Moaddel and Inglehart, 2006a,

2006b), which rests on a well-known *two-dimensional scale of global values and global value change. It is based on the statistical technique of factor analysis* of up to over twenty key WVS variables from the originally more than 900 survey items in the WVS. Factor analysis, as we highlighted in our methodological introduction, reduces variables to the underlying statistical dimensions, and is based on a mathematical procedure, implemented today on most advanced statistical software packages for social statistics, like IBM SPSS (here versions 20 and 21).

Several essays developed the critique of Inglehart's methodology (Hadenius, and Teorell, 2005; Haller, 2002; Haller and Hadler, 2006; Steenkamp, and Geyskens, 2012; furthermore Au, 2000; Davis, Dowley, and Silver, 1999; Li, Zinn, Chick, Absher, Graefe, and Hsu, 2007; Sacchi, 1998; Van de Vijver, and Poortinga, 2002).

The two Inglehart dimensions are: *(1) the Traditional/ Secular-Rational dimension and (2) the Survival/Self-expression dimension.* These two dimensions also explain more than 70 percent of the cross-national variance in a factor analysis of ten indicators, and each of these dimensions is strongly correlated with scores of other important variables.

So let us look at Inglehart's theoretical main pillars. To understand his theory, we best should look at an important paragraph written in Inglehart and Baker, 2000:

> 'In the course of human history, thousands of societies have existed, most of which are now extinct. These societies had a vast range of characteristics. Infanticide was common in hunting and gathering societies, but became rare in agrarian societies; homosexuality was accepted in some preindustrial societies; and women are believed to have dominated political and social life in some preindustrial societies. Although the full range of "traditions" is diverse, a mainstream version of preindustrial society having a number of common characteristics can be identified. All of the preindustrial societies for which we have data show relatively low levels of tolerance for abortion, divorce, and homosexuality; tend to emphasize male dominance in economic and political life, deference to parental authority, and the importance of family life, and are relatively authoritarian; most of them place strong emphasis on religion. Advanced industrial societies tend to have the opposite characteristics. It would be a gross oversimplification to assume that all known preindustrial societies had similar characteristics, but one can meaningfully contrast the cultural characteristics of industrial societies with those of this mainstream version of preindustrial society.' (Inglehart and Baker, 2000: 23-24)

In that quoted path-breaking essay, which has become a true classic of contemporary global sociology, they also go on to say that the two mentioned dimensions explain 70 percent of the total cross-national variation among 10 variables. The factor scores generated by these 10 items are highly correlated with factor scores from earlier research. In a statistical Table in that landmark article, Inglehart and Baker, 2000 also show the results from a factor analysis of variables with 165,594 *World Values Survey* respondents (Inglehart and Baker, 2000, their Table 1). As expected, the factor loadings are considerably lower than those at the national level, and are reproduced here below.

Traditional values are defined by:

- God is very important in respondent's life 0.70
- It is more important for a child to learn obedience and religious faith than independence and determinations (the so-called autonomy index from WVS data) 0.61

- Abortion is never justifiable 0.61
- Respondent has strong sense of national pride 0.60
- Respondent favors more respect for authority 0.51

Self-expression values are defined by:

- Respondent gives priority to self-expression and quality-of-life (measured by the four-item materialist/post-materialist values index, documented in the WVS data) 0.59
- Respondent describes self as very happy 0.58
- Respondent has signed and would sign a petition 0.59
- Homosexuality is always justifiable 0.54
- You should be trusting people 0.44

Inglehart and Baker, 2000 also maintain that in traditional societies a main goal in life is to make one's parents proud and one must always love and respect one's parents, regardless of how they behave. Conversely, parents must do their best for their children even if their own well-being suffers. People in traditional societies idealize large families, and they actually have them (high scores on this dimension are strongly correlated with high fertility rates):

> "Yet although the people of traditional societies have high levels of national pride, favor more respect for authority, take protectionist attitudes toward foreign trade, and feel that environmental problems can be solved without international agreements, they accept national authority passively: They seldom or never discuss politics. In preindustrial societies the family is crucial to survival. Accordingly, societies at the traditional pole of this dimension reject divorce and take a pro-life stance on abortion, euthanasia, and suicide. They emphasize social conformity rather than individualistic striving, believe in absolute standards of good and evil, support deference to authority, and have high levels of national pride and a nationalistic outlook. Societies with secular-rational values have the opposite preferences on all of these topics. The survival/self-expression dimension taps a syndrome of trust, tolerance, subjective well-being, political activism, and selfexpression that emerges in postindustrial societies with high levels of security. At the opposite extreme, people in societies shaped by insecurity and low levels of well-being, tend to emphasize economic and physical security above all other goals, and feel threatened by foreigners, by ethnic diversity and by cultural change. This leads to an intolerance of gays and other out-groups, an insistence on traditional gender roles, and an authoritarian political outlook. A central component of this dimension involves the polarization between materialist and postmaterialist values. Extensive evidence indicates that these values tap an intergenerational shift from an emphasis on economic and physical security toward an increased emphasis on self-expression, subjective well-being, and quality-of-life concerns." (Inglehart and Baker, 2000)

This cultural shift, Inglehart and Baker, 2000 say, is found throughout advanced industrial society; it emerges among birth cohorts that have grown up under conditions in which survival is taken for granted. These values are linked with a growing emphasis on environmental protection, the women's movement, and rising demands for participation in decision-making in economic and political life. But as our data in Chapter 10 show, recent experiences in the "North Atlantic arena" under the pressure of the current economic crisis clearly contradict the contention of the Inglehart/Baker 2000 essay that self-expression values will go on to become increasingly widespread in almost all advanced industrial societies. Instead of the predicted extension, we witness an implosion of self-expression values in the West.

According to Inglehart and Baker, 2000, societies that emphasize survival values show relatively low levels of subjective well-being, report relatively poor health, are low on interpersonal trust, relatively intolerant of out-groups, are low on support for gender equality, emphasize materialist values, have relatively high levels of faith in science and technology, are relatively low on environmental activism, and relatively favorable to authoritarian government. Societies high on self-expression values tend to have the opposite preferences on these topics:

> "When survival is uncertain, cultural diversity seems threatening. When there isn't "enough to go around," foreigners are seen as dangerous outsiders who may take away one's sustenance. People cling to traditional gender roles and sexual norms, and emphasize absolute rules and familiar norms in an attempt to maximize predictability in an uncertain world. Conversely, when survival begins to be taken for granted, ethnic and cultural diversity become increasingly acceptable- indeed, beyond a certain point, diversity is not only tolerated, it may be positively valued because it is interesting and stimulating. In advanced industrial societies, people seek out foreign restaurants to taste new cuisine; they pay large sums of money and travel long distances to experience exotic cultures. Changing gender roles and sexual norms no longer seem threatening. The past few decades have witnessed one of the most dramatic cultural changes that has occurred since the dawn of recorded history - the emergence of new gender roles enabling women to enter the same occupations as men. Polarization over new gender roles is strikingly evident in the survival/self-expression dimension: One of its highest-loading issues involves whether men make better political leaders than women. In the world as a whole, a majority still accepts the idea that men make better political leaders than women, but this view is rejected by growing majorities in advanced industrial societies and is overwhelmingly rejected by the younger generation within these societies. Equal rights for women, gays and lesbians, foreigners, and other out-groups tend to be rejected in societies where survival seems uncertain and increasingly accepted in societies that emphasize self-expression values." (Inglehart and Baker, 2000)

In another important publication, Inglehart and Welzel (2010a) highlighted their understanding of modernization theory and the modernization process in an illuminating way:

> "The core concept of modernization theory is that economic development produces systematic changes in society and politics. If so, one should find pervasive differences between the beliefs and values of people in low-income and high-income societies. The World Values Survey and European Values Study ([...] WVS/EVS) provide evidence that the transition from agrarian to industrial society produces one set of changes, and the rise of postindustrial societies produces another set of changes in peoples' values and motivations. Analyses of WVS/EVS data reveal two major dimensions of cross-cultural variation: a traditional versus secular-rational values dimension and a survival versus self-expression values dimension. These two dimensions tap scores of attitudinal variables, and are robust enough that researchers obtain similar results using various combinations of these variables.
>
> Theoretically, the traditional/secular-rational dimension reflects changes linked with the transition from agrarian to industrial society, associated with bureaucratization, rationalization, and secularization. *Accordingly, the publics of agrarian societies emphasize religion, national*

pride, obedience and respect for authority, while the publics of industrial societies emphasize secularism, cosmopolitanism, autonomy, and rationality. With the emergence of postindustrial society, unprecedented levels of prosperity and the advent of the welfare state bring high levels of existential security. When survival is insecure, it tends to dominate people's life strategies. But the younger birth cohorts of these societies have grown up taking survival for granted, allowing other goals to become more prominent. This trend is reinforced by the fact that in knowledge societies, one's daily work requires individual judgment and innovation, rather than following routines prescribed from above. Both factors bring increasing emphasis on self-expression. The survival versus self-expression dimension reflects polarization between emphasis on order, economic security, and conformity and emphasis on self-expression, participation, subjective well-being, trust, tolerance, and quality of life concerns. In recent decades the publics of virtually all rich countries have gradually moved toward increasing emphasis on self-expression values, but the relative positions of given countries have been remarkably stable." (Inglehart and Welzel, 2010a)

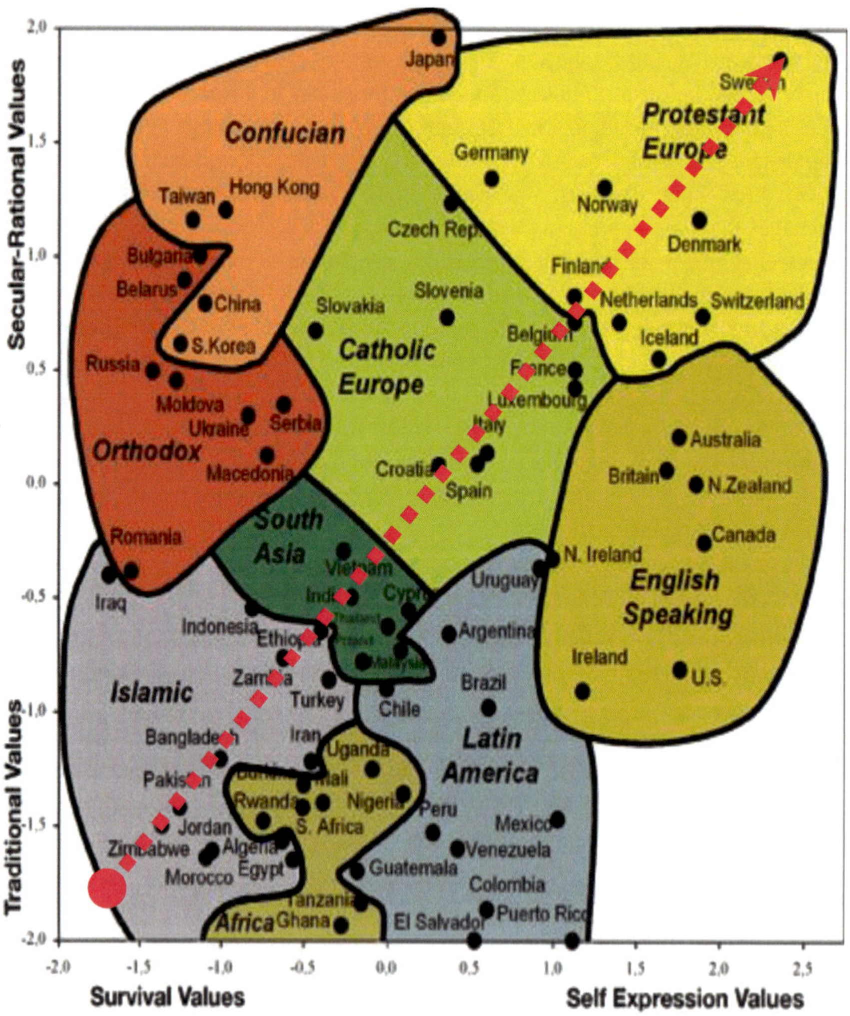

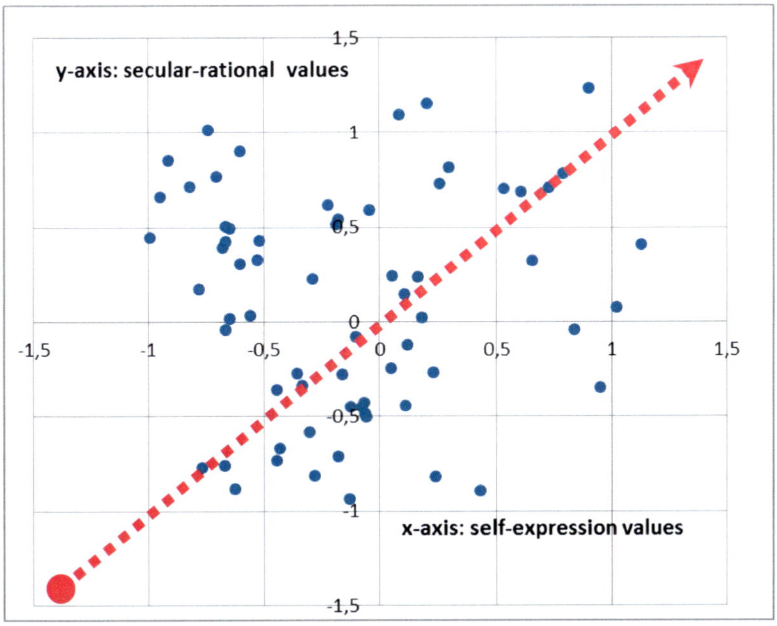

Source: adapted from Ronald Inglehart and Christian Welzel, "Changing Mass Priorities: The Link Between Modernization and Democracy." *Perspectives on Politics,* June 2010 (vol. 8, No. 2) page 554. Graph 1 is the latest published version of the Inglehart/Welzel map; earlier data, including a number of countries NOT included in Graph 1, are to be found at the supplementary data source in doc. format at the bottom of the WVS Internet page at http: //www. worldvaluessurvey.org/wvs/articles/folder_ published/article_base_54

Graph 4.4. Map of global human values according to Inglehart and associates (our adaption).

We emphasize that our own empirical approach to the analysis of human values and their interrelationship among each other revealed many results, which partially contradict, correct or extend the Inglehartian findings, and in some ways re-iterate the interpretations, forwarded by Hofstede, Davidov, and Schwartz (2006). In contrast to all three hitherto dominant explanations, we introduce the dimension of the shadow economy, acceptancy of corruption, and overall moral decay. And as we see in the results, this consideration of the dimension of economic permissiveness results in a significant redrawing of the global maps of human values.

The Inglehart dimensions are based upon *World Values Survey* data from at least 145.000 interview partners around the globe.[30] For Inglehart and his associates, the rise of rational-secular values is an important element in socio-economic and democratic development. Self-expression values, as opposed to survival values, give high priority to environmental protection, tolerance of diversity and rising demands for participation in decision making in economic and political life. For Inglehart, there is a dramatic shift in child-rearing values, from emphasis on hard work toward emphasis on imagination and tolerance as important values to teach a child in the course of socio-economic development. Societies that rank high on self-expression values also tend to rank high on interpersonal trust. The culture of trust and tolerance are crucial, Inglehart emphasizes throughout his recent works, to democracy. Self-

[30] http: //www. worldvaluessurvey. org/wvs/articles/folder_published/article_base_54.

expression (x-axis) and secularism (y-axis) and are but two sides of the same coin – modernity. The Inglehart School assumes the following famous global *"map of human values"*. In our version of the famous graph, we check the Inglehart map data with the data, available to us from the electronic publicly available download facilities from the *World Values Survey* website. The two graphs are practically identical.

But in many ways, the map should and could be reverted, because *secular values* should be seen on an *x-axis* of global value development, and *self-expression* values on the *y-axis.*

Table 4.5 ranks the countries of the world according to their secularism/traditionalism scale according to the results of waves 1-4 of the *World Values Survey*. Tanzania, Puerto Rico and Jordan are the least secular countries of the world, while Sweden, Japan and the Czech Republic are the most secular countries of the world.

Table 4.6 now ranks the countries of the world according to their survival/self-expression scale based on the results of waves 1-4 of the *World Values Survey*. New Zealand, Australia and the United States are the most self-expression oriented countries of the world, while all of the five most survival oriented and least self-expression oriented countries of the world were of Orthodox Christian cultural heritage: Moldova, Ukraine, Russia, Belarus and Romania.

Table 4.5. Secular values according to Inglehart

	Traditional/Secular RATIONAL VALUES WVS 1-4/2006
Sweden	1,234
Japan	1,152
Czech Republic	1,093
Estonia	1,011
Latvia	0,902
Russian Federation	0,852
Germany	0,812
Netherlands	0,780
Bulgaria	0,766
France	0,730
Belarus	0,714
Switzerland	0,706
Norway	0,704
Finland	0,688
Ukraine	0,658
Slovenia	0,617
South Korea	0,591
Croatia	0,543
Slovakia	0,516
Serbia and Montenegro	0,503
Lithuania	0,494
Moldova	0,444
Hungary	0,429
Taiwan Province of China	0,422
New Zealand	0,408
Armenia	0,390
Albania	0,329
Great Britain	0,321
Macedonia, Republic of	0,304
Spain	0,239
Italy	0,235
Bosnia and Herzegovina	0,226

	Traditional/Secular RATIONAL VALUES WVS 1-4/2006
Romania	0,172
Israel	0,147
Australia	0,075
Kyrgyzstan	0,032
Uruguay	0,020
Azerbaijan	0,018
Canada	-0,042
Georgia	-0,042
India	-0,080
Viet Nam	-0,125
Argentina	-0,247
Mexico	-0,267
Poland	-0,275
Singapore	-0,278
Indonesia	-0,339
United States	-0,346
Turkey	-0,360
Chile	-0,431
Dominican Republic	-0,444
Philippines	-0,450
Brazil	-0,452
Peru	-0,489
South Africa	-0,506
Uganda	-0,586
Bangladesh	-0,671
Egypt	-0,713
Algeria	-0,734
Pakistan	-0,763
Zimbabwe	-0,772
Nigeria	-0,816
Venezuela	-0,822
Jordan	-0,882
Puerto Rico	-0,894
Tanzania, United Republic Of	-0,937

Table 4.6. Self-expression according to Inglehart

	Survival/Self-Expression Values WVS 1-4/2006
New Zealand	1,130
Australia	1,025
United States	0,951
Sweden	0,902
Canada	0,841
Netherlands	0,794
Switzerland	0,732
Great Britain	0,661
Finland	0,610
Norway	0,536
Puerto Rico	0,431
Germany	0,298
France	0,259
Venezuela	0,243
Mexico	0,235
Japan	0,206

Table 4.6 (Continued)

	Survival/Self-Expression Values WVS 1-4/2006
Uruguay	0,185
Italy	0,165
Viet Nam	0,124
Dominican Republic	0,113
Israel	0,111
Czech Republic	0,086
Spain	0,057
Argentina	0,052
South Korea	-0,041
South Africa	-0,054
Peru	-0,061
Chile	-0,065
Philippines	-0,071
India	-0,101
Brazil	-0,124
Tanzania, United Republic Of	-0,125
Singapore	-0,158
Croatia	-0,173
Egypt	-0,175
Slovakia	-0,184
Slovenia	-0,220
Nigeria	-0,276
Bosnia and Herzegovina	-0,288
Uganda	-0,300
Indonesia	-0,332
Poland	-0,353
Bangladesh	-0,426
Algeria	-0,441
Turkey	-0,443
Hungary	-0,518
Albania	-0,525
Kyrgyzstan	-0,555
Macedonia, Republic of	-0,600
Latvia	-0,600
Jordan	-0,621
Azerbaijan	-0,644
Lithuania	-0,645
Serbia and Montenegro	-0,662
Taiwan Province of China	-0,663
Georgia	-0,664
Pakistan	-0,669
Armenia	-0,678
Bulgaria	-0,701
Estonia	-0,737
Zimbabwe	-0,765
Romania	-0,780
Belarus	-0,819
Russian Federation	-0,912
Ukraine	-0,947
Moldova	-0,993

Our choropleth maps – Maps 4.7 and 4.8 – designed with Inglehart's data, further highlight these aspects of the Inglehart global map of human values. Choropleth maps show global statistics on an easily readable real world map. They allow important insights into the essence of social scientific theories. Even at the price of oversimplification they can show to audiences confronted with theories where a given phenomenon is strongest and where a given phenomenon is weakest. Projected onto such world choropleth maps, it is clear that Inglehart's secular values, based on the *World Values Survey*, waves 1-4, are strongest in Sweden, the Czech Republic, Estonia, Japan, and indeed the entire former USSR, China, Korea, France, Germany, and the Benelux countries.

Poland, the United States, but also Canada, the UK, Australia and New Zealand, as well as Spain, Italy and several other European countries are less secular than the leading secularist countries, with Turkey, Poland, and most of Latin America belonging to a middle range of the continuum, and most of the Muslim nations, Venezuela and Zimbabwe being the least secular countries in the world.

Self-expression in the *World Values Survey* studies, waves 1-4, was highest in the highly developed Western Anglo-Saxon democracies, Scandinavia, and in the Netherlands and Switzerland. Most of the other continental European countries exhibited lower rates of self-expression. Latin America, several Muslim nations and Japan were to be found in the middle ranges of self-expression. Lowest rates of self-expression were to be found in the Orthodox countries and/or the countries of former Communism.

In Graph 4.5, we further take the results of Inglehart's calculations at their face value. Here, we draw a scatterplot of our own in the style of Graph 4.4, but with reverted axes. There are hundreds of good historical and sociological reasons why secularization and not self-expression values are the prime movers of the modernization process. Also historically, secularization and not postmaterialistic self-expression values first developed in the Western countries from the 19th to the late 20th Century.

As to be expected from the results of standard factor analysis without oblique factor rotation (see Chapter 3), the bi-variate correlation between the two Inglehart factors is very small. But there is an interesting wave structure in the relationship between secularism (in our Graph 4.5 projected on the x-axis) and self-expression (in our Graph 4.5 projected on the y-axis). With rising secularization, first there is a certain implosion and only then a rise in self-expression values, considered by Inglehart and his associates to be so important for *"effective democracy"* (see below). There is then a rise of self-expression values, in order to implode again at later stages of development, and only at very high levels of secularization, self-expression rises in a linear fashion. This wave-structure of modernization, secularization and self-expression is perhaps an important qualification of Inglehart's theory.

In Table 4.7 we now document these non-linear developments; the trend values and their residuals, used in Graph 4.4. Again taking Inglehart's sociology at its face value, let us assume that secularization is the prime mover of value change. We all of a sudden discover then that the real problematic cases with exceptionally low self-expression rates are not the majority Muslim countries, but the countries of the Christian Orthodox cultural tradition, including the former USSR. In Table 4.7, which lists the distances of the dots (country values) in Graph 4.5 from the non-linear trend line, Egypt and its neighbor Israel have about the same residuals. All Muslim countries in the sample (with the exception of post-Soviet Kyrgyzstan and Azerbaijan)

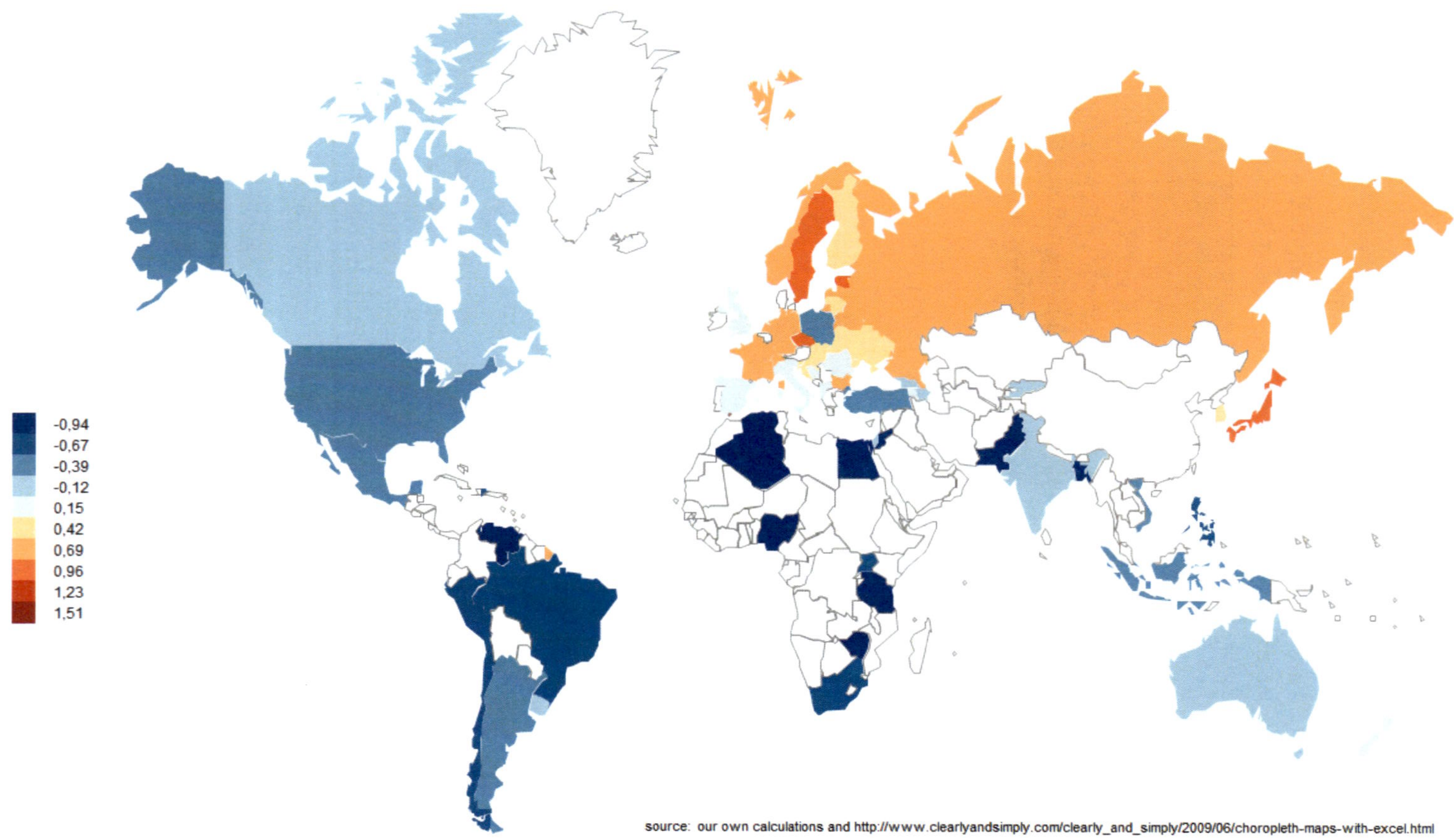

Map 4.14. Inglehart: Secular Values. Data from the WVS waves 1-4.

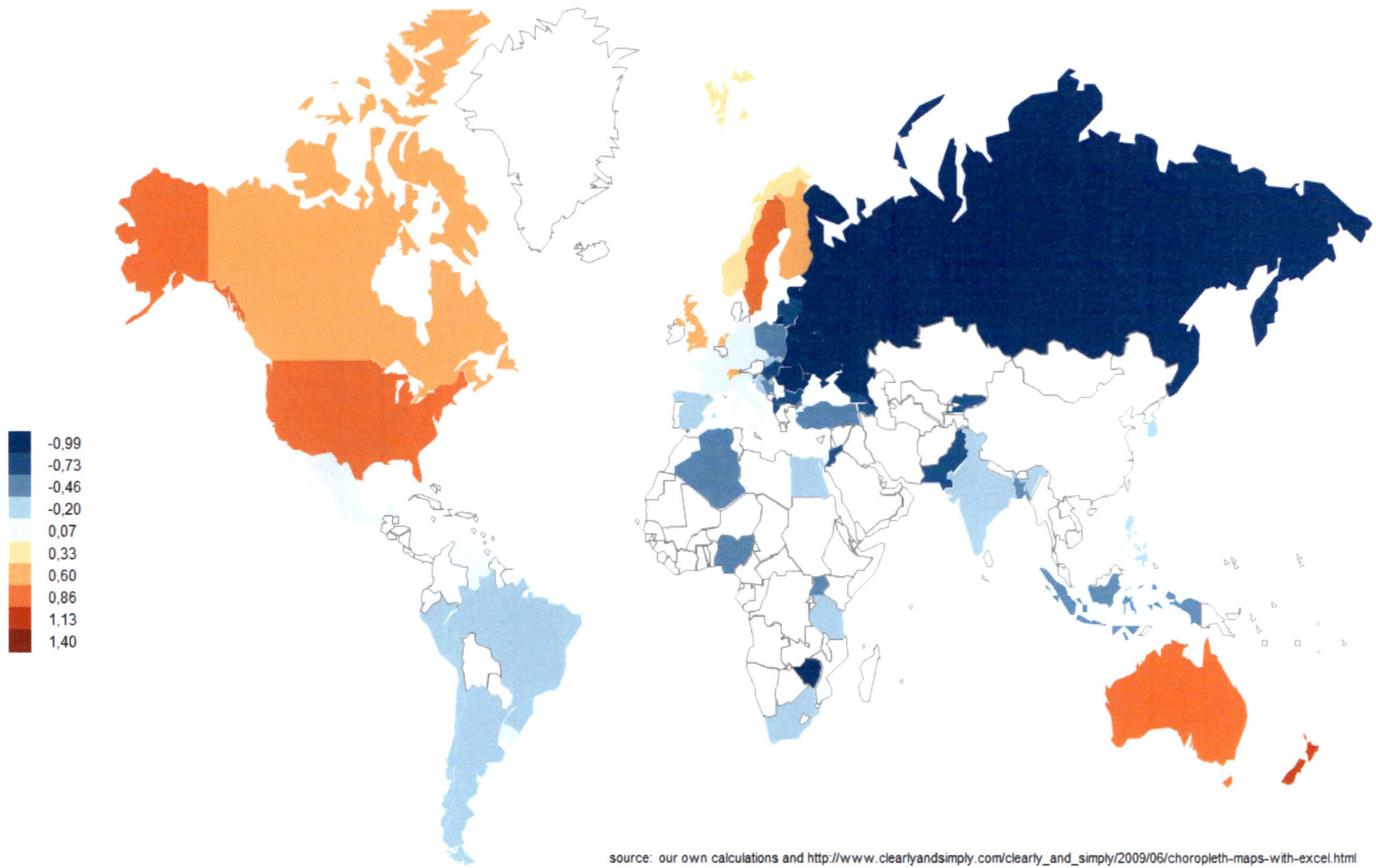

Map 4.15. Inglehart: Self-Expression Values. Data from the WVS waves 1-4.

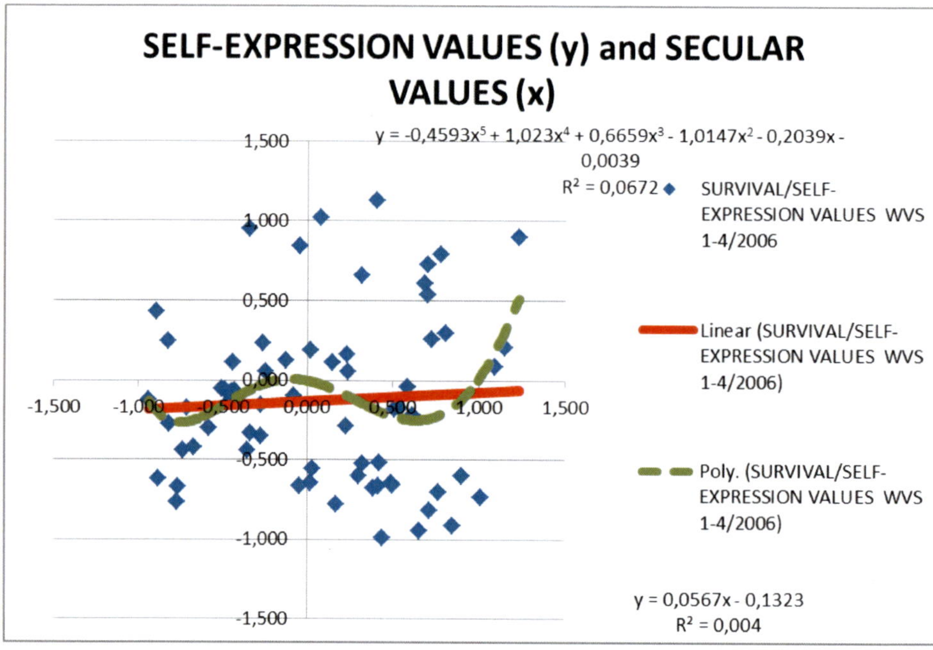

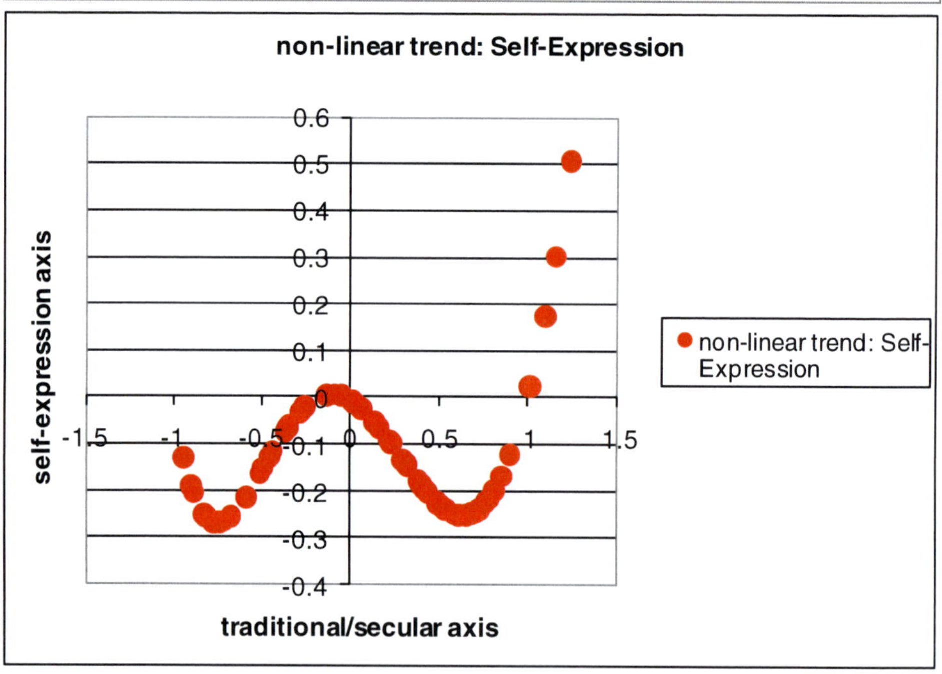

Graph 4.5. Secular values (x-axis) and self-expression values (y-axis) in the *World Values Surveys*, 1-4, 2006.

Source: adapted from Ronald Inglehart and Christian Welzel, "Changing Mass Priorities: The Link Between Modernization and Democracy." *Perspectives on Politics,* June 2010 (vol. 8, No. 2) page 554. Their Graph 1 is one of the latest published versions of the Inglehart/Welzel map; earlier data are to be found at the supplementary data source in doc. format at the bottom of the WVS Internet page at http: //www. worldvaluessurvey. org/wvs/articles/folder_published/article_base_54

Table 4.7. The two waves of self-expression – results from our non-linear re-analysis of the Inglehart global map of human values

	Traditional/secular rational values WVS 1-4/2006	**Survival/Self-Expression Values WVS 1-4/2006**	non-linear trend: Self-Expression	Residual: Self-Expression
New Zealand	0,4080	1,1300	-0,1876	1,3176
Australia	0,0750	1,0250	-0,0246	1,0496
United States	-0,3460	0,9510	-0,0655	1,0165
Netherlands	0,7800	0,7940	-0,2182	1,0122
Switzerland	0,7060	0,7320	-0,2456	0,9776
Finland	0,6880	0,6100	-0,2492	0,8592
Canada	-0,0420	0,8410	0,0029	0,8381
Great Britain	0,3210	0,6610	-0,1425	0,8035
Norway	0,7040	0,5360	-0,2461	0,7821
Puerto Rico	-0,8940	0,4310	-0,1926	0,6236
France	0,7300	0,2590	-0,2391	0,4981
Germany	0,8120	0,2980	-0,1993	0,4973
Venezuela	-0,8220	0,2430	-0,2523	0,4953
Sweden	1,2340	0,9020	0,5087	0,3933
Mexico	-0,2670	0,2350	-0,0286	0,2636
Italy	0,2350	0,1650	-0,0964	0,2614
Dominican Republic	-0,4440	0,1130	-0,1240	0,2370
South Korea	0,5910	-0,0410	-0,2496	0,2086
Uruguay	0,0200	0,1850	-0,0084	0,1934
Israel	0,1470	0,1110	-0,0532	0,1642
Spain	0,2390	0,0570	-0,0985	0,1555
Viet Nam	-0,1250	0,1240	0,0047	0,1193
South Africa	-0,5060	-0,0540	-0,1645	0,1105
Peru	-0,4890	-0,0610	-0,1533	0,0923
Egypt	-0,7130	-0,1750	-0,2667	0,0917
Argentina	-0,2470	0,0520	-0,0212	0,0732
Croatia	0,5430	-0,1730	-0,2399	0,0669
Philippines	-0,4500	-0,0710	-0,1279	0,0569
Chile	-0,4310	-0,0650	-0,1157	0,0507
Slovakia	0,5160	-0,1840	-0,2320	0,0480
Slovenia	0,6170	-0,2200	-0,2523	0,0323
Tanzania, United Republic of	-0,9370	-0,1250	-0,1312	0,0062
Brazil	-0,4520	-0,1240	-0,1292	0,0052
Nigeria	-0,8160	-0,2760	-0,2552	-0,0208
Uganda	-0,5860	-0,3000	-0,2145	-0,0855
Czech Republic	1,0930	0,0860	0,1742	-0,0882

	Traditional/secular rational values WVS 1-4/2006	**Survival/Self-Expression Values WVS 1-4/2006**	non-linear trend: Self-Expression	Residual: Self-Expression
Japan	1,1520	0,2060	0,3027	-0,0967
India	-0,0800	-0,1010	0,0056	-0,1066
Singapore	-0,2780	-0,1580	-0,0331	-0,1249
Bangladesh	-0,6710	-0,4260	-0,2553	-0,1707
Algeria	-0,7340	-0,4410	-0,2695	-0,1715
Bosnia and Herzegovina	0,2260	-0,2880	-0,0917	-0,1963
Indonesia	-0,3390	-0,3320	-0,0617	-0,2703
Hungary	0,4290	-0,5180	-0,1975	-0,3205
Poland	-0,2750	-0,3530	-0,0318	-0,3212
Turkey	-0,3600	-0,4430	-0,0731	-0,3699
Albania	0,3290	-0,5250	-0,1468	-0,3782
Pakistan	-0,7630	-0,6690	-0,2694	-0,3996
Jordan	-0,8820	-0,6210	-0,2061	-0,4149
Lithuania	0,4940	-0,6450	-0,2245	-0,4205
Serbia and Montenegro	0,5030	-0,6620	-0,2277	-0,4343
Macedonia, Republic of	0,3040	-0,6000	-0,1334	-0,4666
Taiwan Province of China	0,4220	-0,6630	-0,1943	-0,4687
Bulgaria	0,7660	-0,7010	-0,2250	-0,4760
Latvia	0,9020	-0,6000	-0,1217	-0,4783
Zimbabwe	-0,7720	-0,7650	-0,2683	-0,4967
Armenia	**0,3900**	**-0,6780**	**-0,1787**	**-0,4993**
Kyrgyzstan	**0,0320**	**-0,5550**	**-0,0114**	**-0,5436**
Belarus	**0,7140**	**-0,8190**	**-0,2437**	**-0,5753**
Azerbaijan	**0,0180**	**-0,6440**	**-0,0079**	**-0,6361**
Georgia	**-0,0420**	**-0,6640**	**0,0029**	**-0,6669**
Ukraine	**0,6580**	**-0,9470**	**-0,2525**	**-0,6945**
Romania	**0,1720**	**-0,7800**	**-0,0647**	**-0,7153**
Russian Federation	**0,8520**	**-0,9120**	**-0,1694**	**-0,7426**
Estonia	**1,0110**	**-0,7370**	**0,0247**	**-0,7617**
Moldova	**0,4440**	**-0,9930**	**-0,2043**	**-0,7887**

are even ahead of the three Baltic EU-member states Lithuania, Latvia and Estonia in their self-expression development. Judged by the size of its residual self-expression, the EU-member country Poland is ranked about equally with Turkey, and is even behind several Muslim countries, i.e., judging by the advancement of secularization, Poland has a lower self-expression than several Muslim countries. Such comparisons could be continued almost endlessly, all showing that prospects for the Muslim countries in the framework of Inglehart's sociology must be considered as more positive than commonly thought.

Maps 4.16 and 4.17 further highlight these tendencies with the help of choropleth maps.

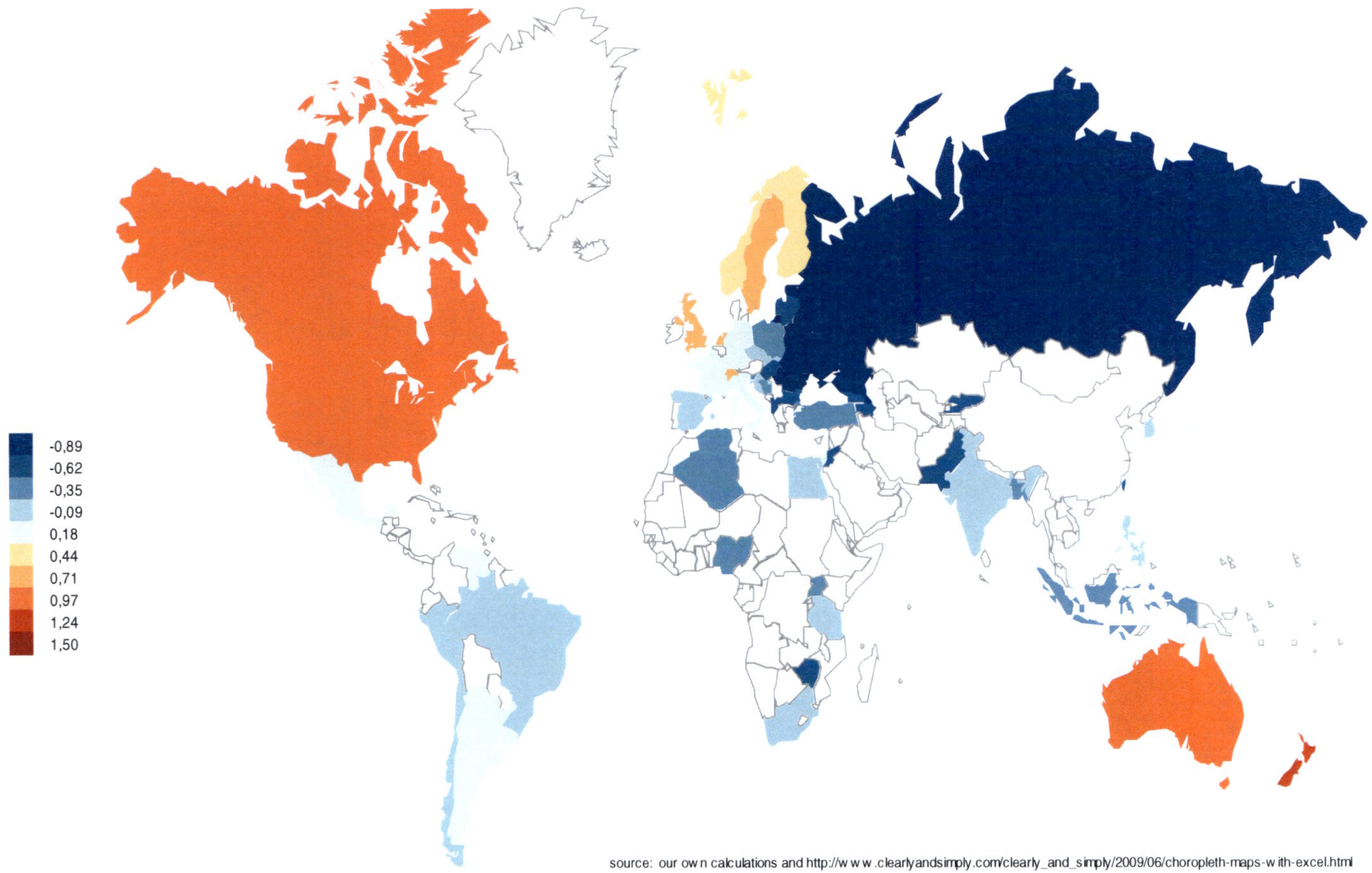

source: our own calculations and http://www.clearlyandsimply.com/clearly_and_simply/2009/06/choropleth-maps-with-excel.html

Map 4.16. Redrawing the Inglehartian map – the residuals from the linear function secularization→ selfexpression.

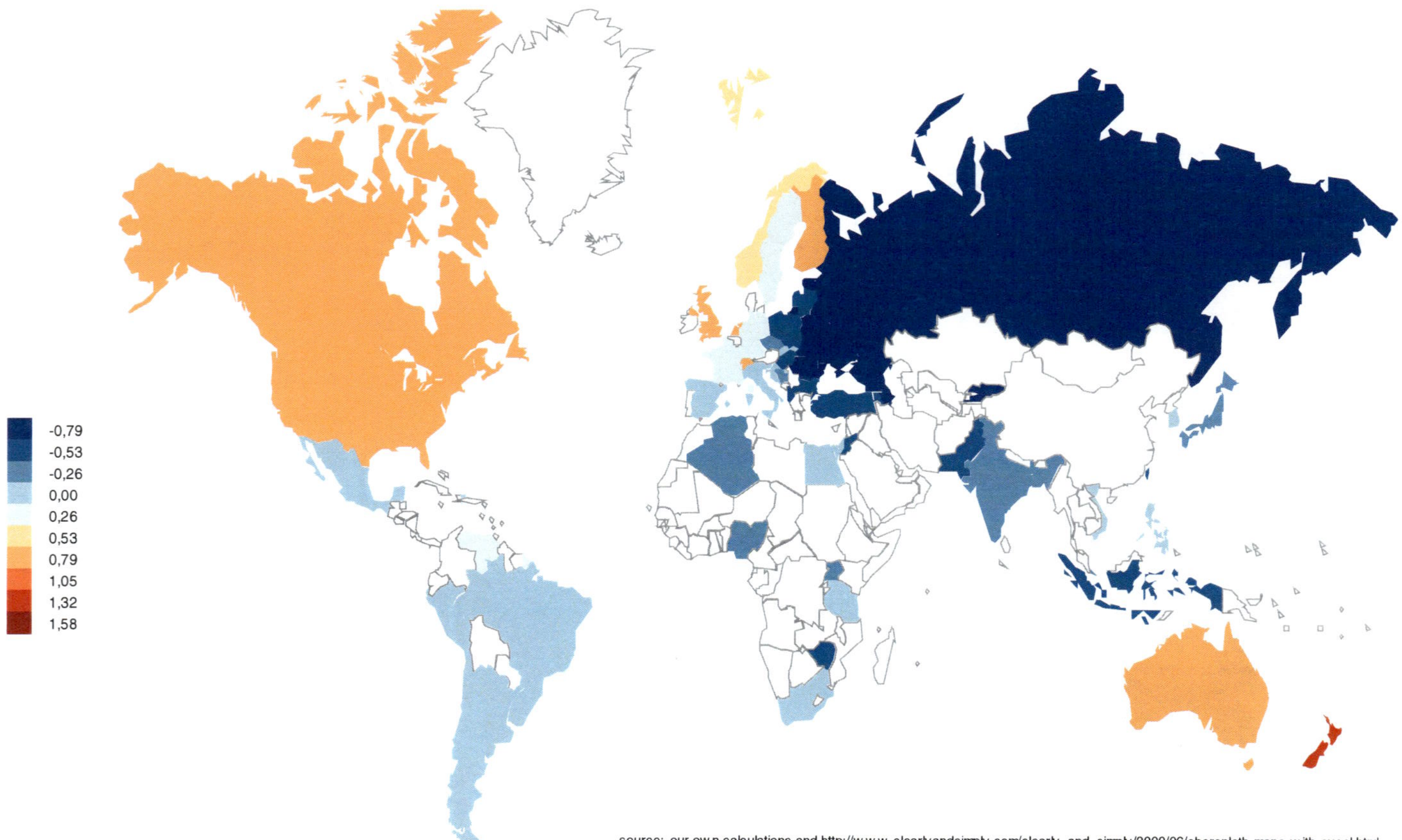

source: our own calculations and http://www.clearlyandsimply.com/clearly_and_simply/2009/06/choropleth-maps-with-excel.html

Map 4.17. Redrawing the Inglehartian map – the residuals from a polynomial function of the fifth degree secularization→ selfexpression (see also Table 4.7).

Policy pointer:

Inglehart developed an interpretation of global value change which rests on a well-known two-dimensional scale of global values and global value change. It is based on the statistical technique of factor analysis of up to some twenty key *World Values Survey* variables. The two Inglehart dimensions are: (1) the Traditional/ Secular-Rational dimension and (2) the Survival/Self-expression dimension. These two dimensions explain more than 70 percent of the cross-national variance in a factor analysis of ten indicators, and each of these dimensions is strongly correlated with scores of other important variables. For Inglehart and Baker, 2000, all of the preindustrial societies show relatively low levels of tolerance for abortion, divorce, and homosexuality; tend to emphasize male dominance in economic and political life, deference to parental authority, and the importance of family life, and are relatively authoritarian; most of them place strong emphasis on religion. Advanced industrial societies tend to have the opposite characteristics.

When survival is uncertain, cultural diversity seems threatening. When there isn't "enough to go around," foreigners are seen as dangerous outsiders who may take away one's sustenance. People cling to traditional gender roles and sexual norms, and emphasize absolute rules and familiar norms in an attempt to maximize predictability in an uncertain world. Conversely, when survival begins to be taken for granted, ethnic and cultural diversity become increasingly acceptable - indeed, beyond a certain point, diversity is not only tolerated, it may be even positively valued because it is seen as interesting and stimulating. In advanced industrial societies, people seek out foreign restaurants to taste new cuisine; they pay large sums of money and travel long distances to experience exotic cultures. Changing gender roles and sexual norms no longer seem threatening.

Tanzania, Puerto Rico and Jordan are the least secular countries of the world, while Sweden, Japan and the Czech Republic are the most secular countries of the world.

New Zealand, Australia and the United States are the most self-expression oriented countries of the world, while all of the five most survival oriented and least self-expression oriented countries of the world were of Orthodox Christian cultural heritage: Moldova, Ukraine, Russia, Belarus and Romania.

Our choropleth maps designed with Inglehart's data further highlight these aspects of the Inglehart global map of human values.

We also further take the results of Inglehart's calculations at their face value and draw a scatterplot of our own. While the bi-variate correlation between the two factors is indeed very small, there is an interesting wave structure in the relationship between secularism and self-expression. With rising secularization, first there is a certain implosion and only then a rise in self-expression values.

We also document these non-linear trend values and their residuals. We discover that from the point of view of self-expression, the real problematic cases are not presented by Muslim countries, but by the countries of the Christian Orthodox cultural tradition.

In Chapter 10, we will look at the changes in self-expression over time. We show with Inglehart's own data that self-expression values are imploding over time in leading Western countries. In addition, direct measurements undertaken by the *World Values Survey,* wave 2010-2014, asking respondents whether they assign importance to the value of self-expression in child education, directly contradict the Inglehart scores (negative correlation, around 20% of variance explained).

Chapter 5

TOWARDS A NEW POLITICAL GEOGRAPHY OF HUMAN VALUES

5.1. RE-ANALYSIS OF INGLEHART'S DATA

Convincing, as Inglehart's theory and empirics might appear at first sight, several essays questioned Inglehart's way of combining the analyzed variables into his dimensions or the linkage between his dimensions and democratic development (Hadenius, and Teorell, 2005; Haller, 2002; Haller and Hadler, 2006; Steenkamp, and Geyskens, 2012; furthermore Au, 2000; Davis, Dowley, and Silver, 1999; Li, Zinn, Chick, Absher, Graefe, and Hsu, 2007; Sacchi, 1998; Van de Vijver, and Poortinga, 2002). In our view, the most important theoretical and at the same time empirical problem is the following: principle component analysis or factor analysis with orthogonal standard rotation of factors (as is the current SPSS default option) is a convenient, but not always best way to reduce the relationships in a statistical correlation matrix between variables.

It is of paramount importance to be of the "small print" here – what relationships exist between the factors? Which relationships are being allowed by the model? No correlations between the factors? The statistical method chosen by Inglehart to reduce the complexity of the different components allows for no correlations between the chosen factors. It is the standard varimax rotation. Today, this technique is increasingly being substituted by better and more advanced methods, like the promax rotation, which exactly allows such correlations between the factors (Finch, 2006). To make matters worse, Inglehart's choice of the *WVS* data did not always use the items, which are the best available items for a maximum number of countries. His analyses are based on a maximum of 146789 global interviews reflecting 22 variables; while we thought it more appropriate to base our analysis on a maximum of 180041 global interviews for 30 variables. Our analysis is based on the *World Values Survey* data roll-out entitled: *"wvs1981_2008_v20090914.sav",* which was freely available at the WVS website. Any person around the globe in the possession of an IBM/SPSS 21 license, and an internet connection to download the freely available WVS data should be able to arrive at exactly the same results as we do.

Box 5.1. The overall message of this Chapter

In this Chapter, we present the results of our re-analysis of the entire openly available global *World Values Survey* data base. We use the advanced statistical multivariate analysis technique of Promax factor analysis, which allows for correlations between factors, and which is available via the IBM-SPSS statistical package XXI to the global public. We eliminated missing values by listwise deletion. There were 92289 interview partners around the globe with complete data for all the 30 variables of our research design. We worked with the very best documented *World Values Survey* items.

We used the background variables

- Age
- Highest educational level attained
- Sex (Gender) [in multivariate analysis: female] (1=male; 2=female)

The main opinion and value dimensions were

- Competition good or harmful
- Confidence: Armed Forces
- Confidence: The Press
- How important is God in your life
- How often do you attend religious services?
- Immigrant policy (prevent people from coming)
- Important child qualities: determination and perseverance
- Important child qualities: feeling of responsibility
- Important child qualities: hard work
- Important child qualities: imagination
- Important child qualities: independence
- Important child qualities: obedience
- Important child qualities: religious faith
- Important child qualities: thrift, and saving money and things
- Important child qualities: tolerance and respect for other people
- Important child qualities: unselfishness
- Justifiable: avoiding a fare on public transport
- Justifiable: cheating on taxes
- Justifiable: claiming government benefits even if one is not entitled to them
- Justifiable: someone accepting a bribe
- Most people can be trusted [highest numerical value: you just can't be too careful])
- Neighbors: Immigrants/foreign workers
- Neighbors: People of a different race
- Political system: having a democratic political system
- Satisfaction with your life
- Self-positioning in political scale (scale 1-left to 10-right)
- University is more important for a boy than for a girl

Our model explains 47.89% of the total variance of all the 30 variables. In Table 5.5 of this Chapter, we highlighted the relationships between the original 30 variables and the newly derived factor analytical dimensions:

a) Economic permissiveness
b) Traditional religion
c) Racism
d) Higher education for the younger generation (education gap between the generations)
e) Distrust of the army and the press
f) Authoritarian character
g) Tolerance and respect
h) The 'ego' company (i.e., the rejection of obedience and unselfishness as values in education)
i) [Predominantly] female rejection of the market economy and democracy

We also look at the trajectory of global society by analyzing the factor scores from Table 5.7 of this Chapter along the path of the Human Development Indicator of the UNDP ("human security indicator", used by Inglehart and his associates):

- Traditional religion diminishes in its societal role, as human security is rising
- Tolerance and respect, distrust of the army and the press, the "ego-company" orientation, and the [predominantly] female rejection of the market economy and democracy all rise with human security
- There is an inverted U-shaped relationship between human security and permissiveness/pessimism, racism, the education gap between the generations, and the authoritarian character

Economic permissiveness clearly captures the dimension of lawlessness, moral-ethical decay and the shadow economy, so prominent in contemporary economic theory of growth. This dimension is completely lacking in the hitherto published analyses by Hofstede, Schwartz/Davidov and Inglehart. In statistical terms it is the most important of all the resulting factors.

Traditional religion is linked in a very complex way to the absence of economic permissiveness.

We present Choropleth maps of human values across the globe, and also show the regional implications of our analysis.

In our research design, we worked with listwise deletion of missing values. At the end of the day, there were 92289 persons around the globe with complete data for all the 30 variables. We worked with the very best documented *World Values Survey* items. Our "new" nine factors all make very much 'sense' and are completely free from problematic theoretical assumptions. We included all 30 original indicators, based on the original more than 180.000 WVS interviews, into the original principal components and later the promax factor analysis. Our research design was thus intended to be more straightforward and simple than Inglehart's. In addition, it uses a more advanced and up to date statistical methodology. To explain the mathematical-statistical details of the Inglehart approach, let us quote here at length from Inglehart and Welzel (2010a), where the two authors very concisely explain the intricate details of their own factor analytical approach:

> "These two dimensions (i.e., Secular Values and Self-Expression Values) explain more than 70 percent of the cross-national variance in a factor analysis of ten indicators—and each of these dimensions is strongly correlated with scores of other orientations. The Traditional/Secular-rational values dimension reflects the contrast between societies in which religion is very important and those in which it is not. A wide range of other orientations are closely linked with this dimension. Societies near the traditional pole emphasize the importance of parent-child ties and deference to authority, along with absolute standards and traditional family values, and reject divorce, abortion, euthanasia, and suicide. These societies have high levels of national pride, and a nationalistic outlook. Societies with secular-rational values have the opposite preferences on all of these topics. The second major dimension of crosscultural variation is linked with the transition from industrial society to post-industrial societies, which brings a polarization between Survival and Selfexpression values. The unprecedented wealth that has accumulated in advanced societies during the past generation means that an increasing share of the population has grown up taking survival for granted. Thus, priorities have shifted from an overwhelming emphasis on economic and physical security toward an increasing emphasis on subjective well-being, self-expression and quality of life. A central component of this emerging dimension involves the polarization between Materialist and Postmaterialist values, reflecting a cultural shift that is emerging among generations who have grown up taking survival for granted. Self-expression values give high priority to environmental protection, tolerance of diversity and rising demands for participation in decision making in economic and political life. These values also reflect mass polarization over tolerance of out-groups, including foreigners, gays and lesbians and gender equality. The shift from survival values to self-expression values also includes a shift in child-rearing values, from emphasis on hard work toward emphasis on imagination and tolerance as important values to teach a child. And it goes with a rising sense of subjective well-being that is conducive to an atmosphere of tolerance, trust and political moderation. Finally, societies that rank high on self-expression values also tend to rank high on interpersonal trust. These two dimensions tap scores of variables [...]. *For technical reasons, early work using factor analysis derived the dimensions from ten of these variables; more recent work uses a 12-item additive index to measure Survival/Self-expression values. Regardless of how it is constructed, the resulting cross-cultural map is so robust that it produces very similar results.* Moreover, using a completely different way of measuring basic values, different types of samples and a different type of dimensional analysis, Schwartz (2006) finds very similar transnational groupings among 76 countries." (Inglehart and Welzel, 2010a)

In contrast to Inglehart, we also include background variables, like gender, age and education. Thus, our analysis is not anymore a "gender-free zone":

Our re-analysis is based on 68 countries, and there is a good number of Muslim societies or predominantly Muslim societies among them [the Organization of Islamic Cooperation members Albania; Azerbaijan; Bangladesh; Bosnia and Herzegovina; Indonesia; Jordan; Kyrgyzstan; Mali; Morocco; Tanzania; Turkey; Uganda]. The smallest n for any of our items is more than 180.000 persons and our total analysis is based finally on none the less than 92289 persons with complete data covering all the 30 variables from around the globe. No substitution of missing values had to be carried out. The SPSS XXI statistics program with the default options for factor analysis/promax rotation was applied. Independent of culture, religion or moral convictions, any researcher around the globe could arrive at the same results as we did.

Table 5.1. The research designs compared

	Inglehart and associates	Our re-analysis
choice and naming of the dimensions	ex ante, perhaps controversial, for example that survival values (and not tradition values) include the divorce and abortion item and the limits on selling foreign goods item, while the item on incomes and jobs, currently listed in the traditional values dimensions, very plausibly might be listed under the survival dimension as well	ex post, as suggested by the empirical results; the results were compared to the literature
factor analytical design	standard factor analysis using orthogonal varimax rotation (does not allow for correlations between the factors)	more advanced versions of factor analysis, allowing for relationships between the factors (Promax rotation)
inclusion of socio-economic background variables like age, education in the model	no	yes
n of countries, considered in the analysis	65	68
number of variables in the original analysis	22 variables/later reduced to 10	30
results are based on items, available for at least the following number of respondents	146789	180041
% of total variance explained at the individual analysis level	26% (traditional values) + 13% (survival values)	47. 89%

Table 5.2. The choice of our variables from the *World Values Survey*

Variable label	Interpretation of the variable according to the highest numerical value	n =
Age	Age	247978
Competition good or harmful	Competition good or harmful	203976
Confidence: Armed Forces	[No] Confidence: Armed Forces	231665
Confidence: The Press	[No] Confidence: The Press	236529
Highest educational level attained	Highest educational level attained	230283
How important is God in your life	How important is God in your life	240112
How often do you attend religious services	[Never attend religious services. Scale:] How often do you attend religious services	238981

Table 5.2. (Continued)

Variable label	Interpretation of the variable according to the highest numerical value	n =
Immigrant policy (prevent people from coming)	Immigrant policy (prevent people from coming)	187066
Important child qualities: determination and perseverance	Important child qualities: determination and perseverance	247782
Important child qualities: feeling of responsibility	Important child qualities: feeling of responsibility	255656
Important child qualities: hard work	Important child qualities: hard work	253331
Important child qualities: imagination	Important child qualities: imagination	252238
Important child qualities: independence	Important child qualities: independence	255656
Important child qualities: obedience	Important child qualities: obedience	255656
Important child qualities: religious faith	Important child qualities: religious faith	253503
Important child qualities: thrift, and saving money and things	Important child qualities: thrift, and saving money and things	255656
Important child qualities: tolerance and respect for other people	Important child qualities: tolerance and respect for other people	255656
Important child qualities: unselfishness	Important child qualities: unselfishness	252238
Justifiable: avoiding a fare on public transport	Justifiable: avoiding a fare on public transport	224394
Justifiable: cheating on taxes	Justifiable: cheating on taxes	232012
Justifiable: claiming government benefits even if one is not entitled to them	Justifiable: claiming government benefits	230882
Justifiable: someone accepting a bribe	Justifiable: someone accepting a bribe	243824
Most people can be trusted [highest numerical value: you just can't be too careful])	Lack of social capital (Most people can be trusted [highest numerical value: you just can't be too careful])	246798
Neighbors: Immigrants/foreign workers	[Rejecting] Neighbors: Immigrants/foreign workers	225868
Neighbors: People of a different race	[Rejecting] Neighbors: People of a different race	231410
Political system: having a democratic political system	Political system: (It's very bad] having a democratic political system	193889
Satisfaction with your life	Satisfaction with your life	252679
self-positioning in political scale (scale 1-left to 10-right)	[Right wing] self-positioning in political scale (scale 1-left to 10-right)	180041
Sex (Gender) [in multivariate analysis: female] (1=male; 2=female)	Sex (Gender) [in multivariate analysis: female] (1=male; 2=female)	252941
University is more important for a boy than for a girl	Rejecting sexist position: University is more important for a boy than for a girl	207655

The WVS data we used correspond to 88.96% of the total global population and 84.75% of the world's Muslim population. The complete data matrix for the re-analysis of the Inglehart world map of human values is based on the sample of 56.89% of the global population and 56.16% of the global Muslim population. Table 5.2 listed the final variables of our investigation. Table 5.3 now lists the indicators, their data range, and the original wording in the *World Values Survey* questionnaire. Our readers should thus be in a better position to be able to follow our naming of the different factors, which reproduce the underlying correlation matrix for the data.

Table 5.3. Description of the variables of our model according to the *World Values Survey*

Indicator	range	maximum numerical value	wording
Age	15 to 65	65+	V237. - This means you are __ __ years old.
Competition good or harmful	1 to 10	10 (harmful)	V119. - Now I'd like you to tell me your views on various issues. How would you place your views on this scale? 1 means you agree completely with the statement on the left; 10 means you agree completely with the statement on the right; and if your views fall somewhere in between, you can choose any number in between. Sentences: Competition is good. It stimulates people to work hard and develop new ideas vs. Competition is harmful. It brings the worst in people
[No] Confidence: Armed Forces	1 to 4	4 (none at all)	V132. - I am going to name a number of organizations. For each one, could you tell me how much confidence you have in them: is it a great deal of confidence, quite a lot of confidence, not very much confidence or none at all? The armed forces
[No] Confidence: The Press	1 to 4	4 (none at all)	V133. - I am going to name a number of organizations. For each one, could you tell me how much confidence you have in them: is it a great deal of confidence, quite a lot of confidence, not very much confidence or none at all? The press
Highest educational level attained	1 to 9	9 University with degree/Higher education - upper-level tertiary certificate	V238. - What is the highest educational level that you have attained? (use functional equivalent of the following, in given society; IF STUDENT, CODE HIGHEST LEVEL HE/SHE EXPECTS TO COMPLETE):
How important is God in your life	1 to 10	10 (very important)	V192. - How important is God in your life? Please use this scale to indicate- 10 means very important and 1 means not at all important.
[Never attend religious services. Scale:] How often do you attend religious services	1 to 7	7 (never, practically never)	V186. - Apart from weddings, funerals and christenings, about how often do you attend religious services these days?
Immigrant policy	1 to 4	4 (prohibit people from coming)	V124. - How about people from other countries coming here to work. Which one of the following do you think the government should do?
Important child qualities: determination and perseverance	0 to 1	1 (important)	determination and perseverance
Important child qualities: feeling of responsibility	0 to 1	1 (important)	feeling of responsibility
Important child qualities: hard work	0 to 1	1 (important)	hard work

Table 5.3. (Continued)

Indicator	range	maximum numerical value	wording
Important child qualities: imagination	0 to 1	1 (important)	imagination
Important child qualities: independence	0 to 1	1 (important)	independence
Important child qualities: obedience	0 to 1	1 (important)	obedience
Important child qualities: religious faith	0 to 1	1 (important)	religious faith
Important child qualities: thrift, and saving money and things	0 to 1	1 (important)	thrift, and saving money and things
Important child qualities: tolerance and respect for other people	0 to 1	1 (important)	tolerance and respect for other people
Important child qualities: unselfishness	0 to 1	1 (important)	unselfishness
Justifiable: avoiding a fare on public transport	1 to 10	10 (always justifiable)	V199. - Please tell me for each of the following statements whether you think it can always be justified, never be justified, or something in between, using this card. (Read out statements. Code one answer for each statement). Avoiding a fare on public transport
Justifiable: cheating on taxes	1 to 10	10 (always justifiable)	V200. - Please tell me for each of the following statements whether you think it can always be justified, never be justified, or something in between, using this card. (Read out statements. Code one answer for each statement). Cheating on taxes if you have a chance
Justifiable: claiming government benefits even if one is not entitled to them	1 to 10	10 (always justifiable)	V198. - Please tell me for each of the following statements whether you think it can always be justified, never be justified, or something in between, using this card. (Read out statements. Code one answer for each statement). Claiming government benefits to which you are not entitled
Justifiable: someone accepting a bribe	1 to 10	10 (always justifiable)	V201. - Please tell me for each of the following statements whether you think it can always be justified, never be justified, or something in between, using this card. (Read out statements. Code one answer for each statement). Someone accepting a bribe in the course of their duties
Lack of social capital (Most people can be trusted [highest numerical value: you just can't be too careful])	1 to 2	2 (you can't be too careful)	V23. - Generally speaking, would you say that most people can be trusted or that you need to be very careful in dealing with people?

Indicator	range	maximum numerical value	wording
[Rejecting] Neighbors: Immigrants/foreign workers	0 to 1	1 (mentioned)	On this list are various groups of people. Could you please sort out any that you would not like to have as neighbors?
[Rejecting] Neighbors: People of a different race	0 to 1	1 (mentioned)	On this list are various groups of people. Could you please sort out any that you would not like to have as neighbors?
Political system: [It's very bad] having a democratic political system	1 to 4	4 (very bad)	V151. - I'm going to describe various types of political systems and ask what you think about each as a way of governing this country. For each one, would you say it is a very good, fairly good, fairly bad or very bad way of governing this country? Having a democratic political system
Satisfaction with your life	1 to 10	10 (satisfied)	V22. - All things considered, how satisfied are you with your life as a whole these days? Please use this card to help with your answer.
[Right wing] self-positioning in political scale (scale 1-left to 10-right)	1 to 10	10 (right wing)	V114. - In political matters, people talk of "the left" and "the right." How would you place your views on this scale, generally speaking?
Sex (Gender) [in multivariate analysis: female] (1=male; 2=female)	1 to 2	2 (female)	V235. - Sex (Gender)
Rejecting sexist position: University is more important for a boy than for a girl	1 to 4	4 (strongly disagree)	V62. - For each of the following statements I read out, can you tell me how much you agree with each. Do you agree strongly, agree, disagree, or disagree strongly? A university education is more important for a boy than for a girl

Table 5.4 provides an overview of the sample sizes of *World Values Survey* respondents with complete data for all the chosen 30 variables.

Table 5.4. The Sample size of *World Values Survey* respondents with complete data for all 30 variables of our investigation

Country/region	n =
Albania	1297
Azerbaijan	860
Argentina	2031
Australia	2750
Bangladesh	1033
Armenia	1154
Bosnia and Herzegovina	1809
Brazil	2161
Bulgaria	953
Belarus	861
Canada	2832
Chile	2109
Taiwan	1820
Cyprus	932
Czech Republic	835
Dominican Republic	276

Country/region	n =
Ethiopia	950
Estonia	685
Finland	1552
Georgia	2042
Germany	2953
Ghana	718
Guatemala	748
Hong Kong	785
Hungary	478
India	2358
Indonesia	1112
Italy	597
Jordan	300
South Korea	1156
Kyrgyzstan	883
Latvia	772
Lithuania	489
Mali	595
Mexico	2299
Moldova	1809
Morocco	278
New Zealand	1022
Nigeria	1494
Norway	2016
Peru	1864
Philippines	2174
Poland	543
Puerto Rico	1378
Romania	1433
Russian Federation	821
Slovakia	777
Viet Nam	1830
Slovenia	1181
South Africa	6086
Zimbabwe	667
Spain	2336
Sweden	1639
Switzerland	910
Thailand	1452
Trinidad and Tobago	597
Turkey	945
Uganda	759
Ukraine	1186
Macedonia	1176
Tanzania	671
United States	3317
Burkina Faso	725
Uruguay	1075
Venezuela	1634
Serbia and Montenegro	1975
Zambia	775
Serbia	559
Total	**92289**

So our re-analysis of Inglehart's theories is based on 68 countries. The smallest n for any of our single variables is slightly more than 180.000 persons. For 92289 persons around the globe, there are complete data for all the 30 variables available in the WVS database (http://www.wvsevsdb.com/wvs/WVSData.jsp) at once.[31]

5.2. A "Light" Version of the Results and Their Immediate Consequences for the General Public: The Kuznets Curves of Human Values along the Paths of Human Development

Since the content of Chapter 5.3, presenting the factor analytical model in great detail will be only of special interest to the social scientific audience of this book, we present here for the general public a "light" version of the results and their immediate consequences.

We already highlighted that our model, based on the usual standard criteria of factor analysis and the promax rotation, explains 47.89% of the total variance of all the 30 variables. For an analysis of original survey data from more than 90.000 global representative interview partners, this percentage must be considered as high.

In Table 5.5, we highlight the relationships between the original 30 variables and the newly derived factor analytical dimensions. As it is well-known from the methodological literature on the subject, the naming of the different factors might sound arbitrarily. The best proposed solution, suggested by the literature, is to provide readers with a full unbiased original account of all the loadings from the rotated structure matrix and also to mention the table of the correlations between the newly derived factors (to cite some classical and contemporary literature on the subject, see: Child, 2006; Diena et al., 2005; Finch, 2006; Jackson and Borgatta, 1981; Kim and Mueller, 1978a, 1978b; Thompson, 2004; Ueberla, 1970). We think that our naming of the factors, based on the structure of the loadings, represents a good choice, compatible with the results of a large body of world-values-related literature, quoted in our bibliography. But as in any factor analysis, the naming of the factors is indeed subjective and open to criticism. But we claim that from the given lists of results, the choice taken by us is a good one.

We also should highlight that a loading (i. e., correlation coefficient of a factor with an underlying variable) of, say, 0.400 only represents 16% of variance of the original variable explained, and 0.500 only implies a percentage of 25% of explained variance, while 0.600 already implies 36% of explained variance, and 0.900 would imply 81% of total variance explained et cetera. Readers are invited to compare the loadings with our proposed naming of the factors:

a) economic permissiveness
b) traditional religion
c) racism
d) higher education for the younger generation (education gap between the generations)
e) distrust of the army and the press
f) authoritarian character
g) tolerance and respect

[31] WVS FIVE WAVE AGGREGATED FILE 1981-2005 (NEW) available at http: //www. wvsevsdb. com/wvs/WVSData. jsp

h) the 'ego' company (i. e., the rejection of obedience and unselfishness as values in education)
i) [predominantly] female rejection of the market economy and democracy

Let us now look at the trajectory of the factor scores along the path of Human Development (measured by the well-known UNDP Human Development Indicator). Norris and Inglehart, 2004 link the UNDP Human Development Index directly with the concept of *"human security"*. They maintain that human security will be increasing with a rising UNDP Human Development Index.

By tracking the progress of the value indicators along the axis of the Human Development Indicator, we discover that value change in the societies of this world is far from linear. Our new factors, best representing the core items of the *World Values Survey*, as a rule present a curve-linear relationship with the UNDP-Human Development Index (UNDP HDI). This is an important new qualification in the debate about global development.

We should add here that there is not yet a widely accepted definition of the concept of Human Security. While definitions vary, their common ground remains the individual, not the state. There is a clear difference between the individual security and the state security.

According to the 2009 UNDP Human Development Report,

> "Most contemporary Arab writers express the belief in an unbreakable bond between individual human security and national security on the one hand and human security and external military threats on the other. Other writers believe that the authoritarian state is the source of the greatest threats to human security in the region (...) However, such writers do not maintain that the achievement of human security rules out state security. Rather, they hold that the achievement of individual human security will have the effect of transforming the authoritarian state into one that respects the rule of law. This shift may require, among other things, the reform of existing security apparatuses or possibly even the establishment of new security institutions." (UNDP, Arab Human Development Report, 2009).

It is noteworthy that the UNDP's report, drawing on the concerns of the Arab thinkers, defines Human Security as *"the liberation of human beings from those intense, extensive, prolonged, and comprehensive threats to which their lives and freedom are vulnerable."*

However, while the UNDP's report relies primarily on the internationally recognized dimensions of human security, as identified in the 1994 global Human Development Report, it acknowledges that *"any identification of threats to Arab human security needs to reflect the status of the region as a highly diverse area which, through most periods of its history, has been the object of conflict among the world's superpowers."*

Applying this type of reasoning, we might expect the following to happen in the process of modernization:

- Traditional religion diminishes in its societal role, as human security is rising
- Tolerance and respect, distrust of the army and the press, the "ego-company" orientation, and the [predominantly] female rejection of the market economy and democracy all rise with human security

- There is an inverted U-shaped relationship between human security and permissiveness/pessimism, racism, the education gape between the generations, and the authoritarian character

Graph 5.1a and 5.1b highlight these strongly non-linear relationships:

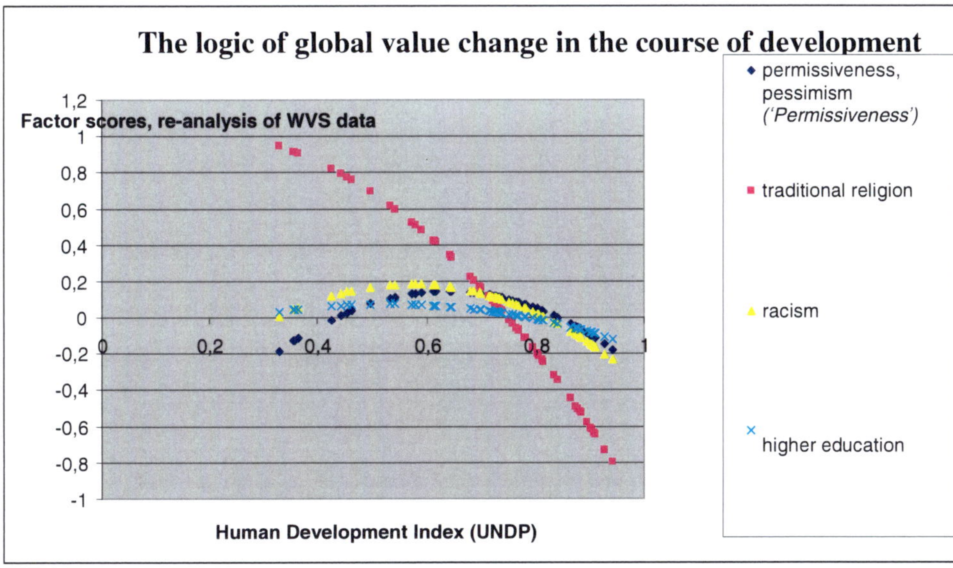

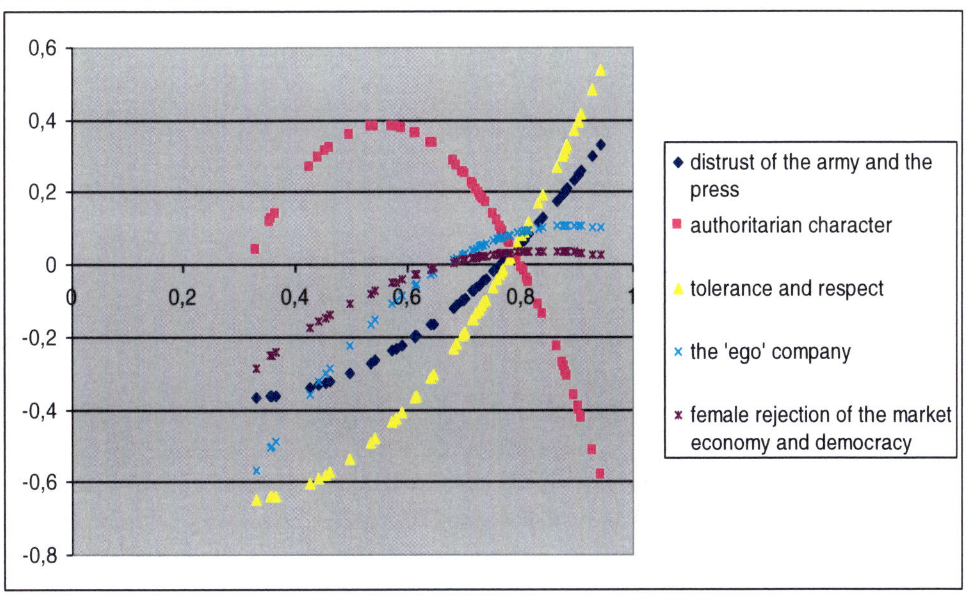

Graph 5.1a and Graph 5.1b. The global trajectory of human values along the scale of the Human Development Index.

In Inglehart's model, the two dimensions – traditional values and survival values - are independent from one another, whereas the promax model allows for correlations between the factors.

We will now look at the factor loadings above +- 0.500 and we will try to debate the similarities, but also differences in the research results with earlier approaches.

Economic permissiveness has its higher loadings with the WVS items

- Justifiable: cheating on taxes
- Justifiable: avoiding a fare on public transport
- Justifiable: someone accepting a bribe
- Justifiable: claiming government benefits even if one is not entitled to them

Traditional religion expresses the higher factor loadings on the items

- How important is God in your life
- Important child qualities: religious faith
- Negative loading [Never attend religious services. Scale:] How often do you attend religious services

Racism

[Rejecting] Neighbors: Immigrants/foreign workers

- [Rejecting] Neighbors: People of a different race
- Immigrant policy (prevent people from coming)

Higher education of the younger generation (education gap between the generations)

- Highest educational level attained
- Rejecting sexist position: University is more important for a boy than for a girl
- Negative loading Age
- Negative loading Important child qualities: thrift, and saving money and things

Distrust of the army and the press

- [No] Confidence: The Press
- [No] Confidence: Armed Forces
- Negative loading: [Right wing] self-positioning in political scale (scale 1-left to 10-right)

Authoritarian character

- Lack of social capital (Most people can be trusted [highest numerical value: you just can't be too careful])
- Important child qualities: hard work

- Important child qualities: obedience
- Negative loading: Important child qualities: imagination
- Negative loading: Important child qualities: independence

Tolerance and respect + post-materialism

- Important child qualities: tolerance and respect for other people
- Important child qualities: feeling of responsibility
- Rejecting sexist position: University is more important for a boy than for a girl
- Negative loading: Important child qualities: hard work

The 'ego' company (rejection of obedience + unselfishness)

- Important child qualities: feeling of responsibility
- Important child qualities: independence

- Negative loading: Important child qualities: unselfishness
- Negative loading: Important child qualities: obedience

[Predominantly] female rejection of the market economy and democracy

- Sex (Gender) [in multivariate analysis: female] (1=male; 2=female)
- Competition good or harmful
- Political system: (It's very bad] having a democratic political system

Traditional religion reflects very well the results from all the three mentioned major research traditions in cross-national value research today. This is a factor, whose relationship with these research traditions is very clear, and whose significance can be easily interpreted also from the viewpoint of Hofstede's, Inglehart's and Schwartz/Davidov's theory.

Economic permissiveness clearly captures the dimension of lawlessness, moral-ethical decay and the shadow economy, so prominent in contemporary economic theory of growth. It is completely lacking in the hitherto published analyses by Hofstede, Schwartz/Davidov and Inglehart. Yet it is the most important of all the resulting factors today. To have overlooked this dimension is a major still existing shortcoming of contemporary cross-national quantitative value research. As Graph 5.1a and Graph 5.1b show, this factor presents a clear U-shaped function of the development level, so well-known from economic literature ever since the discoveries of Simon Kuznets (1955) about the relationship between income levels and economic inequalities. We foresee a culmination of crises tendencies, manifesting themselves in lawlessness, racism, the educational generation gap, and the authoritarian character, all at middle development levels, coinciding with the maximum points of income inequality in a nation's history. As the power of traditional religion fades away in modern developed society, not only positive processes (like the increase in tolerance) set in: we also witness a rising distrust in the army and the press, and a growing and only slowly decreasing level of [predominantly] female rejection of the market economy and democracy.

Inglehart expects that with a rising level of capitalist development, the values of *'self-expression'* gain upper hand, and that the religious element is basically a human answer to the problem of a lack of security. If human security increases, religion slowly recedes. But economic permissiveness, racism, the educational generation gap, and the authoritarian character all first increase with rising levels of development and increasing secularization. Only at very high levels of development racism and economic permissiveness diminish again. In a cultural shift, well explainable by Inglehart's theories, tolerance increases as the social power of religion recedes; but as traditional religion diminishes, distrust of the army and the press, and the [predominantly] female rejection of the market economy and democracy increase. So we are far less optimistic than the hitherto existing theories.

Comparable to Hofstede's LTO-factor, we also detect a factor, which we call 'the ego company', which is especially strong in the countries of East Asia.

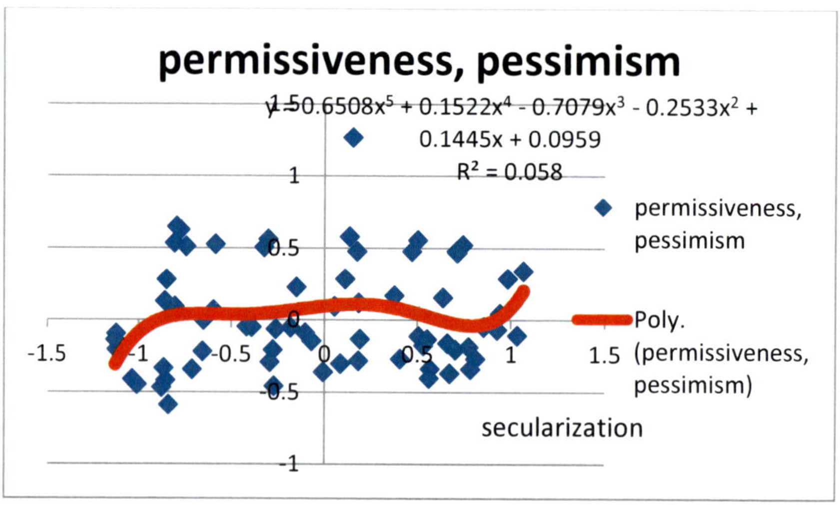

X-axis: secularization (= our factor "traditional religion"*-1); y-axis: economic permissiveness.

Graph 5.2. Economic permissiveness (lowest value: Bangladesh, highest value: Serbia, y-axis) as a function of secularization (lowest value: Nigeria; highest value: Russia, x-axis) in our own theoretical system .

Traditional religion is linked in a very complex way to the absence of economic permissiveness. An interesting question for future research would be which countries are exceptional performers ("residuals") on each stage of secularization.

5.3. The Final Factor Analytical Model : A Presentation for the Specialists

The general public might skip over this Chapter and directly go to Chapter 5.4. We will start our presentation of the final factor analytical results by looking at the correlations between the nine factors. The final diagram about the noticeable interrelationships between the various factors of our model is described in Graph 5.4:

The methodological literature already referred to above suggests that the assumption of non-correlation between the factors, inherent in Inglehart's varimax-rotation model, is unrealistic. We have highlighted the correlations above 0.100 in our Table in **bold letters,** to allow our readers a closer look at the underlying relationships between the factors.

The following Tables and Graphs are directed towards the specialists. The general public might simply gaze over the following Tables and Graphs, while the specialists are invited to take a closer look at the following materials.

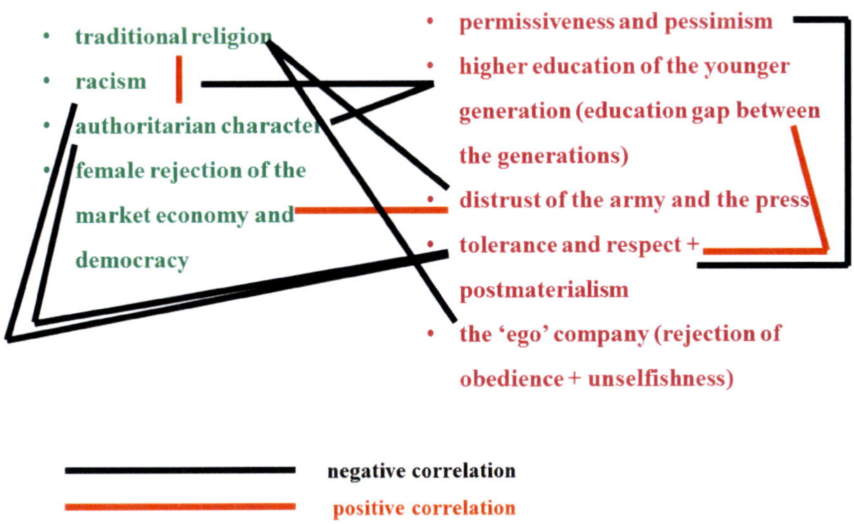

Graph 5.3. Final summarizing model, based on the correlations between the nine factors.

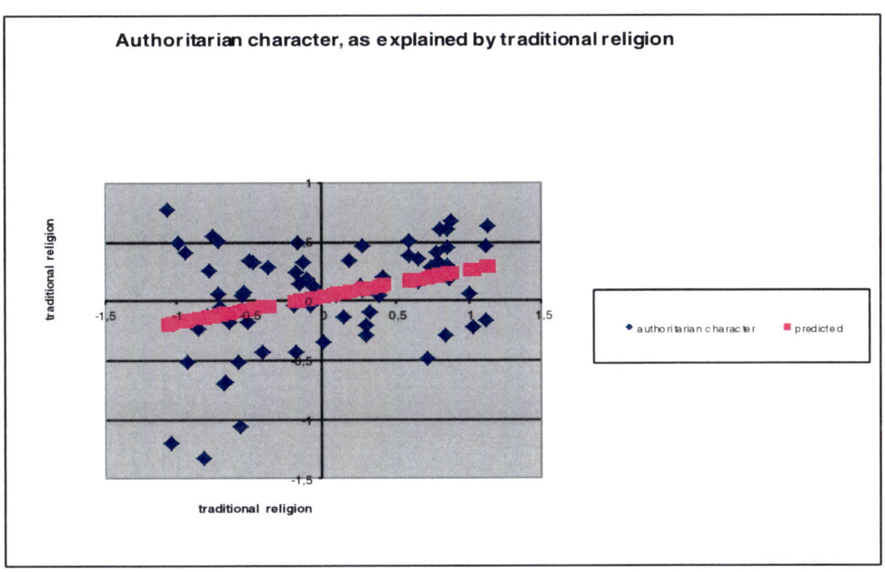

Graph 5.4.

Policy pointer:

Far from conforming to the expectations of contemporary atheism and secularism, the correlations of our factors, reflecting the individual responses of more than 90.000 global representative interview partners around the globe show that the effects of traditional religion nowhere exceed 5% of variance explained. Even if traditional religion is weakly and positively associated with the "authoritarian character", this association is much weaker than expected by many and leaves 95% of the variance unexplained. Religious or highly religious societies with a remarkable low or only medium authoritarian character are[32]

Ethiopia
Bangladesh
Indonesia
Jordan
Thailand
United States
Mexico
Tanzania
Poland
Puerto Rico
Dominican Republic

More religious societies with remarkable authoritarian tendencies are:

Zimbabwe
Trinidad and Tobago
Uganda
Turkey
Nigeria
Brazil
Burkina Faso
Peru
Philippines
Ghana
South Africa

Relatively secular societies with more than usual authoritarian tendencies are:

Russian Federation
Belarus
Czech Republic
Latvia
Estonia
Ukraine

[32] Microsoft EXCEL Trend analysis, based on the data of this work. The scatterplot between the two variables is the following:

Armenia
Bulgaria
Lithuania
Hungary
Serbia and Montenegro
Spain
Taiwan
Viet Nam

The correlations between the most important factors are the following:

	economic permissiveness	**traditional religion**
traditional religion	-0,051	
racism	0,067	0,063
higher education of the younger generation (education gap between the generations)	0,010	-0,036
distrust of the army and the press	0,058	**-0,122**
authoritarian character	0,055	**0,190**
tolerance and respect + post-materialism	**-0,175**	-0,090
the 'ego' company (rejection of obedience + unselfishness)	-0,046	**-0,173**
[predominantly] female rejection of the market economy and democracy	*0,093*	*0,034*

The methodological literature in Chapter 3 of this book already highlighted the importance of the *Eigen values* as a defining mathematical benchmark of a principal components/factor analysis. Graph 5.4 shows the *Eigen value*s of our investigation; with the first two factors way past any linear continuation of the factors 10 – 29, shown in the x-axis.

Thus eight factors,

- economic permissiveness
- traditional religion
- racism
- higher education of the younger generation (education gap between the generations)
- distrust of the army and the press
- authoritarian character
- tolerance and respect + post-materialism
- the 'ego' company (rejection of obedience + unselfishness)

are not only well above *Eigen value* 1.0, but also way above the linear continuation of the *Eigen values* of factors 9 – 19. Following the simple standard procedure of analyzing the *Eigen values,* we suggest to treat the results for factor nine with some caution at least. Although its *Eigen value* is still above 1.0, its numerical vicinity to the low *Eigen values* of factors 10 – 19 is clearly visible. In all, our model explains some 47% of the total variance of the correlation matrix of the data for 30 variables from more than 90.000 interview partners of the *World Values Survey*, with two factors explaining more than 17% of the total variance in between them already – economic permissiveness, and traditional religion, the two defining processes of global values and global value change today.

We have highlighted all factor loadings from the rotated structure matrix (*Promax rotation* according to the SPSS 20 with *Kaiser normalization)* in different typing and shadings in order to facilitate our readers to arrive at their own independent opinions about our results (provided that they think that the original *World Values Survey* data are credible and unbiased).

We should emphasize the point that the correlations between the factors are not correlations between aggregations at the country level but reflect the correlations between the factors, to be extracted from the data *at the individual level* of the more than 90. 000 interview partners across countries and across cultures. Under such conditions, correlations between factors of more than +-.100 are already to be considered high. Nevertheless, we have to emphasize that the relationships between the factors with *Eigen values* of 2.0 or above (economic permissiveness, and traditional religion), with the rest of the nine factors under consideration here in no way sufficiently firmly confirm widespread expectations of contemporary secularism and atheism directed against the religious phenomenon.

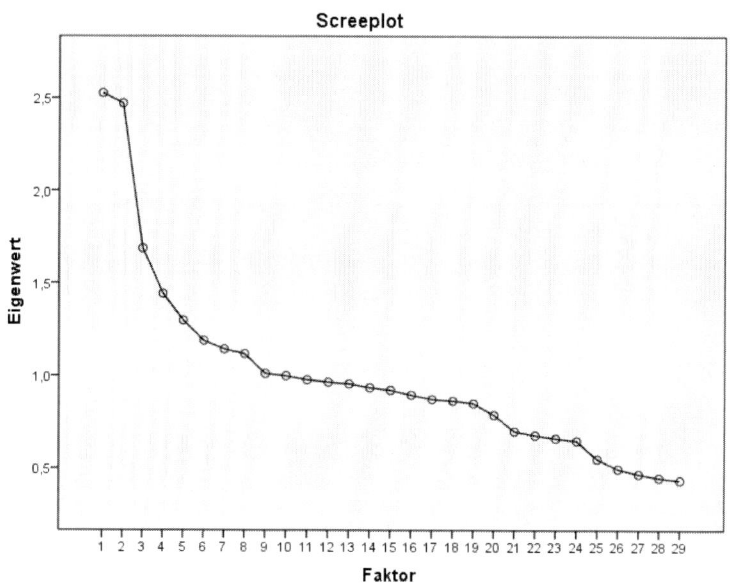

	Eigen value	% of variance explained	Cumulated percentage
economic permissiveness	2,526	8,711	8,711
traditional religion	2,472	8,523	17,234
racism	1,688	5,822	23,056
higher education of the younger generation	1,442	4,974	28,029
distrust of the army and the press	1,298	4,475	32,504
authoritarian character	1,189	4,099	36,604
tolerance and respect	1,143	3,942	40,545
the 'ego' company	1,118	3,854	44,399
[predominantly] female rejection of the market economy and democracy	1,012	3,489	47,888

Graph 5.4. Screeplot for our factor analysis.

Table 5.5. Structure matrix – Promax with Kaiser normalization

	Economic permissiveness	**Traditional religion**	**Racism**	**Higher education of the younger generation (education gap between the generations)**	**Distrust of the army and the press**	**Authoritarian character**	**Tolerance and respect + post-materialism**	**The 'ego' company (rejection of obedience + unselfishness)**	***[Predominantly] female rejection of the market economy and democracy***
Age	-0,169	-0,071	-0,026	**-0,680**	-0,022	-0,037	0,195	0,114	*-0,135*
Competition good or harmful	0,155	-0,059	0,020	-0,163	0,064	-0,093	-0,076	-0,101	***0,556***
[No] Confidence: Armed Forces	0,079	-0,087	-0,054	0,157	**0,757**	-0,035	0,014	0,032	*0,111*
[No] Confidence: The Press	-0,016	-0,090	-0,076	-0,020	**0,759**	-0,010	0,120	0,039	*-0,007*
Highest educational level attained	-0,041	-0,098	-0,094	**0,648**	0,078	-0,195	0,165	0,158	*-0,136*
How important is God in your life	-0,039	**0,813**	0,035	-0,026	-0,123	0,200	-0,076	-0,106	*0,054*
[Never attend religious services. Scale:] How often do you attend religious services	0,018	**-0,800**	-0,064	0,033	0,138	-0,119	0,108	0,080	*-0,021*
Immigrant policy (prevent people from coming)	0,003	0,089	***0,302***	-0,191	0,107	0,227	0,097	0,044	*0,095*
Important child qualities: determination and perseverance	0,002	-0,232	0,015	0,282	-0,003	-0,104	0,114	0,114	*-0,192*
Important child qualities: feeling of responsibility	-0,043	-0,186	-0,044	0,071	-0,009	-0,028	***0,462***	***0,408***	*0,071*
Important child qualities: hard work	0,007	0,033	-0,023	-0,029	-0,200	***0,437***	**-0,500**	0,256	*-0,148*
Important child qualities: imagination	0,000	-0,098	-0,022	0,159	0,082	**-0,613**	-0,074	0,005	*0,032*
Important child qualities: independence	-0,020	-0,133	0,057	0,207	0,067	**-0,508**	0,007	***0,353***	*0,031*
Important child qualities: obedience	0,030	0,268	0,035	-0,158	-0,043	***0,300***	-0,201	**-0,562**	*0,031*
Important child qualities: religious faith	-0,028	**0,741**	0,041	-0,065	-0,038	0,166	-0,082	-0,169	*-0,007*
Important child qualities: thrift, and saving money and things	0,006	**-0,069**	0,080	-0,338	-0,097	0,212	-0,282	0,260	*0,170*

Table 5.5. (Continued)

	Economic permissiveness	**Traditional religion**	**Racism**	**Higher education of the younger generation (education gap between the generations)**	**Distrust of the army and the press**	**Authoritarian character**	**Tolerance and respect + post-materialism**	**The 'ego' company (rejection of obedience + unselfishness)**	***[Predominantly] female rejection of the market economy and democracy***
Important child qualities: tolerance and respect for other people	-0,064	-0,029	-0,104	-0,009	0,026	0,033	**0,623**	-0,032	*-0,072*
Important child qualities: unselfishness	-0,008	-0,013	-0,024	0,063	-0,016	-0,107	0,131	**-0,644**	*0,031*
Justifiable: avoiding a fare on public transport	**0,779**	-0,059	-0,002	0,079	0,066	0,021	-0,078	-0,026	*0,112*
Justifiable: cheating on taxes	**0,791**	-0,103	0,020	0,047	0,084	0,018	-0,089	0,021	*0,026*
Justifiable: claiming government benefits even if one is not entitled to them	**0,716**	0,006	0,059	-0,003	0,017	0,041	-0,112	-0,053	*0,132*
Justifiable: someone accepting a bribe	**0,736**	0,015	0,070	0,001	0,006	0,049	-0,132	-0,024	*0,058*
Lack of social capital (Most people can be trusted [highest numerical value: you just can't be too careful])	0,037	0,149	0,073	0,085	0,164	**0,562**	-0,238	0,058	*0,121*
[Rejecting] Neighbors: Immigrants/foreign workers	0,039	0,025	**0,845**	-0,064	-0,076	0,045	-0,113	0,018	*0,003*
[Rejecting] Neighbors: People of a different race	0,043	0,059	**0,827**	-0,055	-0,115	0,026	-0,161	0,006	*-0,008*
Political system: (It's very bad] having a democratic political system	0,175	-0,106	0,097	-0,087	0,136	0,246	-0,239	-0,009	***0,458***
[Right wing] self-positioning in political scale (scale 1-left to 10-right)	0,005	0,193	0,164	-0,063	***-0,300***	0,045	-0,035	0,095	*-0,145*
Sex (Gender) [in multivariate analysis: female] (1=male; 2=female)	-0,079	0,165	-0,045	0,167	-0,021	-0,011	0,216	0,061	***0,661***
Rejecting sexist position: University is more important for a boy than for a girl	-0,153	-0,064	-0,182	***0,384***	0,087	-0,160	***0,399***	-0,042	*0,177*

Table 5.6. The correlations between the factors

				Matrix of components				
	Economic permissiveness	**Traditional religion**	**Racism**	**Higher education of the younger generation (education gap between the generations)**	**Distrust of the army and the press**	**Authoritarian character**	**Tolerance and respect + post-materialism**	**The 'ego' company (rejection of obedience + unselfishness)**
traditional religion	-0,051							
racism	0,067	0,063						
higher education of the younger generation (education gap between the generations)	0,010	-0,036	**-0,103**					
distrust of the army and the press	0,058	**-0,122**	-0,069	0,084				
authoritarian character	0,055	**0,190**	0,064	**-0,192**	-0,042			
tolerance and respect + post-materialism	**-0,175**	-0,090	**-0,144**	**0,119**	0,072	**-0,276**		
the 'ego' company (rejection of obedience + unselfishness)	-0,046	**-0,173**	0,034	0,058	-0,028	0,014	-0,004	
[predominantly] female rejection of the market economy and democracy	*0,093*	*0,034*	*0,018*	*0,030*	***0,117***	*0,045*	*-0,005*	*-0,027*

Table 5.7. The country means for the different new factors – mapping the new map of global values on earth

	Economic permissiveness	Traditional religion	Racism	Higher education of the younger generation (education gap between the generations)	Distrust of the army and the press	Authoritarian character	Tolerance and respect + post-materialism	The 'ego' company (rejection of obedience + unselfishness)	*[Predominantly] female rejection of the market economy and democracy*
Albania	0,094	-0,056	-0,137	-0,201	0,214	0,114	-0,149	-0,060	*-0,229*
Argentina	-0,090	0,101	-0,369	-0,103	0,557	0,050	0,067	0,037	*0,147*
Armenia	0,476	-0,471	-0,022	0,237	-0,060	0,332	-0,596	0,287	*-0,030*
Australia	-0,340	-0,569	-0,387	0,004	0,275	-0,516	0,554	-0,241	*-0,166*
Azerbaijan	0,283	-0,112	-0,135	0,297	0,204	0,183	-0,504	0,683	*-0,169*
Bangladesh	-0,589	0,839	1,712	-0,022	-0,854	-0,286	-0,626	0,457	*-0,463*
Belarus	0,520	-0,746	-0,413	-0,035	0,000	0,555	-0,508	0,323	*0,010*
Bosnia and Herzegovina	-0,304	-0,087	0,000	0,010	-0,128	-0,044	-0,083	-0,143	*-0,172*
Brazil	0,525	0,583	-0,373	-0,071	-0,037	0,503	-0,153	-0,429	*0,346*
Bulgaria	-0,188	-0,776	-0,037	0,028	-0,282	0,257	-0,258	0,196	*-0,037*
Burkina Faso	0,128	0,859	-0,440	-0,225	-0,243	0,450	-0,825	-0,710	*-0,228*
Canada	-0,284	-0,183	-0,424	0,038	0,018	-0,437	0,499	-0,112	*-0,003*
Chile	0,225	0,147	-0,139	-0,131	0,223	-0,131	0,209	-0,316	*0,457*
Cyprus	-0,146	0,068	0,040	0,154	-0,026	0,056	0,194	-0,204	*-0,017*
Czech Republic	0,286	-0,986	0,112	-0,448	0,180	0,496	-0,134	0,266	*-0,079*
Dominican Republic	-0,220	0,653	-0,079	0,920	0,273	0,159	0,282	0,064	*-0,030*
Estonia	0,041	-0,944	-0,110	0,095	0,081	0,406	-0,231	0,560	*-0,108*
Ethiopia	-0,343	0,711	-0,114	0,250	0,429	-0,493	-0,759	-0,062	*0,137*
Finland	-0,162	-0,664	-0,046	-0,168	-0,059	-0,680	0,811	-0,018	*0,187*
Georgia	-0,054	0,179	-0,080	0,208	-0,075	0,335	-0,228	0,510	*-0,178*
Germany	-0,068	-0,928	-0,256	-0,268	0,563	-0,521	0,629	0,529	*0,132*
Ghana	-0,098	1,120	0,097	0,054	-0,688	0,461	-0,464	-0,584	*-0,402*
Guatemala	0,535	0,806	-0,354	0,125	0,310	0,236	-0,120	-0,546	*0,424*
Hong Kong	-0,048	-0,856	1,801	-0,193	0,008	-0,236	-0,575	-0,164	*0,222*
Hungary	0,467	-0,716	0,256	-0,132	0,353	0,057	0,455	0,074	*0,075*
India	-0,064	0,260	0,704	-0,044	-0,824	0,133	-0,464	-0,100	*-0,436*
Indonesia	-0,412	1,032	0,591	0,268	-0,366	-0,218	0,017	0,182	*-0,110*
Italy	-0,362	0,005	-0,240	0,093	0,153	-0,346	0,579	0,041	*-0,032*
Jordan	-0,207	1,118	1,329	0,095	-0,824	-0,160	0,058	-0,695	*-0,403*
Kyrgyzstan	0,117	-0,185	0,047	0,433	0,041	0,239	-0,409	0,419	*0,175*
Latvia	0,478	-0,712	-0,115	0,046	0,319	0,511	-0,144	0,561	*0,018*
Lithuania	0,169	-0,376	0,167	0,047	-0,011	0,289	-0,360	0,487	*0,080*
Macedonia	-0,134	-0,191	0,362	0,056	0,508	-0,035	0,458	0,002	*0,058*

	Economic permissiveness	**Traditional religion**	**Racism**	**Higher education of the younger generation (education gap between the generations)**	**Distrust of the army and the press**	**Authoritarian character**	**Tolerance and respect + post-materialism**	**The 'ego' company (rejection of obedience + unselfishness)**	***[Predominantly] female rejection of the market economy and democracy***
Mali	0,509	0,743	0,087	-0,454	-0,619	0,278	-0,879	-0,453	*-0,129*
Mexico	0,505	0,322	0,208	0,075	0,093	-0,097	-0,137	-0,407	*0,307*
Moldova	0,579	-0,136	-0,098	-0,034	0,221	0,322	-0,385	0,359	*0,184*
Morocco	-0,419	0,855	-0,184	-0,086	-0,208	0,291	-0,232	-0,068	*-0,463*
New Zealand	-0,371	-0,675	-0,419	-0,085	0,125	-0,702	0,601	-0,090	*-0,323*
Nigeria	-0,139	1,126	0,040	0,371	0,012	0,640	-0,700	-0,489	*-0,483*
Norway	-0,271	-0,813	-0,288	0,165	0,151	-1,324	0,989	0,165	*-0,088*
Peru	0,067	0,594	-0,188	0,327	0,432	0,387	-0,072	-0,297	*-0,018*
Philippines	0,624	0,775	0,249	0,076	-0,397	0,405	-0,543	-0,009	*0,270*
Poland	-0,051	0,390	-0,080	-0,449	0,093	0,040	0,299	-0,057	*0,496*
Puerto Rico	-0,335	0,865	-0,216	0,231	-0,043	0,181	0,472	-0,163	*0,005*
Romania	-0,209	0,275	0,111	0,008	-0,279	0,084	-0,083	0,514	*-0,374*
Russian Federation	0,339	-1,070	-0,251	-0,174	0,099	0,777	-0,462	0,363	*0,259*
Serbia	1,266	-0,152	0,285	0,078	0,445	0,143	-0,194	0,100	*0,140*
Serbia and Montenegro	-0,178	-0,534	-0,112	-0,037	0,229	0,076	-0,101	0,106	*-0,322*
Slovakia	0,476	-0,176	-0,016	-0,411	-0,042	0,494	-0,139	0,221	*-0,105*
Slovenia	0,156	-0,639	0,029	0,017	0,384	-0,182	0,279	0,100	*0,068*
South Africa	-0,003	0,649	0,124	0,102	0,010	0,350	-0,120	-0,098	*0,022*
South Korea	-0,114	-0,507	0,563	0,272	-0,026	-0,182	-0,431	0,963	*0,376*
Spain	-0,135	-0,554	-0,372	-0,265	0,264	0,039	0,075	-0,158	*0,126*
Sweden	-0,109	-1,036	-0,443	0,045	0,311	-1,205	1,067	0,133	*-0,137*
Switzerland	-0,404	-0,565	-0,364	0,067	0,253	-1,067	0,968	0,296	*-0,153*
Taiwan	-0,206	-0,705	0,191	-0,140	0,325	-0,057	0,180	0,563	*0,210*
Tanzania	-0,448	1,008	0,034	-0,032	-1,063	0,056	-0,437	-0,564	*-0,337*
Thailand	0,561	0,300	0,722	-0,739	0,093	-0,297	-0,051	-0,334	*0,382*
Trinidad and Tobago	0,090	0,804	-0,307	-0,253	0,460	0,601	0,037	-0,256	*0,060*
Turkey	-0,460	0,271	0,300	0,026	-0,218	0,463	-0,217	0,024	*0,083*
Uganda	0,279	0,849	-0,051	0,540	-0,545	0,607	-0,503	-0,478	*-0,256*
Ukraine	0,553	-0,504	-0,219	-0,016	-0,020	0,345	-0,435	0,230	*0,080*
United States	-0,292	0,292	-0,250	-0,050	-0,076	-0,207	0,348	-0,038	*-0,155*
Uruguay	-0,270	-0,406	-0,311	-0,225	0,274	-0,427	0,442	-0,455	*0,422*
Venezuela	-0,047	0,417	0,142	0,159	-0,303	0,199	0,170	-0,369	*0,202*
Viet Nam	-0,341	-0,786	0,496	-0,266	-1,833	-0,122	-0,408	0,173	*-0,292*
Zambia	0,648	0,795	0,403	0,323	-0,086	0,329	-0,584	-0,343	*-0,012*
Zimbabwe	-0,468	0,876	-0,063	0,049	-0,071	0,678	-0,405	-0,704	*-0,144*

Table 5.8. The global map of the major world denominations[33] in the *World Values Survey* project

N	Religious denomination	**Economic permissiveness**	**Traditional religion**	**Racism**	**Higher education of the younger generation (education gap between the generations)**	**Distrust of the army and the press**	**Authoritarian character**	**Tolerance and respect + post-materialism**	**The 'ego' company (rejection of obedience + unselfishness)**	***[Predominantly] female rejection of the market economy and democracy***
75239	**Total number**	**-0,014**	**0,208**	**0,016**	**-0,025**	**-0,048**	**0,037**	**-0,019**	**-0,013**	***0,004***
26269	Roman Catholic	0,062	0,294	-0,112	-0,016	0,024	0,097	0,038	-0,098	*0,109*
12652	Protestant	-0,179	0,249	-0,093	-0,004	0,025	-0,137	0,221	-0,048	*-0,070*
10726	Orthodox	0,110	-0,075	-0,022	0,015	0,082	0,189	-0,157	0,235	*-0,027*
9602	**Muslim**	**-0,144**	**0,484**	**0,300**	**0,014**	**-0,249**	**0,134**	**-0,347**	**-0,025**	***-0,182***
2660	Buddhist	0,238	-0,069	0,561	-0,494	-0,109	-0,198	-0,105	0,025	*0,310*
2256	Hindu	-0,082	0,310	0,622	0,014	-0,661	0,161	-0,374	-0,099	*-0,338*
1270	The Church of Sweden	-0,117	-1,056	-0,430	-0,019	0,272	-1,169	1,066	0,141	*-0,146*
973	Armenian Apostolic Church	0,445	-0,330	0,000	0,244	-0,077	0,302	-0,567	0,250	*-0,013*
783	Ancestral worshipping	-0,349	-0,888	0,791	-0,178	-1,926	-0,134	-0,479	0,313	*-0,379*
599	Independent African Church (e.g., ZCC, Shembe, etc.)	0,019	0,934	0,251	-0,100	-0,358	0,402	-0,356	-0,201	*0,158*
536	Pentecostal	-0,280	1,181	-0,055	0,402	-0,001	0,503	-0,314	-0,388	*-0,276*

[33] It is important to emphasize that these results are only unweighted world averages as they emerge from the *World Values Survey* project. These data say nothing about the value structures of given individuals across the globe and in individual countries adhering to the denominations, listed in our Table. For each denomination, there would be wide standard deviations, which we did not list in Table 5.8 due to reasons of the limitations of space.

Seven factors under consideration here are a new input for the entire global debate about human values:

- the strength of economic permissiveness,
- the clear proof of the existence of a factor describing racism,
- the generation gap in education, connected with value transformation processes,
- the existence of a joint political distrust factor directed against important institutions of society - the army and the press. This factor is also relevant for some highly developed countries, and not only countries in the periphery and semi-periphery of the world system, recovering from years of military dictatorship (like Argentina, Chile *et cetera)*
- the clear proof of an "authoritarian personality" factor, and a
- re-emergence of the Hofstede dimension of "long term orientation", called here "the ego company".

Two factors bear great semblance to the results, achieved by Inglehart and his associates –

- traditional religion and
- tolerance + respect and post-materialism.

Our readers are also invited to look at Table 5.7 for the aggregate country results and at Table 5.8 for the "global map" of human religious denominations. Table 5.8 and its results about factor scores for the major global denominations are way past of what Huntington, 1996 had to say on the clash between civilizations. Of course, we should remind our readers that results based on 500 interview partners have an error margin of up to +-4.4% (see also Graph 2.1, confidence interval 95%).

Policy pointer:

The seven factors under consideration here are a new input for the entire global debate about human values:

- the strength of economic permissiveness,
- the clear proof of the existence of a factor describing racism,
- the generation gap in education, connected with value transformation processes,
- the existence of a joint political distrust factor against both the army and the press, also existing in some countries, considered until recently as highly developed,
- the clear proof of an "authoritarian personality" factor, and a
- re-emergence of the Hofstede dimension of "long term orientation", called here "the ego company".

Two factors bear great semblance to the results, already achieved by Inglehart and his associates:

- traditional religion and
- tolerance + respect and post-materialism.

5.4. The New Choropleth Maps of Global Human Values

Graph 5.1 depicted the Inglehart/Welzel map of human values. The choropleth maps, Maps 5.1 to 5.9, designed by us on the basis of the preceding data, now focus on our analysis and its geographical implications.

Economic permissiveness most strongly affects large parts of Latin America (except Venezuela, Peru, and the countries of the Southern Cone), some, but not all nations of Africa, most of the former USSR and some other countries of East Central and Southeastern Europe, and Thailand and the Philippines. Medium levels of economic permissiveness are to be found – among others – in India, in Germany, Sweden, the Czech Republic and Poland, in Venezuela, Peru, and the countries of the Southern Cone of Latin America, in South Africa, in Ghana *et cetera,* while there were some really shining results to be reported from several African countries. The Anglo-Saxon Western democracies had very good performances on this indicator.

In our factor analysis, *traditional religion* was highest in most of the nations of Africa, and the countries of Latin America with the highest poverty rates (Brazil, Peru, Guatemala, Dominican Republic, Mexico), in the Arab speaking Muslim countries and in Poland. It is noteworthy that Turkey, India and the United States are classified in the same league of religiosity, while the real bastions of secularism are Scandinavia, Russia, Germany and the Czech Republic.

Racism is lowest in most of the old democracies of the West and in the South of Latin America, while in Venezuela, Mexico, in several African countries, in Eastern Europe and in Turkey, middle levels of racism are to be observed. The worst performances to be recorded are to be encountered in Bangladesh, Jordan, and South Korea, Indonesia, and India. Thailand and Vietnam also have higher racism-scores.

The *generational education gaps* are lowest in Thailand, Mali, Poland, the Slovak Republic and the Czech Republic. In most developed western democracies, generational education gaps are also low. In Uganda, Nigeria, Zambia, Peru and Kyrgyzstan we find the highest values on this score.

Distrust of the army and the press is highest in Germany, Spain, Australia, Argentina, Uruguay, Peru, Hungary, Latvia, Ethiopia, Guatemala and the Dominican Republic. Russia and several countries in East and Central Europe, but also Canada, Mexico, Italy, Switzerland, Norway, Nigeria, South Africa, Thailand, New Zealand, South Korea, Kyrgyzstan are in the same league of medium skepticism regarding the army and the press. The United States, Brazil, Venezuela, Turkey, Finland, several countries of Central and Southeastern Europe, Morocco, Burkina Faso, Zambia, and Zimbabwe already exhibit lower levels of skepticism in their armed forces and the press. The highest level of trust in the army and the press is to be found in Vietnam, followed by Tanzania, Jordan, India and several nations in Africa and Indonesia.

The *authoritarian character* is highest in Russia, and tends to be high in many countries of East Central Europe and in several developing countries of Africa, Asia, Brazil and Peru. Notable exceptions are Indonesia, which is in the same league as the United States, Uruguay, Chile, Mexico, Bangladesh and Thailand, and also Ethiopia. Lowest rates of the authoritarian character are to be registered in Scandinavia and in Switzerland.

Tolerance, respect and post-materialism are highest in Scandinavia, Switzerland and New Zealand, followed by Germany and then Australia, Canada, Italy, Hungary, Macedonia and Uruguay. Considering the very clear correlations of tolerance and respect and post-materialism with *"effective democracy"* (see Chapter 10, below), the low levels of tolerance and respect and post-materialism in Russia and most of the former Communist countries of Eastern Europe bode ill for the future of democracy in those countries. Also in India, Kyrgyzstan, South Korea, India and Bangladesh, the low values of tolerance, respect and post-materialism have to be recalled.

The *"ego company"* bears a lot of resemblance with Hofstede's "long term orientation" factor. Predictably, it is highest in South Korea, but also the high values in Russia, Germany, the Baltic Republics, Azerbaijan, Romania, Moldavia and Bangladesh have to be observed. Lowest values for the "ego company factor" are to be encountered in several African countries. Also several states in Latin America, East and Southeast Asia are on the lower ranks of this factor. Also most of the developed western democracies and the Muslim nations are not especially characterized by the "ego company".

The *[predominantly] female rejection of the market economy* and democracy is highest in Poland, Guatemala, Chile, Uruguay, Thailand and South Korea, and also Russia Brazil and the Philippines: these are all countries, which in one form or the other had the experience of authoritarian governments in the not too distant past in the 1970s or 1980s, and where distrust against democracy and the market economy today above all is a distrust along gender lines. Interestingly enough, also Argentina, Ethiopia and Kyrgyzstan belong to this pattern, while Finland, Mexico, and Germany are interesting "outlayers" further to be studied in future political science research. In the following, we list the factor definitions and the countries with the highest and lowest factor scores. Members of the Organization of Islamic Cooperation (OIC) are marked in **bold**.

- Economic permissiveness

Definition:

Justifiable: cheating on taxes
Justifiable: avoiding a fare on public transport
Justifiable: someone accepting a bribe
Justifiable: claiming government benefits

Best practice of avoiding it:
Bangladesh
Zimbabwe
Turkey
Tanzania
Morocco

Worst practice of avoiding it:
Serbia
Zambia
Philippines

Moldova
Thailand

- Traditional religion

Definition:

How important is God in your life
Important child qualities: religious faith
Negative loading: never attend religious services

Highest values:
Nigeria
Ghana
Jordan
Indonesia
Tanzania

Lowest values:
Russian Federation
Sweden
Czech Republic
Estonia
Germany

- Racism

Definition:

[Rejecting] Neighbors: Immigrants/foreign workers
[Rejecting] Neighbors: People of a different race
Immigrant policy (prevent people from coming)

Best practice of avoiding it:
Sweden
Burkina Faso
Canada
New Zealand
Belarus

Worst practice of avoiding it:
Hong Kong
Bangladesh
Jordan
Thailand
India

- Generational education gaps

Definition:

Highest educational level attained
Rejecting sexist position: University is more important for a boy than for a girl
Negative loading Age
Negative loading important child qualities: thrift saving money and things

Highest values:
Dominican Republic
Uganda
Kyrgyzstan
Nigeria
Peru

Lowest values:
Thailand
Mali
Poland
Czech Republic
Slovakia

- Distrust of the army and the press

Definition:

[No] Confidence: The Press
[No] Confidence: Armed Forces
Negative loading: [Right wing] self-positioning in political scale (scale 1-left to 10-right)

Best practice of avoiding it:
Viet Nam
Tanzania
Bangladesh
Jordan
India

Worst practice of avoiding it:
Germany
Argentina
Macedonia
Trinidad and Tobago
Serbia

- The authoritarian character

Definition:

Lack of social capital (Most people can be trusted [highest numerical value: you just can't be too careful])
Important child qualities: hard work
Important child qualities: obedience
Negative loading: Important child qualities: imagination
Negative loading: Important child qualities: independence

Best practice of avoiding it:
Norway
Sweden
Switzerland
New Zealand
Finland

Worst practice of avoiding it:
Russian Federation
Zimbabwe
Nigeria
Uganda
Trinidad and Tobago

- Tolerance, respect and post-materialism

Definition:

Important child qualities: tolerance and respect for other people
Important child qualities: feeling of responsibility
Rejecting sexist position: University is more important for a boy than for a girl
Negative loading: Important child qualities: hard work

Best practice:
Sweden
Norway
Switzerland
Finland
Germany

Worst practice:
Mali
Burkina Faso
Ethiopia
Nigeria
Bangladesh

- The “ego company”

Definition:

Important child qualities: feeling of responsibility
Important child qualities: independence
Negative loading: Important child qualities: unselfishness
Negative loading: Important child qualities: obedience

Highest values:
South Korea
Azerbaijan
Taiwan
Latvia
Estonia

Lowest values:
Burkina Faso
Zimbabwe
Jordan
Ghana
Tanzania

- The rejection of the market economy and democracy

Definition:

Sex (Gender) [in multivariate analysis: female] (1=male; 2=female)
Competition good or harmful
Political system: (It's very bad] having a democratic political system

Best practice of avoiding it:
Nigeria
Bangladesh
Morocco
India
Jordan

Worst practice of avoiding it:
Poland
Chile
Guatemala
Uruguay
Thailand

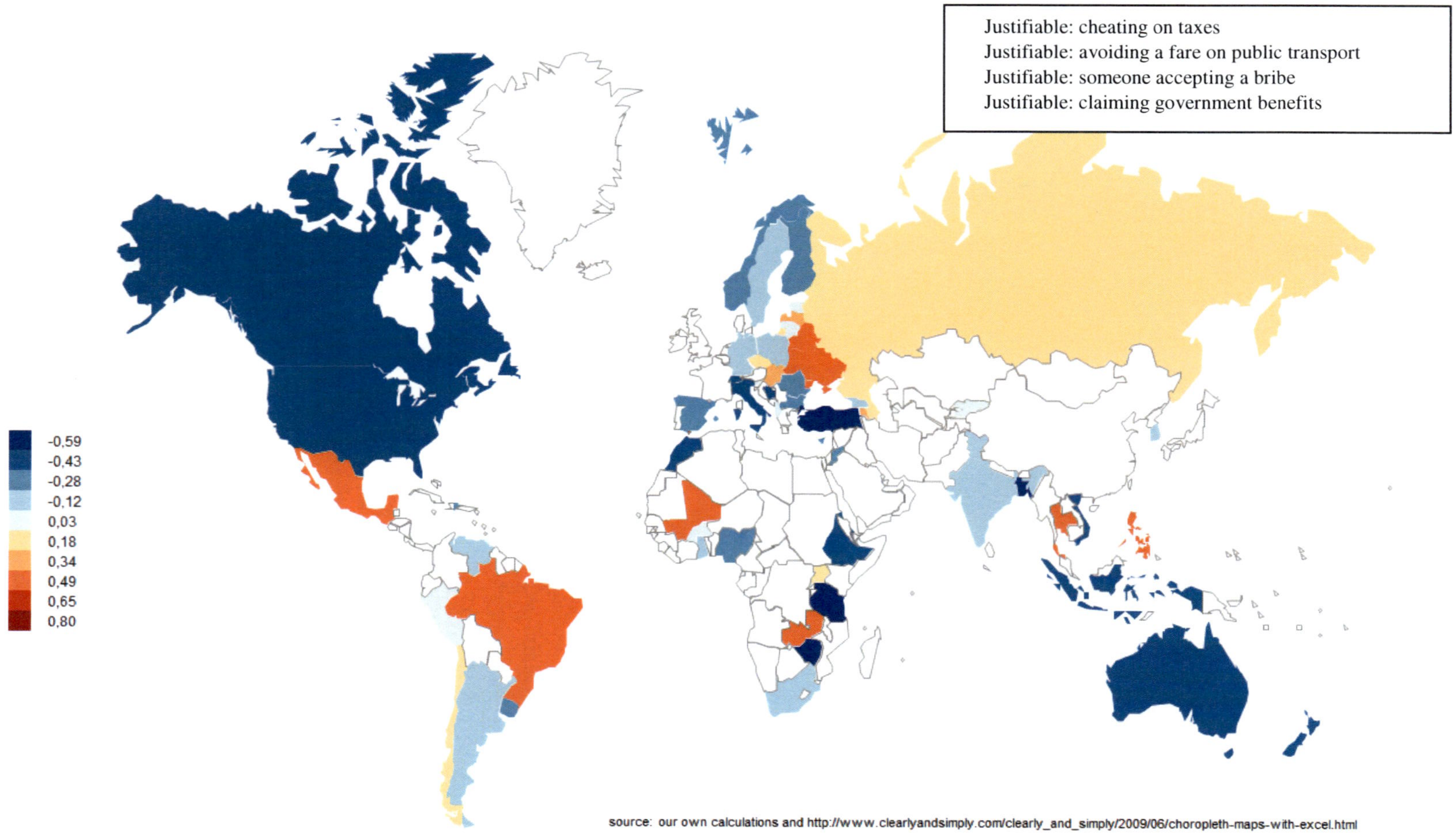

Map 5.1. Economic permissiveness.

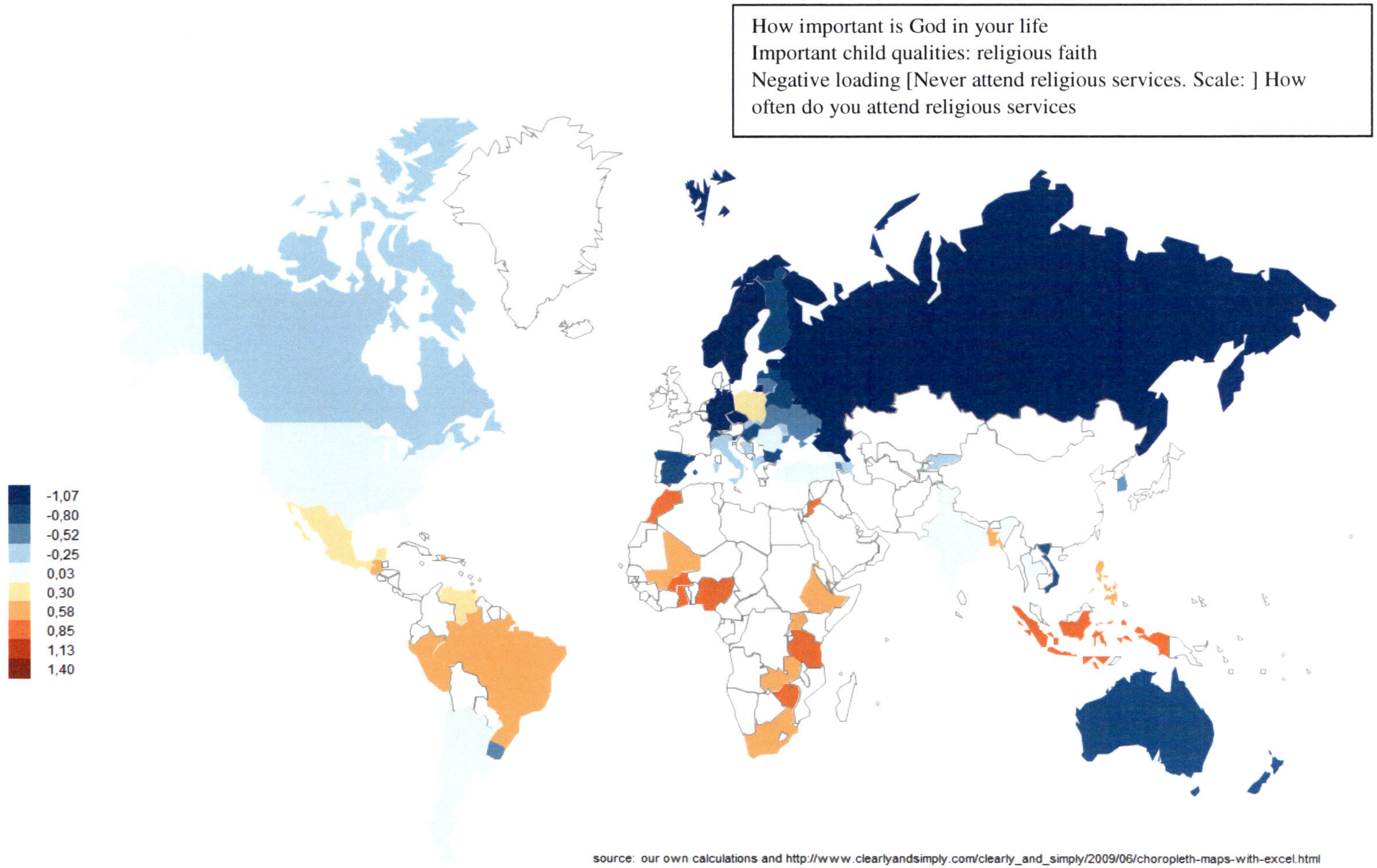

Map 5.2. Traditional religion.

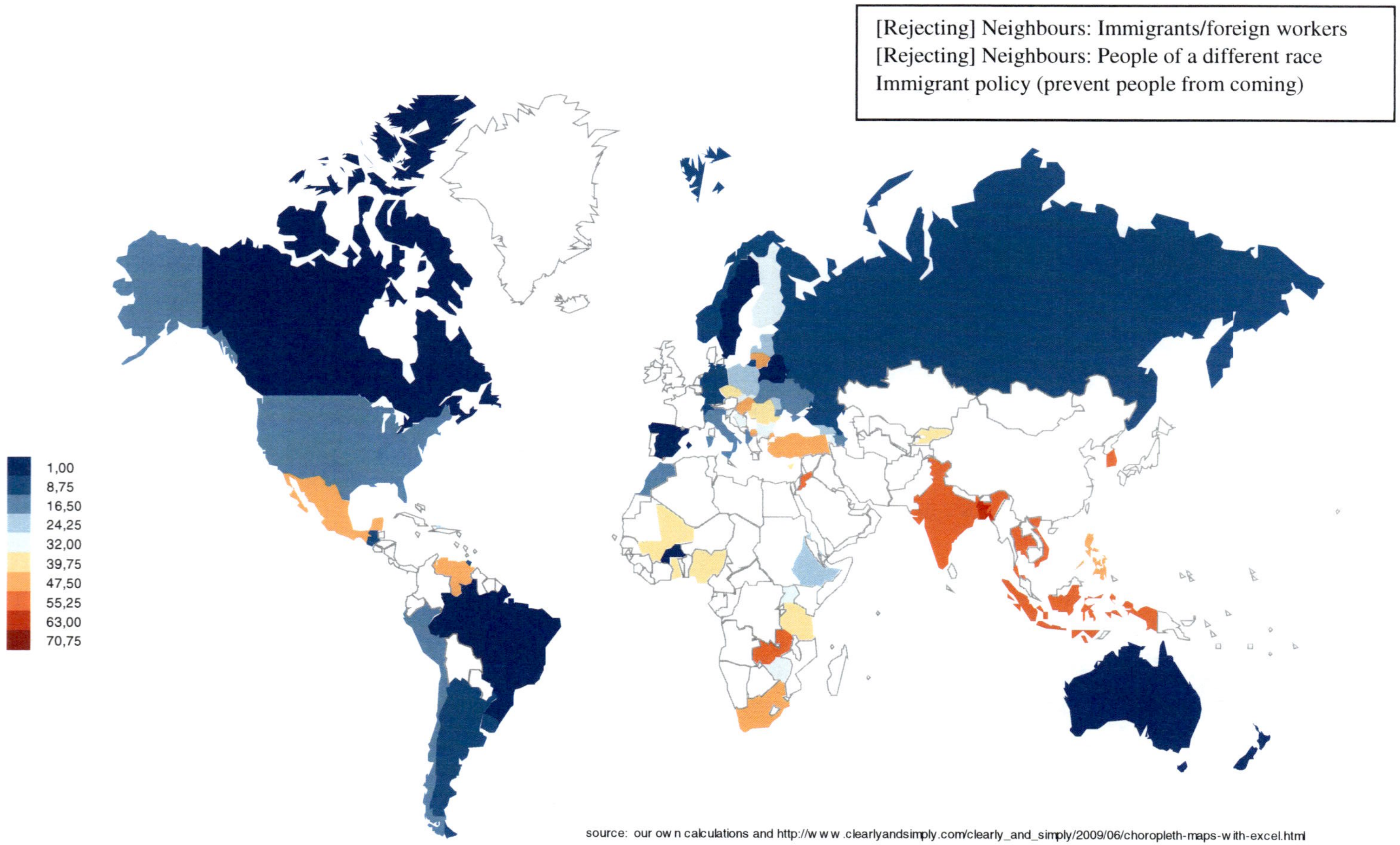

Map 5.3. Racism (global rank, avoiding racism).[34] Highest racism: marked in red; lowest racism: marked in blue.

[34] The indicator did not produce meaningful maps with the original values, so we transformed it to a ranking.

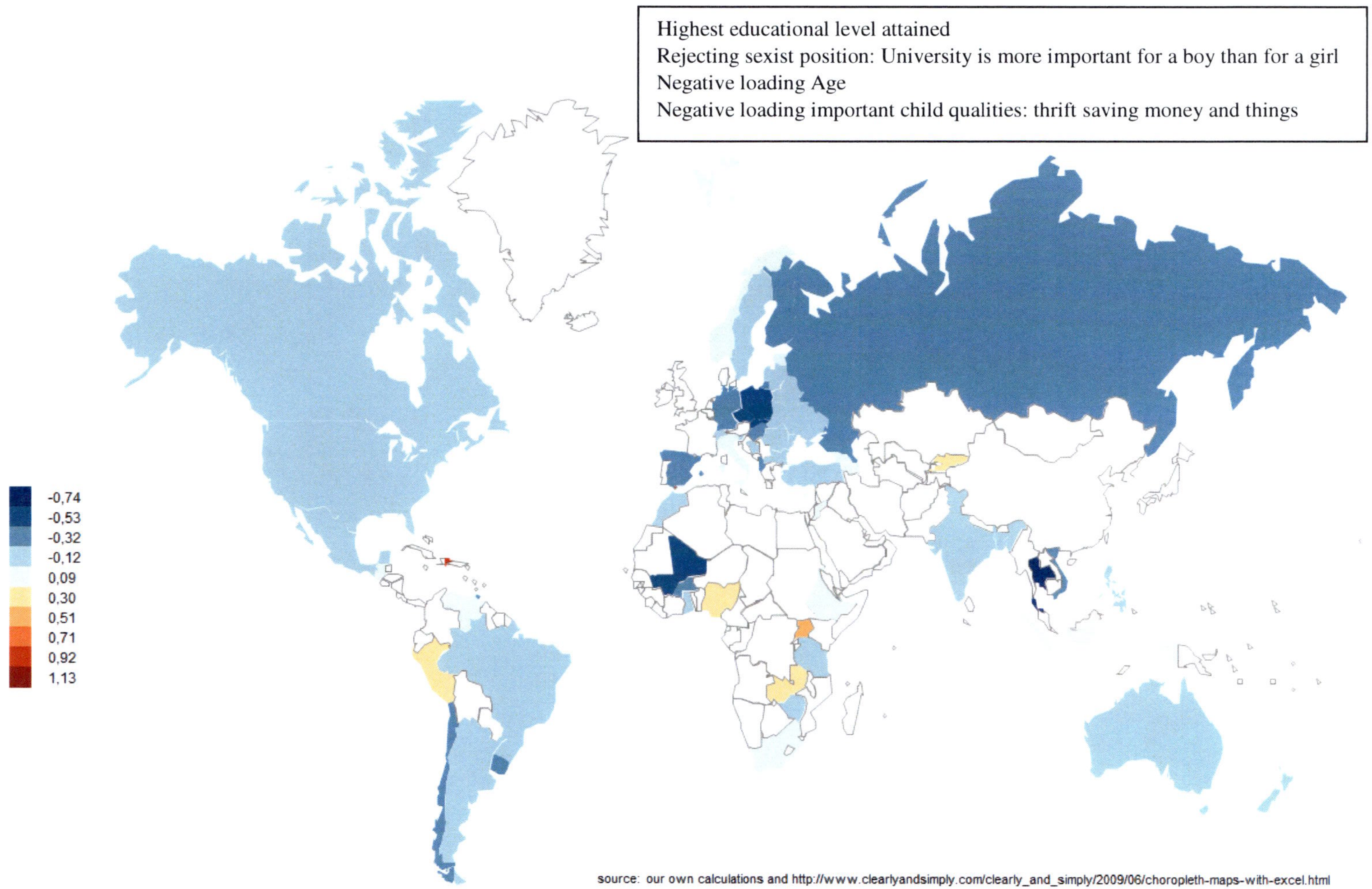

Map 5.4. Higher education of the younger generation (education gap between the generations).

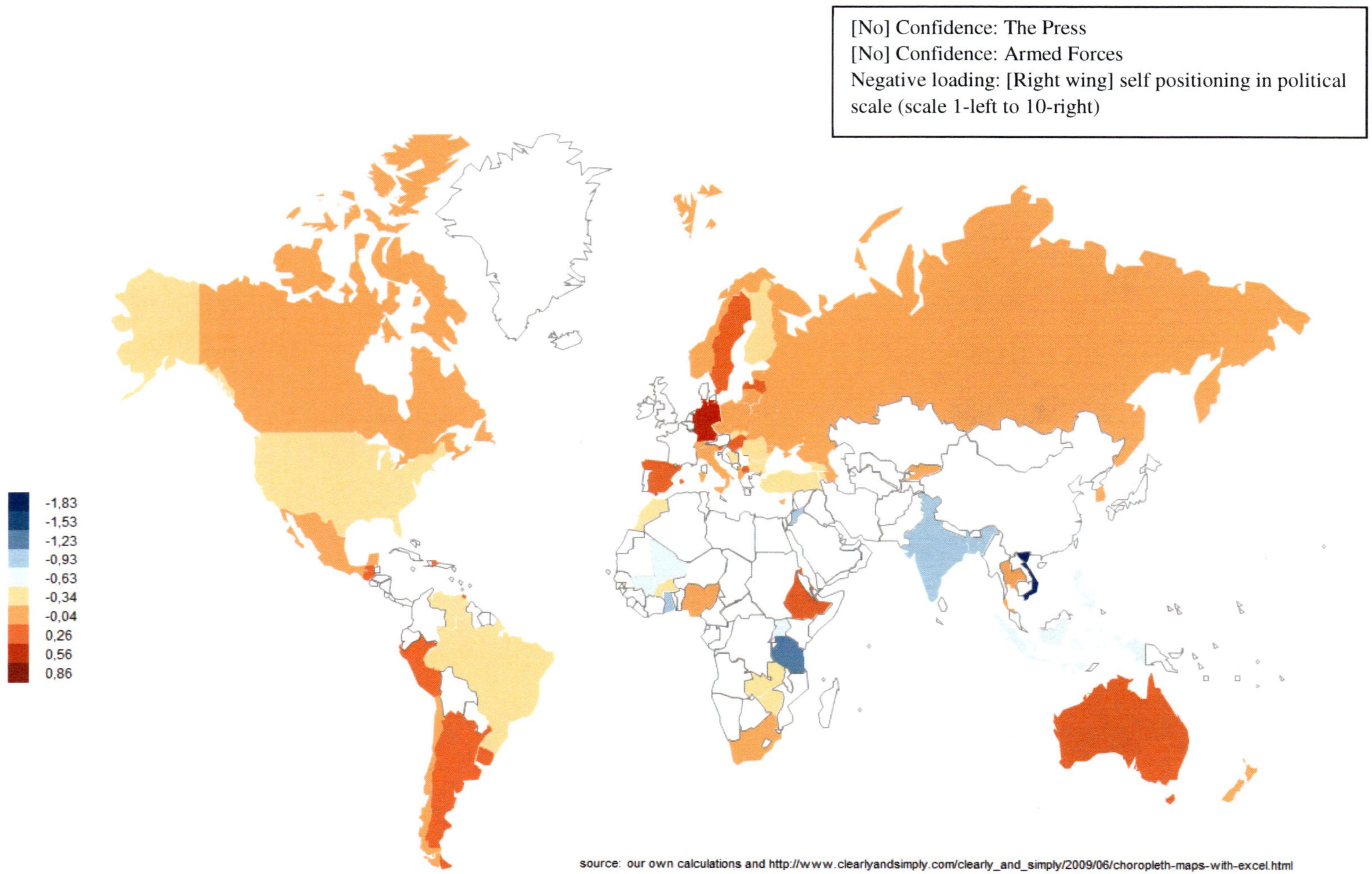

Map 5.5. Distrust of the army and the press.

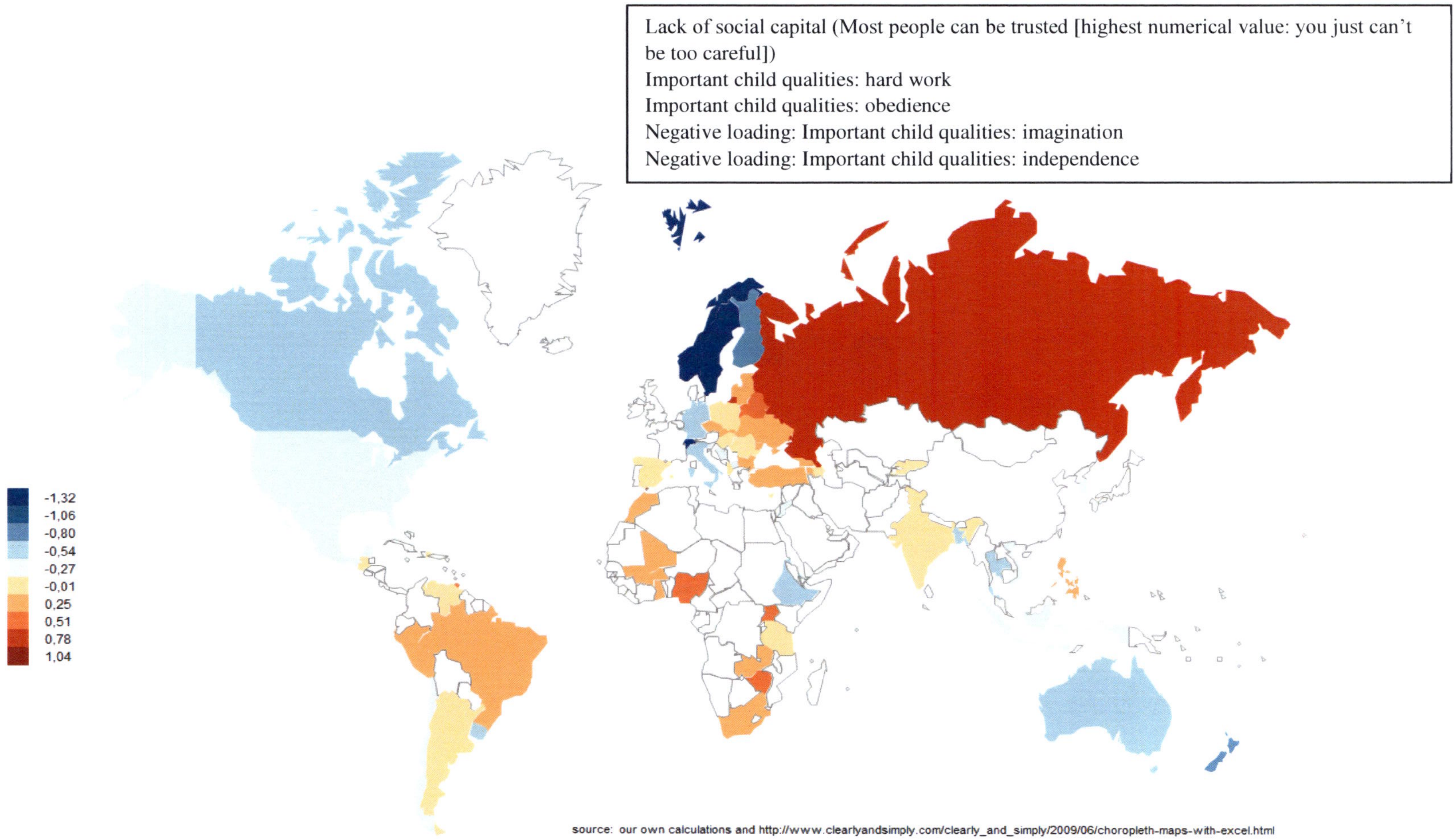

Map 5.6. Authoritarian character.

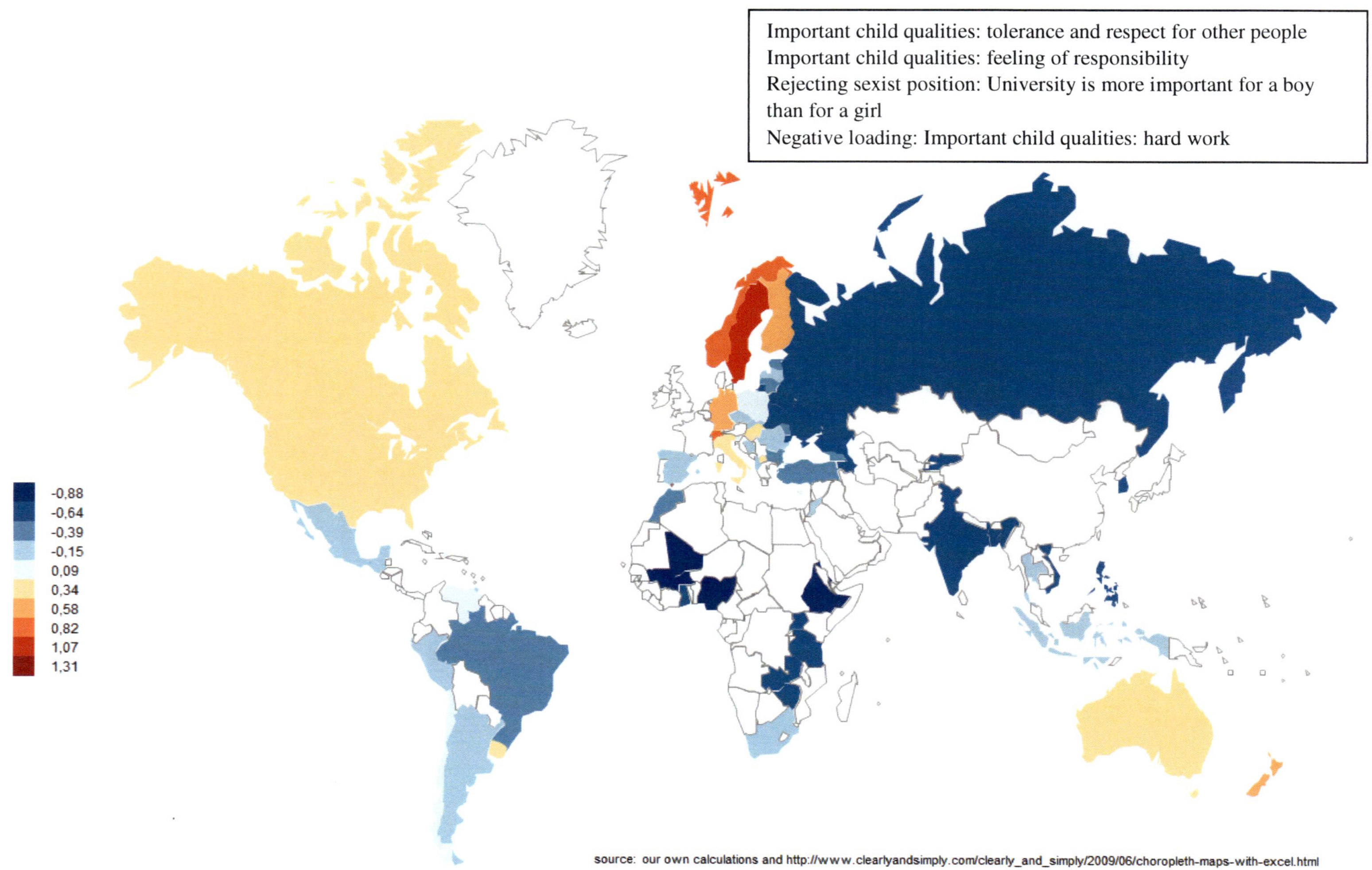

Map 5.7. Tolerance and respect + post-materialism.

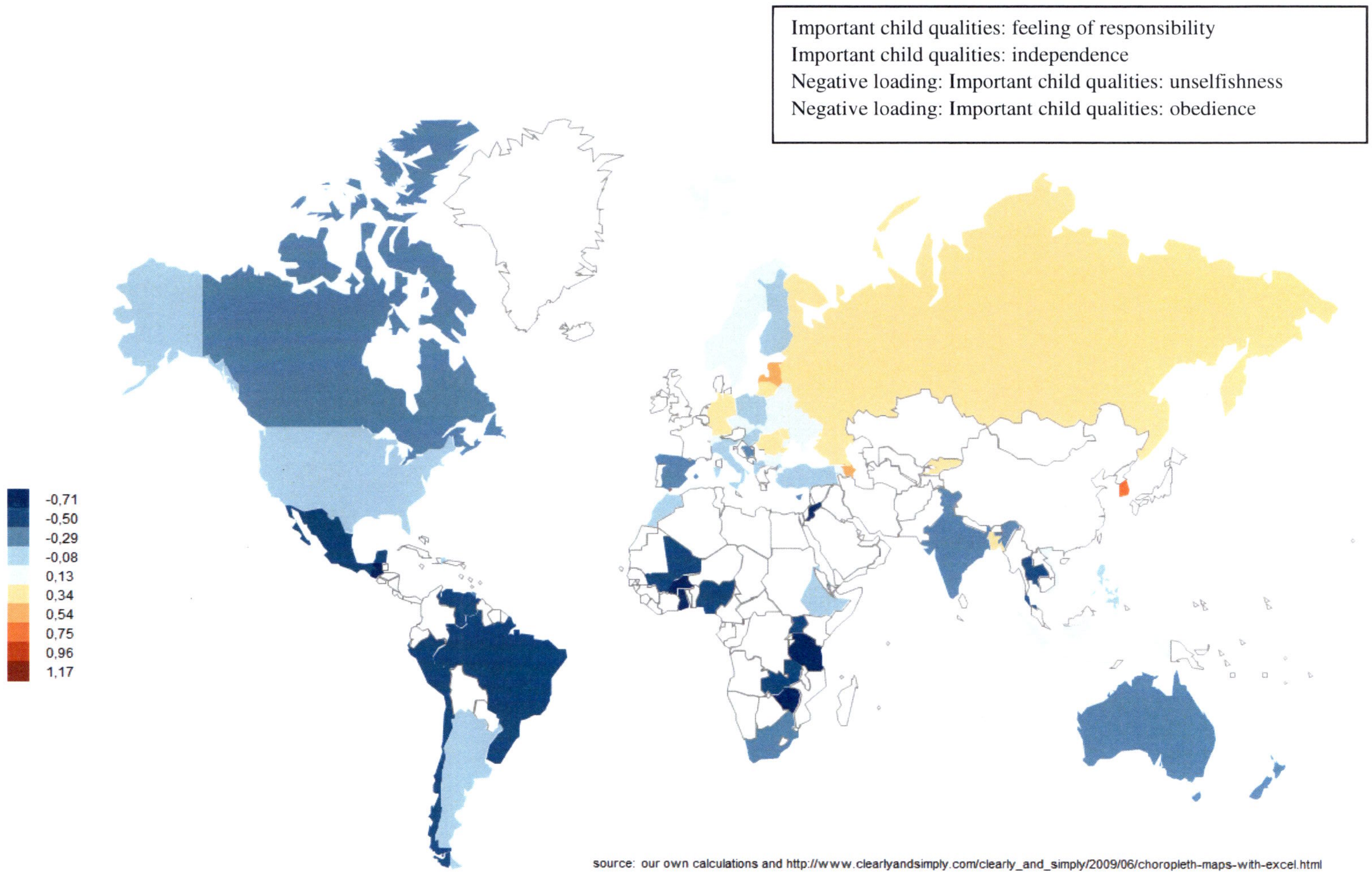

Map 5.8. The 'ego' company.

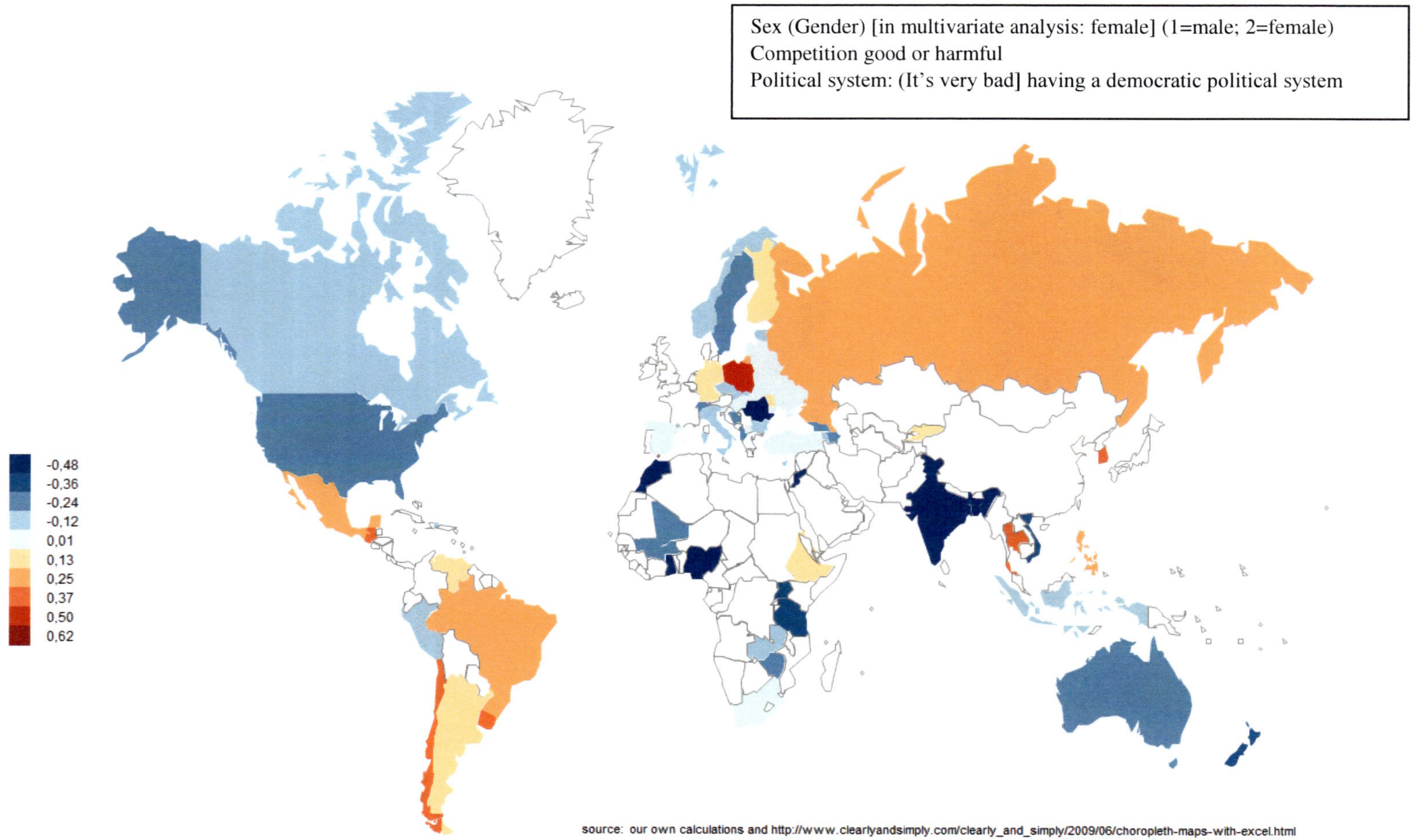

Map 5.9. [Predominantly] female rejection of the market economy and democracy.

5.5. SOME CONCLUSIONS FROM THE COUNTRY FACTOR SCORES

Looking back on Chapters 5.3 and 5.4 of this book, one might summarize for the general readership of this publication that factor analysis (Promax version) of the 30 variables presented in Table 5.6 yielded nine underlying dimensions (factors), also presented in the quick overview of Table 5.9. Our results suggested the following nine factors to be relevant:

Table 5.9. The main results of the principal components analysis – *Eigen values* and percentage of variance explained

		Eigen value	% of variance explained	cumulated % of total variance explained
1)	economic permissiveness	2,526	8,711	8,711
2)	traditional religion	2,472	8,523	17,234
3)	racism	1,688	5,822	23,056
4)	higher education of the younger generation (education gap between the generations)	1,442	4,974	28,029
5)	distrust of the army and the press	1,298	4,475	32,504
6)	authoritarian character	1,189	4,099	36,604
7)	tolerance and respect + post-materialism	1,143	3,942	40,545
8)	the ‘ego’ company (rejection of obedience + unselfishness)	1,118	3,854	44,399
9)	[predominantly] female rejection of the market economy and democracy	1,012	3,489	**47,888**

Graph 5.2 and 4.3 as well as our maps for the nine newly derived scales on a global level revisit the old debate, started by Huntington, 1993 and 1996. Graph 5.2 and 4.3 explain to us, just where there are the main differences and where are the main similarities between the Muslim/Orthodox global population and the average global citizen. But as the country results (factor scores), reproduced in Table 5.7 explain to us quite well, closer inspection tells us that things turn out to be more complicated than most social science approaches today would predict. Global rank 7 (Latvia), rank 9 (Czech Republic), rank 10 (Slovakia), rank 14 (Estonia), rank 24 (Lithuania) and rank 26 (Bulgaria) on the global *authoritarianism scale* are held by NATO Western allies in Europe, and Catholic Latin American countries are among the global leaders of the authoritarianism scale. And interestingly enough, there are Muslim or predominantly Muslim societies like Bangladesh, and Indonesia, which are less authoritarian than several key Western allies – the United States, Slovenia, and South Korea; and Jordan is less authoritarian than Spain or Poland.

So what can we say about religious global value differences in the light of Huntington’s theories? Based on our data, we could say with great caution that there is a certain trend towards racism and traditional religion in too many places in the global Muslim community (*Umma),* and a lack of the values of tolerance and respect in too many places in the global Muslim community (*Umma),*, all compared to the global community of humankind.

The world of global Orthodoxy is characterized also in too many places by a very strong tendency towards the 'ego company' (rejection of the educational values of unselfishness and obedience) and the authoritarian character, and a lack of the values of tolerance and respect and traditional religiosity. Future research would have to highlight not only the global denominational means, but also the global standard deviations, to arrive at more meaningful results.

We used the terms "great caution" because even if a racist trend exists in too many places in several Muslim societies, it is not a characteristic feature of "Islamic culture" as such. Nor is it in the Western countries, although we know that racism exists there too. The religious traditions of the major global denominations are very clear on this subject, and also the Noble Quran is clear about this issue. There is no difference between human beings based on race. For God, all are equal. The only difference is the degree to which they are religiously committed:

> {O mankind, indeed We have created you from male and female and made you peoples and tribes that you may know one another. Indeed, the most noble of you in the sight of Allah is the most righteous of you. Indeed, Allah is Knowing and Acquainted. }[Quran 49: 13]

In the Islamic understanding, God's plan was to make us different types of people so that we could realize our diversity and put forth every effort to lead one another to the Almighty. Being different only obliged us to work as one people to promote the good and forbid the wrong. Another verse of the Quran says:

> {...Had Allah willed, He would have made you one nation [united in religion], but [He intended] to test you in what He has given you; so race to [all that is] good. To Allah is your return all together, and He will [then] inform you concerning that over which you used to differ. } [Quran 5: 48]

Considering these Quranic verses, racism is a form of the fanaticism of the Pre-Islamic era *(Jahiliyya)* that Islam condemned and fought against it. Islam teaches that people are from Adam, and Adam is from dust and there is no superiority of anyone over another except in terms of Taqwa (piety and righteousness). This is shown in the Hadith of the Prophet Muhammad (S):

> O people! The Lord is one, and your Father is one, there is no privilege for an Arab over a non-Arab, nor for a non-Arab over an Arab, nor for a red-skinned over a black person, nor for a black person over a red, but by piety. The best among you in God's sight is the most pious. Did I make the message enough clear? They said: Yes, O Messenger of Allah. He said: those who heard this have to inform the absent. (*Narrator: Jabir bin Abdullah. Al-Albani. Saheeh al-Targhib, 2964).*

There is a very great variation of these results, documented in Table 5.7 and in our country maps. And we have to consider as well that phenomena of racism and traditional religion, lack of the values of tolerance and respect, all are dependent on the development level. Since the majority of Muslim nations are underdeveloped countries, it is to be expected that their performance could improve over time.

5.6. Adorno's F-Scale and Global Economic Growth in the Post-2008 Crisis Era

Among the results, which have a direct consequence for the explanation of economic growth rates since the global economic crisis of 2008, we find our factor *"authoritarian character"*. It emerges that with other statistical predictors of economic growth or social development being constant, the *"authoritarian character"* is not conducive to economic growth in our era. Our concept of the *"authoritarian character"* takes some elements from the classical sociological work by Adorno et al. (1950), which defined the *"authoritarian personality"* as having high average scores on the following variables.[35] Scholarly interest in the F-scale has been enormous ever since the main English language publication Adorno et al., 1950, among the most influential studies being Fahrenberg and Steiner, 2004; Flere, 1991; Meloen, Van Der Linden, and De Witte, 1996; Ray, 1985; Ray, and Lovejoy, 1990; Rubinstein, 1995, Rusby, 2010.

Today, some elements, based on the *World Values Survey*, are completely different from the original Adorno *"F-scale"*, which was intended by its authors to be a measurement scale to assess the potentials of authoritarianism in modern society after the horrors of Nazism and Fascism in Europe:

1. Obedience and respect for authority are the most important virtues children should learn.
2. A person who has bad manners, habits, and breeding can hardly expect to get along with decent people.
3. If people would talk less and work more, everybody would be better off.
4. The business man and the manufacturer are much more important to society than the artist and the professor.
5. Science has its place, but there are many important things that can never be understood by the human mind.
6. Every person should have complete faith in some supernatural power whose decisions he obeys without question.
7. Young people sometimes get rebellious ideas, but as they grow up they ought to get over them and settle down.
8. What this country needs most, more than laws and political programs, is a few courageous, tireless, devoted leaders in whom the people can put their faith.
9. No sane, normal, decent person could ever think of hurting a close friend or relative.
10. Nobody ever learned anything really important except through suffering.
11. What the youth needs most is strict discipline, rugged determination, and the will to work and fight for family and country.
12. An insult to our honor should always be punished.
13. Sex crimes, such as rape and attacks on children, deserve more than mere imprisonment; such criminals ought to be publicly whipped, or worse.
14. There is hardly anything lower than a person who does not feel a great love, gratitude, and respect for his parents.

[35] http: //www. zonalatina. com/Zldata387. htm

15. Most of our social problems would be solved if we could somehow get rid of the immoral, crooked, and feebleminded people.
16. Homosexuals are hardly better than criminals and ought to be severely punished.
17. When a person has a problem or worry, it is best for him not to think about it, but to keep busy with more cheerful things.
18. Nowadays more and more people are prying into matters that should remain personal and private.
19. Some people are born with an urge to jump from high places.
20. People can be divided into two distinct classes: the weak and the strong.
21. Some day it will probably be shown that astrology can explain a lot of things.
22. Wars and social troubles may someday be ended by an earthquake or flood that will destroy the whole world.
23. No weakness or difficulty can hold us back, if we have enough will power.
24. It is best to use some prewar authorities in Germany to keep order and prevent chaos.
25. Most people don't realize how much our lives are controlled by plots hatched in secret places.
26. Human nature being what it is there will always be war and conflict.
27. Familiarity breeds contempt.
28. Nowadays when so many different kinds of people move around and mix together so much, a person has to protect himself especially carefully against catching an infection or disease from them.
29. The wild sex life of the old Greeks and Romans was tame compared to some of the goings-on in this country, even in places where people might least expect it.
30. The true American way of life is disappearing so fast that force may be necessary to preserve it.

Table 5.10 explains the factor loadings for our own scale of authoritarianism, achieved by promax rotation. We define the *authoritarian character* by the following *five factor loadings* equal or above the absolute value of .30:

Table 5.10. The authoritarian character

	authoritarian character
Most people can be trusted (you can't be too careful)	**0,562**
Important child qualities: hard work	**0,437**
Important child qualities: obedience	**0,300**
Political system: Having a democratic political system (reject)	0,246
Immigrant policy (prohibit people from coming)	0,227
Important child qualities: thrift, and saving money and things	0,212
How important is God in your life	0,200
Important child qualities: religious faith	0,166
Justifiable: someone accepting a bribe	0,049
Neighbors: Immigrants/foreign workers (reject)	0,045
Self-positioning in political scale (right)	0,045
Justifiable: claiming government benefits even if one is not entitled to them	0,041
Important child qualities: tolerance and respect for other people	0,033
Neighbors: People of a different race (reject)	0,026
Justifiable: avoiding a fare on public transport	0,021

	authoritarian character
Justifiable: cheating on taxes	0,018
Confidence: The Press (no confidence)	-0,010
Sex (Gender) (female)	-0,011
Important child qualities: feeling of responsibility	-0,028
Confidence: Armed Forces (no confidence)	-0,035
Age	-0,037
Competition good or harmful (harmful)	-0,093
Important child qualities: determination and perseverance	-0,104
Important child qualities: unselfishness	-0,107
How often do you attend religious services (never)	-0,119
University is more important for a boy than for a girl (reject)	-0,160
Highest educational level attained	-0,195
Important child qualities: independence	**-0,508**
Important child qualities: imagination	**-0,613**

Table 5.11. The authoritarian tendencies in major world denominations (analyses about global denominations with a WVS sample size of below 500 should be considered with utmost care and do not reflect population/country weighted averages but are simple based on global overall WVS samples for the different denominations)

	n =	authoritarian character
Pentecostal	536	0,50
Independent African Church (e.g., ZCC, Shembe, etc.)	599	0,40
Armenian Apostolic Church	973	0,30
Orthodox	10726	0,19
Hindu	2256	0,16
Muslim	9602	0,13
Roman Catholic	26269	0,10
Ancestral worshipping	783	-0,13
Protestant	12652	-0,14
Evangelical	2586	-0,16
Buddhist	2660	-0,20
The Church of Sweden	1270	-1,17

Adorno and associates expected a strong tendency of their scale – reflecting conventionalism, authoritarian submission, authoritarian aggression, anti-intellectualism, anti-intraception, superstition and stereotypedness, power and "toughness", destructiveness and cynicism, projectivity, and exaggerated concerns over sex – to be encountered among the political right (hence also the name "F-scale" for the "F" in the word "fascism"), while there is hardly any empirical connection today between our version of the *World Values Survey*-based authoritarianism scale and the conventional right-left political spectrum. The three religion variables used in our analysis – importance of God in life, the desire for religious education of children, and religious service attendance rate, have just 4% or less of variance in common with the authoritarianism phenomenon.

Table 5.11 shows the average tendencies towards the authoritarian character among the major global denominations. It is simply wrong to attempt to portray adherents of the Muslim religion as more authoritarian as adherents of several Eastern religions, Orthodox and Greek Catholics as well as other Churches inspired by Christianity. The difference between global Roman Catholics and Muslims on this scale is minimal; while several different Protestant Churches are the real strongholds of the spirit of an enlightened education and also social capital (=trust in other people).

Policy pointer:
"Modernity" (ln GDP per capita^2) in global society is characterized by

- less traditional religion and
- more tolerance and respect

and the negative phenomena of

- more distrust in the army and the press and
- an increasingly ego-centric culture
- the female rejection of democracy and the market economy

In global society, the paths to modernity are indeed not conflict-free, and at middle levels of development we generally reach a climax of societal

- racism
- the generation gap in education
- Economic permissiveness
- the authoritarian character

On average, the Arab States reached a Human Development Index of 0.641 in 2011. Our analysis suggests that the Arab nations are currently undergoing a critical phase in their development. Based on the trajectories of our global values analysis in comparison to the Human Development Index (UNDP HDI), we could expect a more optimistic scenario for the countries of the Arab world, since from the level of UNDP HDI = 0.600 onwards economic permissiveness will decrease, the weight of religious traditionalism will decline, racism already reached its climax and will decline, the education gap between the generations will decline sharply, tolerance and respect will increase, egoism will decrease, and the trajectory of the authoritarian character and the [predominantly] female rejection of the market economy and the press will not increase sharply. Decision makers and democratic civil society will be however confronted with one major negative phenomenon: distrust of the army and the press will increase with rising human development

Graph 5.5 and following now depicts the development of human value scales along the trajectory of different development levels, as measured by the UNDP Human Development Index in global society. While traditional religiosity clearly diminishes with increasing Human Development, distrust of the army and the press, tolerance and respect, and the 'ego company' increase with increasing UNDP levels, and there are U-shaped trade-offs between the UNDP

Human Development levels and Economic permissiveness, racism, comparative higher education levels of the younger generation, the authoritarian character, and the [predominantly] female rejection of the market economy and democracy.

All this suggests that "modernity" in global society would be characterized by what most sociologists interpret as the phenomena of

- less traditional religion and
- more tolerance and respect

and the negative phenomena of

- more distrust in the army and the press and
- an increasingly ego-centric culture
- the female rejection of democracy and the market economy

In global society, the paths to modernity are indeed not conflict-free, and at middle levels of development we generally reach a climax of societal

- racism
- the generation gap in education
- Economic permissiveness
- the authoritarian character

On average, the Arab States reached a Human Development Index of 0.641 in 2011. In the framework of the general tendencies to be deduced from our theory, we can cautiously predict that Arab nations indeed are currently undergoing the most critical phase in their development. In the light of our analysis based on the trajectories of our global values analysis in comparison to the Human Development Index (UNDP HDI), we can expect for the countries of the Arab world the following to happen: from the level of UNDP HDI = 0.600 onwards, also in the Muslim countries economic permissiveness will decrease, the weight of religious traditionalism will decline, racism already reached its climax and will decline, the education gap between the generations will decline dramatically, tolerance and respect will increase, egoism will decrease, and the trajectory of the authoritarian character and the [predominantly] female rejection of the market economy and the press will not increase sharply anymore. Decision makers and democratic civil society will be however confronted with one negative phenomenon. From the level of UNDP HDI = 0.600 onwards, distrust of the army and the press will increase with rising human development (see also Graphs 5. 21 ff.).

Now, our data also can be used to re-analyze the drivers and bottlenecks of global economic growth during the current world economic slump, which began in 2007/2008. Our data are based on the well-known IMF World Economic Outlook data series, and represent the average economic growth rates 2008, 2009, 2010 and 2011.

Based on stepwise multiple regressions and the macro-quantitative world system analysis "Corvinus University" data set,[36] we can show that under due consideration of the non-linear

[36] http: //www. uni-corvinus. hu/index. php?id=47854.

tradeoffs between economic development levels and subsequent growth rates ("convergence effects"), market size (% of world population) is an important driver of contemporary growth. The authoritarian character is one of its main stumbling blocks. Our analysis is based on n = 62 countries with complete data and our regression results are based on the SPSS, version XX, and the default options for the multiple regression procedure (i.e., forward regression).

Table 5.12. The determinants of global economic growth during the current economic crisis (mean economic growth rates, 2008, 2009, 2010, 2011) according to IMF data

	Regression Coefficient B	Standard error	Beta	T	Sig.
Constant	-0,029	18,796		-0,002	0,999
% world population	**0,251**	**0,115**	**0,199**	**2,181**	**0,033**
2000 Economic Freedom Score	0,055	0,038	0,175	1,463	0,149
ln GDP per capita	2,458	4,362	0,917	0,564	0,575
ln GDP per capita ^2	-0,280	0,258	-1,818	-1,086	0,282
authoritarian character	**-1,893**	**0,770**	**-0,291**	**-2,457**	**0,017**

N = 62 countries, adj. R^2 = 52. 1%; F = 14. 244; error p = .000. Economic growth data are IMF data, and the data matrix is freely available from http://economics.uni-corvinus.hu/index.php?id=47854

Policy pointer:

Among the results, which have a direct consequence for the explanation of different economic growth rates since the global economic crisis since 2008, we find our factor "authoritarian character". It emerges that with other statistical predictors of economic growth or social development being constant, the "authoritarian character" is not conducive to economic growth in our era. Our concept of the "authoritarian character" only takes some elements from the classical sociological work by Adorno et al. (1950).

We define the authoritarian character by the following five factor loadings equal or above the absolute value of .30:

Most people can be trusted (you can't be too careful)	+0,562
Important child qualities: hard work	+0,437
Important child qualities: obedience	+0,300
Important child qualities: independence	-0,508
Important child qualities: imagination	-0,613

Based on stepwise multiple regressions and the "Corvinus University" data set (http://economics.uni-corvinus.hu/index.php?id=47854) we can show that under due consideration of the non-linear tradeoffs between economic development levels and subsequent growth rates ("convergence effects"), market size (% of world population) is an important driver of contemporary growth. The authoritarian character is one of its main stumbling blocks. Our analysis is based on n = 62 countries with complete data and our regression results are based on the SPSS, version XX, and the default options for the multiple regression procedure (i.e., forward regression).

5.7. The Trajectories of Global Value Change along the Paths of Human Development. Some Further Considerations

In the following, we will further analyze the trajectories of global value change along the paths of development. The Inglehart approach assumes that values largely depend on human security, and that human security is well measured by the UNDP Human Development Index. Now we will analyze the trajectory of value change of the OIC-members, i.e., the Muslim world, and global society separately. Such an analysis wields astonishing results and interesting predictions for the Muslim world at the same time. Knowing the UNDP Human Development Index of a given Arab or Muslim country, we can predict the probable path of value development during the next decade, provided that the UNDP Human Development Index will move in a given, predictable direction.

Based on the trade-off between human values according to the *World Values Survey* and the UNDP Human Development Index, we find that in Muslim countries there will be several *optimistic tendencies at work*, which could imply that under proper policies being pursued, Muslim countries could overcome with resilience the crisis of modernity. Yes, unfortunately, there is a U-shaped phenomenon of racism in these countries which is strongest at middle UNDP Human Development Index development levels, and there is a plateau curve of human egoism, and an inter-generational education gap, which are all strongest at middle development levels, just as in world society at large. But traditionalist religion, racism, the educational generation gap, the ego company, they all will diminish in their influence over society, when Muslim countries in general and Arab countries in particular will increase their UNDP Human Development Index. Economic permissiveness will decrease in weight, just as tolerance and respect will increase with rising development levels. These empirically robust relationships are one of the reasons for our optimism for the future of the democracy movement in the Muslim world.

Our statistics also imply that the crisis of modernization in the countries of the world system is at times more dramatic than in the countries of the OIC.

We have to note as well that the authoritarian character and the female rejection of democracy and the market economy are at a relatively constant level along the development trajectory.

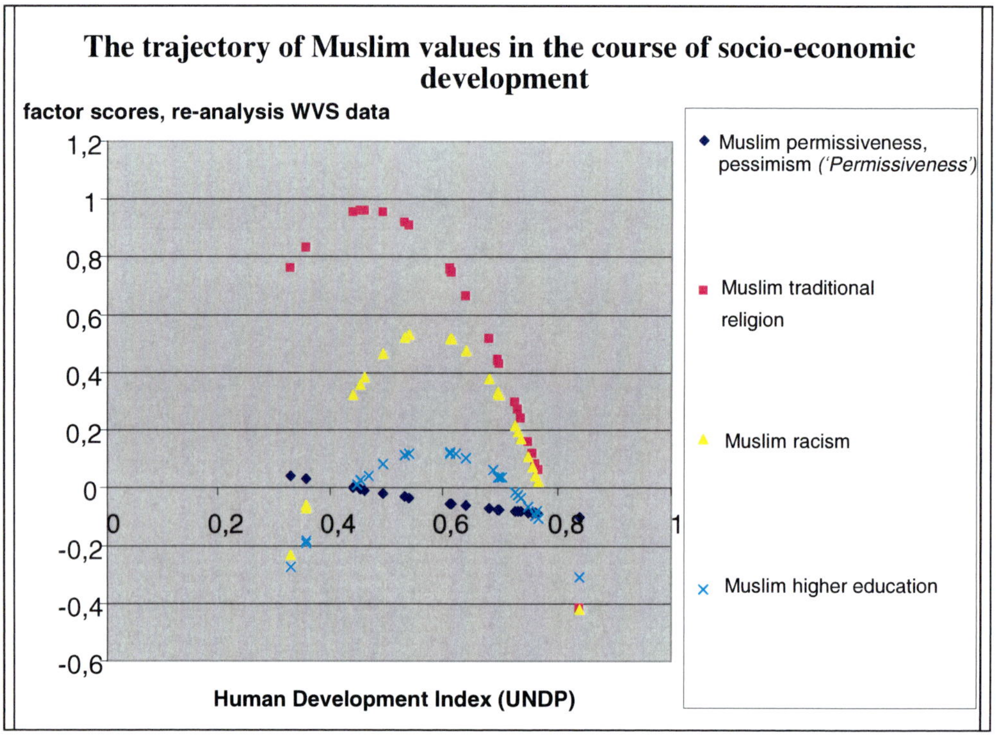

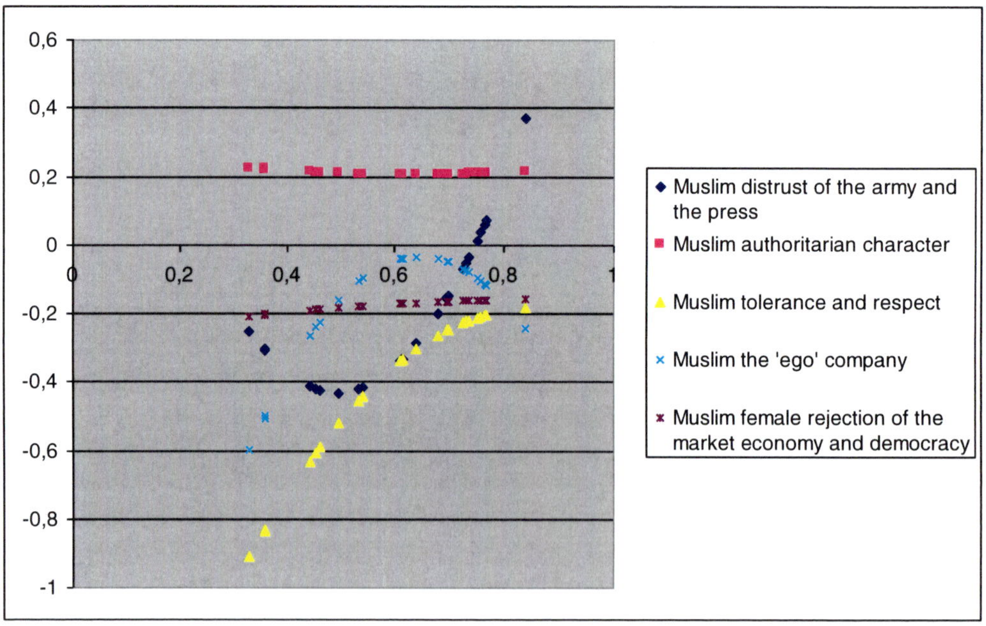

Graphs 5.5 and Graph 5.6. predicting the future trajectory of the global *Umma:* permissiveness, traditional religion, educational gap, distrust of the army and the press, authoritarian character, tolerance and respect, the "ego company", female rejection of markets and democracy as a function of the UNDP Human Development Index in Muslim societies.

5.8. The Performance of Muslim Countries along These Trajectories

To our great surprise, the trajectories of our development paths show that Muslim societies would have at least the potential of relatively smoother transitions to a mature, enlightened and humanistic society than the countries of the world system as a whole. How can we dare to advance such an optimistic proposition, which is counter to the explicit or implicit conclusions from most other published social science research around our globe? Our results generally imply that especially the rich and wealthy Muslim societies of our globe have the human and material abilities to initiate democratic reforms towards a full parliamentary democracy, if they did not do so already (perhaps in the framework of a Constitutional Monarchy, British style) at relatively low risk-levels. It should be also noted that one possible framework could be that of a Constitutional Monarchy, British style.[37]

Already 50 years ago, the American political scientist Karl Wolfgang Deutsch had the following message for the countries like the Arab world today, facing the challenges of modernization:

> "[...] Deliberate political and economic intervention into the social mobilization process, on the other hand, might open up some more hopeful perspectives. Such intervention should not aim at retarding economic and social development, in the manner of the policies of the regime of Prince Metternich in Austria during much of the first half of the 19th century. Those policies of slowing down social mobilization and economic development in the main only diminished the capabilities of the government, paved the way to domestic failures and international defeats and were followed over the course of three generations by the persistent backwardness and ultimate destruction of the state. A more promising policy might have to be, on the contrary, one of active intervention in favor of more rapid and more balanced growth; a somewhat more even distribution of income, related more closely to rewards for productive contributions rather than for status and inheritance; the more productive investment of available resources; and a sustained growth in the political and administrative capabilities of government and of ever wider strata of the population. The crude model outlined above may have some modest usefulness in surveying and presenting in quantitative terms some of the magnitudes and rates of change that would be relevant for understanding the basic problems of such a more constructive policy in developing countries. Somewhat as the economic models of the late Lord Keynes drew attention to the need of keeping the national rates of spending and investment in a country in balance [...]" (Deutsch, 1961: 505-506].

a) Economic permissiveness is highest at low levels of Muslim Human Development, while for global society, it reaches a climax at the current average UNDP Human Development Index level of the Arab world. Thus it can be said that development in the Muslim world promises to be more corruption free and shadow economy free than in the rest of the countries of our globe, once these Muslim countries reach the current UNDP Human Development Index level of the Arab world.
b) Traditionalist religion and racism both in the Muslim world and in global society indeed are evolving in an inverted U-shaped pattern along the trajectory of

[37] This parallel is all the more interesting, since Her Britanic Majesty the Queen is also the Head of the global Anglican Church.

development, but the climax levels at middle development stages are in fact lower in Muslim countries than in global society

c) The generational education gap is a problem for Muslim societies too, but still the climax levels are higher in global society
d) The distrust level concerning the army and the press is certainly a problem for global Muslim development, but still, at high levels of the UNDP Human Development Index, these Muslim societies will reach distrust levels which are in fact lower than in global society
e) Yes indeed, there is also a problem with the authoritarian character, but in global society the climax level of the authoritarian character phenomenon is far worse than in Muslim countries
f) At high levels of Human Development, societal egoism and the female rejection of democracy and the market economy are relatively lower than in other mature societies of our globe
g) What is a problem, indeed, is the future trajectory of Muslim tolerance and respect. Muslim societies – in accordance with the prescriptions of the Noble Quran, have to learn to become open towards immigration and multiculturalism; and to forgo – particularly in the Gulf –the old and useless tradition of "sponsorship", and to be more respectful toward human rights (included the rights of foreign workers).

> 'Let there be no compulsion in religion. Truth has been made clear from error. Whoever rejects false worship and believes in God has grasped the most trustworthy handhold that never breaks. And God hears and knows all things.' (Quran 2: 256)
>
> 'If it had been your Lord's will, all of the people on Earth would have believed. Would you then compel the people so to have them believe?' (Quran 10: 99)
>
> 'So if they dispute with you, say 'I have submitted my whole self to God, and so have those who follow me.' And say to the People of the Scripture and to the unlearned: 'Do you also submit yourselves?' If they do, then they are on right guidance. But if they turn away, your duty is only to convey the Message. Allah observes the affairs of His servants.' (Quran 3: 20)
>
> Mankind! We created you from a male and female, and made you into peoples and tribes so that you might come to know each other. The noblest among you in God's sight is that one of you who best performs his duty. God is All-Knowing, All-Aware. (The Qur'an, 49: 13)
>
> 'The Messenger's duty is but to proclaim the Message.' (Quran 5: 99)

According to Al Kadhi (1995), *"Muslims believe in all of the prophets of the Bible. If we were to study the stories of the prophets of the Bible we would find that they all came from the general area of Egypt, Palestine, Syria, and the surrounding regions. Muhammad (PBUH), the Prophet of Islam, also came from this locale. He was born in the Arabian Peninsula and that is where the message of Islam first came into the world. However, just as the religion of Moses was not an 'Arab' religion simply because it came from Egypt, and just as Christianity was not an 'Arab' religion simply because it came from Palestine, so too Islam is not an Arab religion but the final message of God to all humanity. In fact, Arabs make up less than 20% of the estimated 1.2 billion Muslims in the world today."*

And sure enough, as he also emphasized, "Muslims do not hate Christians nor harbor ill will against them. This is a common misconception in the West about Muslims." Muslims are told in the Qur'an:

> "... and nearest among them (mankind) in love to the believers (Muslims) will you find those who say 'we are Christians': because among these are men devoted to learning and men who have renounced the world, and they are not arrogant". (Qur'an, 5: 82).

"The people of the book" (Jews are Christians) are also pointed out by God in the Qur'an to receive the following treatment:

> "And argue not with the people of the book unless it be in [a way] which is better [than mere arguing], except with such of them that do wrong, and say [to them]: 'We believe in that which has been revealed to us and revealed to you, and our God and your God is One, and to Him we have submitted (literally: 'We are Muslims')." *(Qur'an, 29: 46).*

In the following, we chart for each of our empirically derived value indicators their trajectories in comparison to the Human Development Index already achieved – each time comparing the OIC members (=Muslim countries) with global society at large. For one, our results reveal a surprising new dimension to the entire debate about human values, going back to the discoveries of the great American economist Simon Kuznets and his insights about the development crisis of inequality being most profound at middle development levels.

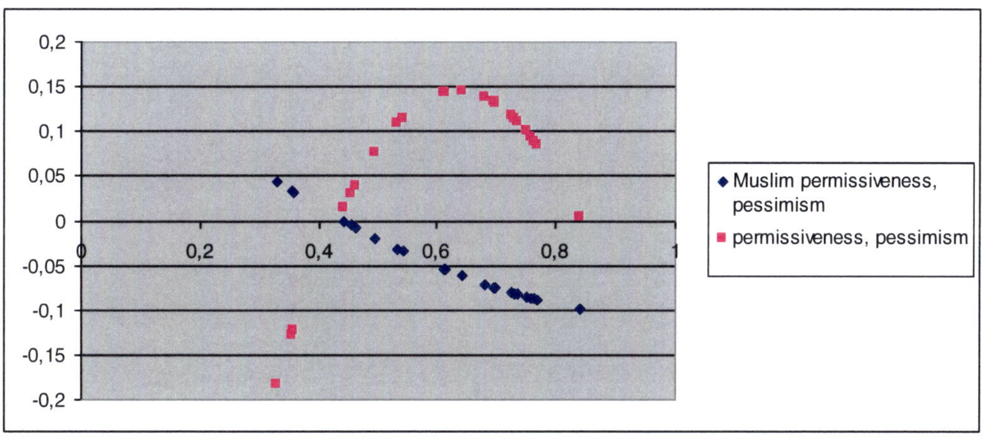

Graph 5.7.

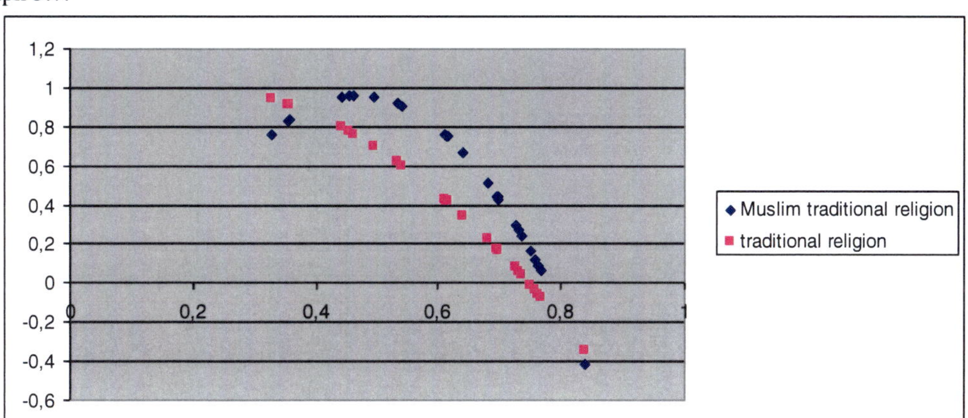

Graph 5.8.

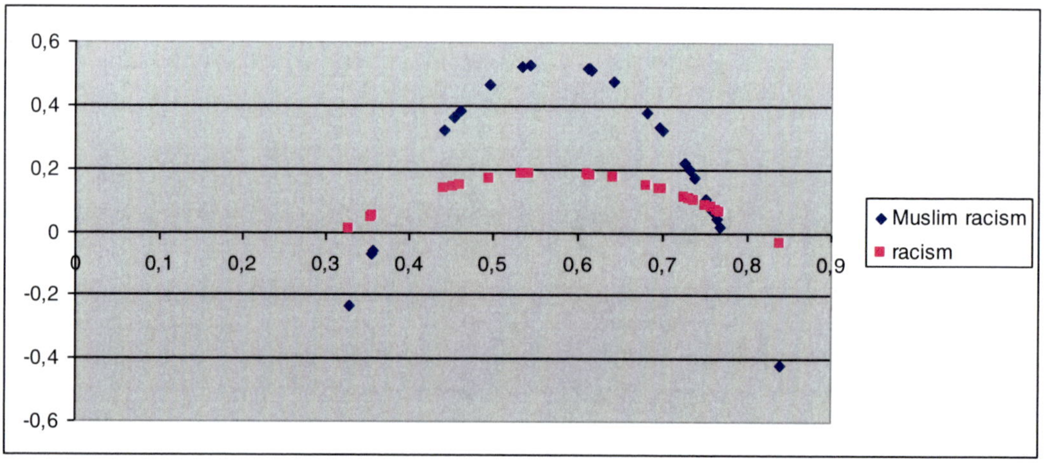

Graph 5.9.

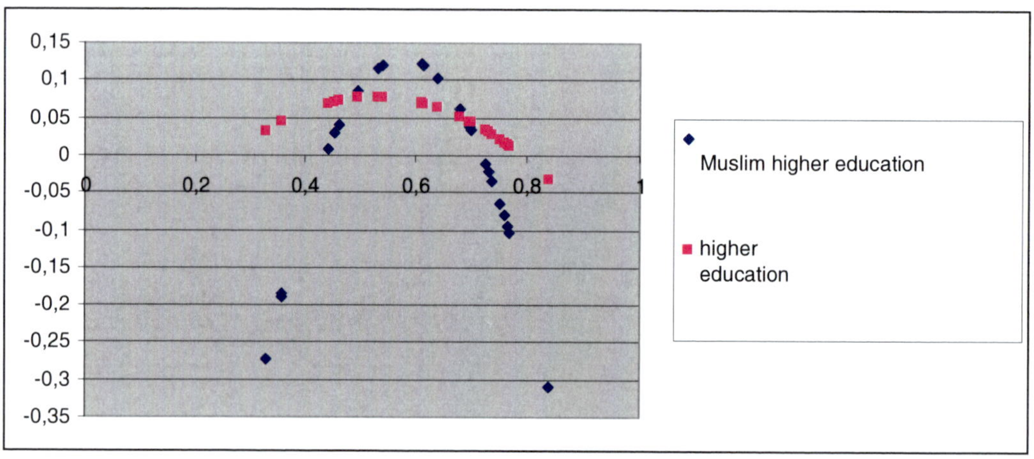

Graph 5.10.

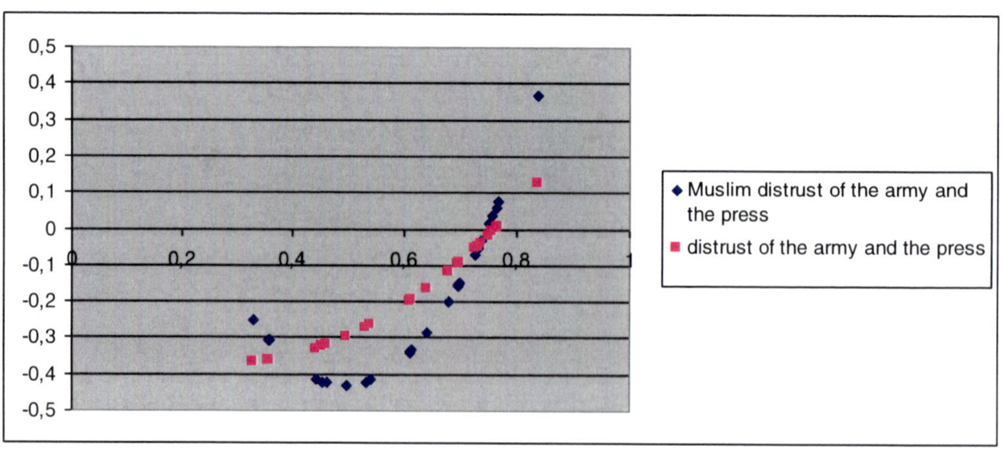

Graph 5.11.

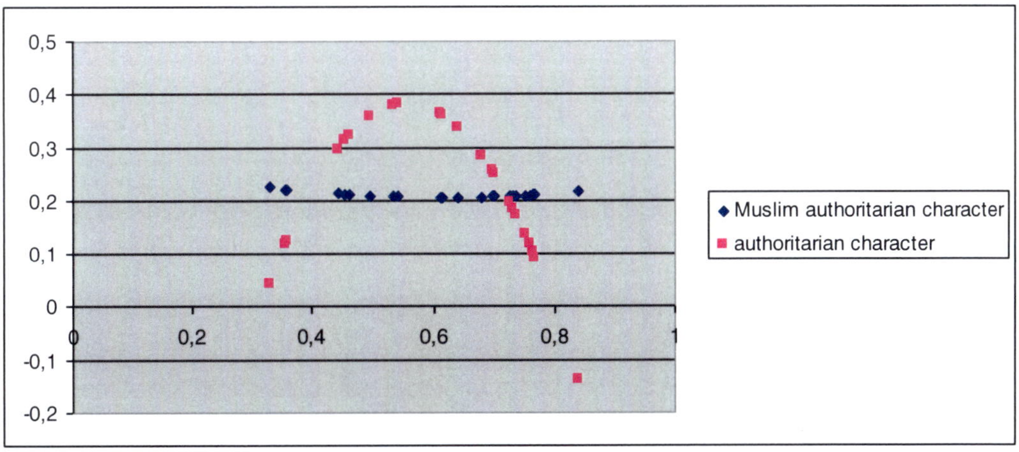

Graph 5.12.

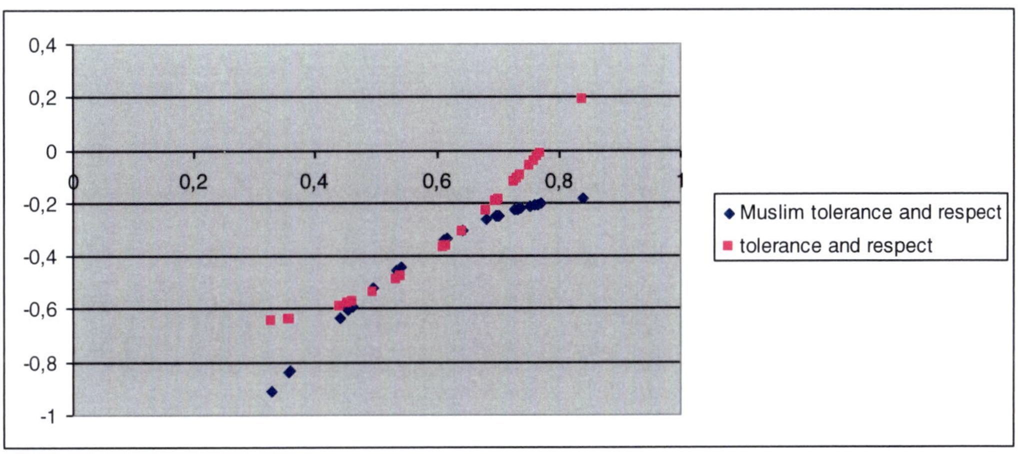

Graph 5.13.

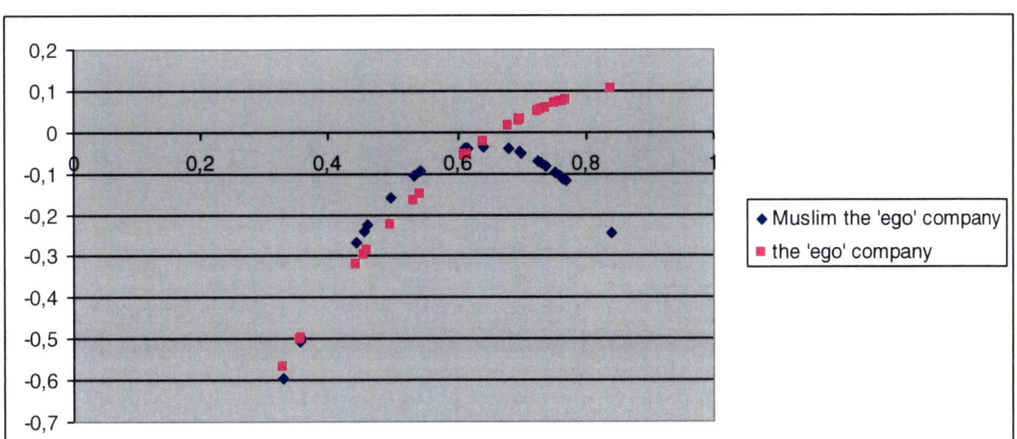

Graph 5.14.

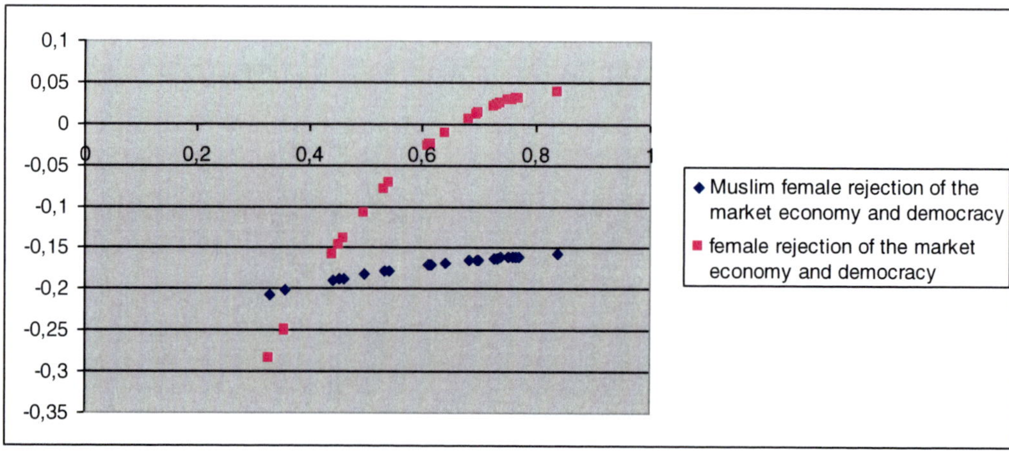

Graph 5.15.

Graph 5.7 to Graph 5.15. Predicting the future trajectory of global society and the global *Umma:* permissiveness, traditional religion, educational gap, distrust of the army and the press, authoritarian character, tolerance and respect, the "ego company", female rejection of markets and democracy as a function of the UNDP Human Development Index in global society and in Muslim societies

Like the curve-linear effect between the income level and income inequality, generally known by the term 'the Kuznets curve' in global cross-national research nowadays, traditional religion, racism, the gaps in higher education between the different generations, the authoritarian character in Muslim societies clearly present a curve-linear effect. These phenomena cannot be separated from the overall process of the transition of a country from a rural to an urban society. This process implies the concomitant processes of the transition of the employment structure of a society from the primary to the secondary and the tertiary sector. We now invite our readers to look at our summary of global value change in Muslim and global society along the paths of achieved human development.

Policy pointer:

The Inglehart approach assumes that values largely depend on human security, and that human security is well measured by the UNDP Human Development Index. We analyze the trajectory of value change of the OIC-members, i. e., the Muslim world, and global society separately. Such an analysis wields astonishing results and interesting predictions for the Muslim world at the same time. Knowing the UNDP Human Development Index of a given Arab or Muslim country, we can predict the probable path of value development during the next decade, provided that the UNDP Human Development Index will move in a given, predictable direction.

Based on the trade-off between human values according to the *World Values Survey* and the UNDP Human Development Index, we find that in Muslim countries there will be several optimistic tendencies at work, which could imply that under proper policies being pursued, Muslim countries could overcome with resilience the crisis of modernity.

Yes, unfortunately, there is a U-shaped phenomenon of racism in these countries which is strongest at middle UNDP Human Development Index development levels, and there is a plateau curve of human egoism, and an inter-generational education gap, which are all strongest at middle development levels, just as in world society at large.

But traditionalist religion, racism, the educational generation gap, the ego company, they all will diminish in their influence over society, when Muslim countries in general and Arab countries in particular will increase their UNDP Human Development Index.

Economic permissiveness will decrease in weight, just as tolerance and respect will increase with rising development levels. These empirically robust relationships are one of the reasons for our optimism for the future of the democracy movement in the Muslim world.

Our statistics also imply that the crisis of modernization in the countries of the world system is at times more dramatic than in the countries of the OIC.

Our results also imply that especially the rich and wealthy Muslim societies of our globe have the human and material abilities to initiate democratic reforms towards a full parliamentary democracy, if they did not do so already (perhaps in the framework of a Constitutional Monarchy, British style) at relatively low risk-levels. Already 50 years ago, the American political scientist Karl Wolfgang Deutsch had the following message for the countries like the Arab world today, facing the challenges of modernization:

"[...] Deliberate political and economic intervention into the social mobilization process, on the other hand, might open up some more hopeful perspectives. Such intervention should not aim at retarding economic and social development, in the manner of the policies of the regime of Prince Metternich in Austria during much of the first half of the 19th century. Those policies of slowing down social mobilization and economic development in the main only diminished the capabilities of the government, paved the way to domestic failures and international defeats and were followed over the course of three generations by the persistent backwardness and ultimate destruction of the state." (Deutsch, 1961: 505-506].

5.9. The Discovery of a Global Value Development Index and Regional Value Differences at the Sub-National Level

In the following, we will use the factor analytical scores of this Chapter to calculate a **new Global Value Development Index,** which uses the measurement scales (factors) of our work.

Our global value development index combines law-abiding and social capital, avoiding racism; trust of the army and the press; no authoritarian character; a high degree of tolerance and respect + post-materialism; and a female acceptance of the market economy and democracy.

Hopefully the index, based on our factor analytical model, will be recognized by religious and non-religious readers alike as a measurement scale which expresses the true degree of development of a civil society of a country. Among the factors of Chapter 5, we exclude the extent of traditional religion in the country, because this factor reflects the religious choice of an individual. It also excludes the generational educational gap. It also neglects the factor "ego company", where different cultural codes of global society might sharply diverge on the assessment whether it constitutes a "good" or "bad phenomenon".

Thus, our average Global Value Development Index country score combines

- avoiding economic permissiveness;
- avoiding racism;

- avoiding distrust of the army and the press;
- avoiding the authoritarian character;
- tolerance and respect + post-materialism; and
- avoiding a [predominantly] female rejection of the market economy and democracy.

The weight, given to each factor, corresponds to the Eigen values listed in Graph 5.4 of this work.

Our country results show that the five best ranked countries of our entire globe are all western democracies with a solid historical anchoring of their societies in the traditions of the Enlightenment – Switzerland, Norway, Sweden, New Zealand, and Australia. But we already find among the next five countries Canada, the two developing countries Vietnam and Tanzania, and the EU-member countries Italy (predominantly Roman Catholic, with a long history of liberal Catholicism since the Second Vatican Council) and Finland (predominantly Protestant). Our global value development index, ranks the predominantly Muslim nation of Morocco twelfth – just behind the United States of America – and still ahead the Latin American democracy Uruguay and the EU-country Germany, to be followed by Bosnia and Indonesia. In our Table, we highlighted the member countries of the Organization of Islamic Cooperation in green color, while the member countries of the European Union are marked in blue color.

Table 5.13. The rankings of the countries of the world on a new Global Value Development Index (OIC member countries and Bosnia and Hercegovina are marked in green colors)

Country	**Value Development Index**	**Global Rank Value Development Index**
Switzerland	3,84	1
Norway	3,77	2
Sweden	3,41	3
New Zealand	3,33	4
Australia	2,57	5
Canada	2,50	6
Vietnam	2,38	7
Tanzania	2,23	8
Italy	2,23	9
Finland	2,11	10
United States	2,06	11
Morocco	1,50	12
Uruguay	1,44	13
Germany	1,08	14
Bosnia and Herzegovina	1,07	15
Indonesia	0,91	16
Romania	0,89	17
Spain	0,54	18
Cyprus	0,51	19
Dominican Republic	0,50	20

Table 5.13. (Continued)

Country	**Value Development Index**	**Global Rank Value Development Index**
Bulgaria	0,34	21
Ghana	0,31	22
Zimbabwe	0,26	23
Ethiopia	0,08	24
Turkey	0,06	25
Venezuela	0,03	26
Jordan	0,01	27
Argentina	0,00	28
Poland	-0,07	29
Georgia	-0,11	30
Bangladesh	-0,20	31
India	-0,20	32
Albania	-0,36	33
Macedonia	-0,43	34
Slovenia	-0,47	35
Burkina Faso	-0,51	36
Estonia	-0,66	37
Chile	-0,69	38
South Africa	-0,79	39
Nigeria	-0,81	40
Peru	-0,94	41
Uganda	-0,95	42
Trinidad and Tobago	-1,04	43
Korea, South	-1,28	44
Kyrgyzstan	-1,36	45
Azerbaijan	-1,37	46
Lithuania	-1,53	47
Hungary	-1,69	48
Slovakia	-1,76	49
Brazil	-1,77	50
Czech Republic	-1,81	51
Mali	-1,83	52
Belarus	-1,87	53
Ukraine	-1,99	54
Guatemala	-2,00	55
Mexico	-2,10	56
Armenia	-2,13	57
Latvia	-2,22	58
Russia	-2,27	59
Moldova	-2,59	60
Thailand	-2,85	61
Philippines	-2,86	62
Zambia	-3,25	63

Policy pointer:

Our global value development index combines law-abiding and social capital, avoiding racism; trust of the army and the press; no authoritarian character; a high degree of tolerance and respect + post-materialism; and a female acceptance of the market economy and democracy.

The weight, given to each factor, corresponds to the Eigen values listed in Graph 5.4 of this work.

Our country results show that the five best ranked countries of our entire globe are all western democracies with a solid historical anchoring of their societies in the traditions of the Enlightenment – Switzerland, Norway, Sweden, New Zealand, and Australia. But we already find among the next five countries Canada, the two developing countries Vietnam and Tanzania, and the EU-member countries Italy (predominantly Roman Catholic, with a long history of liberal Catholicism since the Second Vatican Council) and Finland (predominantly Protestant). Our global value development index, ranks the predominantly Muslim nation of Morocco twelfth – just behind the United States of America – and still ahead the Latin American democracy Uruguay and the EU-country Germany, to be followed by Bosnia and Indonesia. In our Table, we highlighted the member countries of the Organization of Islamic Cooperation in green color, while the member countries of the European Union are marked in blue color.

While in general terms, our analysis is quite optimistic about the civil society foundations for a stable democracy for several Muslim countries, including Morocco, Bosnia, Indonesia, Turkey and Jordan, our analysis is fairly pessimistic for the former communist countries and successor states of the former Soviet Union, predominantly Muslim and non-Muslim alike. Our materials also show that the euro-centric assumptions by European decision makers, which pushed European Union Enlargement ahead of democratic consolidation after the fall of Communism in Eastern Europe, are utterly wrong. Table 5.13 of this study provides readers with a first overview of the highest and lowest placed regions in the current European Union members and the EU-accession countries, ranked again by an average Value Development Index score. Finally, we also show country ranks of the Global Value Development Index.

Our map shows that the euro-centric assumptions by European decision makers, which pushed European Union Enlargement ahead of democratic consolidation after the fall of Communism in Eastern Europe and the Balkans, are utterly wrong. The degree of development of a democratic civil society [characterized by law-abiding and social capital, avoiding racism; trust of the army and the press; no authoritarian character; a high degree of tolerance and respect + post-materialism; and a female acceptance of the market economy and democracy] is very poorly developed in several of the countries, admitted into the European Union in 2004 and after. Our choropleth map of global value development (Map 5.10) then summarizes the results of Table 5.14 at a glance.

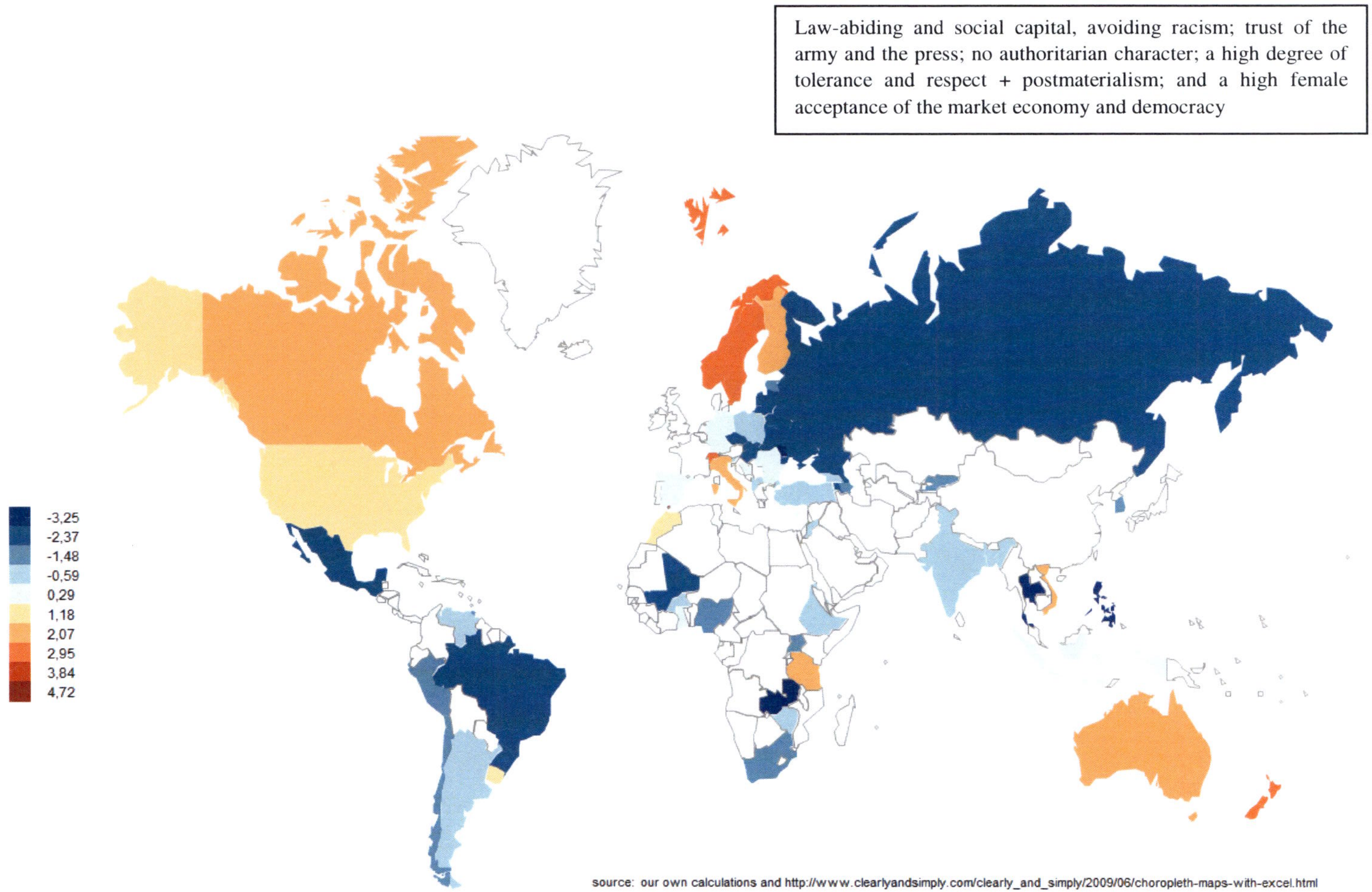

Map 5.10. Combined global value development index (avoiding permissiveness, racism, distrust of the army and the press, authoritarian character, [predominantly] female rejection of the market economy and democracy; and practicing the values of tolerance and respect (weighted by the Eigen values of the promax factor analytical model).

Once more, it emerges that the political geography of global values in the 21st Century is far away from the imaginary of culturalist scientists and populist politicians alike: even for neighboring countries, culturalist explanation mechanisms utterly fail. Just compare Uruguay and Brazil, both predominantly Roman Catholic; Italy and Hungary, both predominantly Roman Catholic, Morocco and Mali, both predominantly Muslim, or for that matter, Morocco and Spain, just separated by the Straits of Gibraltar; Tanzania and Zambia, two neighboring African countries, and Vietnam and Thailand, two Asian neighboring countries. While Uruguay, Italy, Morocco, Tanzania and Vietnam are real frontrunners in value development, we find that Brazil; Hungary, Mali, Zambia, and Thailand are real laggards in global value development. These phenomena hold independently of the attained development level of a country, measured by the Human Development Index of the UNDP. All of a sudden we discover how exceptional countries like *Uruguay, Italy, Morocco, Tanzania and Vietnam* really are. Global sociological research would do well to focus on factors which contributed towards their performance.

The *World Value Survey* data also permit the research community to analyze the results not only at the national level, but also at the regional level, where the interviews were recorded.

The idea that global values are often distributed in the nations of the world in a highly regionally contradictory pattern is relatively new in the research literature on the subject. First studies in this direction were published, among others, by Alm, Martinez-Vazque, and Torgler, 2006; Bonini, 2008; Freitag, 2003; Howard, 2002; and Torgier and Schneider, 2007.

Our regional evaluations of the *World Values Survey* data are all available from https://uibk.academia.edu/ArnoTausch/Documentation-for-books-and-articles where readers will find the file entitled: *"Regional aspects of socio-religious values according to the World Values Survey"* and *"The global analysis of feminism and its regional implications for the Muslim world"*.

Of the global regions with more than 30 interview partners each, it emerges for example that the **30 most permissive and pessimistic social climates** are located in:

ZM: Copperbelt Province	Zambia
CS: South East Serbia	Serbia and Montenegro
CS: Central West Serbia	Serbia and Montenegro
MD: Vulkaneshtskij	Moldova
MD: Autonomous Gaugasian Republic	Moldova
UA: Zhitomyr oblast	Ukraine
RO: Arges	Romania
BF: Sahel Region	Burkina Faso
UA: Kirovograd oblast	Ukraine
GH: Northern	Ghana
IN: Assam	India
ML: Mopti	Mali
AM: Ararat Marz	Armenia
UA: Southern	Ukraine
UA: Central	Ukraine

BY: Gomel oblast	Belarus
ML: Sikasso	Mali
LV: Ziemelu reg.	Latvia
UG: Kampala	Uganda
GT: Altiplano/Centro	Guatemala

The **30 superstars of law-abiding and social trust** (=highest negative loadings on the economic permissiveness factor) we find in the following regions of the world:

BD: Kishoreganj	Bangladesh
TR: East Central Anatolia	Turkey
BD: Feni	Bangladesh
BD: Mymensingh	Bangladesh
TR: Western Black Sea	Turkey
BD: Sylhet	Bangladesh
TR: Western Marmara	Turkey
BD: Chittagong	Bangladesh
BD: Rangpur	Bangladesh
BD: Habiganj	Bangladesh
GE: Samegrelo	Georgia
ZW: Masvingo	Zimbabwe
BD: Brahmanbaria	Bangladesh
ID: Lampung	Indonesia
ZW: Mashonaland West	Zimbabwe
ET: Addis Ababa	Ethiopia
ID: East Java	Indonesia
TR: Eastern Black Sea	Turkey
ZW: Midlands	Zimbabwe
BD: Nator	Bangladesh
VN: northwest	Vietnam
BD: Sirajgonj	Bangladesh
MA: Marrakech-Tensift	Morocco
IN: Punjab	India
TR: Eastern Marmara	Turkey
ID: Dareah Istimewa Yogyakarta	Indonesia
VN: north central	Vietnam
TR: Western Anatolia	Turkey
BD: Dhaka	Bangladesh
ZW: Mashonaland Central	Zimbabwe
BD: Barisal	Bangladesh

The 30 **most traditional religious regions of the world** are to be found in:

BD: Mymensingh	Bangladesh
GH: Brong Ahafo	Ghana
ZM: Western Province	Zambia
BD: Sylhet	Bangladesh
BD: Kishoreganj	Bangladesh
BD: Brahmanbaria	Bangladesh
BD: Comilla	Bangladesh
NG: West	Nigeria
BD: Habiganj	Bangladesh
NG: Middle belt	Nigeria
GH: Eastern	Ghana
GH: Ashanti	Ghana
GH: Northern	Ghana
BD: Rangpur	Bangladesh
NG: North	Nigeria
ID: Lampung	Indonesia
ID: Central Java province	Indonesia
NG: East	Nigeria
GH: Greater Accra	Ghana
NG: Lagos	Nigeria
BF: Central North	Burkina Faso
ID: Banten	Indonesia
GH: Volta	Ghana
ID: West java province	Indonesia
BF: North Region	Burkina Faso
GH: Central	Ghana
UG: Mbale	Uganda
ID: East Java	Indonesia
UG: Tororo	Uganda
ZW: Mashonaland West	Zimbabwe

Equally important, it emerges that the **30 most secularist regions of the entire world** are to be found in:

DE: Ost-Berlin	Germany
RU: Far East	Russia
CZ: Severoèeský kraj - North Bohemia -	Czech Republic
RU: Western Siberia	Russia
DE: Mecklenburg-Vorpommern	Germany

RU: Volga; Povolzskij	Russia
DE: Sachsen	Germany
DE: Brandenburg	Germany
DE: Sachsen-Anhalt	Germany
BG: Lovech	Bulgaria
SE: Skåne	Sweden
EE: Isa-Virumaa	Estonia
UA: Vinnytsia oblast	Ukraine
EE: Tartumaa	Estonia
DE: Thueringen	Germany
EE: Laane-Viruma	Estonia
SE: Vasterbotten	Sweden
UA: Kharkiv oblast	Ukraine
UA: Luhansk oblast	Ukraine
VN: northwest	Vietnam
RU: Volgo-Vyatki	Russia
SE: Västra Götaland	Sweden
CZ: Jihoèeský kraj - South Bohemia -	Czech Republic
RU: North Caucasus	Russia
SE: Vasternorrland	Sweden
SE: Stor Stockholm	Sweden
CZ: Západoèeský kraj - West Bohemia -	Czech Republic
RU: Urals	Russia
DE: Berlin	Germany
BG: Razgrad	Bulgaria
SE: Uppsala	Sweden

The 30 most **racist and xenophobic cultures** of the world are to be found in the following regions:

BD: Sirajgonj	Bangladesh
IN: Jharkhand	India
BD: Brahmanbaria	Bangladesh
BD: Kishoreganj	Bangladesh
BD: Feni	Bangladesh
BD: Nator	Bangladesh
BD: Habiganj	Bangladesh
BD: Sylhet	Bangladesh
BD: Chittagong	Bangladesh
BD: Barisal	Bangladesh
ID: Banten	Indonesia

BD: Dhaka	Bangladesh
GH: Eastern	Ghana
TH: The South	Thailand
IN: Assam	India
BD: Mymensingh	Bangladesh
BD: Comilla	Bangladesh
TR: Central Anatolia	Turkey
IN: Madhya Pradesh	India
IN: Rajasthan	India
TH: The North	Thailand
IN: West Bengal	India
IN: Orrisa	India
IN: Bihar	India
VN: northwest	Vietnam
KR: Jeonbuk / North Jeolla	Korea-South
ID: DKI	Indonesia
ZM: Copperbelt Province	Zambia
VN: southeast	Vietnam
ID: Dareah Istimewa Yogyakarta	Indonesia

The **least racist and xenophobic cultures** of the world are to be found in the following regions of the world:

BF: Central West	Burkina Faso
MD: Floreshtskij	Moldova
BF: Central North	Burkina Faso
BF: Central Plateau	Burkina Faso
ES: Asturias	Spain
ZW: Midlands	Zimbabwe
ZW: Masvingo	Zimbabwe
BY: Mogilev oblast	Belarus
SE: Stor Stockholm	Sweden
CA: Saskatchewan	Canada
UA: Hmelnytsk oblast	Ukraine
AR: Rosario	Argentina
CA: Manitoba	Canada
BF: Boucle du Mouhoun	Burkina Faso
SE: Skåne	Sweden
RO: Dolj	Romania
UY: Fray Bentos	Uruguay
SE: Uppsala	Sweden

NZ: Wellington	New Zealand
CA: Alberta	Canada
AL: Tirana	Albania
BY: Brest oblast	Belarus
CA: Nova Scotia	Canada
NZ: Hawkes Bay	New Zealand
CA: Ontario	Canada
SE: Vasterbotten	Sweden
MD: Kantemirskij	Moldova
NO: Oslo and Akershus	Norway
NO: Nordland	Norway
CA: British Columbia	Canada
BR: RJ	Brazil

Such comparisons for the hopefully flourishing future research field of regional analyses of human values could be continued endlessly. On a European level, one should not underestimate the long-term implications of such a discourse.

It also emerges, for example that even in highly developed overseas democracies, regional value differences are considerable. Differences between the deeply religious "Bible Belt" in the US South and the relatively secular New England are but one example. Secular Western Turkey quickly catches up with other European regions concerning the "Westernization" of values, while Central Anatolia lags behind. Table 5.13 provides us with a first overview of the highest and lowest placed regions in the current European Union member states and the EU-accession countries, ranked again by an average Value Development Index score, which combines, as before

- avoiding economic permissiveness;
- avoiding racism;
- avoiding distrust of the army and the press;
- avoiding the authoritarian character;
- tolerance and respect + post-materialism; and
- avoiding a [predominantly] female rejection of the market economy and democracy.

The regional index is again built by weighting the regional factor scores according to the *World Values Survey* results by the *Eigen values* of each of these factors.

Tables 5.14 and 5.15 suggest huge national and regional differences in the European Union, which will increase even more after the next proposed round of EU-enlargement. Table 5.15 could be an important checklist for European decision makers in future EU-enlargement negotiations, pinpointing just where the real strengths and weaknesses of candidates in global value development really are situated.

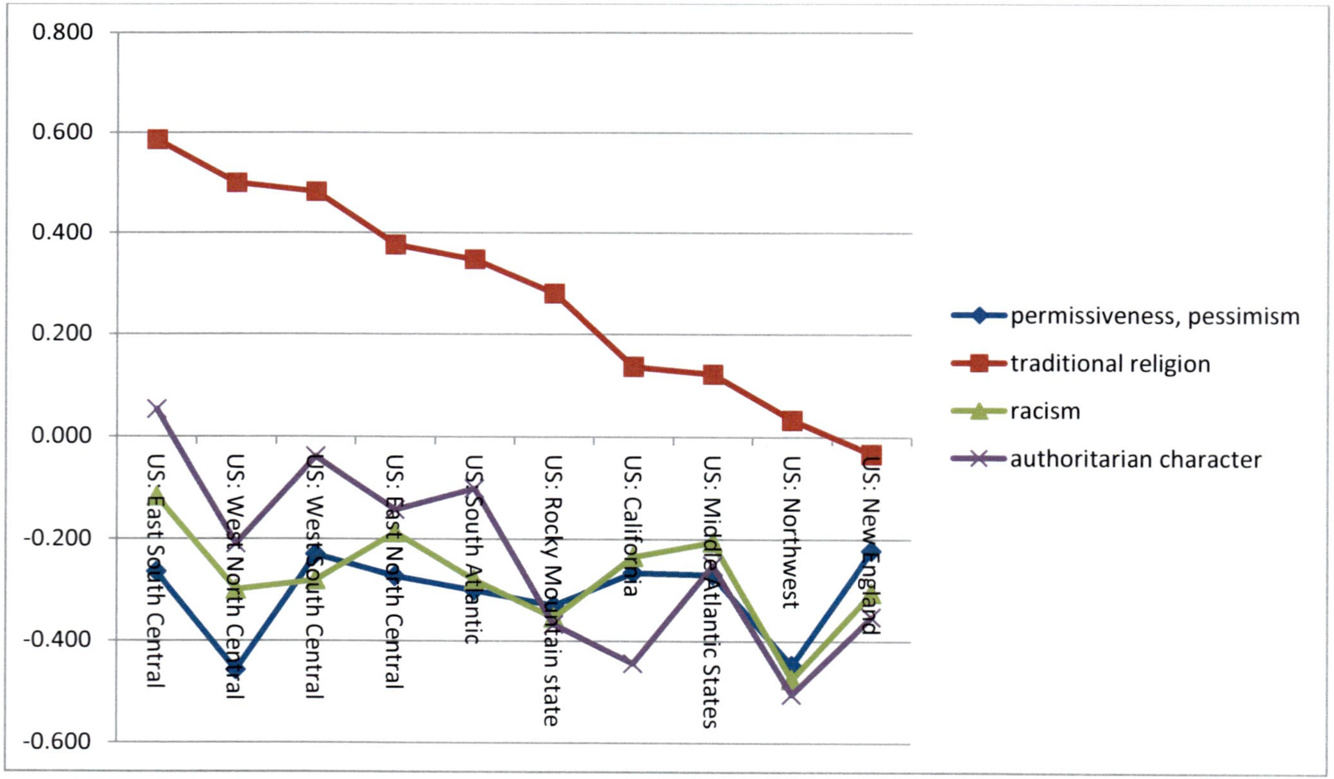

Graph 5.16. The regional value dimension – nine factor model for the different regions in the US.

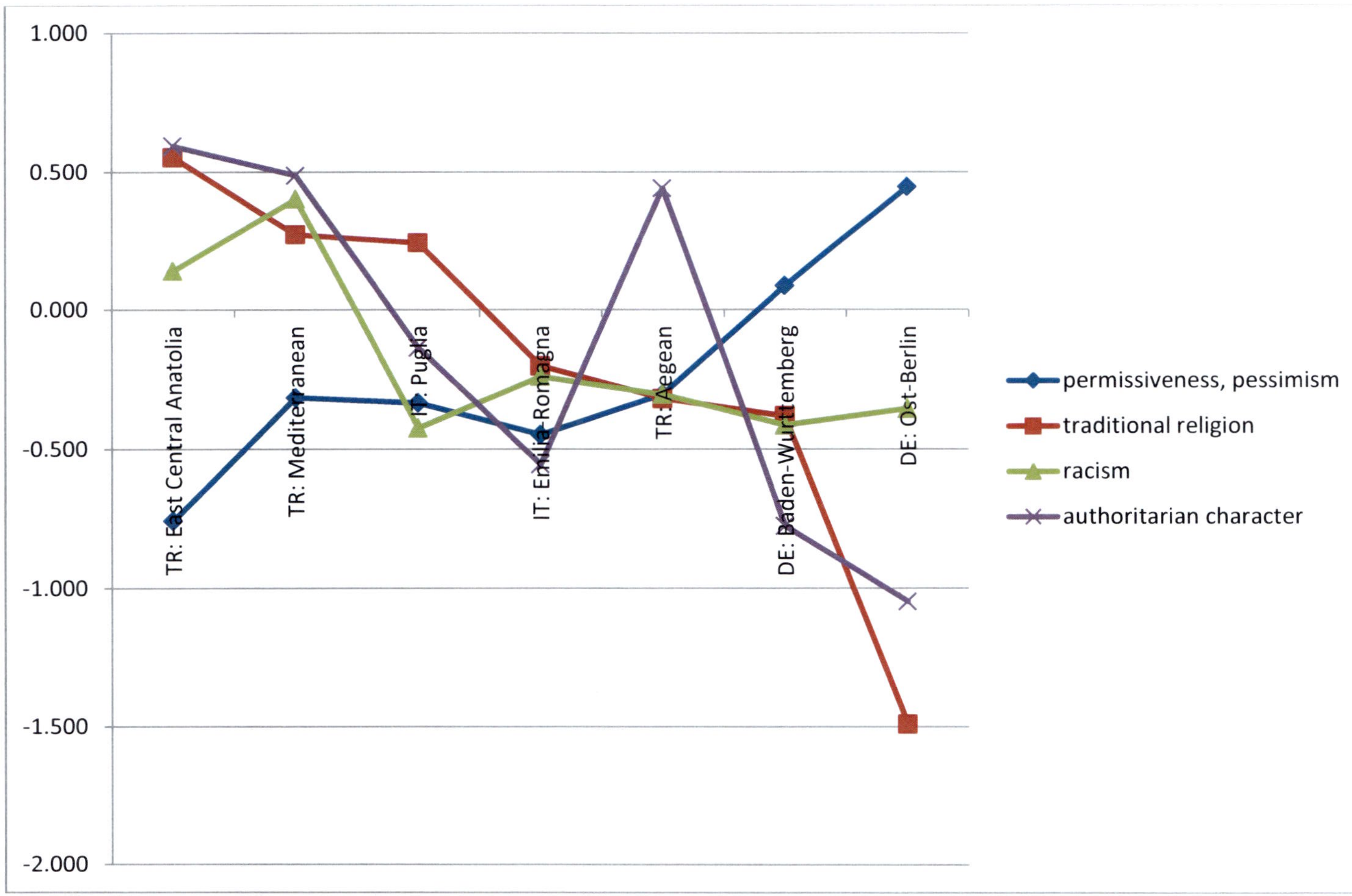

Graph 5.17. The regional value dimension – nine factor model for some European regions.

Table 5.14. The rankings of the best and the worst ranked regions in the European Union and the candidate and potential candidate countries

Region where the interview was conducted	Country	avoiding economic permissiveness	avoiding racism	avoiding distrust of the army and the press	avoiding the authoritarian character	tolerance and respect + post-materialism	avoiding a [predominantly] female rejection of the market economy and democracy	Regional development index	World Rank	Rank in the enlarged EU
SE: Skåne	Sweden	0,745	0,928	0,309	0,999	0,971	0,584	0,756	3	1
FI: Lappi	Finland	0,764	0,910	0,473	0,824	0,893	0,468	0,722	10	5
SE: Jonkoping	Sweden	0,756	0,880	0,238	0,781	0,979	0,528	0,694	21	9
RO: Dolj	Romania	0,889	0,928	0,487	0,294	0,686	0,859	0,690	22	10
DE: Rheinland-Pfalz	Germany	0,659	0,893	0,282	0,743	0,910	0,510	0,666	33	13
IT: Emilia-Romagna	Italy	0,866	0,843	0,281	0,586	0,790	0,569	0,656	43	14
PR: Montana	Portugal	0,860	0,906	0,330	0,430	0,681	0,583	0,632	71	23
CS: Central Serbia	Serbia and Montenegro	0,807	0,887	0,381	0,325	0,618	0,629	0,608	97	36
CY: Kyrenia	Cyprus	0,830	0,760	0,460	0,446	0,585	0,525	0,601	108	42
ES: Castilla Leon	Spain	0,894	0,891	0,330	0,387	0,626	0,447	0,596	115	49
IT: Campania	Italy	0,770	0,834	0,255	0,486	0,786	0,354	0,581	138	58
TR: Eastern Black Sea	Turkey	0,926	0,740	0,446	0,138	0,693	0,534	0,579	141	60
PR: Metropolitana	Portugal	0,742	0,792	0,360	0,317	0,809	0,394	0,569	157	68
SI: Osrednja Slovenska	Slovenia	0,635	0,799	0,204	0,546	0,743	0,463	0,565	169	73
BG: Ruse	Bulgaria	0,809	0,862	0,459	0,256	0,550	0,427	0,561	178	78
EE: Tartumaa	Estonia	0,808	0,691	0,431	0,263	0,504	0,626	0,554	193	84
MK: Ohridski	Macedonia	0,780	0,728	0,152	0,446	0,687	0,468	0,544	222	99
AL: Tirana	Albania	0,644	0,923	0,217	0,411	0,483	0,550	0,538	233	104
DE: Thueringen	Germany	0,750	0,818	0,109	0,474	0,714	0,335	0,533	244	110
CZ: Prague	Czech Republic	0,547	0,759	0,295	0,263	0,693	0,619	0,530	257	115
AL: Center	Albania	0,614	0,743	0,355	0,312	0,546	0,602	0,529	258	116
RO: Arges	Romania	0,212	0,735	0,617	0,433	0,424	0,736	0,526	266	121
HU: Central-Hungary	Hungary	0,528	0,680	0,220	0,346	0,814	0,510	0,516	289	130
FI: Vaasan	Finland	0,733	0,669	0,292	0,468	0,733	0,190	0,514	295	132
BG: Sofia-province	Bulgaria	0,677	0,762	0,395	0,323	0,471	0,453	0,513	298	134

Region where the interview was conducted	Country	avoiding economic permissiveness	avoiding racism	avoiding distrust of the army and the press	avoiding the authoritarian character	tolerance and respect + post-materialism	avoiding a [predominantly] female rejection of the market economy and democracy	Regional development index	World Rank	Rank in the enlarged EU
SK: Bratislava County	Slovakia	0,425	0,747	0,336	0,303	0,699	0,522	0,505	323	138
LT: Vilnius	Lithuania	0,699	0,782	0,363	0,258	0,454	0,447	0,501	335	140
SI: Pomurska	Slovenia	0,588	0,775	0,233	0,375	0,664	0,366	0,500	337	142
MK: Poloski	Macedonia	0,661	0,842	0,077	0,274	0,604	0,498	0,493	362	149
HU: South-Danubian	Hungary	0,639	0,710	0,260	0,307	0,702	0,285	0,484	374	153
ES: Pais Vasco	Spain	0,505	0,878	0,175	0,388	0,550	0,402	0,483	378	155
EE: Isa-Virumaa	Estonia	0,603	0,839	0,286	0,204	0,411	0,544	0,481	383	157
LV: Daugavpils	Latvia	0,707	0,888	0,242	0,126	0,418	0,500	0,480	386	158
CY: Limassol	Cyprus	0,675	0,723	0,390	0,299	0,585	0,188	0,477	396	160
SK: Northern Slovakia	Slovakia	0,471	0,769	0,338	0,199	0,507	0,506	0,465	419	167
TR: Central Anatolia	Turkey	0,821	0,317	0,509	0,318	0,361	0,400	0,455	436	168
LT: Kaunas	Lithuania	0,551	0,711	0,337	0,323	0,423	0,308	0,442	465	171
CZ: Západoèeský kraj - West Bohemia -	Czech Republic	0,499	0,642	0,278	0,195	0,448	0,490	0,426	482	172
LV: Ziemelu reg.	Latvia	0,355	0,881	0,169	0,193	0,423	0,488	0,418	488	174
CS: South East Serbia	Serbia and Montenegro	0,040	0,662	0,244	0,262	0,505	0,309	0,337	509	176

Table 5.15. Country ranks of global values

	avoiding economic permissiveness	traditional religion	avoiding racism	generational education gaps, a growing acceptance of female higher education and the rejection of thrift	avoiding the distrust of the army and the press	avoiding the authoritarian character	tolerance and respect + post-materialism	the 'ego' company (rejection of obedience + unselfishness)	avoiding the rejection of the market economy and democracy
Albania	**43**	**35**	**24**	**56**	**47**	**33**	**37**	**37**	**12**
Argentina	32	32	9	49	66	28	23	30	51
Armenia	54	46	37	11	22	48	62	16	31
Australia	12	51	6	38	54	7	8	49	17
Azerbaijan	**50**	**37**	**25**	**7**	**46**	**38**	**57**	**2**	**16**
Bangladesh	**1**	**11**	**66**	**40**	**3**	**13**	**63**	**10**	**3**
Belarus	59	58	5	43	31	62	58	14	37
Bosnia and Herzegovina	**14**	**36**	**39**	**36**	**17**	**24**	**29**	**44**	**15**
Brazil	60	21	7	46	26	60	38	57	61
Bulgaria	23	59	36	33	12	42	44	20	28
Burkina Faso	45	8	2	57	14	55	66	67	13
Canada	16	41	3	32	35	9	9	43	35
Chile	48	31	23	50	49	20	18	52	66
Cyprus	25	33	42	16	27	30	19	48	33
Czech Republic	51	65	48	64	45	59	33	17	27
Dominican Republic	19	18	32	1	52	36	16	28	30
Estonia	40	64	28	20	37	54	42	5	24
Ethiopia	10	17	27	10	61	8	65	38	49
Finland	24	53	35	53	23	5	4	34	54
Georgia	35	30	30	13	20	49	41	8	14
Germany	33	63	15	62	67	6	5	6	48
Ghana	31	2	46	27	6	56	55	64	6
Guatemala	61	12	11	17	55	40	31	62	65
Hong Kong	37	62	67	55	32	14	60	47	57
Hungary	53	57	55	51	59	31	12	27	43
India	34	29	63	44	4	34	54	42	4
Indonesia	6	4	62	9	10	15	26	21	23

	avoiding economic permissiveness	traditional religion	avoiding racism	generational education gaps, a growing acceptance of female higher education and the rejection of thrift	avoiding the distrust of the army and the press	avoiding the authoritarian character	tolerance and respect + post-materialism	the 'ego' company (rejection of obedience + unselfishness)	avoiding the rejection of the market economy and democracy
Italy	9	34	18	21	44	11	7	29	29
Jordan	21	3	65	19	5	19	24	65	5
Kyrgyzstan	44	42	44	3	36	41	49	11	52
Latvia	56	56	26	30	57	61	36	4	38
Lithuania	47	44	51	29	30	44	45	9	44
Macedonia	**28**	**43**	**58**	**26**	**65**	**25**	**11**	**32**	**40**
Mali	58	16	45	66	7	43	67	58	22
Mexico	57	24	53	24	39	22	34	56	60
Moldova	64	38	29	42	48	46	46	13	53
Morocco	5	9	22	48	16	45	43	39	2
New Zealand	8	54	4	47	42	4	6	40	9
Nigeria	26	1	43	4	34	65	64	61	1
Norway	17	61	14	14	43	1	2	23	26
Peru	41	20	21	5	62	52	28	51	32
Philippines	65	15	54	23	9	53	59	33	59
Poland	36	23	31	65	38	27	15	36	67
Puerto Rico	13	7	20	12	24	37	10	46	36
Romania	20	27	47	37	13	32	30	7	7
Russian Federation	52	67	16	54	41	67	53	12	58
Serbia	**67**	**39**	**56**	**22**	**63**	**35**	**39**	**25**	**50**
Slovakia	55	40	38	63	25	58	35	19	25
Slovenia	46	52	40	35	60	18	17	26	42
South Africa	39	19	49	18	33	51	32	41	39
South Korea	29	48	61	8	28	17	50	1	62
Spain	27	49	8	60	51	26	22	45	47
Sweden	30	66	1	31	56	2	1	24	21
Switzerland	7	50	10	25	50	3	3	15	19
Taiwan	22	55	52	52	58	23	20	3	56
Tanzania	4	5	41	41	2	29	52	63	8

Table 5.15. (Continued)

	avoiding economic permissiveness	traditional religion	avoiding racism	generational education gaps, a growing acceptance of female higher education and the rejection of thrift	avoiding the distrust of the army and the press	avoiding the authoritarian character	tolerance and respect + post-materialism	the 'ego' company (rejection of obedience + unselfishness)	avoiding the rejection of the market economy and democracy
Thailand	63	25	64	67	40	12	27	53	63
Trinidad and Tobago	42	13	13	59	64	63	25	50	41
Turkey	**3**	**28**	**57**	**34**	**15**	**57**	**40**	**31**	**46**
Uganda	49	10	34	2	8	64	56	60	11
Ukraine	62	47	19	39	29	50	51	18	45
United States	15	26	17	45	19	16	14	35	18
Uruguay	18	45	12	58	53	10	13	59	64
Venezuela	38	22	50	15	11	39	21	55	55
Viet Nam	11	60	60	61	1	21	48	22	10
Zambia	66	14	59	6	18	47	61	54	34
Zimbabwe	2	6	33	28	21	66	47	66	20

So, our analysis is fairly pessimistic for the former communist countries and successor states of the former Soviet Union, predominantly Muslim and non-Muslim alike. They excel hardly anywhere by an overwhelmingly positive performance. The history of Communism which began to be implemented in Russia in 1917 destroyed the religious fabric of society and left a hyper-authoritarian society in place. *Russia's global value performance percentile performance scores,* which might be interpreted as a serious question mark about Russia's future trajectory, are the following (see below). The best value performance of Russia is avoiding the distrust of the army and the press, where Russia is only at standardized rank 61 among a hypothetical 100 nations. Avoiding economic permissiveness, Russia is only at standardized rank 78 among a hypothetical 100 nations *et cetera, et cetera.* Regaring traditional religion and avoiding the authoritarian character, Russia has the lowest values on earth.

avoiding the distrust of the army and the press	61%
avoiding economic permissiveness	78%
tolerance and respect + post-materialism	79%
avoiding the rejection of the market economy and democracy	87%
traditional religion	100%
avoiding the authoritarian character	100%

The degree of development of a democratic civil society, characterized by law-abiding and social capital, avoiding racism; trust of the army and the press; no authoritarian character; a high degree of tolerance and respect + post-materialism; and a female acceptance of the market economy and democracy is very poorly developed in several of the countries, admitted into the European Union in 2004 and after.

Just how important the dimension of tolerance, respect and postmaterialism is for *"effective democracy"* and how important the closeness to religion (the traditional religion factor) is for economic growth, we realize when we look at the bi-variate scatterplots. *"Effective democracy"* is defined with Alexander, Inglehart and Welzel, 2011 as the *combination of civil rights with the absence of corruption* (Graph 5.18). The overall Value Development Index shows an interesting trade-off with "effective democracy": the non-linear effect is certainly stronger than the linear effect. Tolerance and respect and postmaterialism explains almost 2/3 of effective democracy, while the closeness to religion explains more than ¼ of economic growth, 2008-2011:

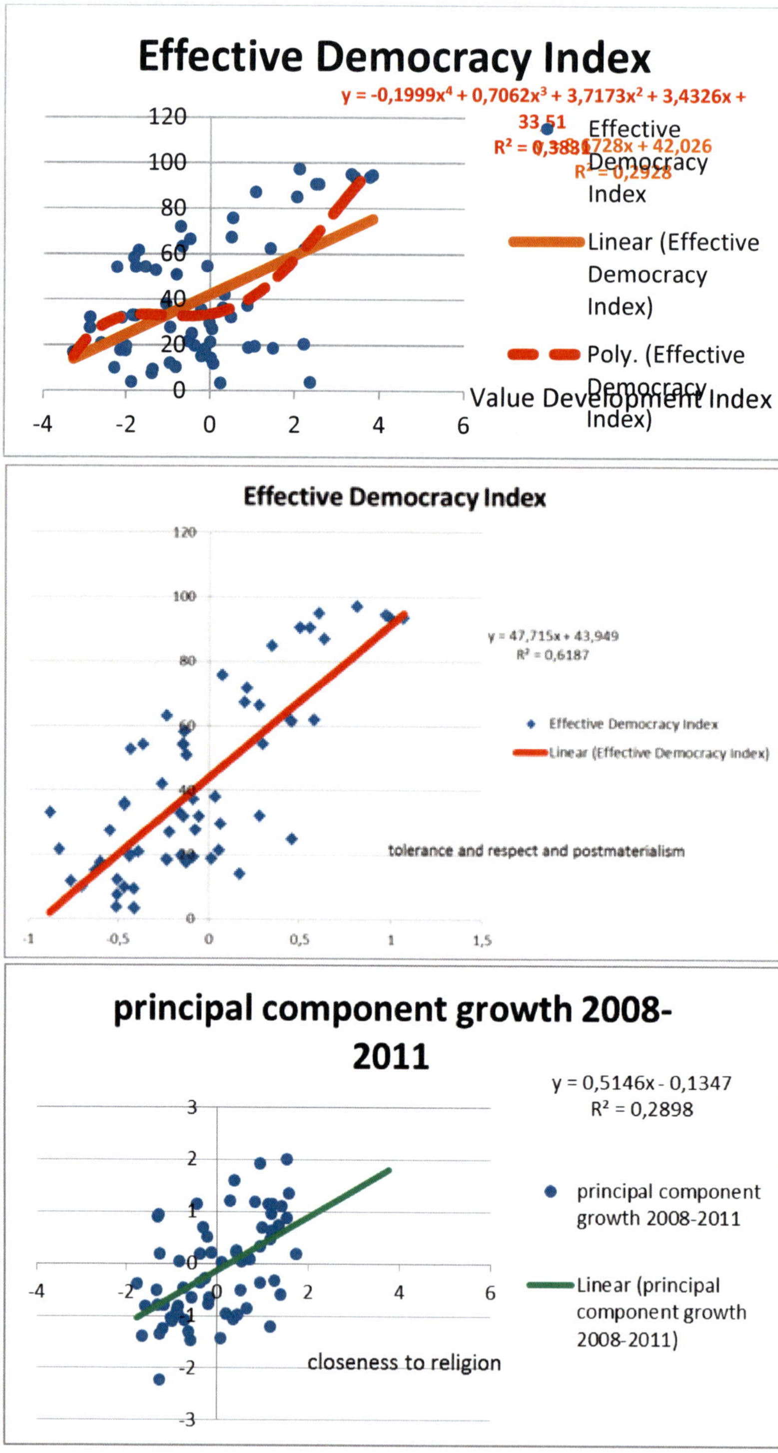

Source: our own calculations from Alexander, Inglehart and Welzel, 2011 and the results of this work. The Alexander, Inglehart and Welzel, 2011 data are freely available from http://ips.sagepub.com/content/33/1/41.refs.

Graph 5.18. Value development and closeness to religion as drivers of *"effective democracy"* and economic growth.

Finding the necessary combination between liberal values and religion will be an important task for democratic forces around the world.

Our materials support the verdict that the member countries of the European Union and the European Commission should carefully weigh the costs and benefits of further enlargements, also in terms of the value balances in comparison to the world's leading democracies and the ascending democracies in Latin America and other regions of the world, which conform much better to the essence of the values of the Enlightenment. What is also at stake was recently spelt out by the Turkish political scientist Prof. Suat Kiniklioglu, who remarked recently:

> "An examination of the intense social media discourse in the aftermath of the coup in Egypt as well as the growing failure in Syria illustrates the increasingly blurred lines between the Turkish nation and the ummah as a unit of analysis in the minds of Turkey's conservatives/Islamists. Atılgan Bayar, who is a consultant to the staunchly pro-government news channel A Haber, announced in August 2013 that he accepts Turkish Prime Minister Recep Tayyip Erdoğan as the "caliph of the earth," a title used by Ottoman sultans after Yavuz Sultan Selim. Indeed, since the Gezi Park protests, we have seen a more pronounced identification of Erdoğan as the head of an imagined ummah rather than merely as the head of the Turkish government. In view of Erdogan's domestic troubles, pro-government commentators have established a link between Erdogan's political fate and that of Muslims everywhere. Erdogan's political fortunes are linked to the fate of Muslims from Syria to Egypt, from Myanmar to Africa. This sort of linkage is, of course, astutely constructed from within to be made use of in the domestic political battle.
>
> When Erdoğan is elevated to the role of head of the ummah, any sort of criticism of Turkish domestic or foreign policy is equated to being a "pawn of the Zionists," the US or other foreign entities. Alternatively, one who longs for the old authoritarian state, has no foreign policy vision or cannot foresee a stronger role for Turkey is dismissed as a Kemalist. *However, Turkey's conservatives/Islamists are entertaining an extremely dangerous and irresponsible political discourse for domestic political ends. Many conservative voters are unaware of Turkey's real capacity, the situation on the ground in the region or perceptions developed in the last two years vis-à-vis Turkey. They have bought into the grandiose narrative that portrays Turkey as the leader of the Islamic ummah.* Exploiting such a religious and unrealistic foreign policy narrative for domestic political objectives is destined to lead to enormous disappointments. Worse, Turkey's reputation and international standing is damaged by this irresponsible exploitation of the Turkish public at large. It is telling that Turkey's conservatives/Islamists desperately feel the need to manipulate the public discourse in such an irresponsible fashion."[38]

---

[38] http://www.todayszaman.com/columnists/suat-kiniklioglu_336109-nation-or-ummah.html

Chapter 6

THE GLOBAL ANALYSIS OF FEMINISM AND ITS REGIONAL IMPLICATIONS FOR THE MUSLIM WORLD: VOICES FROM QUANTIATIVE SOCIAL SCIENCE AND FROM A RELIGIOUS MUSLIM VIEWPOINT

In the following Chapter, we will re-analyze the entire question of feminism, religion and Islam anew.

6.1. MUSLIM FEMINISM – THE VIEW OF A QUANTITATIVE SOCIAL SCIENTIST

In the following analysis, we particularly try to provide global scholarship with a promax factor analysis of the interrelationships between religion, gender, feminism and social background variables, like age and education.

Box 6.1. The overall message of this Chapter

In this Chapter, we re-analyze the question of Islam and feminism, based on an analysis of all respondents from the *World Values Survey*. The Muslim population covered in our survey comprises representatives of 62.6% of the Muslim population of our globe.

The data comprised the following variables:

- Age
- Education level (recoded)
- Highest educational level attained
- How important is God in your life
- How often do you attend religious services (never?)
- Important child qualities: religious faith
- Jobs scarce: Men should have more right to a job than women (reject)
- Sex (Gender)

- University is more important for a boy than for a girl (reject)
- Acceptancy of woman as a single parent

The respondents (all denominations) comprised n = 173231 representative global citizens in 83 countries and territories. After Promax factor analysis, the three factors explained 53.8% of total variance. While the distance to religious practice is explained to some 4% by the education level (correlation between the two factors is 0.192), one can say with certainty that there is no real sharp contradiction between religion and feminism on a global scale. And while gender determines feminist convictions, contained in our analysis to some 40%, it is also evident that feminist convictions are not only held by women, but also increasingly by enlightened men, non-Muslims and Muslims alike. Interestingly enough, our data also show that people supporting typical feminist contentions, like female access to tertiary education and jobs even at a time of crisis (Factor 3), are not necessarily too strongly in support of the acceptancy of women as a single parent (factor loading 0.352, i.e., 12.39% of variance explained), which is a form of household organization so common now in Western countries (the argument is that marriage is an outdated institution *et cetera)*. Support for single parenthood by women is rather an expression of the distance towards religion around the globe (factor loading of 0.431, i.e., 18.58% of variance explained).

The correlations between the factors suggest that the often described contradiction between feminism and religion hardly exists today, because the squared correlation coefficients indicate that the common variance between the religious and the feminist dimension is below 5%. The correlation between secularism and feminism is only 0.106, i.e., only 1.12% of the variance of feminism is explained by secularism, a long distance from the often militant contentions of the critics of religion around the globe about the incompatibility between any form of religion and feminism.

We also present the Choropleth maps of global feminism and analyze the regional implications of our analysis for the Muslim countries in the analysis. For the first time in global literature, we present data on the regional aspects of feminism in such countries as Saudi-Arabia.

Influential *World Values Survey* oriented studies on the issue of global feminism were, among others, Cao and Stack, 2010; Hayes, McAllister, and Studlar, 2000; and Inglehart, and Norris, 2003b, 2003c, and the studies Inglehart, 2002, 2003c, 2007; Inglehart and Baker, 2000; Inglehart and Norris, 2003a, 2003b, 2003c, 2010, Inglehart and Welzel, 2003, and 2005. Most of these studies strongly believed that the real difference between the West and the Muslim world is not the commitment towards democracy, which is now very strong in the Muslim world, but different views on gender and family issues. By re-analyzing the data from n = 173231 representative global citizens in 83 countries and territories, we are permitted to compare the average values in different regions of the world. The large samples allow us for the first time in cross-national value research to draw comparisons also on the regional level for many countries around the world.

Policy pointer:

To our great surprise, we learn that instead of talking about gender and politics in "the West" and in "the Muslim Orient", there is rather a very wide range of variations within the two civilizations, making underdeveloped regions in the West much more similar to the "average Orient", while at the same time showing the already existing advances in feminism and values of gender justice in the more highly developed regions of the Muslim world.

Our Table 6.2 destroys the myth of value "monolithic" Muslim countries, and instead, a starling variety of regional convictions and value patterns emerge, even in countries, where Western knowledge hitherto had little knowledge about them, like Saudi-Arabia.

The Muslim countries contained in our analysis are Albania; Algeria; Azerbaijan; Bangladesh; Burkina Faso; Egypt; Indonesia; Iran; Jordan; Kyrgyzstan; Mali; Morocco; Nigeria; Pakistan; Saudi Arabia; Turkey; and Uganda. They already make up none the less than an estimated 62.55% of the entire Muslim population of our globe. In addition, our analysis is also representative of the large Muslim minorities in China, Ethiopia, India and Russia, whose population comprise an additional estimated 17.55% of the global *'Umma'*. Our analysis is thus the largest ever undertaken analysis on gender relations and religion in the world and in Muslim countries. A peculiarity of the *World Values Survey* Project must be taken into consideration here: when you introduce only a small additional number of variables into the multivariate analysis you often risk reducing the number of individuals with complete data often dramatically. This is mainly due to large cross-country differences in the actual questionnaires fielded in the different countries. In the present analysis, we kept the variables to a minimum. Nevertheless we hope to be able to tap with them the dimensions of religion, gender, feminism and social background variables, like age and education:

- Age
- Education level (recoded)
- Highest educational level attained
- How important is God in your life
- How often do you attend religious services (never?)
- Important child qualities: religious faith
- Jobs scarce: Men should have more right to a job than women (reject)
- Sex (Gender)
- University is more important for a boy than for a girl (reject)
- Woman as a single parent

Our factor analysis explains almost 54% of the total variance. Our analysis is based on none the less than 173231 representative individuals in 83 countries.

Our analysis wielded three factors with an *Eigen value* of higher or equal to one.

One can say with certainty that there is no real sharp contradiction between religion and feminism on a global scale. Distance to religious practice has only 4% of the variance in common with the education level (correlation between the two factors is 0.192). Gender determines feminist convictions to some 40%. Furthermore, it is also evident that feminist convictions are not only held by women, but also increasingly by enlightened men. Interestingly enough, our data also show that people supporting typical feminist contentions, like female access to tertiary education and jobs even at a time of crisis (Factor 3), are not necessarily too strongly in support of secularism or the acceptancy of women as a single parent (factor loading 0.352, i.e., 12.39% of variance explained). Single parenthood is a form of household organization very common in Western countries. The argument is that marriage is an outdated institution *et cetera*. Support for single parenthood by women is rather an expression of the distance towards religion (factor loading of 0. 431, i.e., 18.58% of variance explained), and not so much an affirmation of feminism

Table 6.1.a. Feminism and religion re-analyzed. Data from n = 173231 representative global citizens in 83 countries and territories

	initial	**extraction**
Important child qualities: religious faith	1,000	0,544
Jobs scarce: Men should have more right to a job than women	1,000	0,387
Woman as a single parent	1,000	0,281
University is more important for a boy than for a girl	1,000	0,459
How often do you attend religious services	1,000	0,570
How important is God in your life	1,000	0,666
Sex (Gender)	1,000	0,438
Age	1,000	0,190
Highest educational level attained	1,000	0,930
Education level (recoded)	1,000	0,921

Table 6.1.b. Feminism and religion re-analyzed. Cumulative variance explained

	Eigen values	**% of variance**	**cumulated percentage**
1,000	2,401	24,009	24,009
2,000	1,768	17,677	41,686
3,000	1,216	12,162	53,848

Table 6.1.c. Feminism and religion re-analyzed. The factor loadings

	distance to traditionalist forms of religion	**Generation gap in educational level**	**feminism**
Important child qualities: religious faith	**-0,736**	-0,093	-0,087
Age	0,115	***-0,391***	-0,034
How important is God in your life	**-0,814**	-0,132	-0,032
How often do you attend religious services (never)	**0,751**	0,081	0,035
Education level (recoded)	0,200	**0,959**	0,108
Highest educational level attained	0,204	**0,964**	0,113
Woman as a single parent	***0,431***	0,100	***0,352***
Jobs scarce: Men should have more right to a job than women (reject)	0,213	0,154	**0,600**
Sex (Gender)	-0,155	-0,119	**0,606**
University is more important for a boy than for a girl (reject)	0,111	0,173	**0,669**

The correlations between the factors suggest that the often described contradiction between feminism and religion hardly exists today. The squared correlation coefficients indicate that the common variance between the religious and the feminist dimension is below 5%. The correlation between secularism and feminism (see Table 6.1.d) is only 0.106, i.e., only 1.12% of the variance of feminism is explained by secularism. This is a long distance from the often militant contentions of the critics of religion around the globe about the incompatibility between religion and feminism.

Table 6.1.d. Feminism and religion re-analyzed. Correlation matrix of components

	correlation matrix of the components		
	distance to traditionalist forms of religion	**Generation gap in educational level**	**feminism**
distance to traditionalist forms of religion	1,000	0,192	0,106
Generation gap in educational level	**0,192**	1,000	0,109
feminism	**0,106**	**0,109**	1,000

n = 173231 respondents with complete data (global sample = 257597 representative respondents).

In Table 6.1.e we present the country results of our analysis. The countries belonging to the Organization of Islamic Cooperation (OIC) are marked in green color. There is a startling variety of feminisms and secularisms in the world today. The feminism among the population in predominantly Muslim Bosnia is ranked equally with that in Mexico and Guatemala, and the people of predominantly Muslim Albania are as feminist as the populations in neighboring Romania and the Czech Republic. Albania, the Czech Republic and Romania all had a very harsh communist rule after 1945. Feminism in Turkey is at the same level of development as the one in China and Belarus, *et cetera.*

Policy pointer:

Regarding feminism's concerns, the predominantly Muslim country of Indonesia is not very different from the neighboring predominantly Catholic country of the Philippines. Simple facts of concomitant neighboring geography – and not the works of Professor Samuel Huntington - explain to us why distance to religion in Bosnia and Herzegovina is at about an equal level as in neighboring Serbia and Macedonia; and why the distance to religion in former communist predominantly Muslim Albania is about at the same level as in former communist predominantly Muslim Kyrgyzstan and Azerbaijan, with similar values also for former communist and predominantly Orthodox Moldova.

Turkey's religiosity or distance to religion depends to a large extent on the region of Turkey which you analyze, with the country averages comparable to those of the United States, Colombia, Poland, Mexico, or India. The deeply Roman Catholic population of El Salvador is at the same level of religiosity as the population of Saudi Arabia, and they all are in the same league of religiosity as Catholic Guatemala, predominantly protestant Zambia, and Muslim Algeria and Iran.

Table 6.1.e. Feminism and religion re-analyzed (country results)

OIC members are marked in green color

Country/region	distance to traditionalist forms of religion	Generation gap in educational level	feminism
Albania	0,264	-0,342	0,138
Algeria	-0,682	0,270	-0,403
Andorra	1,059	0,295	0,371
Argentina	0,095	-0,366	0,189
Armenia	0,409	0,367	-0,436
Australia	0,693	0,352	0,121
Azerbaijan	0,139	0,506	-0,435
Bangladesh	-0,817	-0,215	-0,853
Belarus	0,721	0,083	-0,204
Bosnia and Herzegovina	0,295	0,055	0,202
Brazil	-0,479	-0,315	0,390
Bulgaria	0,854	0,015	0,274
Burkina Faso	-0,866	-0,741	-0,416
Canada	0,282	0,128	0,388
Chile	0,000	-0,105	0,186
China	0,727	-0,352	-0,134
Colombia	-0,189	0,392	0,492
Cyprus	0,100	0,271	0,236
Czech Republic	1,259	-0,210	0,131
Dominican Republic	-0,347	1,061	0,468
Egypt	-0,885	-0,352	-0,596
El Salvador	-0,638	-0,204	0,480
Estonia	1,104	0,120	-0,001
Ethiopia	-0,562	-0,196	0,366
Finland	0,760	-0,098	0,327
France	1,062	-0,283	0,413
Georgia	0,045	0,348	-0,050
Germany	1,040	-0,113	0,276
Ghana	-1,048	-0,597	-0,234
Great Britain	0,828	0,271	0,242
Guatemala	-0,662	-0,019	0,199
Hong Kong	1,175	0,016	0,110
Hungary	0,882	-0,160	0,292
India	-0,246	-0,064	-0,489
Indonesia	-0,992	0,392	-0,301
Iran	-0,699	0,112	-0,677
Italy	0,058	0,137	0,255
Japan	0,955	0,418	0,299
Jordan	-1,008	-0,092	-0,620
Kyrgyzstan	0,164	0,464	-0,210

Country/region	**distance to traditionalist forms of religion**	**Generation gap in educational level**	**feminism**
Latvia	0,773	0,329	0,065
Lithuania	0,467	0,157	0,102
Macedonia	0,336	-0,021	0,198
Mali	-0,788	-0,768	-0,598
Mexico	-0,202	-0,016	0,218
Moldova	0,148	0,124	-0,050
Morocco	-0,799	-0,721	-0,673
Netherlands	1,033	-0,129	0,385
New Zealand	0,943	0,325	0,447
Nigeria	-0,979	0,172	-0,447
Norway	1,089	0,312	0,488
Pakistan	-1,180	-0,372	-0,361
Peru	-0,365	0,198	0,113
Philippines	-0,721	0,214	-0,347
Poland	-0,195	-0,269	0,268
Puerto Rico	-0,574	0,529	0,299
Romania	-0,175	0,025	0,176
Russian Federation	0,900	0,223	-0,093
Rwanda	-0,776	-0,792	-0,110
Saudi Arabia	-0,646	0,472	-0,812
Serbia	0,337	0,144	0,356
Serbia and Montenegro	0,641	-0,116	0,001
Singapore	-0,336	-0,129	-0,065
Slovakia	0,489	-0,113	0,130
Slovenia	0,703	-0,058	0,223
South Africa	-0,495	0,012	0,111
South Korea	0,531	0,704	0,024
Spain	0,655	-0,436	0,277
Sweden	1,270	0,330	0,484
Switzerland	0,699	0,266	0,360
Taiwan	0,724	0,259	-0,195
Tanzania	-0,899	-0,282	0,116
Thailand	-0,107	-0,512	0,083
Trinidad and Tobago	-0,616	-0,337	0,235
Turkey	-0,237	-0,334	-0,153
Uganda	-0,749	0,102	0,079
Ukraine	0,629	0,206	0,006
United States	-0,147	0,256	0,240
Uruguay	0,570	-0,463	0,586
Venezuela	-0,256	0,064	0,193
Viet Nam	0,851	-0,431	-0,212
Zambia	-0,669	-0,082	0,015
Zimbabwe	-0,915	-0,537	-0,111

6.2. Women in Islam – A Voice from a Religious Muslim Viewpoint

One of the most often encountered prejudices in the West against Islam as such in the 21st Century is to portray the Muslim world as a region where there is no significant feminist movement or that no significant rethinking of the religious traditions in the light of gender justice has taken place. However, this is not true.

One cannot deny first, that women in many Muslim countries are indeed and in fact still put in an inferior condition *vis-à-vis* men, if they are not subjected to them. But such is also the condition of people without gender consideration in countries where violations of public liberties and freedoms of opinion and expression are "routine."

In the Arab world, the rise of women is to be achieved as part of society's advancement towards freedom as advocated by the third and the fourth Arab Human Development Reports (2004, 2005). Arab countries can achieve this goal by eradicating all infringements on human dignity and specifically by guaranteeing full citizenship and the enjoyment of all human rights for all women on an equal footing with men.

Policy pointer:
A feminist movement has been active among both local and global Muslim women since many years, as we can infer from the life and works of the Bengali Roquia Sakhawat Hussain (1880 – 1932), the Azerbaijani Hamida Javanshir (1873 –1955), the Lebanese May Ziadeh (1886-1941), the Pakistani Zaib-un-Nissa Hamidullah (1921 – 2000), the Iranian Farideh Mashini (d. 2012), the Egyptian Nawal al-Saadawi, the Moroccan Fatema Mernissi, the French Fadela Amara and Rama Yade, the Sudanese Lubna Ahmed al-Hussein, the American Saleemah Abdul-Ghafur and Amina Wadud, and the Canadian Raheel Raza, to mention just a few examples.

Margot Badran"offered a concise definition of Islamic feminism gleaned from the writings and work of Muslim protagonists as a feminist discourse and practice that derives its understanding and mandate from the Qur'an, seeking rights and justice within the framework of gender equality for women and men in the totality of their existence. Islamic feminism explicates the idea of gender equality as part and parcel of the Qur'anic notion of equality of all insan (human beings) and calls for the implementation of gender equality in the state, civil institutions, and everyday life. It rejects the notion of a public/private dichotomy (by the way, absent in early Islamic jurisprudence or fiqh) conceptualising a holistic umma in which Qur'anic ideals are operative in all space." (Badran, 2006: http: //www. countercurrents. org/gen-badran100206. htm)

Another Muslim woman explained why she is feminist, thus:

> "While Islam has undoubtedly granted women many rights—some of which were radical for much of the world in the 7th century, such as the rights to divorce, consent in marriage, education, and financial independence—many Muslim women around the world are denied those rights in practice. That these rights were "radical" for the 7thcentury is significant: one would think that this is an indication that our rights should be "radical" in all times. What Islamic feminism does is to help us deal with this tension of the existence of women's rights in

theory but their denial in practice" (http: //feminismandreligion. com/2013/08/02/why-i-am-an-islamic-feminist-by-orbala/)

A single blog (i.e., Muslim Feminists: http: //muslimfeminists. tumblr. com/) presented as many as 17 links to Muslim women organizations, only in the USA, such as:

- American Muslim Women's Association: http: //www. amwa. us/
- American Society for Muslim Advancement (ASMA): http: //www. asmasociety. org/
- Muslims for Progressive Values USA (MPVUSA): http: //www. mpvusa. org/
- Arab-American Family Support Center: http: //www. aafscny. org/
- AWAAM: Arab Women Active in the Arts and Media: http: //www. awaam. org/
- Karamah: Muslim Women Lawyers for Human Rights: http: //www. karamah. org/
- Femin Ijtihad: http: //feminijtihad. com/
- Milwaukee Muslim Women's Coalition: http: //mmwconline. org/
- Muslim Women in the Arts (MWIA): http: //mwiarts. wordpress. com/
- Muslim Women Resource Center: http: //www. mwrcnfp. org/
- Muslim Women's Fund: http: //www. womensfundingnetwork. org/the-network/member/muslim-womens-fund
- Muslim Women's League: http: //www. mwlusa. org/
- Muslimah Compass: http: //www. muslimahcompass. org/
- Muslim Women's Giving Circle: http: //muslimwomensgiving. org/
- Sisters United in Human Service: http: //www. sistersunited. org/
- Sufi Women Organization: http: //www. sufiwomen. org/
- Women In Islam, Inc.: http: //www. womeninislam. org/

Policy pointer:
The Muslim women movements are not a simple import; many could be probably considered to be an extension of a social women's movement born in the early 20th Century, whose leaders have been advocating women education and emancipation. We cited a few examples previously; but men also have participated to this movement: Ahmed Amin (1886-1954: Egypt); Tahir Haddad (1899- 1935: Tunisia), and so on.

Away from some Western stereotypes about gender issues in the world of Islam, and assuming that religious creeds are a determining factor in social interaction, there is no difference in Islam between men and women regarding their behavior in this world, their rights and responsibilities. The Noble Quran is more than clear about this:

> "Every soul will be (held) in pledge for its deeds" (Quran 74: 38).
>
> For Muslim men and women— for believing men and women, for devout men and women, for true men and women, for men and women who are patient and constant, for men and women who humble themselves, for men and women who give in charity, for men and women who fast (and deny themselves) for men and women who guard their chastity and for men and women who engage much in Allah's praise— for them has Allah prepared forgiveness and great reward. *(33: 35)*

As Patel (1997) observed,

> "A number of Islamic virtues are mentioned here, but the primary message of this ayah is that these virtues are applicable to both, women as well as men. Both sexes have human rights and duties to an equal degree, and the rewards of the Hereafter are available to men and women alike. Each individual will be judged according to his or her deeds. Gender is simply not an issue in this matter."

Subsequently, we think that a moderate discourse avoiding the outright condemnation of religion as the cause of sexism will find a favorable echo in the Muslim world of today. The reason for this is simple. As we will show here, Islam's position regarding women should be distinguished from the history of oppression that women have undergone everywhere on the globe, and in the region, too. Feminism is today accepted in all societies. Certainly, in the majority of countries of the Muslim world, radical feminism will be a trend that meets with difficulties, and like all radical social movements, is marginalized.

Woman, according to the Noble Quran, are not blamed for Adam's first mistake. Both were jointly wrong in their disobedience to God, both repented, and both were forgiven (Quran 2: 36, 7: 20 – 24). In one verse in fact (20: 121), Adam specifically, was blamed. (Badawi, 1971) In this specific context, there is a *"spiritual equality of sexes"*, (Patel: 1997; Khan: 2008), which a number of verses show in the Noble Quran, such as:

> And their Lord has accepted of them, and answered them: Never shall I suffer to be lost the work of any of you, be he male or female: you are members, one of another ... [AI 'Imran 3: 195]
>
> Whoever works righteousness, man or woman, and has Faith, verily, to him will We give a new Life, and life that is good and pure, and We will bestow on such their reward according to the best of their action. [al-Nahl 16: 97]
>
> Say: Come, I will rehearse what Allah has (really) prohibited you from: join not anything as equal with Him; be good to your parents; kill not your children on a plea of want - We provide sustenance for you and for them - come not nigh to shameful deeds, whether open or secret; take not life, which Allah has made sacred, except by way of justice and law: thus does He command you, that you may learn wisdom. [al-an 'am 6: 151]
>
> When the female (infant) buried alive, is questioned - for what crime she was killed [al-Takwir 81: 8-9]
>
> When news is brought to one of them, of (the birth of) a female (child), his face darkens and he is filled with inward grief! With shame does he hide himself from his people, because of the news that he has had! Shall he retain it on (sufferance and) contempt, or bury it in the dusts Ah! What an evil (choice) they decide on. [al-Nahl 16. 58-59]

We should note that sexual relations in Islam (as in the other monotheistic religions) are permitted only within marriage.

> "And among His signs is this: That He created mates for you from yourselves that you may find rest, peace of mind in them, and He ordained between you love and mercy. Lo, herein indeed are signs for people who reflect." (Quran 30: 2 1).

However, according to Islamic Law (several Hadiths of the Prophet), women cannot be forced to marry anyone without consent.

Indeed, Islam does not accept as right the position of some feminists who

> "seek to steer women away from marriage and childbearing, as if women were unable to find happiness in these. It is a sign of arrogance on the part of feminists to make such assumptions. Even worse, they contend that those women who have found happiness and security in marriage are only fooling themselves." (Patel, 1997)

For the Noble Quran, marriage between a man and a woman is the only acceptable way:

> And Allah has made for you mates (and companions) of your own nature, and made for you, out of them, sons and daughters and grandchildren, and provided for you sustenance of the best: will they then believe in vain things, and be ungrateful for Allah's favours? [al-Nahl 16: 72] (See also al'Baqarah 2: 187; al-Shura 42: 11; al-Nisa '4: 1)

Contemporary Muslim authors think that extremist currents in the Western feminist movement were a reaction against the oppression of women in the respective societies. In his *essay The Subjection of Women (1869)*, John Stuart Mill wrote:

> We are continually told that civilization and Christianity have restored to the woman her just rights. Meanwhile the wife is the actual bondservant of her husband; no less so, as far as the legal obligation goes, than slaves commonly so called.

There is indeed a long history of oppression regarding women in ancient and modern civilizations. Several Muslim authors addressing the status of women in Islam compare it to their status in different other civilizations. According to Badawi (1971):

> "In Hindu scriptures, the description of a good wife is as follows: 'a woman whose mind, speech and body are kept in subjection, acquires high renown in this world, and, in the next, the same abode with her husband.'
>
> In Athens, women were not better off than either the Indian or the Roman women.
>
> Athenian women were always minors, subject to some male - to their father, to their brother, or to some of their male kin. Her consent in marriage was not generally thought to be necessary and 'she was obliged to submit to the wishes of her parents, and receive from them her husband and her lord, even though he were stranger to her.'
>
> A Roman wife was described by an historian as: 'a babe, a minor, a ward, a person incapable of doing or acting anything according to her own individual taste, a person continually under the tutelage and guardianship of her husband.'
>
> In the Encyclopedia Britannica, we find a summary of the legal status of women in the Roman civilization:
>
> In Roman Law a woman was even in historic times completely dependent. If married she and her property passed into the power of her husband ... the wife was the purchased property of her husband, and like a slave acquired only for his benefit. A woman could not exercise any civil or public office, could not be a witness, surety, tutor, or curator; she could not adopt or be adopted, or make will or contract. Among the Scandinavian races women were: under perpetual tutelage, whether married or unmarried. As late as the Code of Christian, at the end of the 17th

> Century, it was enacted that if a woman married without the consent of her tutor he might have, if he wished, administration and usufruct of her goods during her life."

This history of oppression was therefore legally justified and nourished by women's own ignorance and forced illiteracy.

Women in the pre-Islamic Arab society also were no less oppressed. Baby-daughters were often killed. Islam delivered them from their predicament and improved considerably their condition, supporting their right to seek knowledge. Prophet Muhammad (PBUH) said:

> "Seeking knowledge is mandatory for every Muslim".

However, even after the apparition of three monotheistic religions,

> "The woman in Europe and all over the world was looked upon as a mere nonentity. She formed the theme of many a discourse of the learned 'scholars' and 'philosophers' who wrangled among themselves over questions such as these: Has woman got a soul or not? If yes, then what precisely is the nature of her soul, is it human or animal? Supposing she does possess a human soul then what social and human position should she occupy in relation to man? Is she born as a slave to man or does she hold a position slightly superior to that of a slave?" (Kutub, 1984).

Some Muslim writers argued that

> "This position of woman in Europe remained unaltered during the periods of serfdom and feudalism. She in her ignorance was blandished sometimes by luxury and license and at times was content to live as animals-eating, drinking, bearing children, giving birth to others, and working day and night. When the industrial revolution took place in Europe it brought in its wake the worst possible sufferings for woman yet experienced by her throughout the history of mankind." (Kutub, 1984)

The whole family life was upset when women and children were forced to work in factories in order to get a living. The weakening of the family ties under the burden of material needs would lead gradually to a disintegration of the family. Women and children more than men were ruthlessly exploited by the system and underpaid. The First World War added more misery to this dark picture. Millions of young men in Europe and the USA had to fight. Many lost their lives. Millions of widowed women had to survive by their own means.

In the eyes of Muslim writers, this was a deplorable condition:

> "To go and work in the factories, however, required that woman should ignore her moral temperament as well as her feminine nature which bad now become a positive hurdle in earning her living. Moreover, factory-owners did not want working hands merely; they wanted to satisfy their lust as well. The helplessness of woman now promised them an excellent opportunity of which they fully availed themselves. The woman thus had to discharge a two-fold duty: work in the factories, and try her level best to please her employer." (Kutub, 1984)

Moreover, wages paid to women were far less than those paid to men doing the same work. The dual standards in law favored men over women. Patel (1997) reminds us that

> "As late as the nineteenth century, oppressive marriage laws were still restricting women with regard to earnings. In the event of a divorce, women were further humiliated by being denied access to their children and being cut off from any source of maintenance. The divorce laws were heavily biased in favour of men."

In their book, *Marriage East and West*, David and Vera Mace (1960) wrote:

> Let no one suppose, either, that our Christian heritage is free of such slighting judgments. It would be hard to find anywhere a collection of more degrading references to the female sex than the early Church Fathers provide. Lecky, the famous historian, speaks of (these fierce incentives which form so conspicuous and so grotesque a portion of the writing of the Fathers ... woman was represented as the door of hell, as the mother of all human ills. She should be ashamed at the very thought that she is a woman. She should live in continual penance on account of the curses she has brought upon the world. She should be ashamed of her dress, for it is the memorial of her fall. She should be especially ashamed of her beauty, for it is the most potent instrument of the devil). One of the most scathing of these attacks on woman is that of Tertullian: Do you know that you are each an Eve? The sentence of God on this sex of yours lives in this age: the guilt must of necessity live too. You are the devil's gateway: you are the unsealer of that forbidden tree; you are the first deserters of the divine law; you are she who persuades him whom the devil was not valiant enough to attack. You destroyed so easily God's image, man. On account of your desert – that is death – even the Son of God had to die). Not only did the church affirm the inferior status of woman, it deprived her of legal rights she had previously enjoyed. *(Mace, 1960; Badawi, 1971)*

Under these circumstances, women's revolt was just normally expected.

> "The earliest feminist campaigners demanded an end to the double standard of sexual morality, but this did not mean that they sought an overall lowering of moral standards: the early feminists saw chastity not as oppressive, but as both natural and necessary," says Patel (1997).

However, this revolt, though absolutely justified, engaged into a radical path with the *The New York Feminist Manifesto* of 1971, which states:

> "Radical feminism recognises the oppression of women as a fundamental political oppression wherein women are categorised as an inferior class based upon their sex. It is the aim of radical feminism to organise politically to destroy this sex class system. As radical feminists we recognise that we are engaged in a power struggle with men, and that the agent of our oppression is man in so far as he is identified with and carries out the supremacy privileges of the male role. For while we realise that the liberation of women will ultimately mean the liberation of men from their destructive role as oppressor, we have no illusions that men will welcome this liberation without struggle. Radical feminism is political because it recognises that a group of individuals - men - have set up institutions throughout society to maintain this power."

Some of those radical feminists argued that

> "in order to make a complete commitment to feminism, a woman has to be or become a lesbian. A leading radical feminist, Bunch believed that only lesbians can be serious feminists, and that lesbianism is best understood as a revolutionary rejection of all males and male-defined institutions. Adrienne Rich suggested that compulsory heterosexuality is the central social structure perpetuating male domination." (Patel, 1997).

It is obvious; first, that such a revolt was primarily against the subjection of women in the Western world. Feminism initially was not concerned with Islam or the Islamic view of gender, as Muslims were not much visible on the Western social and political stage at the time. Second, Islam, which is essentially against all kinds of radicalism and fanaticism, would not support such a struggle.

The fundamental principal of Islam holds that woman is a human being; and she has a soul similar to that of man.

> O mankind! Reverence your Guardian-Lord Who created you from a single person, created, of like nature, his mate and from them twain scattered (like seeds) countless men and women;— fear Allah, through Whom ye demand your mutual (rights) and (reverence) the wombs (that bore you): for Allah ever watches over you. ` (Quran, 4: 1)

In Islam, men and women being equal in their origin are thus entitled to similar rights: the right to life, honor, property and respect. The same laws are applicable to women and men alike.

> "O ye who believe! let not some men among you laugh at others: it may be that the (latter) are better than the (former): nor let some women laugh at others: it may be that the (latter) are better than the (former): nor defame nor be sarcastic to each other, nor call each other by (offensive) nicknames: Ill-seeming is a name connoting wickedness, (to be used of one) after he has believed: And those who do not desist are (Indeed) doing wrong" (Al-Hujraat: 49: 11)
>
> "Men and women are equal in their rights to realize their material needs in the world including similar rights to hold property, and dispose of it as they should wish. They are free to mortgage, to give it in lease, or bequeath it, sell or buy it or exploit it for his or her own benefit." (Kutub, 1984)
>
> "From what is left by parents and those nearest related there is a share for men and a share for women, whether the property be small or large, — a determinate share."[Quran: 4: 7]
>
> "And in no wise covet those things in which Allah hath bestowed His gifts more freely on some of you than on others: to men is allotted what they earn, and to women what they earn: but ask Allah of His bounty: for Allah hath full knowledge of all things." [Quran: 4: 32]

Policy pointer:

However, while acknowledging a perfectly equal status as human beings for both men and women, and treating them as equals, entitled to equal rights Islam does however differentiate between man and woman with regard to their special functions in life. (Kutub, 1984).

The Noble Quran thus states:

> "And they (women) have rights similar to those (of men) over them, and men are a degree above them." (*Quran 2: 228).*
>
> Such a decree is called *Quiwama* (maintenance and protection). This refers to that natural difference between the sexes which entitles the weaker sex to protection. It implies no superiority or advantage before the law.
>
> "Men are appointed guardians over women, because of that in respect of which Allah has made some of them excel others, and because the men spend of their wealth. So virtuous women are obedient and safeguard, with Allah's help, matters the knowledge of which is shared by them with their husbands." (4: 35)

Most feminist antagonistic positions against Islam in the West and among the Arab left-wing are founded on the issue of *Quiwama*. Why should men, they ask, enjoy the right of "protecting" women, while protection paves the way to exploitation and abuse and while women, being financially autonomous, can perfectly assure their self-protection?

Yet, man's role of leadership in relation to his family does not mean the husband's dictatorship over his wife. Islam emphasizes the importance of taking counsel and mutual agreement in family decisions. (Badawi, 1971) The Quran gives us an example:

> "…If they (husband wife) desire to wean the child by mutual consent and (after) consultation, there is no blame on them…" (Quran 2: 233)

Radical feminists would not accept the concept of *Quiwama*, though. According to Shulamith Firestone, *"Nature produced the fundamental inequality which was later consolidated and institutionalised in the interests of men."* That is why she states that *"The end goal of feminist revolution must be not just the elimination of male privilege but of the sex distinction itself. Genital differences between human beings would no longer matter culturally. The tyranny of the biological family would be broken and with it the psychology of power."* (Patel, 1997)

Contemporary Muslim thinkers do not support such extremism that considers men as the core of evil. Nor do they condone its logic result, which is homosexuality. The Quranic view (which, by the way, coincides with that of the Hebrew and the Christian "Old Testament" Bible, e.g., Leviticus 20:13) is that homosexuality is an abomination and that those who indulge in it are "committing excesses". This includes both male and female homosexuals.

No impartial feminist may hold Islam as such responsible for any abuse or injustice committed against women. Actually,

> "Muslims are admonished to treat women equitably. The Quran forbids them to inherit women and abuse them, sexually or otherwise, as was the custom prior to the advent of Islam and as is still practiced in many societies." (Patel, 1997)
>
> O you who believe! You are forbidden to inherit women against their will. Nor should you treat them with harshness, that you may take away part of the dower you have given them - except when they have been guilty of open lewdness; on the contrary live with them on a footing of kindness and equity. If you take a dislike to them it may be that you dislike a thing, and Allah brings about through it a great deal of good. *[Quran, 4: 19]*

The Prophet Muhammad (PBUH) enhanced this message of kindness and fairness by several stances reported by credible biographers, such as Bukhari, Muslim, Abu Dawud, Tirmidhi and others. We will close the discussion of this issue with some examples from this Islamic Tradition:

- Mu'awiah ibn Haidah relates: I asked the Holy Prophet: What is the right of a wife against her husband? He said: Feed her when you feed yourself, cloth her when you cloth yourself, do not strike her on her face, do not revile her and do not separate yourself from her except inside the house. (Abu Daud)
- Abu Hurairah relates that the Holy Prophet said: The most perfect of believers in the matter of faith is he whose behaviour is best, and the best of you are those who behave best towards their wives. (Tirmidhi)
- Abdullah ibn Amr ibn 'As relates that the Holy Prophet said: The world is but a provision and the best provision of the world is a good woman. (Muslim)
- Aysha relates: A woman came to me begging with her two daughters; I could not find anything except a single date which I gave her. She divided it between her daughters and did not herself eat any of it. Then she got up and left. When the Holy Prophet came, I told him of it. He said: one who is tried with daughters and treats them well will find that they will become his shield from the Fire. (Bukhari and Muslim)
- Abu Shuraih Khuwailad ibn Amr Khuzai relates that the Holy Prophet said: Allah, I declare sinful any failure to safeguard the rights of two weak ones: orphans and women. (Nisai)
- Abu Hurairah relates that the Holy Prophet said: Of the dinar you spend in the cause of Allah; the dinar you spend in procuring the freedom of a slave; the dinar you give away in charity to the poor and the dinar you spend on your wife and children, the highest in respect of reward is the one you spend on your wife and children. (Muslim)
- Sa'ad ibn Abi Waqqas relates in the course of a long hadith that the Holy Prophet said: Whatever you might spend seeking thereby the pleasure of Allah will have its reward, even that which you put in the mouth of your wife. (Bukhari and Muslim)
- "Many women have come to the family of Muhammad complaining about their husbands ... Those husbands are not the best of you".
- "By assisting your wives in their household duties, you will receive the reward of sadaqah (charity)"
- "A believer must not hate a believing woman; if he dislikes one of her characteristics, he should be pleased with another"
- "When a woman breast feeds, for every gulp of milk she will receive a reward as if she had granted life to being, and when she weans her child, the angels pat her on the hack saying, 'Congratulations!' All your past sins have been forgiven, now start all over again":
- "O women! Remember that the pious among you will enter paradise before the pious men". "During pregnancy until the time of childbirth, and until the end of the suckling period, a woman earns reward similar to that of the person who is guarding the borders of Islam."

6. 3. THE CHOROPLETH MAPS OF GLOBAL FEMINISM

Let us now return to the quantitative social statistical argument. Our Choropleth maps of feminism are based on more countries but fewer variables than the choropleth maps and the factor analytical results of Chapter 5. Distance to traditional (istic) forms of religion is highest in Scandinavia and in some of the highly industrialized countries of Western and Central Europe with a long-run history of political secularization, like France, Germany, the Netherlands, and Estonia; and distance to traditionalist forms of religion is also high in Japan, Australia, New Zealand, in Russia and China, in the United Kingdom, Hungary and Bulgaria.

Spain, Uruguay, the Ukraine, Lithuania and the Slovak Republic are in the "next league" of secularization, while the populations in Argentina, Canada, and Italy are more secular than the United States, Turkey, Mexico, Chile, Venezuela, Colombia, India, Poland and Romania, which are forming the next "secularization league". The next group of countries is a group of already more religious nations of mainly Western Christianity – Peru, Brazil and South Africa. Most of the populations in the predominantly Muslim countries in the sample, including Algeria, Morocco, Mali, Saudi Arabia Iran and Bangladesh, belong to the next group and are as religious or secular as the populations in the predominantly Catholic countries Guatemala, El Salvador, and the Philippines, and are also on the same level as populations in the African countries Uganda and Zambia. According to our analysis, the most traditionally religious countries on earth are Ghana, Nigeria, Tanzania and Zimbabwe, as well as the predominantly Muslim countries Egypt, Jordan, Pakistan and Indonesia.

In Chapter 5 we already highlighted the factor "educational generation gap". Our analysis would have to repeat many of the statements, already made there. We concentrate on the comparative aspects of our analysis for the countries, which were not included in the analysis of Chapter 5. Saudi Arabia certainly belongs to the group of countries, experiencing the strongest "clash of educational levels" between generations, together with the Dominican Republic, South Korea, Colombia, Azerbaijan, Kyrgyzstan, Japan and Indonesia. Finally, our choropleth map of feminism (Map 6.3) also shows several interesting results. One is the high degree of feminism achieved not only in Scandinavia, New Zealand and France, but also in El Salvador, the Dominican Republic, Colombia and Uruguay. A striking result is also the fact that Ethiopia and Brazil are already in the same feminism league as most of the Western democracies in the North Atlantic arena. Ethiopia and Brazil are even ahead of Australia and South Africa. In this map, also other surprising and notable results emerge, like the position of Indonesia. Indonesia is in the same league of feminism as Turkey, Belarus and China. Concerning feminism, Saudi Arabia, Iran and Bangladesh are real laggards. Our book would not like to indulge into too many predictions, but these data have to be combined with the knowledge about the generation gap in education, which would imply that in the next years there will be considerable societal clashes and confrontations over gender policy in Saudi-Arabia and Iran.

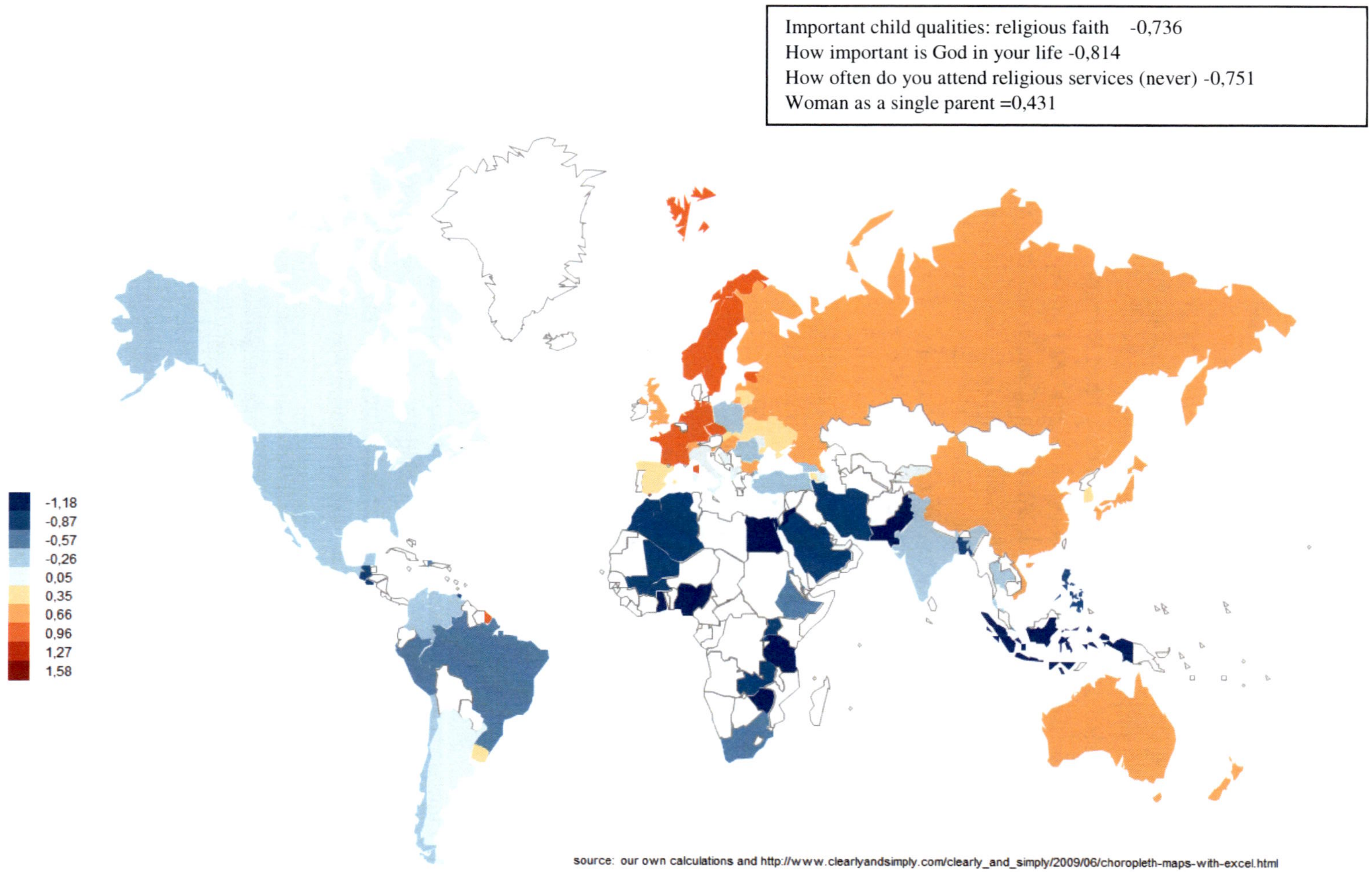

Map 6.1. Distance to traditionalist forms of religion.

Age- 0,391
Education level (recoded) - 0,959
Highest educational level attained - 0,964

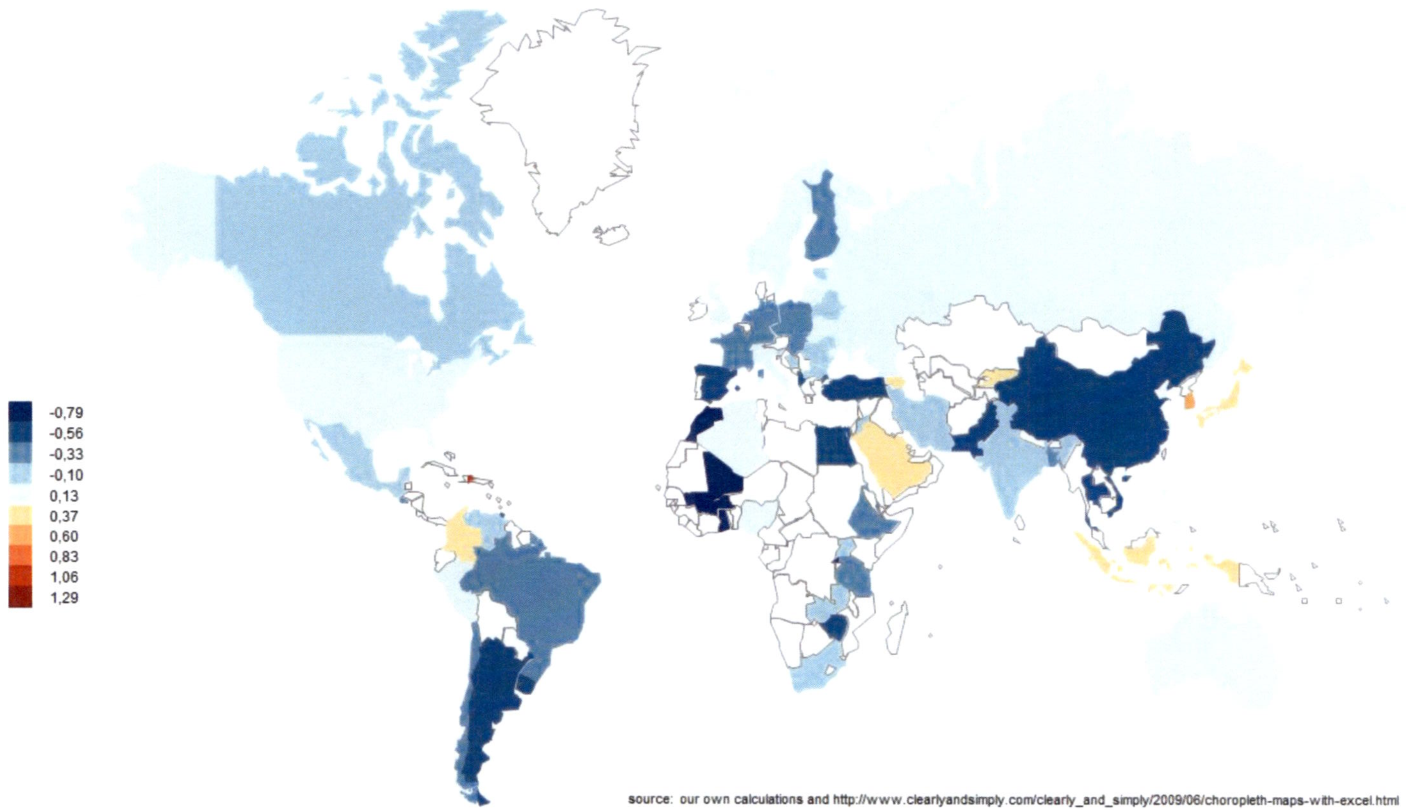

Map 6.2. Generation gap in educational level.

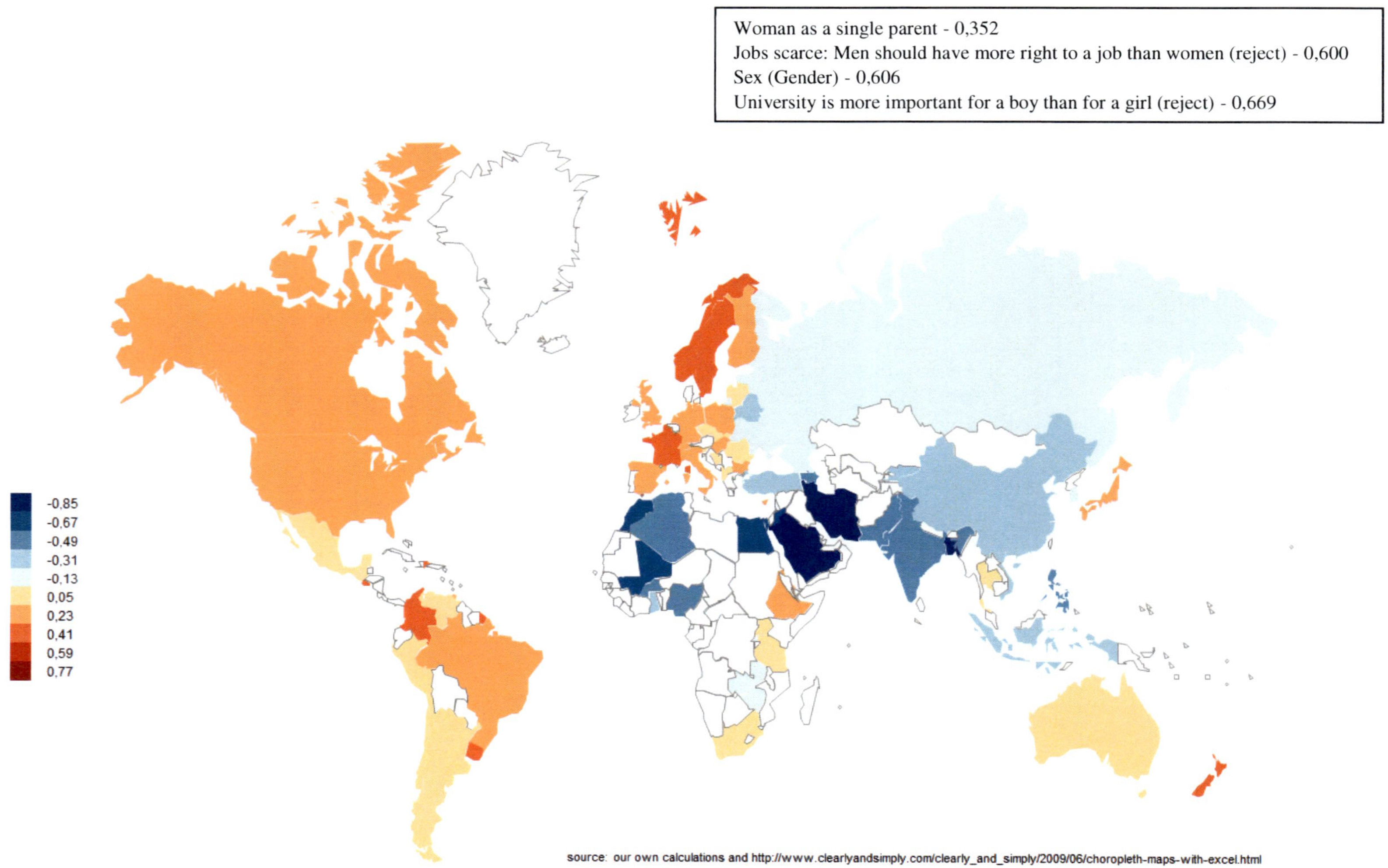

Map 6.3. Feminism.

6.4. THE REGIONAL GEOGRAPHY OF MUSLIM FEMINISM

In the following Table (Table 6.2), we present the factor scores of our analysis at a regional level. For reasons of space, we mention here only regions in Muslim countries which never had undergone a communist development history with its notorious policies of "scientific atheism". Interested readers and researchers are referred to our website[39] where we made available all the factor scores from all the regions of the world with available data. Our Table is ordered by the amount of feminism in the entire core center of the Muslim world. We also document for each region the development history and denominational characteristics of each region.

Using the facilities of the Microsoft EXCEL program, our readers can easily estimate that the most *secular regions* in the center of the Muslim world outside the orbit of (former) Communism are the Muslim Kemalist Turkish regions of the Aegean (with a factor score of 0.171); the Muslim Shia Iranian region of Kurdistan (factor score of 0.110) and the Muslim Kemalist Turkish region of South Eastern Anatolia, largely populated by Turkey's Kurdish minority (factor score: -0.154).

The secularism factor scores of some of the more pronouncedly secular Muslim regions range between +0.171 to -0.389:

Turkey	TR: Aegean
Iran	IR: Kordestan
Turkey	TR: South Eastern Anatolia
Turkey	TR: Mediterranean
Saudi Arabia	SA: Hail
Saudi Arabia	SA: Qassim
Bangladesh	BD: Feni
Bangladesh	BD: Tangail

This corresponds exactly to the following results in predominantly non-Muslim highly or medium developed countries, which are members of the OECD or the European Union. The list of these regions includes not only poorer regions of the European Union and some new member countries of the OECD, but also regions in Canada and the United States:

Secularism factor scores in OECD or European Union regions ranging between +0.171 to -0.389:

Romania	RO: Mutenia
Canada	CA: Nova Scotia
Romania	RO: Galati
USA	US: New England
Canada	CA: Saskatchewan
USA	US: Northwest
Canada	CA: Newfoundland
Chile	CL: Zona Metropolitana
Italy	IT: Lombardia
Cyprus	CY: Nicosia

[39] https: //www. academia. edu/5255298/The_global_analysis_of_feminism_and_its_regional_ implications_ for_the_Muslim_world.

Romania	RO: Moldova
Chile	CL: Norte
Romania	RO: Oltenia
Romania	RO: Bucuresti
USA	US: Middle Atlantic States
USA	US: California
Cyprus	CY: Famagusta
Italy	IT: Veneto
Italy	IT: Lazio
Canada	CA: New Brunswick
Romania	RO: Timis
Italy	IT: Sicilia
Romania	RO: Banat
Cyprus	CY: Paphos
Chile	CL: Centro
Romania	RO: Dobrogia
Romania	RO: Sibiu
Canada	CA: Prince Edward Island
Romania	RO: Prahova
USA	US: Rocky Mountain state
Italy	IT: Campania
Romania	RO: Arad
USA	US: South Atlantic
Romania	RO: Buzau
Chile	CL: Sur
Mexico	MX: Centro
Italy	IT: Puglia
Romania	RO: Dolj
Romania	RO: Transylvania
USA	US: East North Central
Mexico	MX: Norte
Mexico	MX: Sur
Romania	RO: Iasi
Cyprus	CY: Larnaca
Romania	RO: Valcea
Romania	RO: Crisana-Maramures (region)
Romania	RO: Arges
USA	US: West North Central
USA	US: West South Central
Romania	RO: Brasov
Cyprus	CY: Limassol

The very high educational gaps of more than 0.500 between the generations in the predominantly Muslim regions in:

Indonesia	ID: Dareah Istimewa Yogyakarta
Indonesia	ID: South Kalimantan
Indonesia	ID: DKI
Indonesia	ID: West Nusa Tenggara
Algeria	DZ: Comune 9
Pakistan	PK: Urban Sindh
Iran	IR: Lorestan
Saudi Arabia	SA: Abha \ Khamis
Saudi Arabia	SA: Tabuk
Indonesia	ID: Jakarta province
Algeria	DZ: Comune 4
Indonesia	ID: Banten
Indonesia	ID: East Java
Iran	IR: Mazandaran
Indonesia	ID: Lampung
Iran	IR: Ilam
Saudi Arabia	SA: Dammam \ Khobar
Saudi Arabia	SA: Riyadh
Saudi Arabia	SA: Qassim
Iran	IR: Boyer ahmad

well correspond to the range of results to be observed in the **following regions in core-Western countries**:

Korea-South	KR: Tawjeon / Daejeon
Korea-South	KR: Kwangju / Gwangju
Korea-South	KR: Seoul
Norway	NO: Oslo and Akershus
Korea-South	KR: Taegu / Daegu
Sweden	SE: Stor Stockholm
New Zealand	NZ: Coastal-North Otago
Sweden	SE: Uppsala
Portugal	PR: Sur
Korea-South	KR: Kyeongbuk / North Gyeongsang
Portugal	PR: Montana
Korea-South	KR: Kyeonggi / Gyeonggi Do
Korea-South	KR: Kyowygi Do
Korea-South	KR: Jeonnam / South Jeolla
Korea-South	KR: Pusan / Busan
New Zealand	NZ: Wellington
Korea-South	KR: Kangwon / Gangwon Do
Sweden	SE: Vasterbotten
Germany	DE: Ost-Berlin
Korea-South	KR: Kyeongnam / South Gyeongsang
GB (UK)	GB: Eastern
Sweden	SE: Skåne
Korea-South	KR: Chungnam / South Chungcheong
Australia	AU: New South Wales

Portugal	PR: Oeste
Sweden	SE: Halland
Latvia	LV: Kurzeme
Romania	RO: Bucuresti
Germany	DE: West-Berlin
Japan	JP: Kanto
Cyprus	CY: Paphos
Portugal	PR: Centro
Korea-South	KR: Chungbuk / North Chungcheong
Sweden	SE: Västra Götaland
Norway	NO: South Eastern Norway
Portugal	PR: Metropolitana
Latvia	LV: Daugavpils

The **feminism factor loadings of the predominantly Muslim regions in**

Bangladesh	BD: Kishoreganj
Turkey	TR: Aegean
Turkey	TR: Eastern Marmara
Turkey	TR: Eastern Black Sea
Indonesia	ID: Bengkulu province
Indonesia	ID: West Nusa Tenggara

are in the **same range of the development of feminism in core-western regions of:**

Finland	FI: Etelä-Savo
Korea-South	KR: Taegu / Daegu
Finland	FI: Kymen
Germany	DE: Brandenburg
Bulgaria	BG: Varna
Czech R	CZ: Støedoèeský kraj - Central Bohemia -
Spain	ES: C Valenciana
Korea-South	KR: Seoul
Germany	DE: Thueringen
Korea-South	KR: Chungbuk / North Chungcheong
Mexico	MX: Norte
Romania	RO: Brasov
Estonia	EE: Parnumaa
Romania	RO: Arges
Latvia	LV: Kurzeme
Spain	ES: Aragon
Hungary	HU: East Hungary
Korea-South	KR: Kyeongbuk / North Gyeongsang
Italy	IT: Veneto
Mexico	MX: Sur
Cyprus	CY: Limassol
Czech R	CZ: Jihoèeský kraj - South Bohemia -
Hungary	HU: North Hungary

Korea-South	KR: Kyowygi Do
USA	US: East North Central
Chile	CL: Sur
Korea-South	KR: Kangwon / Gangwon Do
Australia	AU: Victoria (Vic)
GB (UK)	GB: London
Slovenia	SI: Obalno-Kraska
Romania	RO: Sibiu
Czech R	CZ: Jihomoravský kraj - South Moravia -
Estonia	EE: Tartumaa
Italy	IT: Puglia
Slovakia	SK: Central Slovakia
Spain	ES: Navarra
Korea-South	KR: Pusan / Busan
Australia	AU: Western Australia (WA)
Australia	AU: Tasmania (Tas)
Korea-South	KR: Kyeonggi / Gyeonggi Do
Australia	AU: South Australia (SA)
Slovakia	SK: Northern Slovakia
Australia	AU: New South Wales and ACT (NSW&ACT)
New Zealand	NZ: Southland
Bulgaria	BG: Razgrad
Korea-South	KR: Chungnam / South Chungcheong
Latvia	LV: Latgale
Chile	CL: Centro
Estonia	EE: Laane-Viruma
Romania	RO: Prahova
Spain	ES: Murcia
Mexico	MX: Centro occidente
Australia	AU: Queensland (Qld)
Romania	RO: Moldova
Czech R	CZ: Severoèeský kraj - North Bohemia -
Korea-South	KR: Incheon
Cyprus	CY: Iskele (Karpas)
Romania	RO: Banat
Romania	RO: Caras-Severin
Germany	DE: Schleswig-Holstein
Latvia	LV: Liepaja
Bulgaria	BG: Haskovo
Cyprus	CY: Paphos
Romania	RO: Transylvania

Policy pointer:

Our regional analysis provides global research and global policy-making with startling insights about the degree to which secularism, a rising educational level of the younger generations, and feminism have become a reality in some regions of the Muslim world.

There is already a significant feminist movement and a significant rethinking of the religious traditions in the light of gender justice in the entire Muslim world, and not just in Muslim post-communist countries. The rise of women is to be achieved as part of society's advancement towards freedom, and Arab countries can achieve this goal by eradicating all infringements on human dignity, and specifically by guaranteeing full citizenship and the enjoyment of all human rights for all women on an equal footing with men.

Table 6.2. Feminism and religion - regional value analysis for the Muslim world ranked by the feminism factor in descending order

Description of the region	Country	Region	n =	distance to traditionalist forms of religion	Generation gap in educational level	feminism
Muslim	Bangladesh	BD: Kishoreganj	60	-1,211	-0,498	0,182
Muslim Kemalist	Turkey	TR: Aegean	131	0,171	0,098	0,135
Muslim Kemalist	Turkey	TR: Eastern Marmara	102	-0,434	0,064	0,120
Muslim Kemalist	Turkey	TR: Eastern Black Sea	49	-0,408	-0,068	0,099
Muslim	Indonesia	ID: Bengkulu province	191	-1,038	0,085	0,053
Muslim	Indonesia	ID: West Nusa Tenggara	57	-1,129	0,778	0,028
Muslim	Bangladesh	BD: Faridpur	50	-1,214	-0,462	-0,002
Muslim Kemalist	Turkey	TR: Western Black Sea	85	-0,529	-0,310	-0,046
Muslim	Indonesia	ID: South Kalimantan	33	-1,192	0,830	-0,088
Muslim Kemalist	Turkey	TR: Western Marmara	47	-0,651	-0,608	-0,091
Arab Muslim, former French colony	Algeria	DZ: Comune 6	67	-0,741	0,437	-0,125
Muslim Kemalist	Turkey	TR: East Central Anatolia	71	-0,558	-0,483	-0,132
Muslim Kemalist	Turkey	TR: Western Anatolia	125	-0,452	-0,145	-0,133
Muslim	Indonesia	ID: East Java	481	-0,974	0,576	-0,135
Muslim	Pakistan	PK: Urban NWFP	148	-1,163	0,353	-0,146
Muslim Kemalist	Turkey	TR: Istanbul	245	-0,448	-0,205	-0,147
Arab Muslim, former French colony	Algeria	DZ: Comune 1	306	-0,709	0,262	-0,164
Muslim	Pakistan	PK: Urban Sindh	212	-0,979	0,704	-0,176
Arab Muslim, former French colony	Algeria	DZ: Comune 4	87	-0,835	0,587	-0,206
Muslim Kemalist	Turkey	TR: South Eastern Anatolia	123	-0,154	-0,673	-0,212
Muslim	Indonesia	ID: Dareah Istimewa Yogyakarta	45	-0,726	1,098	-0,233
Muslim	Indonesia	ID: Jakarta province	192	-0,869	0,620	-0,253
Muslim	Indonesia	ID: Lampung	97	-1,123	0,566	-0,253
Arab Muslim	Egypt	EG: Beni Swaif	75	-1,015	-0,254	-0,268
Muslim	Indonesia	ID: DKI	116	-0,863	0,809	-0,271
Arab Muslim	Egypt	EG: Menofia	121	-1,120	-0,438	-0,279
Arab Muslim	Morocco	MA: Region 13	96	-0,605	-0,584	-0,282
Muslim Kemalist	Turkey	TR: North Eastern Anatolia	49	-0,508	-0,275	-0,290
Arab Muslim	Egypt	EG: Canal zone	174	-0,922	-0,396	-0,290
Muslim Kemalist	Turkey	TR: Mediterranean	152	-0,325	-0,307	-0,303
Muslim	Pakistan	PK: Urban Punjab	525	-1,038	0,171	-0,304
Arab Muslim	Jordan	JO: MADABA	39	-0,904	-0,048	-0,307
Muslim	Pakistan	PK: Urban Baluchistan	46	-1,274	0,084	-0,333
Muslim	Indonesia	ID: Central Java province	593	-0,991	0,232	-0,334
Arab Muslim, former French colony	Algeria	DZ: Comune 2	161	-0,606	0,263	-0,348

Description of the region	Country	Region	n =	distance to traditionalist forms of religion	Generation gap in educational level	feminism
Arab Muslim	Egypt	EG: Sinai	99	-0,949	0,118	-0,352
Arab Muslim	Morocco	MA: Region 10	114	-0,690	-0,575	-0,360
Arab Muslim	Morocco	MA: Doukkla - Abda	95	-0,597	-0,674	-0,364
Arab Muslim	Egypt	EG: Qalubia	197	-0,973	-0,349	-0,371
Arab Muslim, former French colony	Algeria	DZ: Comune 9	31	-0,821	0,760	-0,378
Muslim	Indonesia	ID: Jambi province	189	-1,038	-0,029	-0,390
Muslim	Bangladesh	BD: Coxex Bazar	44	-0,441	-0,602	-0,393
Muslim	Pakistan	PK: Rural Punjab	527	-1,254	-1,070	-0,407
Muslim	Indonesia	ID: Banten	111	-1,044	0,587	-0,413
Arab Muslim	Egypt	EG: Gharbia	174	-0,873	-0,375	-0,417
Arab Muslim	Egypt	EG: Giza	274	-0,862	-0,063	-0,419
Muslim Shia	Iran	IR: North Khorasan	30	-0,905	-0,250	-0,443
Muslim Shia	Iran	IR: Lorestan	121	-0,599	0,678	-0,448
Arab Muslim	Egypt	EG: Alex	401	-0,971	-0,260	-0,449
Arab Muslim	Morocco	MA: Region 9	247	-0,676	-0,632	-0,458
Arab Muslim	Egypt	EG: Fayoum	98	-1,055	-0,771	-0,460
Arab Muslim	Morocco	MA: Region 7	226	-0,729	-0,873	-0,467
Arab Muslim	Jordan	JO: MA´AN	36	-0,793	-0,221	-0,469
Muslim	Bangladesh	BD: Chittagong	113	-0,677	0,254	-0,486
Arab Muslim	Egypt	EG: Cairo	773	-0,887	-0,022	-0,488
Arab Muslim	Saudi Arabia	SA: Dammam \ Khobar	216	-0,509	0,536	-0,489
Arab Muslim, former French colony	Algeria	DZ: Comune 7	175	-0,624	0,305	-0,494
Arab Muslim	Saudi Arabia	SA: Al Madinah	93	-0,977	0,324	-0,497
Muslim	Indonesia	ID: West java province	670	-1,001	0,323	-0,506
Arab Muslim	Jordan	JO: KARAK	77	-0,640	-0,226	-0,508
Arab Muslim	Egypt	EG: Damiatta	50	-0,860	-0,440	-0,512
Muslim Shia	Iran	IR: Fars	269	-0,581	0,421	-0,518
Arab Muslim	Egypt	EG: Kafr El shiekh	97	-0,758	-0,537	-0,519
Muslim	Pakistan	PK: Rural Sindh	203	-1,368	-1,165	-0,532
Muslim Shia	Iran	IR: Ilam	34	-0,760	0,557	-0,545
Muslim	Bangladesh	BD: Feni	40	-0,375	0,411	-0,547
Arab Muslim	Morocco	MA: Region 5	117	-0,808	-0,913	-0,563
Muslim Shia	Iran	IR: Kerman	165	-0,703	-0,065	-0,566
Muslim Shia	Iran	IR: Golestan	94	-0,681	0,321	-0,583
Arab Muslim	Egypt	EG: Behaira	197	-0,809	-0,554	-0,588
Muslim Shia	Iran	IR: Khozestan	173	-0,541	0,176	-0,591
Muslim Shia	Iran	IR: Kordestan	45	0,110	0,329	-0,594
Arab Muslim	Jordan	JO: JARASH	42	-0,803	-0,132	-0,594
Arab Muslim	Egypt	EG: Lower Egypt	1126	-0,882	-0,314	-0,595
Arab Muslim	Morocco	MA: Region 6	63	-0,576	-0,831	-0,596
Arab Muslim	Jordan	JO: TAFELA	50	-0,847	-0,100	-0,609
Arab Muslim	Egypt	EG: Sharkia	224	-0,877	-0,444	-0,617
Muslim Shia	Iran	IR: Hormozgan	86	-0,827	-0,299	-0,620
Arab Muslim	Jordan	JO: MAFRAQ	61	-0,873	-0,111	-0,623
Muslim Kemalist	Turkey	TR: Central Anatolia	75	-0,555	-0,304	-0,624
Muslim Shia	Iran	IR: Tehran	765	-0,591	0,196	-0,630
Muslim Shia	Iran	IR: Ghom	68	-0,733	0,421	-0,632

Table 6.2. (Continued)

Description of the region	Country	Region	n =	distance to traditionalist forms of religion	Generation gap in educational level	feminism
Arab Muslim	Jordan	JO: AMMAN	368	-0,799	-0,061	-0,638
Arab Muslim	Jordan	JO: IRBID	176	-0,815	-0,173	-0,639
Arab Muslim	Jordan	JO: BALQA	76	-0,838	-0,450	-0,641
Muslim Shia	Iran	IR: Ardabil	71	-0,776	0,024	-0,651
Arab Muslim	Egypt	EG: Assuiot	123	-0,807	-0,374	-0,651
Arab Muslim, former French colony	Algeria	DZ: Comune 5	119	-0,590	0,103	-0,654
Muslim Shia	Iran	IR: Isfahan	269	-0,753	0,290	-0,654
Arab Muslim	Saudi Arabia	SA: Hail	70	-0,331	0,282	-0,662
Muslim Shia	Iran	IR: East azarbayjan	287	-0,796	-0,024	-0,672
Muslim Shia	Iran	IR: Gilan	167	-0,699	-0,060	-0,676
Muslim	Pakistan	PK: Rural NWFP	112	-1,438	-1,201	-0,679
Muslim	Bangladesh	BD: Comilla	81	-1,191	0,243	-0,680
Muslim	Pakistan	PK: Rural Baluchistan	43	-1,467	-1,165	-0,683
Arab Muslim	Jordan	JO: AQABA	53	-0,674	0,042	-0,685
Muslim Shia	Iran	IR: Mazandaran	259	-0,767	0,574	-0,687
Arab Muslim, former French colony	Algeria	DZ: Comune 3	114	-0,710	0,001	-0,696
Arab Muslim	Egypt	EG: Aswan	49	-1,038	-0,129	-0,699
Arab Muslim	Morocco	MA: Grand Casablanca	328	-0,903	-0,600	-0,702
Arab Muslim	Egypt	EG: Sohag	123	-0,841	-0,677	-0,705
Muslim Shia	Iran	IR: Khorasan	331	-0,777	-0,174	-0,718
Arab Muslim	Jordan	JO: AJLOON	49	-0,843	-0,546	-0,726
Arab Muslim	Jordan	JO: ZARQA	162	-0,849	-0,445	-0,726
Muslim Shia	Iran	IR: West azarbayjan	177	-0,675	-0,131	-0,732
Muslim Shia	Iran	IR: Ghazvin	79	-0,848	-0,202	-0,744
Muslim	Bangladesh	BD: Nator	35	-0,448	-0,467	-0,749
Arab Muslim	Egypt	EG: Dakahlia	249	-0,885	-0,470	-0,763
Muslim Shia	Iran	IR: Kermanshah	139	-0,692	0,077	-0,764
Muslim Shia	Iran	IR: Sistan and balouchestan	58	-0,962	-0,203	-0,766
Arab Muslim	Egypt	EG: Qena	100	-0,964	-0,552	-0,778
Muslim Shia	Iran	IR: Hamadan	136	-0,809	-0,258	-0,780
Arab Muslim	Morocco	MA: Rabat-Salé-Zemmour	164	-0,850	-0,554	-0,781
Muslim Shia	Iran	IR: Bushehr	49	-0,597	-0,092	-0,784
Arab Muslim	Saudi Arabia	SA: Jeddah	296	-0,810	0,415	-0,788
Muslim	Bangladesh	BD: Rangpur	63	-1,016	0,127	-0,795
Arab Muslim	Saudi Arabia	SA: Tabuk	73	-1,057	0,631	-0,807
Muslim	Bangladesh	BD: Sylhet	75	-1,241	-0,288	-0,823
Arab Muslim	Egypt	EG: Upper egypt	990	-0,794	-0,523	-0,847
Arab Muslim	Saudi Arabia	SA: Makkah	93	-0,750	0,481	-0,862
Muslim Shia	Iran	IR: Yazd	85	-0,813	0,268	-0,862
Arab Muslim	Egypt	EG: Luxor	50	-0,959	-0,555	-0,865
Arab Muslim	Morocco	MA: Marrakech-Tensift	194	-0,969	-0,883	-0,873
Arab Muslim, former French colony	Algeria	DZ: Comune 8	58	-0,590	0,260	-0,884
Muslim	Bangladesh	BD: Gaibandha	48	-0,784	0,281	-0,889
Muslim	Bangladesh	BD: Barisal	105	-0,507	0,234	-0,892

Arab Muslim	Saudi Arabia	SA: Abha \ Khamis	96	-0,704	0,641	-0,909
Muslim Shia	Iran	IR: Boyer ahmad	37	-0,666	0,513	-0,913
Muslim Shia	Iran	IR: Semnan	70	-0,774	0,173	-0,937
Arab Muslim	Saudi Arabia	SA: Riyadh	293	-0,496	0,534	-0,951
Muslim Shia	Iran	IR: Markazi	96	-0,815	-0,420	-0,955
Muslim Shia	Iran	IR: Zanjan	80	-0,979	-0,376	-0,957
Muslim Shia	Iran	IR: Chaharmahal	49	-0,683	0,099	-0,960
Muslim	Bangladesh	BD: Mymensingh	118	-1,295	-0,292	-0,981
Arab Muslim	Morocco	MA: Gharb-Chrarda	102	-0,819	-0,877	-0,989
Muslim	Bangladesh	BD: Dhaka	195	-0,585	0,273	-0,992
Arab Muslim	Morocco	MA: Fès-Boulemane	102	-0,833	-0,758	-1,006
Arab Muslim	Egypt	EG: Menya	148	-0,922	-0,793	-1,032
Arab Muslim	Morocco	MA: Meknès-Tafilalet	120	-0,944	-0,730	-1,046
Muslim	Bangladesh	BD: Jhenaidah	35	-0,723	-0,578	-1,056
Arab Muslim	Morocco	MA: Chaouia-Ouardigha	41	-0,869	-0,875	-1,079
Muslim	Bangladesh	BD: Tangail	70	-0,389	-0,745	-1,096
Arab Muslim	Egypt	EG: Ismailia	49	-0,989	-0,583	-1,110
Muslim	Bangladesh	BD: Sirajgonj	60	-0,484	-1,031	-1,139
Arab Muslim	Saudi Arabia	SA: Taif	38	-0,617	-0,087	-1,172
Arab Muslim	Morocco	MA: Tanger-Tétouan	32	-1,040	-0,799	-1,207
Muslim	Bangladesh	BD: Brahmanbaria	69	-1,137	-1,010	-1,213
Muslim	Bangladesh	BD: Habiganj	47	-1,060	-0,629	-1,266
Arab Muslim	Saudi Arabia	SA: Qassim	98	-0,358	0,521	-1,311
Muslim	Bangladesh	BD: Sherpur	32	-0,747	-0,641	-1,357
Muslim	Bangladesh	BD: Rajshahi	50	-0,605	-1,066	-1,634

Chapter 7

HUMAN VALUES AND THE "ARAB SPRING"

Without question, current global scientific knowledge about mass opinion in the Arab countries is still in its infancy. One of the deficits of earlier *World Values Survey* research was a still existing lack of data for the Arab world. While certainly the global maps, reproduced in this book, are not an entirely blank space for the region anymore, we admit that the *"Arab Opinion Index"* by the *Arab Center for Research and Policy Studies (ACRPS)* in Doha, Qatar, practically neglected by Western scholarship until now, offers startling new perspectives on the dynamics of opinion in the Arab world, which go beyond anything which the Arab data from the *World Values Survey,* wave 1-5, could offer. The *Arab Opinion Index* project is currently the largest of its kind in the world. It covers 12 Arab countries, representing 85 percent of the population of the Arab countries.

To what extent can we still talk about an "Arab spring" as a game changer in the regional order as well as on the level of political values and moral behavior? How can we assess the change introduced by these revolutions and social protest movements? This chapter is dedicated to answering these questions.

7.1. SOME SOCIOLOGICAL MAPS GUIDING TOWARDS AN ANALYSIS OF THE "ARAB SPRING"

In 2011, the old regional order of the Arab world was upset by a series of upheavals that resulted in the overthrow of the heads of states in Tunisia, Egypt (twice), Libya, and Yemen. Civil uprisings have been vented in Bahrain, and Syria. Major protests have broken out in Algeria, Iraq, Jordan, Kuwait, Morocco, and Sudan; and minor protests have occurred in Mauritania, Oman, Saudi Arabia, Djibouti, Western Sahara, and the Palestinian Authority territories.

Box 7.1. The overall message of this Chapter

In this Chapter, we evaluate Arab public opinion with a newly available source, the "*Arab Opinion Index*" by the Arab Center for Research and Policy Studies (ACRPS) in Doha, Qatar, practically neglected by Western scholarship until now. The *Arab Opinion*

Index project is currently the largest of its kind in the world. It covers 12 Arab countries, representing 85 percent of the population of the Arab world. It is thus a larger Arab opinion survey project than any other scholarly effort to estimate Arab opinion. The Index compiles the findings of 16,173 face-to-face interviews with subjects who were drawn from a random, representative sampling of the populations of their countries of origin. The questionnaire was prepared in 2010 and the survey was conducted in the first half of 2011. The findings are freely available from the ACRPS website. Indeed, the Arab Center for Research and Policy Studies clearly reflects an "Arab national viewpoint" in international affairs.

For the purpose of this Chapter, the original data of the *Arab Opinion Index* were weighted by UNDP population figures for the year 2010/2011, so that we can arrive at conclusions about the totality of opinions in the Arab states.

On the one hand, the *Arab Opinion Index* indeed shows the overwhelming support for democracy and change in the region. At the same time, also these data may suggest some basic weaknesses of the civil society support for the structures of democracy.

The liberal heritage, which resulted from the movement of Enlightenment and democracy, implying respect of other religions and civilizations, is part of modern Islamic culture. It does not need to be sown anew, but simply supported through institutions facilitating cooperation, interaction and acculturation.

If the Arab world wants positively to confront the 21st Century and become a full and mature democracy, issues like tolerance, sexism, and the overcoming of authoritarian thought patterns become important for the future of the democratic system. While "Islamophobia" in the West remains to be confronted, xenophobia and discrimination against immigrants and minority groups in Muslim countries have also to be faced. Even if a strong emphasis on religion is expressed, the religious doctrines are invited to start to learn from their own Enlightenment traditions, especially from Muslim philosophy and the great global Enlightenment thinkers of the 18th and the 19th Century.

Theorists and observers are still debating about the real scope and meanings of these social movements, and to what degree they changed the political system, introducing new actors and a new political behavior, or leaving the core of the regimes unchanged while allowing more freedom and flexibility within the political vista.[40]

As some observers noted, *"the regimes' ability to withstand popular pressure depended on whether they retained significant allies in their respective societies. In Egypt, protesters represented a wide range of social classes, excluding only the wealthy elite associated with the regime, the very poor urban sub-proletariat and the peasantry."* (Dalacoura, 2012) This explains why, since the first days of the uprising 2010/2011, Egyptian officials insisted on making clear the "differences" between their country and Tunisia. More than a self-protection reflex, it was actually a reminder to the protesters as well as to the Western allies that Mr. Mubarak was internationally supported and that his downfall would send a lot of negative messages to the entire region, Israel included. (Karoui, 2012: 75)

> "The Tunisian uprising was also the product of wide consensus against the regime, a symptom of the shrinking of the latter's support base during the 2000s. In contrast, Libya,

[40] The Russian world systems scholar and Arabist Professor Andrey Korotayev insists on the fact that Yemen experienced not even a regime change, let alone a real revolution. However, it is indeniable that a revolt, similar to Tunisia and Egypt, occurred in this country.

> Syria, Bahrain and Yemen were divided, with important social and political entities continuing to support the regimes. Yemen's regime was 'less repressive, more broadly inclusive and adaptable', better able to coopt the opposition using an extensive patronage network and to keep open avenues for participation, dissent and negotiation." (Dalacoura, 2012)

As this book is being written, battles rage on the streets of major Syrian cities. A decade ago, many observers would have thought that such an unfolding scenario was simply unthinkable.[41] The regime in Damascus engaged in a desperate attempt to cling on to power supported by Moscow, Tehran and Beijing, - an alliance, which to a liberal Arab observer is not amazing because these supporters of Damascus *"had bombed Chechnya, crushed the students of Tiananmen, or smothered the Green revolt in Iran"* (Karoui, 2012). Assessing the results of the *"Arab Spring"* will be a theme for international social science for months and years to come.

Among the *World-Values Survey* and *Arab Barometer* survey centered studies on Arab opinions, the essays Tessler, 2002 and 2004, were particularly influential in Western scholarship.

The yearly report of the Center for Security Studies (CSS at the ETH Zurich, Switzerland) titled *"Strategic Trends 2011"* made some remarks about the situation in the Middle East and the global and regional power shifts. The overall picture emerging from this analysis is that a *"new multi-polar international order is gradually emerging that is marked by diversity, a lack of leadership, and potentially growing instability."* (Karoui, 2012)

Policy pointer:

In this new order, emerging economies would not be dependent on the West. The US and the EU, still suffering from economic hardships, lack the resources and the political will to provide global stability. The divergences about the Syrian issue have been preceded by similar divergences about the Libyan issue before Western military intervention. Observers have at the time noted that such divergences were a symptom of a crisis which was largely neglected. The domestic problems may hinder Europe from playing a major political role abroad (Karoui, 2012), which relatively explains the refusal of the British Parliament to support a military strike against the regime of Assad, charged of using chemical weapons against its people.

We will now try to evaluate this *"Arab Opinion Index"* of the Arab Center for Research and Policy Studies (ACRPS) in Doha, Qatar.[42] The *Arab Opinion Index* project covers 12 Arab countries, representing 85 percent of the population of the Arab world. It is thus a larger Arab opinion survey project than any other scholarly effort to estimate Arab opinion. The Index compiles the findings of 16,173 face-to-face interviews with subjects who were drawn from a random, representative sampling of the populations of their countries of origin. The questionnaire was prepared in 2010 and the survey was conducted in the first half of 2011. The

---

[41] Russell A. Berman, "European Responses to September 11," *Telos* 121 (Fall 2001): 73-85.

[42] http: //english. dohainstitute. org/release/5083cf8e-38f8-4e4a-8bc5-fc91660608b0. See also the earlier study on this topic by Arno Tausch and Hichem Karoui, *"Les Musulmans: Un cauchemar ou une force pour l'Europe?"* (Paris: L'Harmattan, 2011).

findings are freely available from the ACRPS website. Indeed, the Arab Center for Research and Policy Studies clearly reflects an "Arab national viewpoint" in international affairs.[43]

7.2. The Arab Opinion Index

For the purpose of this Chapter, *the original data of the Arab Opinion Index were weighted by UNDP population figures for the year 2010/2011,*[44] so that we can arrive at conclusions about the totality of opinions in the Arab states. Clearly, the population-rich countries Egypt, Sudan, Algeria, Iraq and Morocco already account for some 62.5% of the total population of the 22 Arab countries. Egypt has about 24 times a bigger population than Mauritania! Looking at Table 7.1 and the changes in the rankings of Arab nations according to the very reliable UNDP "Human Development Index", it emerges that some Arab countries experienced an improvement in their world system position from 2005 to 2011.

Table 7.1. Changes in the world rank of Arab countries according to the UNDP Human Development Index, 2005-2011

Country	2005	2006	2007	2008	2009	2011	improvement or deterioration of relative position in world society, 2005-2011
United Arab Emirates	35	32	31	29	29	29	6
Yemen	147	146	146	146	143	142	5
Tunisia	90	89	89	88	86	87	3
Lebanon	69	70	70	70	67	67	2
Egypt	106	107	107	106	104	104	2
Algeria	89	90	90	89	87	88	1
Morocco	120	119	119	119	119	119	1
Saudi Arabia	54	54	55	55	54	54	0
Jordan	86	88	86	85	85	86	0
Djibouti	152	153	152	152	152	152	0
Sudan*	156	156	157	157	157	157	-1
Iraq	119	120	122	121	121	121	-2
Qatar	32	34	34	35	35	35	-3
Bahrain	37	37	37	37	38	40	-3
Libyan Arab Jamahiriya	56	56	56	57	55	59	-3
Oman	77	80	81	80	80	82	-5
Syrian Arab Republic	104	105	106	107	108	110	-6
Mauritania	141	142	143	143	147	147	-6
Comoros	145	148	149	150	150	151	-6
Kuwait	50	51	54	59	59	60	-10

[43] http: //english. dohainstitute. org/content/2dc3bc49-a4ef-42fe-83d2-806148106b7d.
[44] http: //hdr. undp. org/en/statistics/data/

7.3. Islamism and Salafism

On the one hand, the *Arab Opinion Index* indeed shows the *overwhelming support for democracy and change in the region.* At the same time, also these data may suggest some basic weaknesses of the civil society support for the structures of democracy. However, some observers have another take on this issue. Thus, Katerina Dalacoura argued that the events that unfolded since 2011 in the Arab region suggest, on the contrary, an increasing role of the civil society in the concerned countries.

> "In Egypt's case, there appears to be a continuum between an increasingly active civil society and labour activism in the 1990s and 2000s and the insurrection of 2011. The 'We are all Khaled Said group', the 6 April Movement (established in 2008 in relation to labour protests), Muslim Brotherhood youth, the group around presidential hopeful Muhammad el-Baradei, the 'new left', human rights and other civil society activists, striking workers: all played a role in Mubarak's overthrow. In Bahrain, political and civil society groups which had become powerful in the second half of the 2000s— including the al-Haqq Movement for Liberty and Democracy (more rejectionist than the Shi'i Islamist al-Wifaq and the left-leaning, non-sectarian Wa'ad groups) and the human rights movement, centred on the Bahrain Centre for Human Rights—played a vital role in the rebellion." (Dalacoura, 2012)

Policy pointer:

Support for the separation of religious practices from political and social life is only expressed by 46.6% of the population in the Arab countries, and the separation of religion from politics is only supported by 42.8% of the population. That political freedom and civil liberties are a requirement of democracy is only supported by 36.3%, and that equality and justice among citizens are a requirement of democracy is supported by only 19.5% of the Arab world.

How can we explain these discrepancies?

The revolution that many Arab opposition movements have long waited for in the aftermath of the 1967 defeat (Six-Day War) and in the seventies did not take place. Many decades later, in 2011, it has become a reality. However, this is assumed to be a revolution completely different from that imagined by secular and Islamist 20th Century political movements.

At first glance, observers assumed that the *"Arab spring"* revolutions are a victory for liberal democracy, and maybe because the Arabs have lost their delusions about populist ideologies and dominance by the sword, they finally could find their way humbly to the light. Nasserite and Baathist Pan-Arabism, as well as Marxism, have failed to liberate the Arabs from the impasse. They failed as alternatives on all levels: political, economic, social, and cultural. Only one hope remained: that liberal democracy could achieve what no ideology was able to do for the Arabs. Nevertheless, instead of being a victory for liberalism, the Arab spring brought in the Islamists. Could they deliver on the issue of democracy and human rights? The rise of Islamism in these days of *"Arab Spring"* has reduced the popular revolutions that took place to a bare fight between dictatorial regimes and Islamists.

Is this all what the Arab political culture was able to provide? Arab culture ... maybe not. But ballot boxes ... surely. However, it is important to remember that democracy as a political culture exists in the Arab world. This is also corroborated by recent quantitative research (Tessler and Gao, 2009). The interesting point here is that while studies of democracy and democratic transitions identify a number of attitudes, values and behavior patterns of ordinary citizens associated with a democratic political culture, these norms do not constitute a one dimensional orientation everywhere. Tessler and Gao (2009) showed that the focus on values and patterns of ordinary citizens as indicators of democratic orientation are derived primarily from the study of new democracies. Thus, they proposed the examination of the politically relevant attitudes, values and behavior patterns of citizens in non-democracies, especially those in the Arab world. Against this background, their analysis (preceding the 2011 Arab revolts) used survey data from Jordan, the Palestinian Territories (West Bank and Gaza) and Algeria; three very different Arab polities, to investigate the nature and structure of citizen orientations related to democracy and governance. They concluded that there is a predisposition to accepting democratic values such as pluralism, political tolerance and support for gender equality, as well as involvement composed of political knowledge, political interest, civic participation and interpersonal trust. As they put it, *"That identical patterns should be found in these three countries with very different political, economic and social attributes suggests that this dimensional structure may be present in a broad array of non-democratic countries, including, perhaps, some outside the Arab world."*

Policy pointer:

Arab thinkers do not agree on a single meaning of democracy. The Salafi revivalist prefers to use the Islamic concept of *"Shura"* instead of democracy. The Marxist understands democracy only when it is linked to the dictatorship of the workers. The Nasserite and Baathist Pan-Arab nationalist considered it a secondary objective and did not give it a thought. The liberal meaning has long been obliterated in the succession of dictatorships throughout the Arab world, although several countries have experienced a multiparty system (Iraq, Syria, Lebanon, Egypt, Tunisia. . .) since the first quarter of the 20th Century. Even in the conservative Gulf, demands for political participation began as early as the 1930s.

There are two passages in the Noble Quran that address the issue of *Shura*. *'Imran* (3: 159) says that the Prophet (S) must *"take counsel with them in the conduct of affairs (wa shawirhum fi al-amri)."* The chapter called *al-shura* (42: 37-38) refers to those *"who avoid gross sins and indecencies, and, when angered, are willing to forgive; who obey their Lord, attend their prayers, and conduct their affairs by mutual consent (wa amruhum shura baynahum)."* However, there is no indication that the consultation mentioned in the Noble Quran needs be understood as anything more than a system of advice.

In Islamic jurisprudence, not only the "Shura" (consultancy) is not binding but it is also restrained to the elite, loyal to the ruler *(ahl al-hall wal 'aqd)*. For Sheikh Muhammad Abduh (1849-1905), who was all the same considered as an influential reformist, *"Shura"* means the absence of *"absolute despotism."* Abduh actually believed that the East needed only the rise of a *"strong man"* whom he labelled a *"just despot."* Such a man in his eyes *"remaining just to his people, would achieve in fifteen years what reason alone cannot achieve in fifteen centuries."*

Apparently, such an idea made its way throughout the Arab world in the 20th Century as far as we can infer from the multiplication of autocratic regimes that took over after World

War Two and even before it. Democracy was also suspect in the eyes of their opponents among the Islamist groups. For the latter, the sovereignty of the people which is the core of the democratic rule is at odds with the sovereignty of God *(al-Hakimiyya),* the source of all Islamic laws and governments. Moreover, whatever the virtues of democracy, *"Shura"* is preferable to them since it does not impel imitating the West.

Some of the Islamist groups might have adjusted their views to fit in the new social trends, particularly since the revolutions that toppled the authoritarian regimes in Tunisia and Egypt (January, February 2011). Yet, while these new openings remain to be tested, it should be noted that initially, major Islamist leaders (e.g., Sayyed Qutb, and Abu al-Aala al-Mawdudi) levelled two arguments against democracy, both based on the notion of *Tawheed* (the oneness or unity of God). Firstly, they denounced democracy as *shirk bi-allah* (attributing partners to God). In their eyes, the right to legislate without being bound by a superior Divine authority contradicts the Qur'anic decree: *"The command is for none but God" (12: 40).* Secondly, human sovereignty, they contend, contradicts God's sovereignty *(Hakimiyya).* Whereas in democracy the legislator is the people, according to them, sovereignty *(Hakimiyya)* is the sole prerogative of God, and any usurpation of the legislative authority from God undermines *Tawheed.*

The liberal current might have been the only that took democracy as a serious affair. For the Egyptian author Salama Musa (1887-1958) for example, *"democracy was a system of society before it became a system of government. Nay, it is a system of government only because it is the result of a certain system in society."* Thus, he stressed the necessity for the developing countries to link social and cultural development to the democratic objective. Musa believed that the Arab middle class would grow in importance until it became able to lead the transition to democracy. This is also the stance of many liberals in the Arab world today.

The Egyptian Ahmed Lutfi al-Sayyid (1872-1963), a self-declared disciple of John Stuart Mill, promoted the idea of an independent Egyptian nation, organized according to the principles of liberalism, governed democratically, and linked educationally and culturally to Europe. He wrote that *"a constitution is the nation's right, just as freedom is the right of the individual, and since every individual is born free, it must follow that the nation was also freely established since man is by nature a social animal. This right means the nation's right to govern itself in the way it sees fit, and this is a right to which no one may infringe upon."* (Browers, 2006:86)

Related to this issue is the attitude of Arabs and Muslims towards Western countries, which may be explained in the light of their positions toward the Arab-Israeli conflict. P. A. Furia and R. E. Lucas (2008) for example, suggest that the fact that Arab publics have quite favorable attitudes toward some Western countries (e.g., France), yet very unfavorable attitudes toward others (e.g., the United States), is related to the differences of their foreign policies in the Middle East. Both authors show sympathy to *"those scholars who see amities and enmities as a mark-up on "political" rather than "cultural" or "social" variables (Morgenthau, 1954; Waltz, 1959, 1979; Page and Shapiro, 1992; Holsti, 1996)".* In their own words, *"The assertion that Arab Muslims similarly evaluate Western countries on the basis of their concrete foreign policy actions in the Middle East is of course a well-known alternative to the view that Arab Muslims are hostile to Westerners qua Westerners because of who they are."* (Furia and Lucas, 2008)

They mention Tessler, Nachtwey, and Grant (1999), who already argued that individual Arab attitudes about the Israeli–Palestinian conflict do far more to predict attitudes about regional peace than do sociological factors such as gender. Similarly, in their *"dyad level"* analysis, Furia and Lucas (2006) argue that Arab societies evaluate non-Arab societies *"largely on the basis of relatively recent and specific foreign policy behaviors"* (Furia and Lucas, 2006: 599; see also Telhami, 2002).

7.4. What the "Arab Masses" Really Think and What the Arab-Muslim Attitudes Towards the West Really Are: The Evidence from the *Arab Opinion Index*

Relying on previous research, particularly by Tausch-Karoui (2011), and Furia-Lucas (2008), we can assert that Arab-Muslim attitudes towards the West are not culturally determined, but rather politically influenced. The *essentialist* character of the Clash of civilizations' thesis makes it all the more faulty that it has failed to give an acceptable explanation for the social success of many Arabs and Muslims living in the West as either citizens or residents. Besides, as Islam is by definition a universal monotheistic religion, whose civilization has provided humankind with some of its greatest scientific and artistic achievements; Huntington's essentialism cannot explain the dialectic of civilizations that is still working at the basis of the cultural relationships between the Western and the Islamic societies. Nor can Huntington explain why Arab and Muslim scholars and intelligentsia continue to echo, discuss, and comment Western cultural and scientific output as if it were their own. There is no need to add that since millions of Muslims are living in the West, Islam is not - and cannot be - the "enemy".

Furthermore, we find in the early history of Islam the basis of tolerance toward other people creeds, as illustrated by *Al-Madina Treaty*. Islam initially did not see any problem with pacific coexistence and cooperation with Jews and Christians. When the Prophet Muhammad (S) emigrated to Madina, there were nearly 20 tribes of Jews living in this city and its outskirts. The three largest and most influential tribes in Madina were: *Banu Qainuqua',* allies of the Al-Khazraj tribe, *Banu An-Nadir* and *Banu Quraizah* who allied with the Al-Aws and lived in the suburb of the city. The Prophet Muhammad (S) wished that all those who lived in this city could enjoy peace regardless of their ethnical and religious differences. Thus, he made a treaty which offered security and assurance to those who agreed to adhere to it. The most important provisions of the treaty were the following:

1. The Jews of Bani 'Awf are one community with the believers. Jews will profess their religion and Muslims theirs.
2. Jews shall be responsible for their expenditure, and Muslims for theirs.
3. If attacked by a third party, each shall come to the assistance of the other.
4. Each party shall hold counsel with the other. Mutual relation shall be founded on righteousness; sin is totally excluded.
5. Neither shall commit sins to the prejudice of the other.
6. The wronged party shall be aided.
7. Jews shall contribute to the cost of war so long as they are fighting alongside the believers.
8. Madina shall remain sacred and inviolable for all who join this treaty.

9. Should any disagreement arise between the signatories to this treaty, then God, the All-High and His Messenger shall settle the dispute.
10. The signatories to this treaty shall boycott *Quraysh* commercially; they shall also abstain from extending any support to them.
11. Each shall contribute to defending Madina, in case of a foreign attack, in its respective area.
12. This treaty shall not hinder either party from seeking lawful revenge.

This historic treaty proves that Islam did not consider non-Muslims as enemies as long as they did not attack Muslims. Better, by this treaty the Prophet Muhammad (S) made of citizenship the bond unifying a community, not religion. Jews, who lived in the same city as Muslims, were their allies and co-citizens. Similarly, when Islam was established, the *"people of the Book"* were protected, and we know many of them worked in the administration of the Caliphs.

Policy pointer:

Today, there is indeed a Western Islam, as well as a Middle-Eastern, African, or Asian Islam. There is Shiite and Sunnite Islam, and many other branches, sections, and segments developing from the same tree, since centuries. These different branches and segments have different views about not only how to relate to non-Muslims, but also how to relate to Muslims holding opposite opinions.

Other authors better acquainted with Islam held more subtle views on the perceived hostility toward the West. Although Furia-Lucas (2008) depicted these authors as holding *"a slight variation of the clash thesis",* we think that without being fundamentally essentialist, they have caught an aspect of the reality particularly concerning some contemporary trends of Arab-Muslim thought. Thus, Irshad Manji (2005) argued that it is not so much Muslims in general, but Arab-Muslims influenced by the *"sturdy customs [of] ... Arab warriors,"* who are systematically bellicose and/or hostile to non-Arabs (Manji, 2005: 143). However, if such a hypothesis may find some support throughout history, it should be noted that it does not concern particularly the West, but also Arab neighbors, such as Turks, Iranians, and indeed Israelis since the beginning of the Palestinian tragedy.

For Fareed Zakaria (2003), Muslim hostility to the West is centered in the Middle East, deriving less from Islam than from a continuing tendency of Arabs to embrace *"patriarchy, strongmen [and] romanticism"* and thus to remain *"stuck in primitive political and social arrangements"* (Zakaria, 2003: 131). This hypothesis has been also defended by several modern Arab thinkers. We mention only a few examples: Hisham Sharabi, Burhan Ghalioun, Adonis, Khaldun al-Naqib, George Tarabishi, etc… In our view, it is more related to the new trends of social and political critique of contemporary Arab society and its failure to cope with many aspects of the modern world than to generalizations conventionally called *"anti-Americanism"* or *"hostility toward the West."*

"Social explanations" of such hostility have been advanced by several authors, and they remain indeed more valuable than the *"clash thesis"*. Following Furia-Lucas (2008), we may mention those who suggested that hostility to the West is a mark-up on individual-level *"socioeconomic"* deprivation (Friedman, 1999; Chomsky, 2001; Sardar and Davies, 2002;

Wright, 2002). In this view that reflects some theories explaining rebellion (Lipset, 1968; Gurr, 1970; Skocpol, 1979; Inglehart, 1997), it is assumed that relatively low socioeconomic status is likewise a predictor of Arab-Muslim hostility toward the economically powerful West. This hypothesis is all the more interesting that it may provide an explanation of many acts of revolt and vandalism performed by marginalized angry youth in some Western cities, as happened in France and the United Kingdom in more recent years. Even more interestingly, it can explain the Arab spring revolts, which were not against the West, but against Western-supported local dictators. In this context, it is rather a signal of the blindness of the *"clash thesis"* that people in revolt across the Arab world never chanted anti-Western slogans. Better: in Libya, which Qaddafi has transformed into an anti-Western hub for most of his 40-year rule, people in the streets were chanting *"Vive la France"*, when the NATO strikes against Qaddafi started.

The previous hypothesis should be brought, though cautiously, into comparison with the thesis about the *"clash of generations"*. In our view, the latter is more closely related to the socioeconomic deprivation hypothesis than to the *"clash thesis"* as Furia-Lucas (2008) have suggested under the label of *"clash of generations"* referring to Kaplan, 1994; Friedman, 1999; Wright, 2002; and Urdal, 2006. As with the thesis about socio-economic deprivation leading to revolt, we think that the population boom (essentially concerning Arabs and Muslims), whether local (in the MENA region for example) or global (the West included), conjugated with the absence of good economic prospects for the young generations, and caused systematic hostility toward *"powerful status quo actors in domestic and international politics."*

Finally, we join Furia and Lucas (2008), when they suggest an explanation of hostility toward the West (not all the West, the distinction here is crucial) based on the political factors. We think that there is a sound reason behind the relatively favorable attitudes of Arab and Muslim publics toward some Western countries (e.g., France), yet very unfavorable attitudes toward others (e.g., the United States). Political, rather than social or cultural variables are better tools of understanding here. Furia and Lucas mention for example Tessler, Nachtwey, and Grant (1999), who argued that individual Arab attitudes about the Israeli–Palestinian conflict do far more to predict attitudes about regional peace than do sociological factors such as gender. The assumption that Arab Muslims are hostile to Westerners qua Westerners because of *"who they are,"* does not hold the road, because it remains blind to the fact that Foreign policy behavior is the key here: otherwise concrete actions in the Middle East and reactions to these actions.

On the one hand, 41.5% fully support the takeover of political power by religious people, and 32.2% prefer to deal only with religious people in their personal relationships. On the other hand, the majority (at least in Tunisia and Egypt) support the separation of the religious practices from the social and political life. Clearly, they are against identifying polity with religion. They are also against allowing the religious sheikhs (clergy) to influence the vote of the citizens and the decisions of the government (Karoui, 2012: 299-302).

A very high percentage (85.6%) of the population is declaring itself to be religious or deeply religious. In the two leading countries of the Arab spring (Egypt and Tunisia), the majority defined their political identity, firstly, by religious affiliation (Muslim, Christian. . .), secondly by ethnicity (Arab), thirdly by nationality, and lastly by political affiliation (Karoui, 2012: 288).

But incredibly, 76% of the "Arab masses" are apparently satisfied/very satisfied with their life, and an astonishing further 55.1% are satisfied/very satisfied with economic conditions!

However, this apparent "satisfaction" may be misleading at first glance. As we explained elsewhere (Karoui, 2012: 7-10), the ACRPS questionnaire was purposely intended to be varied and detailed, meaning that knowing the blockades working inside the Arab mind as well as within society, because of uncontrollable elements like: long oppression, fear, inhibition, *et cetera.* ... The interviewees were asked several times the same question: then, as soon as the question becomes more precise, the answer would differ. For example, if you link the question about "satisfaction in life" to "your finances", "the expenses of your family", "your economic condition", "your work", *et cetera...* the answer would be more modulated. Thus, if the first answer about satisfaction in life is generally "praise to Allah for everything", the focus about the financial condition of the family would make this "overall satisfaction" much less obvious.

The sometimes alarming difference between political satisfaction and economic satisfaction reveals that in Tunisia and Egypt, where the current Arab revolutions started, there was a glaring deficit of political versus economic development of -46% (Tunisia) and -43% (Egypt)![45]

Therefore, we refer our readers to Table 7.2 (our overall regional results, calculated with proper population-weights), and to Table 7.3 (country results). In Table 7.3, we highlight some of the country results, which we deem as vital for the development of democratic peace and security in the region, leaving the further interpretation of our results to our readers.

Table 7.2. The drivers and bottlenecks of democratization in the Arab world – population weighted results from the "Arab Opinion Index" (12 countries, 85% of Arab population covered by the analysis)

Possible drivers of future democratic pro-Western development	Weighted percentage results for all the Arab countries
% supporting overthrow of Mubarak	83,1
satisfied/very satisfied with life	76,0
% supporting Tunisian revolution	67,2
% agree: in spite of problems democratic system better	63,7
% satisfied/very satisfied with economic conditions	55,1
great deal of trust in the army	47,4
% agree/strongly agree religious practices must be separated from political & social life	46,6
% agree/strongly agree: best to separate religion from politics	42,8
% saying political freedom and civil liberties requirement of democracy	36,3
% agree strongly: clergy must not influence the way people vote	35,7
% satisfied/very satisfied with political conditions	27,1
% saying equality and justice among citizens requirement of democracy	19,5

45 Michael Mousseau, "Urban poverty and support for Islamist terror: Survey results of Muslims in fourteen countries," *Journal of Peace Research* 48, 1, (2011): 35–47. Mark Tessler, "What Leads Some Ordinary Arab Men and Women to Approve of Terrorist Acts Against the United States?" *Journal of Conflict Resolution* 51, 2 (2007): 305-328 by contrast thinks that the negative image of US foreign policy in the region is the main driver of terror approval.

Table 7.2. (Continued)

Possible bottlenecks for future democratic pro-Western development	Weighted percentage results for all the Arab countries
% saying corruption widespread	49,2
% agree/strongly agree: best if religious people take over public office	41,5
household income does not cover regular expenses	38,6
% prefer to deal with religious people	32,2
% agree/strongly agree: in a democratic system the economic performance runs badly	25,3
% desire to emigrate	22,2
% say partner from a well-known family decisive for parental approval of a marriage	8,7

Data on religion and Arab nationalism	Weighted percentage results for all the Arab countries
% religious/very religious	85,6
% saying: one Arab nation	35,6
% saying neighboring Arab countries most threatening to country's security	11,3

Source: calculated from the *Arab Opinion Index* data, Arab Center for Research and Policy Studies (ACRPS) in Doha, Qatar, available at: http: //english. dohainstitute. org/release/5083cf8e-38f8-4e4a-8bc5-fc91660608b0

Table 7.3. Percentage country results, based on the *Arab Opinion Index* (12 countries, 85% of Arab population covered by the analysis)

	Algeria	Egypt	Jordan	Lebanon	Mauritania	Morocco	Saudi Arabia	Sudan	Tunisia	Yemen	Iraq	Palestinian territories
satisfied/very satisfied with life	76,0	85,0	78,0	72,0	87,0	70,0	88,0	87,0	72,0	69,0	41,0	65,0
% satisfied/very satisfied with economic conditions	62,0	60,0	64,0	50,0	49,0	55,0	75,0	60,0	55,0	34,0	29,0	42,0
household income does not cover regular expenses	32,0	47,0	42,0	36,0	63,0	34,0	18,0	38,0	37,0	53,0	32,0	58,0
% satisfied/very satisfied with political conditions	32,0	17,0	69,0	8,0	42,0	24,0	76,0	38,0	9,0	8,0	7,0	21,0
% desire to emigrate	26,0	8,0	18,0	33,0	19,0	29,0	9,0	50,0	33,0	28,0	13,0	19,0
% saying Israel most threatening to personal security	6,0	65,0	27,0	62,0	3,0	13,0	18,0	37,0	5,0	5,0	8,0	83,0
% saying US most threatening to personal security	15,0	17,0	3,0	3,0	0,4	3,0	4,0	25,0	3,0	2,0	16,0	2,0
% saying Iran most threatening to personal security	0,0	1,0	3,0	4,0	0,0	0,0	10,0	0,3	0,0	1,0	18,0	0,4
% saying Israel most threatening to country's security	15,0	65,0	48,0	70,0	18,0	12,0	30,0	32,0	14,0	22,0	14,0	81,0

	Algeria	Egypt	Jordan	Lebanon	Mauritania	Morocco	Saudi Arabia	Sudan	Tunisia	Yemen	Iraq	Palestinian territories
% saying US most threatening to country's security	20,0	17,0	6,0	5,0	6,0	3,0	6,0	27,0	12,0	19,0	32,0	9,0
% saying Iran most threatening to country's security	0,0	1,0	4,0	8,0	0,0	0,0	28,0	0,5	0,0	10,0	25,0	3,0
% saying neighboring Arab countries most threatening to country's security	17,0	1,0	2,0	5,0	0,0	51,0	3,0	1,0	34,0	13,0	10,0	1,0
% supporting Tunisian revolution	87,0	54,0	55,0	73,0	66,0	71,0	61,0	. .	96,0	81,0	61,0	79,0
% supporting overthrow of Mubarak	89,0	87,0	65,0	79,0	66,0	75,0	. .	. .	90,0	85,0	77,0	82,0
% saying political freedom and civil liberties requirement of democracy	41,0	34,0	39,0	41,0	22,0	17,0	38,0	49,0	29,0	30,0	46,0	32,0
% saying equality and justice among citizens requirement of democracy	22,0	20,0	19,0	13,0	44,0	26,0	17,0	17,0	26,0	17,0	13,0	14,0
% agree/strongly agree: in a democratic system the economic performance runs badly	17,0	21,0	21,0	42,0	15,0	19,0	36,0	40,0	11,0	23,0	28,0	26,0
% agree: in spite of problems democratic system better	56,0	56,0	79,0	83,0	63,0	68,0	64,0	75,0	72,0	74,0	55,0	67,0
perceived level of democracy in the country	5,0	4,9	7,0	4,6	6,0	4,4	5,2	4,3	2,9	3,2	3,3	4,6
great deal of trust in the army	16,0	81,0	83,0	71,0	63,0	24,0	61,0	52,0	70,0	15,0	8,0	. .
% saying corruption widespread	51,0	38,0	48,0	92,0	45,0	57,0	20,0	59,0	75,0	72,0	. .	. .
% religious/very religious	78,0	96,0	96,0	84,0	94,0	73,0	83,0	93,0	69,0	87,0	73,0	90,0
% agree/strongly agree religious practices must be separated from political & social life	44,0	55,0	36,0	83,0	35,0	35,0	38,0	31,0	63,0	36,0	71,0	36,0
% prefer to deal with religious people	18,0	21,0	27,0	16,0	66,0	21,0	22,0	38,0	13,0	21,0	94,0	99,0
% say partner from a well-known family decisive for parental approval of a marriage	4,0	4,0	5,0	5,0	. .	1,0	16,0	3,0	23,0	14,0	28,0	6,0
% agree strongly: clergy must not influence the way people vote	25,0	32,0	45,0	72,0	37,0	34,0	47,0	43,0	35,0	34,0	34,0	32,0

Table 7.3. (Continued)

	Algeria	Egypt	Jordan	Lebanon	Mauritania	Morocco	Saudi Arabia	Sudan	Tunisia	Yemen	Iraq	Palestinian territories
% agree/strongly agree: best if religious people take over public office	31,0	36,0	36,0	20,0	49,0	34,0	53,0	74,0	31,0	37,0	30,0	41,0
% agree/strongly agree: best to separate religion from politics	40,0	52,0	31,0	85,0	27,0	30,0	29,0	25,0	52,0	36,0	74,0	31,0
% saying: one Arab nation	34,0	35,0	41,0	15,0	45,0	34,0	34,0	32,0	20,0	56,0	36,0	43,0
% against recognition of Israel	94,0	78,0	81,0	79,0	72,0	72,0	82,0	90,0	97,0	88,0	86,0	85,0
% Israel's possession of nuclear weapons justifies nuclear proliferation	52,0	60,0	56,0	47,0	50,0	55,0	60,0	76,0	52,0	54,0	61,0	42,0

Source: compiled from the *Arab Opinion Index* data, Arab Center for Research and Policy Studies (ACRPS) in Doha, Qatar, available at: http: //english. dohainstitute. org/release/5083cf8e-38f8-4e4a-8bc5-fc91660608b0.

Judging by the percentage of people who say that political freedom and civil liberties are an absolute requirement of democracy, we arrive at the following hard-core data of the development of a civil society, necessary to support the workings of a democratic system, derived from the *Arab Opinion Index*:

Table 7.4. Real civil society support for democracy in the Arab world (12 countries, 85% of Arab population covered by the analysis)

	% saying political freedom and civil liberties are a requirement of democracy
Sudan	49%
Iraq	46%
Algeria	41%
Lebanon	41%
Jordan	39%
Saudi Arabia	38%
Egypt	34%
Palestinian territories	32%
Yemen	30%
Tunisia	29%
Mauritania	22%
Morocco	17%

Source: calculated from the *Arab Opinion Index* data, Arab Center for Research and Policy Studies (ACRPS) in Doha, Qatar, available at: http: //english. dohainstitute. org/release/5083cf8e-38f8-4e4a-8bc5-fc91660608b0.

Traditionally, an almost unlimited number of well-informed observers, especially from the Western security establishments, would tend to view the roots of radicalism in the region as a mix of the consequences of the demographic situation, unemployment among the young, urbanization, corruption and incompetence of the former governments, *et cetera.*[46]

Such kinds of insights are being taught not only to students in graduate schools of social sciences around the West, but also – more important still – at diplomatic and bureaucratic staff training institutes, and at the staff colleges of the military, the intelligence services and police academies, from Canada to Australia, and from the new Baltic NATO partners to Brazil and Chile. The wisdom that officers and officials are being taught, was neatly summarized in *"Parameters",* the flagship scholarly journal of the U. S. Army War College, by Christopher Henzel, who is indeed very explicit in offering an advice on what the U.S. government [and implicitly also governments allied with America] should do and what it/they should not do. *"Traditional Muslim culture"* is, according to this school of thinking, for all purposes, beyond any question:

> "If mainstream Sunnis come to view the United States as bent on a campaign to weaken or remake traditional Muslim culture, then more and more mainstream Sunni believers will conclude that the revolutionary Salafists they once reviled were right all along. At that point the world really would see the clash of civilizations sought by both al Qaeda and some US pundits." (Henzel, 2005)

The counter-position being advanced here proposes that one should do everything possible to strengthen moderate forces, which indeed favor the separation of religion and politics. The liberal heritage, which resulted from the movement of Enlightenment and democracy, implying respect of other religions and civilizations, is part of the modern Islamic culture. It does not need to be sown anew, but simply supported through institutions facilitating cooperation, interaction and acculturation.

7.5. Some New Evidence on the Basis of the *World Values Survey* Project

Nowadays, Arab politics courses at major universities in the West, more often than not, will also shy away from the perspective of *"weaken or remake traditional Muslim culture".* What is *"traditional Muslim culture"* anyway?

The first tendency was represented by scholars who turned more toward religion as source of power than as a support of knowledge. The second tendency was represented by those often called: the philosophers (like the French intellectuals who paved the way for the 1789 revolution).

The fact that Islam encouraged acculturation and exchange with other cultures and civilization is beyond doubt, although this fact has been – and still is – ignored by the conservatives who rejected and fought progress and free thought.

[46] Alan Richards, „Socioeconomic Roots of Middle East Radicalism" *Naval War College Review,* 40, 4, 2002: 23-38. Retrievable at: http: //www. usnwc. edu/Publications/Naval-War-College-Review. aspx

Policy pointer:

For many experts in Arab-Islamic civilization, this cultural patrimony is to be divided broadly into two distinct tendencies, which we may call today: conservative, and liberal. The first encouraged conservative reproduction of the same. It begat the old fundamentalist tradition (in both the Shiite and Sunnite branches), fighting progress and freedom of thought, always under the pretention of defending the *"true and pure"* religion. Such a tradition is held responsible for the centuries of darkness and backwardness to which the Muslims have been subjected. They were the servants of Tyrants pretending to serve Islam. The second encouraged the intellectual effort *(Ijtihad),* the creative imagination, progress, scientific research, arts, and opened the minds of people to other cultures and civilizations.

The Islamic encounter with Greek philosophy in the ninth century is regarded as *"the locus of one of the earliest sustained interactions with political thought that is neither Arab nor Islamic"* (Browers, 2006). The ancient oriental Christian thought has played a key-role - before and after the incoming of Islam - in introducing Hellenic rational influence to the Near East region, and in trying to adapt it to the religious creeds. It was also an important role in shaping the new Islamic culture through the translation of the great texts of the Greek philosophers and others.

Christian teachers were in touch with the philosophy of their time, *"and so when the Muslims later on did the same, they had to reckon with philosophy as they found it actually living in their own days."* (De Lacy O'Leary, 1922)

The Arab medieval historian al-Masûdi informs us that *"Greek philosophy originally flourished at Athens, but the Emperor Augustus transferred it from Athens to Alexandria and Rome, and Theodosius afterwards closed the schools at Rome and made Alexandria the educational centre of the Greek world."* (Masûdi, transl. 1896: 170, in: De Lacy, 1922). This is important, because from Plato onwards, philosophy had been very largely concerned with theories more or less related to the organization of human societies, and much of what the medieval Muslim philosophers wrote was a commentary of the Greek texts.

For a long time, Jews were less active than Christians in the development of Hellenistic philosophy. Yet, in later times, they participated particularly in medical studies and natural science: *"Hunayn b. Ishaq, his son Ishaq, and his nephew Hubaysh, with some other companions, formed the college of translators established at Baghdad by the Khalif al-Ma'mun to render the Greek and other philosophical and scientific texts into Arabic."* (De Lacy, 1922). Later on, they will transmit the Muslim philosophy of Asia to Spain and Europe. (We further discuss this topic in the next chapter).

Policy pointer:

If the Arab world wants positively to confront the 21st Century and become a full and mature democracy, issues like tolerance, sexism, and the overcoming of authoritarian thought patterns become important for the future of the democratic system. While *"Islamophobia"* in the West remains to be confronted, xenophobia and discrimination against immigrants and minority groups in Muslim countries have also to be faced. Even if a strong emphasis on religion is expressed, the religious doctrines are invited to start to learn from their own Enlightenment traditions, especially of Muslim philosophy, and the great global Enlightenment thinkers of the 18th and the 19th Century.

Seen from such a perspective, and with the free online data analysis version of the *World Values Survey* project,[47] the reasons for the unsatisfactory and incomplete path towards democracy in the Arab world become suddenly clearer. For Table 7.5, we selected from the WVS questions the item, which asks respondents: *"On this list are various groups of people. Could you please sort out any that you would not like to have as neighbors?"*

(V35) People of a different race
(V37) Immigrants/foreign workers
(V39) People of a different religion

In our Table 7.5, we specify the percentages of the total population, corresponding to the different items, the date of the analysis (i. e., *World Values Survey* wave 4 is from 1999/2000; wave 5 is from 2005 to 2008) as well as an average xenophobia rate, and we rank the countries of the world according to their average xenophobia rates. The Arab countries with available data rank very badly on our scale, but also the E.U. member country France; while advanced democracies and also some Latin American new democracies rank very favorably on our Table:[48]

Table 7.5. Large-scale xenophobia is the real problem

	Neighbors: People of a different race	Neighbors: Immigrants/ foreign workers	Neighbors: People of a different religion	WVS survey	average xenophobia rate
Bangladesh	72%	67%	66%	wave 4	68%
Jordan	53%	68%	51%	wave 5	57%
India	49%	39%	49%	wave 5	46%
Iran	32%	60%	38%	wave 5	43%
Viet Nam	42%	45%	42%	wave 5	43%
Saudi Arabia	38%	33%	40%	wave 4	37%
Rwanda	37%	36%	37%	wave 5	37%
Indonesia	33%	36%	35%	wave 5	35%
South Korea	36%	39%	26%	wave 5	34%
Malaysia	21%	57%	23%	wave 5	34%

[47] http: //www. wvsevsdb. com/wvs/WVSAnalizeStudy. jsp
[48] This could be taken as an example of Muslim exclusivism, described by Douglass Pratt, *"Islamic Prospects for Inter-Religious Dialogue: The Contribution of Fethullah Gülen"* retrievable at http: //www. fethullahgulen. org/conference-papers/contributions-of-the-gulen-movement/2469-islamic-prospects-for-inter-religious-dialogue-the-contribution-of-fethullah-gulen. html.

Table 7.5. (Continued)

	Neighbors: People of a different race	Neighbors: Immigrants/ foreign workers	Neighbors: People of a different religion	WVS survey	average xenophobia rate
France[49]	**27%**	**43%**	**30%**	**wave 5**	33%
Thailand	28%	44%	28%	wave 5	33%
Turkey	30%	31%	34%	wave 5	32%
Nigeria	30%	28%	29%	wave 4	29%
Georgia	25%	24%	37%	wave 5	28%
Zambia	30%	28%	27%	wave 5	28%
Morocco	23%	24%	36%	wave 5	28%
Algeria	28%	24%	32%	wave 4	28%
Ghana	23%	26%	24%	wave 5	24%
Mali	22%	25%	24%	wave 5	24%
Moldova	24%	19%	26%	wave 5	23%
Russian Federation	17%	33%	16%	wave 5	22%
Serbia	19%	26%	15%	wave 5	20%
Cyprus	17%	23%	17%	wave 5	19%
Romania	20%	19%	17%	wave 5	19%
Bulgaria	21%	19%	16%	wave 5	19%
Slovenia	18%	21%	16%	wave 5	18%
China	16%	20%	17%	wave 5	18%
Venezuela	16%	18%	17%	wave 4	17%
Ethiopia	17%	15%	16%	wave 5	16%
Ukraine	12%	19%	13%	wave 5	15%
Pakistan	7%	29%	8%	wave 4	15%
Italy	13%	16%	12%	wave 5	14%
Poland	14%	15%	12%	wave 5	13%
Finland	12%	17%	10%	wave 5	13%
Taiwan	8%	24%	5%	wave 5	13%
South Africa	8%	25%	5%	wave 5	13%
Mexico	9%	11%	15%	wave 5	12%
Burkina Faso	10%	12%	12%	wave 5	11%
Germany	9%	16%	5%	wave 5	10%
Chile	10%	11%	7%	wave 5	9%
Spain	9%	8%	7%	wave 5	8%
Great Britain	5%	16%	2%	wave 5	8%

[49] The current socio-religious landscape of xenophobia in France according to the *World Values Survey* wave 5 is the following:

Rejection of neighbors:	Total	Roman Catholic	Muslim
People who have AIDS	22%	43%	56%
Immigrants/foreign workers	22%	44%	39%
Homosexuals	20%	38%	49%
People of a different race	16%	32%	33%
People who speak a different language	15%	31%	34%
People of a different religion	14%	28%	34%
Unmarried couples living together	8%	14%	30%
Sample size n	846	368	38

	Neighbors: People of a different race	Neighbors: Immigrants/ foreign workers	Neighbors: People of a different religion	WVS survey	average xenophobia rate
Uruguay	6%	8%	9%	wave 5	7%
Netherlands	9%	10%	3%	wave 5	7%
Peru	6%	6%	8%	wave 5	7%
Brazil	5%	8%	7%	wave 5	7%
United States	4%	13%	3%	wave 5	7%
Switzerland	6%	8%	5%	wave 5	6%
Norway	4%	8%	3%	wave 5	5%
Australia	5%	6%	3%	wave 5	5%
Andorra	4%	3%	3%	wave 5	3%
Trinidad and Tobago	3%	5%	2%	wave 5	3%
Canada	3%	5%	2%	wave 5	3%
Argentina	3%	4%	2%	wave 5	3%
Sweden	2%	2%	2%	wave 5	2%

Source: our own calculations from *World Values Survey* at http: //www. wvsevsdb. com/wvs/ WVSAnalizeStudy. jsp

Policy pointer:

According to our preliminary hypothesis, one of the main reasons of the trajectory of radicalism, hatred and intolerance, which emerges from our survey, is the dire state of higher education, especially in the social sciences in the region. The highest ranking Arab University or research institute in the recent very comprehensive SCIMAGO/SIR global University ranking report is Cairo University, and it enjoys only a rank at 526 out of all 3290 ranked institutions.[50]

And even this is rather thanks to their respectable performance in science and medicine, and not in the humanities and social sciences.

In this Chapter, we evaluated PEW International, *World Values Survey* (WVS) and other new Arab public opinion data (the "*Arab Opinion Index*" by the Arab Center for Research and Policy Studies [ACRPS] in Doha, Qatar). While the WVS data show lamentable degrees of xenophobia and racism in several Arab countries, the ACRPS data show the overwhelming support for democracy and change in the region. At the same time, the data show basic weaknesses of the civil society support for the structures of democracy.

ACRPS's majority of respondents accepted the rise to power through ballot boxes of a party they did not agree with. However, percentages differed. Positive answers to this question included Tunisia, Egypt, Sudan, Yemen, Mauritania and the Palestinian Territories. The Algerian opinion was divided, while the majority in Jordan and Lebanon opposed. The Lebanese opposition is rather odd, given the fact that this country practiced a multipartisan political life almost exclusively in the Arab world since the end of World War II. However, it also shows the degree of sectarianism (cause of the 15-year civil war) plaguing politics in Lebanon.

---

[50] http: //www. scimagoir. com/ See also the special SCIMAGO/SIR Report on the Middle East academic institutions, ranked in 2011, available from the same website.

It is no less interesting to note that in the same survey, 31 % of respondents in 12 Arab countries claimed to oppose the rise to power of a religiously-based political party. 31% is not a small percentage over this issue if we consider the weight of variables like: tradition, faith, illiteracy, and the lack of a modern political culture.

Concerning the question of separating religion from politics and state, nearly two-thirds of respondents from Lebanon, Iraq and Tunisia, and 56% in Egypt said that they would support the rise to power of a political party seeking to establish a divide between politics and religion. This is a clear majority in countries where religiously-based political parties have been aggressively active recently.

In contrast, less than 50 % in Algeria, Morocco and Yemen showed the same position, while a majority in Sudan, Jordan and The Occupied Palestinian Territories clearly opposed it. Public opinion in Arab countries, according to ACRPS's survey, is in favor of democracy. The majority considers democracy as the best system of governance. Arguments like *"the economy would suffer under a democracy,"* or *"democracy leads to a breakdown in law and order,"* are not popular among the citizens of the region.

Heinsohn and Pipes were right in emphasizing that the available statistics about deadly conflicts, democide and genocide in the second half of the 20th Century suggest that *the number of deaths in the Arab-Israeli conflict since 1950 amounts to 51,000 people, among them 35,000 Arabs and 16,000 Jewish Israelis. By contrast, some 11,000,000 Muslims have been killed since 1948, 90% of them by fellow Muslims, i.e., the 35,000 Arab victims of the Arab-Israeli conflict are 0.3 percent or 1 out of every 315 global Muslim fatalities.*[51] So it is time to start a *"Vergangenheitsbewältigung" (from German: overcoming of the past)* in the Arab world and the Muslim world in general, and expose the crimes of dictators like the Assad clan in Syria and other murderous regimes and conflicts, which killed in all 11 million Muslims. Terrible massacres committed by the Baathist Arab nationalist regimes against their own populations in Syria (Hama)[52] and recently by chemical weapons in Ghouta[53] and in Iraq (Halabja)[54] cost as many or even more lives than the entire Arab-Israeli conflict. Finally, much of the "third worldist" displayed "solidarity" of the political left – or what has remained of it – should "update" its discourse (born in the Cold War) to tune up with the new developments on the world stage. As some observed, long years of Western *"intellectual imperialism"* have produced *"a borrowed consciousness that rendered most social sciences esoteric and irrelevant in much of the Third World."* (Grant and Short, 2002). Not to be alienated from local realities, local social scientists have to offer new theories and research methods corresponding to the societies they study, not entirely borrowed from post-industrial societies.

[51] Gunnar Heinsohn and Daniel Pipes, *"Arab-Israeli Fatalities Rank 49th" FrontPageMagazine. com,* October 8 (2007), retrievable at http: //www. danielpipes. org/4990/arab-israeli-fatalities-rank-49th

[52] *"warplanes and tanks leveled whole districts of the city (of Hama) . . . the evidence clearly suggests that government forces made no distinction between armed insurgents and unarmed civilians . . . the assault on the city represents a clear act of war crimes and murder on a mass scale"* (see Robert Fisk, "Conspiracy of silence in the Arab world" *The Independent* (Saturday, February 10, 2007) retrievable at http: //www. independent. co. uk/voices/commentators/fisk/robert-fisk-conspiracy-of-silence-in-the-arab-world-435762. html. Syrian human rights groups (http: //www. shrc. org/data/aspx/d0/1260. aspx) put the number of victims of the Hama massacre at 40. 000. The regime used cyanide gas against its own citizens, and the dictator who was responsible for it was Hafez al-Assad (1930-2000).

[53] Sarin gas has been identified by the United States as the agent loaded atop small rockets on Aug. 21 and shot into the densely populated suburbs of Damascus, killing more than 1,400 people, according to administration officials.

[54] http: //news. bbc. co. uk/onthisday/hi/dates/stories/march/16/newsid_4304000/4304853. stm and http: //news. bbc. co. uk/2/hi/middle_east/8479115. stm

Chapter 8

ISLAMIC VALUES, KNOWLEDGE, AND MORALS

"Seek knowledge as far as China."
"Seek knowledge from the cradle to the grave."
Hadith of the Prophet Muhammad (S)

In this chapter, we try to find out the common grounds between modern Islam and the rest of the human kind, on the basic assumption that modern culture is universal. Therefore, Islamic societies cannot stay away from it. We show that the belief that Islam is against modernity, the West, and the values humanity has produced since the 18th century, is unfounded and even contradictory to the Islamic vision of knowledge and morals.

Box 8.1. The overall message of this Chapter

In a book about global value change, the question about how Muslims perceive the problem of knowledge and morals is crucial. Actually, it is at the basis of all the issues related to universal values, modernity, democracy, and human rights.

Regardless of how modernity is defined, we can see its effect everywhere in the Arab and Muslim world, in such domains as education, architecture, graphic arts, urban development, political and cultural institutions, *et cetera.* Indeed, the immersion in modernity does not concern exclusively the West. As an Iranian author put it, *"this immersion has expanded to such an extent that presently in the bookstores of most major cities in Iran one encounters the translated works of such thinkers as Arendt, Aron, Carnap, Dewey, Diderot, Habermas, Heidegger, Hume, Jaspers, Kant, Levi-Strauss, Locke, Marcuse, Nietzsche, Pascal, Popper, Rousseau, Bertrand Russell, Sartre, Spinoza, Weber, Wittgenstein, and many others. As the above list indicates, some of the seminal figures of analytic philosophy (logical positivism, linguistic philosophy) and continental philosophy (phenomenology, existentialism, structuralism) have been introduced to Iranian readers"* (Boroujerdi, 1997). And we can surely say the same thing about the bookstores of the major Arab cities.

The belief that Islam is against modernity, against the West, or against the values humanity has produced in its long march since the 18th Century, is naïve and ignorant. Studying critically Western thought does not imply necessarily opposing dogma to science, obscurantism to Enlightenment, irrationality to rationality. After all, medieval Muslim thinkers have dedicated much time and energy to analyzing and commenting texts of Greek philosophers, to see what was adaptable and useful to their own progress. Such a task cannot be achieved without critical thinking. This endeavor is scientifically right. It is proved by the history of philosophy (Hegel, Marx; Frankfort school; M. Foucault, Deleuze, Derrida, etc…) and science.

Muslim thinkers do criticize Western thinkers, in the same measure that Western thinkers criticize each other, not for "religious" reasons, but for inconsistency and logical flaws. Very often, such studies involve interesting questions about the "self" and the "other", science, technology, arts, politics, economy, systems of thought and systems of belief. There is no unanimity among Muslim thinkers concerning the West. If such unanimity were possible, it would rather belong to conspiracy theories, which, since Shakespeare's Julius Caesar, added nothing new to human knowledge.

The December 2012 *Global Religious Landscape Report* from the Pew Research Center's Forum on Religion & Public Life, estimated the number of Muslims around the world at about 1.6 billion, which makes Islam the world's second-largest religious tradition after Christianity. Though 23% of the world's population is affiliated to Islam, contrarily to a broadly-held belief, Arabs do not constitute the majority of Muslims who, consequently, are not living in the Middle East. Nearly two-thirds (62%) of Muslims live in the Asia-Pacific region, according to the Pew Research analysis. In fact, more Muslims live in India and Pakistan [344 million combined] than in the entire Middle East-North Africa region [317 million]; (http: //www. pewresearch. org/fact-tank/2013/06/07/worlds-muslim-population-more-widespread-than-you-might-think/). In 2010, Muslims living in Europe were estimated at around 43.490.000 (which includes Eastern Europe). The world's Muslim population is expected to increase by about 35% in the next 20 years, rising from 1.6 billion in 2010 to 2.2 billion by 2030, according to the new population projections by the Pew Research Center's Forum on Religion & Public Life. The increase is expected to be at a faster rate than the global non-Muslim population (http: //www. pewforum. org/2011/01/27/the-future-of-the-global-muslim-population/).

8.1. How Muslims Perceive the Problem of Knowledge and Morals

In a book about global value change, the question about how Muslims perceive the problem of knowledge and morals is crucial. Actually, it is at the basis of all the issues related to universal values, modernity, democracy, and human rights. If we consider that these are global issues, since modern science and technology as well as world trade are blurring the limits between cultures and civilizations more than distinguishing and pushing them to a clash (as believed by Huntington), we may talk of this global era as a universal civilization.

Policy pointer:

Indeed, when we find "Paris" and "London" in the malls of Doha and Dubai; when young Arabs and Muslims dance the hip hop in the streets of Tunis, Manama, or Cairo; when young Saudi or Palestinian musicians play the rock & roll or the rap; when Egyptian movies imitate so many Hollywood productions; when the Arab actors play Shakespeare, Becket and Pirandello… should we call this "clash of civilizations," globalization, or universal civilization?

Regardless of how modernity is defined, we can see its effect everywhere in the Arab and Muslim world, in such domains as education, architecture, graphic arts, urban development, political and cultural institutions, *et cetera.* Indeed, the immersion in modernity does not concern exclusively the West. As an Iranian author put it, *"this immersion has expanded to such an extent that presently in the bookstores of most major cities in Iran one encounters the translated works of such thinkers as Arendt, Aron, Carnap, Dewey, Diderot, Habermas, Heidegger, Hume, Jaspers, Kant, Levi-Strauss, Locke, Marcuse, Nietzsche, Pascal, Popper, Rousseau, Bertrand Russell, Sartre, Spinoza, Weber, Wittgenstein, and many others. As the above list indicates, some of the seminal figures of analytic philosophy (logical positivism, linguistic philosophy) and continental philosophy (phenomenology, existentialism, structuralism) have been introduced to Iranian readers"* (Boroujerdi, 1997). And we can surely say the same thing about the bookstores of the major Arab cities.

The belief that Islam is against modernity, against the West, or against the values humanity has produced in its long march since the 18th century, is naïve and ignorant. Boroujerdi (1997) pointed out that some of his Western colleagues *"are often surprised to hear that one of the most intense debates in post-revolutionary Iran has been an epistemological debate over the positivist ideas of Popper, Hempel, and Carnap versus the historicism of Nietzsche, Hegel, and Heidegger."* Yet, it is important to add that contemporary Muslim intellectuals did not receive Western knowledge negatively. On the contrary, although many of them have been educated in Western universities, *"a systematic critical study of the West and its various modes of knowledge has been occurring, in Iran since the 1960s"* (Boroujerdi, 1997) and in the Arab world since the early years of the 20th Century.

Studying critically Western thought does not imply necessarily opposing dogma to science, obscurantism to Enlightenment, irrationality to rationality. After all, medieval Muslim thinkers have dedicated much time and energy to analyzing and commenting texts of Greek philosophers, to see what was adaptable and useful to their own progress. Such a task cannot be achieved without critical thinking. This endeavor is scientifically right. It is proved by the history of philosophy (Hegel, Marx; Frankfort school; M. Foucault, Deleuze, Derrida, etc...) and science.

Muslim thinkers do criticize Western thinkers, in the same measure that Western thinkers criticize each other, not for "religious" reasons, but for inconsistency and logical flaws. Very often, such studies involve interesting questions about the "self" and the "other", science, technology, arts, politics, economy, systems of thought and systems of belief. There is no unanimity among Muslim thinkers concerning the West. If such unanimity were possible, it would rather belong to conspiracy theories, which, since Shakespeare's *Julius Caesar,* added nothing new to human knowledge. The other facet of the coin is likely the absence of unanimity in the West concerning the Islamic world. Thus, more and more Westerners today seek a better – i.e., unbiased – understanding of what is going on in the Islamic world.

Policy pointer:

Islam has been a precursor and a major force in promoting scientific endeavor. Unlike medieval Catholicism, Islam did not stifle the spirit of scientific enquiry. Although the first attempts to engage into philosophical argumentation have aimed at a better understanding of the Islamic message, soon Muslim scholars adopted new mental attitudes merging philosophical rational questioning with Islamic precepts and directives, and broadening the span of knowledge to new disciplines previously unknown to the Arabs, like chemistry, physics, mathematics, geometry, etc...

The first authors to write on political subjects were under philosophical influence (Crone, 2004: 166-167). Although their works were lost, they seem to have had greater affinities with *al-'ilm al-madani* (in the sense of *polity or political science).* To be mentioned in this case: al-Kindi (d. 866), al-Sarakhsi (d. 899), and Abu Zayd al-Balkhi (d. 934). But the real founder of Muslim political science is al-Farabi (d. 950). All the philosophers that followed him fell under his influence: ibn Sina (d. 1037), ibn Baja (d. 1139), ibn Tufayl (d. 1185), and ibn Rushd (d. 1198). Other thinkers were more independent: Abu Hatim al-Razi (d. 934), Abu Yaqub al-Sijistani (d. 971), and the Brethren of Purity (a secret society active in the second half of the tenth century).

The early Muslim thinkers shared the early Greek view that membership of a particular polity (the city state or the *Umma)* was "a precondition for human perfection/salvation" (Crone: 171-172). However, neither of them spoke of the "citizen", but of the *"human" (insân)* or the *"inhabitant of the city" (ahl al-madina).* Thus, it seems that *"the modern Arabic word for citizen [muwâtin], in the legal sense of the term (meaning one holding the citizenship or nationality of a sovereign state) was coined around the turn of the twentieth century" (Browers, 33).*

The discussion of the Greek model of democracy started in the works of the classic Muslim philosopher Abu Nasr al-Farabi (870-950), who translated *"democracy"* by *"al-madina al-jama'iyya"* (the social or the collective city).

In his *"al-Madina al-fadila"* (The Virtuous City), al-Farabi, *known to the Arabs* as the *'Second Master' (after Aristotle), "identifies three types of society, which are perfect, and grades these according to size. His ideal virtuous city, which gives its name to the whole volume, is that which wholeheartedly embraces the pursuit of goodness and happiness and where the virtues will clearly abound. This virtuous city is compared in its function to the limbs of a perfectly healthy body. By stark contrast, al-Farabi identifies four different types of corrupt city: these are the ignorant city (al-madina al-jahiliyya), the dissolute city (al-madina al-fasiqa), the turncoat city (al-madina al-mubaddala) and the straying city (al-madina al-dalla). The souls of many of the inhabitants of such cities face ultimate extinction, while those who have been the cause of their fall face eternal torment. In itemizing four corrupt societies, al-Farabi was surely aware of Plato's own fourfold division of imperfect societies in the Republic into timarchy, oligarchy, democracy and tyranny. The resemblance, however, is more one of structure (four divisions) rather than of content." (Richard Netton, 1999).*

In contrast, *"Muslim science only reached its peak under the banner of Islam, and when Islam was at its heights, and never since. Often, as put by Bucaille, The totally erroneous statements made about Islam in the West are sometimes the result of ignorance, and sometimes of systematic denigration." (Zaimeche, 2002).*

Indeed, how can one ignore so many verses in the Noble Quran that called repeatedly the believers to seek knowledge?

> "Say [unto them, O Muhammad]: Are those who know equal to those who know not? But only men of understanding will pay heed." (39: 9).
>
> "In time, we shall show them our signs [ayatina] in the utmost horizons [of the universe] and within their inner selves until it will be manifest unto them that it is the Truth. Doth not thy Lord suffice, since He is Witness over all things." (41: 53).
>
> "And He has subjected to you, as from Him, all that is in the heavens and on earth: behold, in that are signs indeed for those who reflect." (45: 13).

The only difference between the science of the 13th century Islamic world and that of the 19th century Europe, according to Kalin (T. Peters, M. Iqbal, S. N. Haq, 2002) *"turns out to be quantitative, that is, in terms of the accumulation and further specialization of scientific knowledge about the physical world. By the same token, the scientific revolution of the 17th and 18th centuries was a revolution not in the outlook of the modern man concerning nature and the meaning of scientific investigation but in the methodological tools and formulations of the natural sciences. This is how the majority of the 19th century intellectuals would have interpreted the history of science and the rise of modern natural sciences, and this is how the subject is still taught today in the schools in the Islamic world."*

Policy pointer:
Jamal al-Din al-Afghani (1838-97), in his celebrated attack on the 'materialists', i. e., *Haqiqat-i mazhab-i naichiri wa bayan-i hal-i nachiriyan*, translated into Arabic by Muhammad Abduh as *al-Radd 'ala'l-dahriyyin*, was engaged in a self-proclaimed battle of saving science from the positivists, a battle for which he derived support from both the history of Islamic and modern sciences.

He had the following to say in his celebrated response to Renan:

> 'If it is true that the Muslim religion is an obstacle to the development of sciences, can one affirm that this obstacle will not disappear someday? How does the Muslim religion differ on this point from other religions? All religions are intolerant, each one in its way. The Christian religion, I mean the society that follows its inspirations and its teachings and is formed in its image, has emerged from the first period to which I have just alluded; thenceforth free and independent, it seems to advance rapidly on the road of progress and science, whereas Muslim society has not yet freed itself from the tutelage of religion. Realizing, however, that the Christian religion preceded the Muslim religion in the world by many centuries, I cannot keep from hoping that Muhammadan society will succeed someday in breaking its bonds and marching resolutely in the path of civilization after the manner of Western society...No I cannot admit that this hope be denied to Islam.' (Keddie, 1983, in: Kalin 2002)

"Afghani's voice, which was carried on by such figures as Muhammad Abduh, Sayyid Ahmad Khan, Rashid Rida, Muhammad Iqbal, Mehmet Akif Ersoy, Namik Kemal, Said Nursi and Farid Wajdi, was the epitome of the sentiments of the time: modern science is nothing but Islamic science shipped back to the Islamic world via the ports of European Renaissance and Enlightenment. In other words, science is not a culture-specific enterprise, and as such it is not the exclusive property of any civilization." (Kalin, 2002) Afghani put it in the following way:

> 'The strangest thing of all is that our ulama these days have divided science into two parts. One they call Muslim science, and one European science. Because of this they forbid others to teach some of the useful sciences. They have not understood that science is that noble thing that has no connection with any nation, and is not distinguished by anything but itself. Rather, everything that is known is known by science, and every nation that becomes renowned becomes renowned through science. Men must be related to science, not science to men (...)
>
> The father and mother of science is proof, and proof is neither Aristotle nor Galileo. The truth is where there is proof, and those who forbid science and knowledge in the belief that they are safeguarding the Islamic religion are really the enemies of that religion. The Islamic

religion is the closest of religions to science and knowledge, and there is no incompatibility between science and knowledge and the foundation of Islamic faith. ' (*Keddie, 1983, in: Kalin 2002)*

Furthermore, as we have previously hinted, when Islamic thinkers relate knowledge to faith, this means that science (knowledge) is by the same token related to Islamic morals as shown for example in the works of Ahmad Miskawayh (932-1030).

"Islam anyway, makes self-improvement of the individual and the betterment of society part of religious duty, inspiring individuals in all manners and forms. That also included knowledge. Thus, throughout the Muslim land, the search for knowledge and science was undertaken in an effort to improve society as a form of worship." (Zaimeche, 2002). For the author Ahmed Miskawayh, it is on the individual plane that this educative work has first to be performed.

Policy pointer:

In *Tahdhib al-akhlaq wa-tathir al-a'raq* (refinement of character and purification of dispositions) *"Miskawayh makes religion a foundation for training and refinement; and after the supports of faith are established in the boy's soul, he can study books on ethics, then arithmetic and geometry, or whatever can be deduced or proved by rational proofs."* (Jamal al-Din, N. 1994). Miskawayh defines moral happiness as *"the happiness enabling the human being to live happily, in accordance with the requirements of virtue. Thus it was a personal happiness which the human being could reach through intellectual effort, and striving to acquire the sciences which would make his thought inclusive of all the areas and all existent beings, and make him free himself from material things so as to reach the degree of wisdom whereby to grasp human perfection."* (Jamal al-Din, N. 1994).

8.2. Values and Human Security Are Also Related in Classic and Modern Islamic Thought

Values and human security are also related in classic and modern Islamic thought. The Prophet Muhammad (S) was asked: "What is the religion?" He replied: "The religion is morality." When asked, "What is woe (or affliction)?" he replied, "Real affliction is lack of morality."

In another hadith, the Prophet (S) says:

> "Verily, I have been sent to perfect morals."

The link of the belief ('Aqidah) to action is also clear in the Noble Quran:

> " It is not piety that you turn your faces towards East or West - But truly pious is he who believes in Allah and the Last Day, and the angels, and the Book, and the Messengers; and spend of his substance, out of love for Him, upon his near of kin, and the orphans, and the needy, and the wayfarer, and those who ask, and for the freeing of human beings from bondage; and is steadfast in prayer, and gives regular charity; and (truly pious are) they who

keep their promises whenever they promise, and are firm and patient, in pain (or suffering) and adversity, and in times of peril. Such are the people of truth, the God-fearing". [Qur'an 2: 177]

In another verse:

"[Allah is well aware of] those who, when We establish them in the land, remain constant in prayer and give regular charity, enjoin the right and forbid the wrong, with Allah rests the end (decision) of (all) affairs". [Qur'an 22: 41]

Imam Musa b. Ahmad al-Hajjawi (d. 968H/1561CE) - author of the celebrated Hanbali fiqh text, *al-Iqna'* and its abridgement, *Zad a-Mustaqni'* - served as mufti of the Hanbalis in Damascus during his age. He writes as part of his commentary to the adab-poem:

'It has been said that: "The allegory of faith (iman) is as a fortress having five walls. The first [innermost] is made of gold, the second of silver, the third from iron, the fourth of baked bricks and the fifth [outermost wall] from clay bricks. As long as the inhabitants of the fortress are diligent in guarding the clay wall, the enemy will not set its sights on [attacking] the next wall. But if they are negligent in this, they will attack the next wall, then the next, until the entire fortress lies in ruins.

Likewise, faith is defended by five walls: certainty (yaqin), next is sincerity (ikhlas), then comes fulfilling the obligations (ada' al-fara'id), after which are the recommended acts (sunan), and lastly safeguarding beautiful behavior (adab).

Thus, so long as adab is guarded and defended, the devil will not set his sight on it. But if one forsakes it, Satan makes inroads into the sunan, then the fara'id, then ikhlas, and finally yaqin itself."

Moral values are actually related to the five pillars of Islam, which are: testimony of faith, prayer, zakat, fasting, and pilgrimage.

1. **Testimony of faith is the gate to Islam:** accepting Muhammad's message means accepting his Sunna (tradition), teachings and moral values. It is also awareness that the purpose of Prophethood is the perfection of morals. The Prophet (S) said: *"I have been sent only for the purpose of perfecting good morals."* In this context, *"All the Islamic compulsory forms of worship are designed as exercises and training to enable people to acquire correct morals and habits and to live righteously, and to adhere to these virtues till the end, whatever be the changes in their circumstances."*
2. **Prayer prevents commitment of Evil:** when Allah commanded the compulsory prayers to be established, He stated its wisdom in this way:

"Prayer prohibits obscenity and evil." *(Quran: 29: 45)*

In a Hadith Qudsi it is stated:

"I accept the prayers of the person who adopts the policy of humility with it on account of My greatness, obliges My creatures, does not insist on sinning against Me, spends his day in remembering Me, and is kind to the poor, travelers, weak and the suffering people."

3. **Zakat is a means of purification:** The purpose of paying Zakat has been stated in the Quran in the following words:

> "O Prophet! Take alms out of their riches and thereby cleanse them and purify them." (Quran: 9: 103)

For this reason, to pay Zakat has been made compulsory for every eligible Muslim. The Prophet (S) has said:

> "To smile in the company of your brother is charity. To command to do good deeds and to prevent others from doing evil is charity. To guide a person in a place where he can go astray is charity. To remove troublesome things like thorns and bones from the road is charity. To pour water from your jug into the jug of your brother is charity. To guide a person with defective vision is charity for you." (Bukhari)

4. **Fasting is a stepping stone to righteousness:** Quran states the purpose of the imposition of fasting in these words:

> "Fasting has been made compulsory for you, as it was made compulsory for those (followers of earlier prophets) who preceded you, so that you may become righteous." (2: 183)

The Prophet (S) said:

> "Allah does not accept the fasting of those who do not keep away from wicked and obscene behavior." *(Bukhari)*

5. **Hajj is a compulsory moral pilgrimage for whoever is able to perform it:** Nobody should think that to travel to the holy places and perform pilgrimage, which is included in the basic pillars of Islam, is merely a form of worship that has no relation with the world of morality and character. This is a clear misunderstanding. Giving commands about this compulsory worship, Allah states:

> "For Hajj are the months well known. If any one undertakes that duty therein, Let there be no obscenity, nor wickedness, nor wrangling in the Hajj. And whatever good ye do, (be sure) Allah knoweth it. And take a provision (With you) for the journey, but the best of provisions is right conduct. So fear Me, o ye that are wise." (Quran: 2: 197)

The second example is provided by Sheikh ibn Taymiyyah (1263–1328), who writes in his work Al-'Ubudiyah *(Essay on Servitude):*

> Hence in this ummah the issue of shirk is subtler than the noise made by the footsteps of an ant.
>
> According to a sahih hadith, the Prophet (salallahu 'alayhi wasalam) said: "Wretched is the slave of the dirham, and the slave of the dinar, and the slave of velvet and brocade. He is doomed to failure. If he is afflicted with a thorn, may he not be able to remove it. If he is given, he is content, but if he is not given, he becomes angry."
>
> The Prophet (salallahu 'alayhi wasalam) described such a person as being a slave of the dirham and dinar, and of velvet and brocade, and stated that he is doomed to failure; he also

prayed that if such a person were pricked by a thorn (a minor affliction), he would not be able to remove it.

This is the situation of a person who may be afflicted by calamity, but is not able to overcome it, because he is doomed to failure. So he can neither achieve his aim nor rid himself of the thing he hates.

This is the situation of one who is a slave to wealth and material possessions, who was described as being content when he is given something and angry when it is withheld from him. Allah says:

"And among them are men who slander you [the Prophet] in the matter of (the distribution of) alms. If they are given part thereof, they are pleased, but if not, behold! They are indignant!" (al-Tawbah 9: 59)

Neither their pleasure nor their anger is for the sake of Allah.

This is also the situation of a man who is attached to a position of leadership or to a beautiful image, or any other kind of desire. If he gets what he wants, he is happy, but if he does not achieve it, he becomes angry. So he is a slave to whatever it is that he desires, because enslavement is essentially the enslavement of the heart: whatever the heart is enthralled by, is the thing of which the heart is a slave.

For this reason the Arab poet said:

"The slave is free so long as he is content, and the freeman is a slave so long as he is greedy."

"I followed my desires and they enslaved me. If only I had been content with my lot, I would be free."

It is said that greed and longing are like chains around the neck and shackles on the feet. Even if the chain is removed from the neck, the feet are still shackled.

It was reported that 'Umar ibn al-Khattab (radhiAllahu 'anhu) said, "Greed is poverty, and resignation is wealth. When any one of you gives up hope of something, he becomes independent of it."

This is something which anyone can feel in his heart, for when he gives up hope of something, he no longer wants it or longs for it; he no longer feels in need of it or of the one who can help him to get it. If he longs for and hopes for something, then his heart is attached to it, and he feels that he is in need of it and of the one who he thinks can be the means to this end. This is the case in matters of wealth and power and beautiful images or people, etc.

Let us end this chapter with a selection of varied extracts from current research and publications by Muslim authors regarding modern issues:

1. Economics

"Today, the modern system of production of goods exposes its failure in broad daylight. It is a general collapse: a financial collapse (since 2007), an ecological collapse (Fukushima, disrupted climate…), a collapse of the society of work (job insecurity, structural unemployment…), and a collapse of politics (democratic crisis) and of the regulatory State.

If this system has not yet collapsed completely, it is essentially due to credit. In the face of increasing difficulties in the growth of Capital through the production of goods, recourse to credit was inevitable. Credit, which is profit consumed before having been earned, can postpone the moment at which the world-system reaches its limits, but it cannot do away with it.

The "civilization of money" cannot create human and social civilization. The demand of liberal reformists or Keynesian economists that capitalism should be reformed, should start over again and become more just, is an illusion. The problem is not just in the poor distribution of wealth. It is ethical, philosophical and spiritual.

In modern societies, the moral codes are often what remains of the Christian faith, but faith, in western societies that claim to be enlightened, has become a personal affair that makes no commitments; it is no longer a system that structures social life.

This has been described as an advance in the emancipation of western Man. But this way of life today generates schizophrenic individuals who sometimes respect moral values in their private lives and generally recognize the destructive laws of profit and competition only in their professional lives, which will, however, model the lives of others.

If Islam is under debate today, if the attacks against this faith have been multiplied, if they take pleasure in attacking the Prophet (May peace and benediction be upon him and his progeny), it is because Islam continues to define itself as a system of life (dîn) which in a more or less conscious manner clashes head-on with this "civilization of money".

Beyond all the analyses of this iniquitous system that manages all aspects of our lives including our beliefs, the issue of the alternative will come up.

And the alternative to this global, total world system can only be a global, total world system. Islam which proposes a system of life and a vision of death is a global, total and radical alternative world system. Islam proposes a system of belief; it defines its time-space around tawhîd. Because what makes sense is what is placed at the center. But around the Center and in the name of the Center, it organizes a world along the principles of Justice and respect for creation, but always in the name of the Creator and nothing else." *(Makri, 2012; - See more at:* http: //cilecenter. org/areas-of-research/economics/essays/essays-details?articleID=34

2. Bioethics

" (...) Islamic bioethics has a particularly close relationship with the empirical sciences such as psychology and sociology. It has a similar relationship with the normative sciences such as theology and jurisprudence too. Islamic ethics has same main issues as general ethics: "happiness in this world and in the hereafter", "freedom and determination," "good and bad."

In Islamic ethics some foremost ethical methods are employed, such as "logical," discursive dialectical, analogical, transcendental, analytical and hermeneutical (...)

"The fundamental principles of Muslim medical ethics derive from the concentric action of several currents—Hippocratic ethics, classical medical ethics in the holy sources, and juridical authorities:

1. Qur'an 5. 32: "If anyone saved a life, it would be as if he saved the life of the whole people." This is the basic principle of medical practice in Islam, whereby the doctor must not only heal but also prevent, as understood from Qur'an 2. 195: "and make not your own hands contribute to your destruction."
2. An authentic saying of Prophet Muhammad is: "There is no disease that Allah has created, except that He also has created its treatment." This is interpreted as an explicit invitation to doctors to discover cures and defeat pathologies.
3. The juridical principle of necessity maintains: "Pressing needs even allow what is prohibited." Through analogy and extension this arises from various Qur'anic verses, e.g., 6. 145: "Say: .... if a person is forced by necessity, without willful disobedience, nor transgressing due limits, thy Lord is oft-forgiving, most merciful."
4. Once necessity is recognised as the exception to the rule of the right to avoid serious harm, another principle is derived, namely the "lesser evil." Avoiding evil has priority over the

acquisition of good. This is one of the principles traditionally used to justify therapeutic abortion.

5. The principle of the "public good" or "usefulness" (maslaha). The public good remains closely connected with the altruism that has its roots in Qur'an 5. 2: "Help ye one another in righteousness and piety."
6. The sacred character of the human being. God Himself says: "We have honoured the sons of Adam" (Qur'an 17. 70). He wants to promote the human being's wellbeing, harmony and happiness. God says: "God intends every facility for you; He does not want to put you to difficulties."

"I can summarise these principles as follows:

Need and necessity are equivalent. Necessity allows 'prohibited' matters. Injurious 'harm' should be removed. Prevention of evil has priority over obtaining benefit. The greater benefit prevails over the lesser benefit. 15 Islamic ethics, so to speak, employs pragmatic and analogical methods." *(Arda, Berna and Chaim, 2012)*

3. Environment

One day, according to the Hadith (a book of traditional and authenticated accounts of the words and actions of the Prophet Muhammad), the Prophet was traveling from one town to the next with his followers. They were just crossing a river when it became time for prayers. Naturally they used the river to perform the ritual ablutions required before prayer. However, the followers of the Prophet were astonished to see him enter the river with a little bowl. This he filled with water and it was this water with which he performed the ablutions. When asked why, surrounded by a whole river, he took so little water to use, he said that just because there is plenty this does not give us the right to waste or to take more than we really need. (Palmer and Finlay, 2013)

" (...) The Industrial Revolution also proclaimed a new revival of another God: Mammon. Mammon regrettably has no respect for environmental integrity—nor do his followers. The last 250 years have seen a growing decimation of ever more pristine areas of nature to feed the insatiable industrial cuckoo and its resultant consumerism. Forests—particularly tropical forests—have been systematically hewn down, the seas ransacked, the lands made totally dependent on a host of inorganic fertilizers and pesticides for food production. Wastes galore have filled the seas, the rivers, and the lakes, not to mention the landfills.

We must also take note that the "unmatched material progress" of this century that we are often fond of talking about has only been possible for the few: that is, the population of the northern hemisphere and a small minority among the peoples of the South. This is usually translated as less than 25 percent of the world's population consuming over 75 percent of the world's resources.

This rate of consumption by a minority of the human species has caused unparalleled climatic change, ecosystem disintegration, and species extinction (...)

(. . .)Several times in the Qur'an, man is invited to make use of the nourishing goods that Allah has placed on earth for him, but abuse—particularly through extravagance and excess—is strictly forbidden. Sometimes these principles are stated in one breath, so to speak. Sometimes they are stated separately. But the message is the same, as the following verse indicates:

O children of Adam! ... Eat and drink: but waste not by excess for Allah loveth not the wasters. (007. 031).

There are as many invitations to partake of nature as provided for man and for other creatures of the earth as there are for the avoidance of wasteful extravagance. Time and again, Allah reminds us that He loveth not wasters.

> It is He who produceth gardens with trellises and without and dates and tilth with produce of all kinds and olives and pomegranates similar [in kind] and different [in variety]: eat of their fruit in their season but render the dues that are proper on the day that the harvest is gathered. But waste not by excess: for Allah loveth not the wasters. 006. 141 [...] (Mahasneh, 2003)

4. Arts

"The people who acknowledged Islam blended their cultural styles and developed a distinct and uniform style of art best acknowledged as Islamic art. It is said that Islamic art was an outcome of many centuries, extending from Spain to India. However, there are very realistic and authentic views on the emergence and evolution of Islamic art. They maintain that Islamic art emerged at the same time with the very rise of Islam itself. In this correlation, it is essential to refer to some verses of the Quran and few Traditions of the Prophet, which without a doubt endorse the view that Islamic art is inherent in Islam itself:

"We have adorned the lowest heaven with the beauty of the planets."

"We have placed constellations in heaven and made them beautiful for those who look."

In addition to the above Quranic verses, a couple of Traditions of the Prophet (S) would be in order:

"God likes that whenever anyone of you performs any work he should do it in a beautiful manner."

"God is Beautiful and likes Beauty."

There is an inward intimate relationship between Islamic art and Islamic spirituality and as a result Muslim artists think about the sacred art of Islam to be a descent of heavenly Reality upon the earth. Through this creative activity we maintain our symbolic correlation with an Invisible Spiritual Order. "To be spiritual" in the words of George Santayana, "is to live in the presence of the ideal". This "Ideal" is Allah (God) who is the cause of everything. (Khawaja, 2011)

(. . .)The Holy Quran (The Word of Allah) is the source of every art-form according to Muslim artists and aestheticians. They believe that the origin of the sacred calligraphy is the visual incarnation of the Sacred Word. The Divine Pen (al-qalam) wrote the Realities of all things upon the Guarded Tablet as mentioned earlier (. . .)Thus, the art of calligraphy, which reflects on the earthly plane the writing God's Words upon the Guarded Tablet, is considered as the origin of the plastic arts. The Quranic calligraphy owes its genesis to the Islamic revelation and in reality; it is the response of the soul of the Muslims to the Divine Message.

In the field of decorative arts too Muslim artists used the art of writing as an important "ornamental" element to decorate the different religious and non-religious structures as well as to beautify many things of decoration such as cloth, paper, utensils, wood, rubber and glass material etc., on which the Words or Ayat (verse) of the Holy Quran are written, engraved or embossed. (Khawaja, 2011)

5. Islamic Concept of Life and Morality

"The viewpoint of Islam, however, is that this universe is the creation of God Who is One. He created it and He alone is its unrivaled Master, Sovereign and Sustainer. The whole universe is functioning under His Divine Command. He is All-Wise, All-Powerful and Omniscient. He is Subbooh and Quddoos that is, free from all defects, mistakes, weaknesses and faults and pure in every respect). His God-hood is free from partiality and injustice. Man is His creature, subject and servant and is born to serve and obey Him.

The correct way of life for man is to live in complete obedience to Him. It is not for man to determine the mode of worship and obedience; it is for God to decide this. God, being the master, has raised from time to time prophets for the guidance of humanity and has revealed His books through them. It is the duty of man to take the code of his life from these sources of divine guidance. Man is answerable to God for all his actions in life. The time for rendering an account will be in the life-hereafter and not in this world. The short span of worldly life is really an opportunity to prepare for that great test. In this life all efforts of man should be centered on the object of soliciting the Pleasure and Blessings of God in the Hereafter. During this test every person is responsible for all his beliefs and actions. He, with all his faculties and potentialities, is on trial. There will be an impartial assessment of his conduct in life. By a Being Who keeps a complete and correct record not merely of his movements and actions and their influence on all that is in the world from the tiniest speck of dust to the loftiest mountains but also a full record of his innermost ideas and feelings and intentions." (Al-Mawdudi, 1986)

Chapter 9

GRETCHEN, PERMISSIVENESS AND ECONOMIC GROWTH

The question of the connection between religion and economic growth is not new in the social sciences literature. In this chapter, we try to investigate this connection with global relations, especially in Islam. We will show how the purpose of economic activity in Islam is not only material profit, but also worship; while the question about whether the economic crisis is to be linked to the decay of religious values in the West, also deserves further consideration.

Box 9.1. The overall message of this Chapter

We maintain that the decay of religious values in the West is part and parcel of the current downward spiral of the North Atlantic Arena, and that far from being a *"Christian Club"*, which Europe long seized to be, the current European value landscape is characterized above all by a more and more generalized disrespect of the Ten Commandments. We debate the positions by Inglehart and Baker (2000) and Goermez (2012) on secularization, and factor-analyze the self-declared observance of the Western publics of the Ten Commandments. This item was still contained in the first wave of the *World Values Survey,* and unfortunately was dropped in later surveys, while it reveals many insights, which are important to this day.

In statistical terms, we can observe that factor 1 and factor 2 – the two types of non-observance in the mid-1980s - are highly intercorrelated with each other, but that factor 1 has the highest factor loading with the erosion of property rights, and factor 2 has the highest loading with the non-observance of the Command regarding the Sainthood of God's name. Table 9.1 reveals the reality that *"Europe is a club of the erosion of property rights"* and a *"club of the erosion the Sainthood of God's name"* above anything else, and that this phenomenon is strongest in France, Sweden and the Netherlands, and weakest in Ireland, the United States and Malta. Religious observance in the United States of America is indeed different from the rest of Western democracies.

In one of the most touching paragraphs of European world literature, we read in Goethe's *Faust* the following sentences, forever since associated with the human question about the Eternal and the human encounter with religion.

> GRETCHEN: Do you believe in God?
>
> FAUST: My darling, who can (really) say: I believe in God! You may ask priests or wise men, And their answer seems but a mockery Of the questioner to be.
>
> GRETCHEN: So you do not believe?
>
> FAUST: Don't misunderstand me, you lovely sight! Who may name Him, and who declare: I believe in Him. Who can feel and dare to say: I do not believe in Him! The all-embracing one, the all-preserving one, Does He not embrace and preserve You, me, (and) Himself? Does the sky not arch above us up there? Does the earth not lie firm down here? And do not with kind glance the eternal stars rise? Do I not look at you eye to eye, and does not everything press upon your head and heart and weave in eternal mystery Invisible and visible around you? Fill your heart, as big as it is, from that and when you are completely blissful in the feeling, then call it what you like: Call it happiness! Heart! Love! God! I have no name for it! Feeling is everything; (The) name is sound and smoke, Enshrouding heaven's glow.
>
> GRETCHEN: That is all quite fine and good; Much the same thing says the pastor, too Only with slightly different words.
>
> FAUST: It is said everywhere (by) All hearts under the heavenly day, Each in its own language: Why not I in mine?

- From Faust I, lines 3426-3465; by the German poet Johann Wolfgang von Goethe (28 August 1749 – 22 March 1832) http: //german. about. com/library/blgretchenE. htm

In this book, we already have referred to extensively to the theories of Western mainstream sociology and here, most typically, to the works of Ronald F. Inglehart, who is of the opinion that the relationship between religion and economic development leads to occupational specialization, rising educational levels, rising income levels, and eventually brings unforeseen changes in gender roles, attitudes toward authority and sexual norms; declining fertility rates; broader political participation; and less easily led publics. Elites can resist these changes, but in the long run, it becomes increasingly costly to do so and the probability of change rises.

Inglehart and his associates do not exclude, as we highlighted below, the possibility of development crises and development implosions. And yet, their belief in the coherent strength and force of the secularizing modernization process seems to be unbroken in other parts of their work.

Just let us look at this paragraph:

> "[…] Modernization does not automatically bring democracy. Rather, it, in the long run, brings social and cultural changes that make democracy increasingly probable. Simply attaining a high level of per capita GDP does not produce democracy: if it did, Kuwait and the United Arab Emirates would be model democracies. […] *But the emergence of postindustrial society brings certain social and cultural changes that are specifically conducive to democratization.* Knowledge societies cannot function effectively without highly-educated publics who become increasingly accustomed to thinking for themselves. *Furthermore, rising levels of economic security bring a growing emphasis on a syndrome of self-expression values – one that gives high priority to free choice and motivates political action. Beyond a certain*

> *point, accordingly, it becomes difficult to avoid democratization, because repressing mass demands for more open societies becomes increasingly costly and detrimental to economic effectiveness. Thus, in its advanced stages, modernization brings social and cultural changes that lead to the emergence and flourishing of democratic institutions.* The core idea of modernization is that economic and technological development bring a coherent set of social and political changes. A large body of empirical evidence supports this idea. *Economic development is, indeed, strongly linked with pervasive changes in people's beliefs and motivations, and these changes are reshaping the role of religion, job motivations, human fertility rates, gender roles, and sexual norms. And they also bring growing mass demands for democratic institutions and more responsive behavior by elites.* These changes together make democracy increasingly likely to emerge, while also making war less acceptable to publics." (Inglehart and Welzel, 2009)

By contrast, we maintain that the decay of religious values in the West is part and parcel of the current downward spiral of the North Atlantic Arena, and that far from being a *"Christian Club",* which Europe long seized to be in contrast to the famous, often-quoted opinion of a former Turkish Prime Minister Mesut Yilmaz (see below), the current European value landscape is characterized above all by a more and more generalized disrespect of the Ten Commandments.

9.1. THE U-SHAPED RECOVERY OF MUSLIM SOCIETIES SINCE 1800

Open a Western newspaper from the period of the fall of the Berlin Wall to the beginnings of the economic crisis of 2008, you will find gloomy predictions about the economic conditions in the majority of the Arab countries and the Muslim world in general.

The great historian of global capitalism, Andre Gunder Frank, at the late stages of his scientific life came to the conclusion in his 1998 book on global economic history that the days of the Euro-Atlantic arena are counted, and that countries like the United Kingdom become *"underdeveloping" countries,* while the dynamics of the world economy return to where they were always centered until the mid-18th Century: Asia. This global shift has many important conclusions also for the Arab world and the Muslim world/the Indian Ocean arena.

> "[…] from a global perspective Asia and not Europe held center stage for most of early modern history. Therefore the most important question is less what happened in Europe than what happened in the world as a whole and particularly in its leading Asian parts. I render historical events from this much more global perspective and propose to account for 'the Decline of the East' and the concomitant 'Rise of the West' within the world as a whole. This procedure pulls the historical rug out from under the anti-historical/scientific – really ideological – Eurocentrism of Marx, Weber, Toynbee, Polanyi, Braudel, Wallerstein, and most other contemporary social theorists" (Frank, 1998: xv-xvi)

To underline the point that Muslim societies are recovering from a long trough of relative decline in world society, and that they face a long-term U-shaped recovery path, we re-analyze data from the Angus Maddison database,[55] which is the most complete database on real income convergence/divergence in the world over the long term. For the purposes of our book, we use the original version, but we also highlight the fact that there is an alternative, re-calculated version of this database, provided by Robert Barro.[56] We leave it up to our readers to base their own future work on these subjects on either of the two versions of this database. However, we should emphasize here that the differences for the countries under consideration here are really minimal.

Although long-term annual data series about GDP per capita in real purchasing power are available only for Java/Indonesia and the Ottoman Empire/Turkey, we can cautiously draw the following conclusions presented in our following graphs about economic growth in the Arab and the Muslim periphery and semi-periphery of the current world system from the 18th Century onwards dominated, which was dominated by the Western capitalist powers.

Policy pointer

Andre Gunder Frank came to the conclusion in his 1998 book on global economic history that the days of the Euro-Atlantic arena are counted, and that countries like the United Kingdom and the North Atlantic arena in general become "underdeveloping" countries, while the dynamics of the world economy return to where they were always centered until the mid-18th Century: Asia. This global shift has many important conclusions also for the Arab world and the Muslim world/the Indian Ocean arena. To underline the point that Muslim societies are recovering from a long trough of relative decline in world society, and that they face a long-term U-shaped recovery path, we re-analyze data from the Angus Maddison database, which is the most complete database on real income convergence/divergence in the world over the long term.

Although long-term annual data series about GDP per capita in real purchasing power are available only for Java/Indonesia and the Ottoman Empire/Turkey, we can cautiously draw the following conclusions presented in our following graphs about economic growth in the Arab and the Muslim periphery and semi-periphery of the current world system, from the 18th Century onwards dominated by the Western capitalist powers.

Let us start with the optimistic message: all the countries of the region under scrutiny in our graphs currently recover from their relatively weak position into which they were sliding in the 1950s and 1960s and beyond. Albania; Algeria; Egypt; Iran; Jordan; Malaysia; Morocco; Tunisia; all somehow managed to recover from their weak relative position in the world system. Several of the Muslim countries are showing quite a considerable relative ascent.

---

[55] http://www.ggdc.net/maddison/maddison-project/data.htm

[56] http://rbarro.com/data-sets/

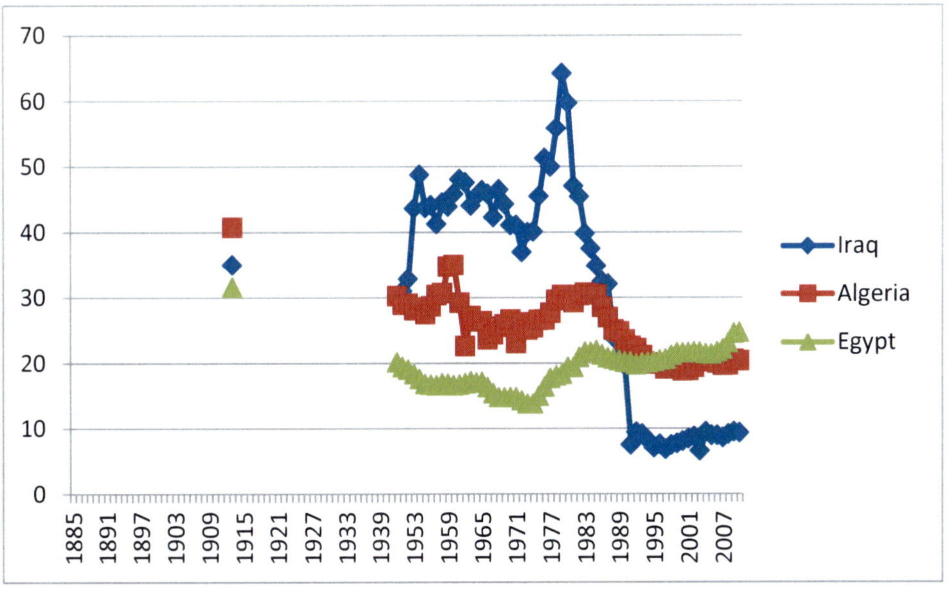

Legend: our own compilations, based on the Maddison data sets, as documented in Bolt and van Zanden, 2013. Calculated from the original data with Microsoft EXCEL 2010.

Graph 9.1. Iraq, Algeria and Egypt – similarities and dissimilarities in their development trajectory (GDP per capita in purchasing power in relationship to the world average of 31 countries).

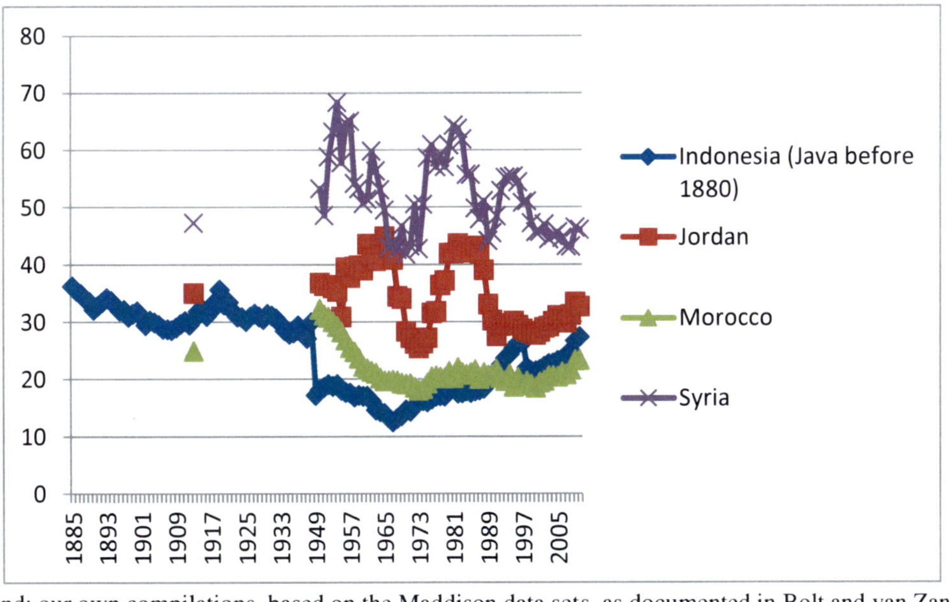

Legend: our own compilations, based on the Maddison data sets, as documented in Bolt and van Zanden, 2013. Calculated from the original data with Microsoft EXCEL 2010.

Graph 9.2. Indonesia, Jordan, Morocco and Syria – similarities and dissimilarities in their development trajectory (GDP per capita in purchasing power in relationship to the world average of 31 countries).

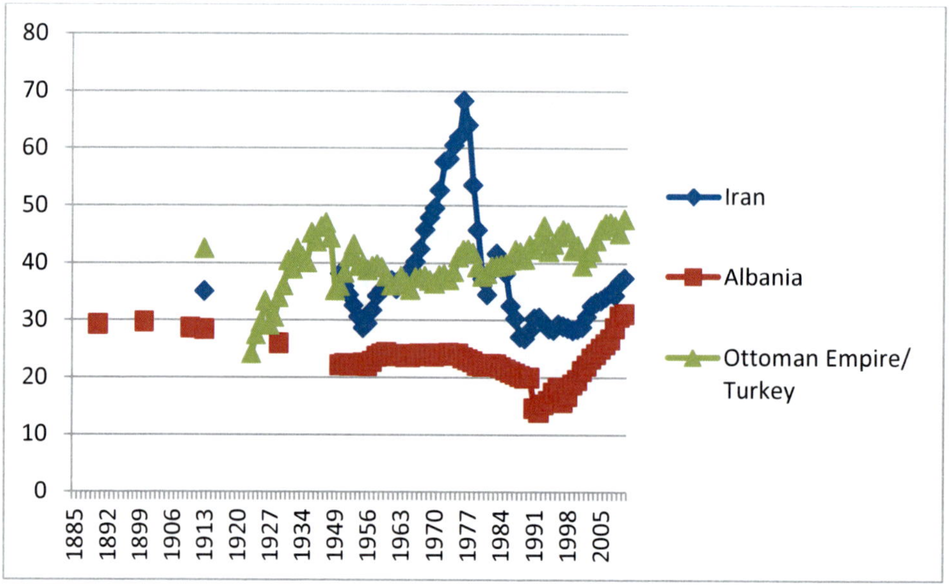

Legend: our own compilations, based on the Maddison data sets, as documented in Bolt and van Zanden, 2013. Calculated from the original data with Microsoft EXCEL 2010

Graph 9.3. Iran, Albania, Turkey – similarities and dissimilarities in their development trajectory (GDP per capita in purchasing power in relationship to the world average of 31 countries).

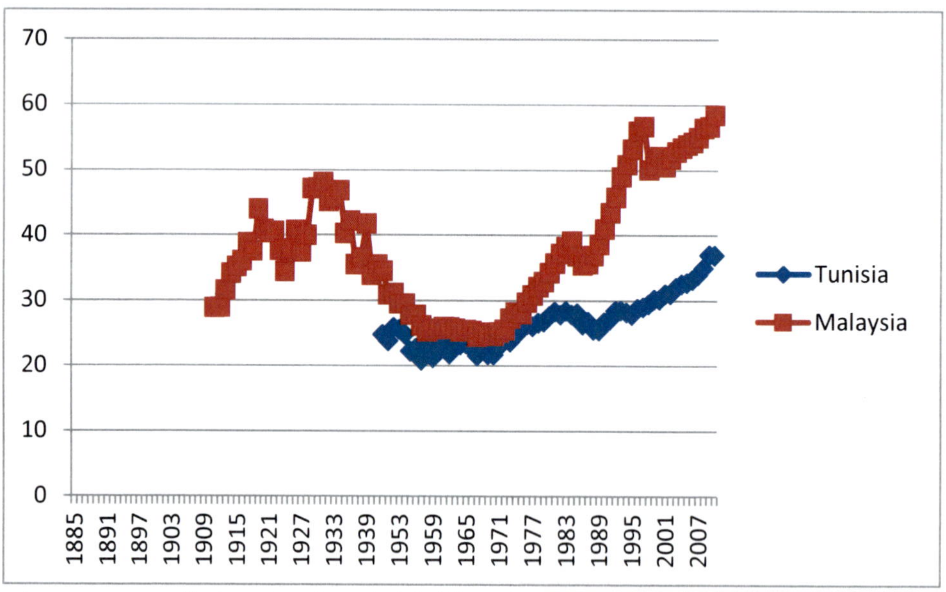

Legend: our own compilations, based on the Maddison data sets, as documented in Bolt and van Zanden, 2013. Calculated from the original data with Microsoft EXCEL 2010.

Graph 9.4. Tunisia and Malaysia – similarities and dissimilarities in their development trajectory (GDP per capita in purchasing power in relationship to the world average of 31 countries).

9.2. THE EROSION OF THE RESPECT FOR PROPERTY RIGHTS IN THE "CHRISTIAN CLUB"

The question about the connection between religion and economic growth has not been ignored in the literature concerned with the Islamic world. The needs created by the specificities of local or global Islamic communities in their day-to-day dealings with non-Muslim societies have urged differential approaches to both economic and financial systems.

In a paper about *"corruption and bribery in Islamic law,"* Arafa (2012) argued that *"the emerging shift in social scientific thought in viewing corruption from 'grease that oils the economic wheel' to a 'menace that undermines economic growth' has brought rational understanding of the phenomenon much closer to Islamic doctrine. Where they differ is with respect to remedial action. The western approach focuses on governance and designing appropriate systems and institutions that gear information and incentives toward minimizing opportunities and enticement for corruption. In short, it emphasizes constraints external to the individual. On the other hand, Islam seeks to go beyond such constraints, and also instill in believers a clear 'second-order' preference for no corrupt behavior. It recommends developing a firm belief in transcendent accountability, stresses character building through practicing moral virtues and shunning vices. In essence, much of the restraint comes from within through a moral renovation. It is the contention that both emphases are important in eliminating corruption and that the followers of Islam and the West can learn from one another."* (Arafa, 2012)

However, there is no recent Islamic model which we may describe as strong in its economy, social institutions, education, *et cetera,* producing enough for its citizens to enjoy a decent life while remaining free from poverty and corrupt behavior. Indeed, *"Muslim thinkers analysed the impact of capitalism on Islam and Muslim societies. They grappled with the new phenomena of capital accumulation, distribution of income, economic backwardness, exploitation of Muslim lands by colonial masters and introduction of capitalist institutions in their countries."* Charles Tripp in his *Islam and the Moral Economy* (2006) studies and synthesizes various responses developed by the Muslim religious scholars, social reformers and political leaders over the last sixty years. But the real problem that Muslim scholars faced since the later part of nineteenth century and throughout the twentieth was *"in translating the ideal Islamic vision into practice. The ideal Muslim, practicing the Islamic values and norms in letter and spirit, could exist in theoretical discussions, religious rhetoric and romantic discourses. When it came to finding a society that is composed of such ideal human beings, the Muslim ummah was unable to present even a single example over the last one-century or more."* (Khan, 2008)

In this context, we document for our readers also the moderate positions of the President of the religious authority in Turkey, Goermez, 2012:

In the Turkish newspaper Hürriyet[57] we read on the 26th of November 1997 the following item:

[57] http: //www. hurriyetdailynews. com/yilmaz-asks-if-europe-is-a-christian-club. aspx?pageID=438&n=yilmaz-asks-if-europe-is-a-christian-club-1997-11-26.

"Prime Minister Mesut Yilmaz stressed, on Tuesday, that *Europe will decide 'whether it will become a Christian club or a multi-cultural organization,'* at the EU enlargement summit that to be held in December in Luxembourg, the Anatolia news agency reported."

Box 9.2. The position, taken up by the Office for Religious Affairs in Turkey (Goermez (2012))

The Muslim world, as it goes through a major socioeconomic transformation, does not offer an Islamic alternative to secularism. Indeed, the culture produced by the Islamic civilization increasingly manifests different forms of secularism, though not in the conventional sense. It will offer an alternative form that will call into question the existing variants of secularism. [...] The failed Soviet regime sought to abolish not only the general teachings but also all aspects of religion from daily life. Undoubtedly, similar practices by different regimes continued to hurt people leading different lifestyles in various parts of the world. While religion was completely removed from daily life in the Eastern bloc countries, it also faced containment in the Western bloc. [...] Religion as the main point of reference to understand human nature, be it Islam or any other religion, has come under attack by rigid applications of laicism. There has been a direct assault to undermine the power of religion and erode the sphere of meaning it represented. Today, it is interesting to see how religion, whose voice can be heard in almost any corner of the world, still remains the only source of hope for humanity despite many such attacks against it. Today, religion rejuvenates itself with its own meaning and existence from South Asia to Eastern Europe and from America to Australia and emerges as an ancient phenomenon that is taken into account within State activities.

[...] Religion continues to manifest itself in a calm manner as the consciousness of the humanity and the heart of life after a relative withdrawal and suppression in modern societies.

Former Turkish PM Mesut Yilmaz was of course basically right in the direction of his argument about Turkish membership in the EU, but at the same time, he was basically wrong by implying that Christianity is still a dominating sociological reality in the Europe (and the West) of today.

To further analyze this question, we factor-analyzed *World Values Survey* data about the observance of the Judeo-Christian Ten Commandments among Western publics. The Ten Commandments also are part and parcel of the traditions of Islam. The Commandments, also in the Islamic tradition, were given to Moses:

"And we ordained laws for him in the tablets in all matters, both commanding and explaining all things, (and said): 'Take and hold these with firmness, and enjoin thy people to hold fast by the best in the precepts: soon shall I show you the homes of the wicked (How they lie desolate).'" (Quran 7: 145)

"When Moses came back to his people, angry and grieved, he said: 'Evil it is that ye have done in my place in my absence: did ye make haste to bring on the judgment of your Lord?' He put down the tablets, seized his brother by (the hair of) his head, and dragged him to him. . ." (Quran 7: 150).

"When the anger of Moses was appeased, he took up the tablets: in the writing thereon was guidance and Mercy for such as fear their Lord." (Quran 7: 154).

The Ten Commandments, as it is presumed to be generally known, are the following:

1. I am the Lord thy God. Thou shalt have no other gods before me
2. Thou shalt not take the name of the Lord thy God in vain
3. Remember the Sabbath day, to keep it holy
4. Honor thy father and thy mother
5. Thou shalt not kill
6. Thou shalt not commit adultery
7. Thou shalt not steal
8. Thou shalt not bear false witness against thy neighbor
9. Thou shalt not covet your neighbor's house
10. Thou shalt not covet your neighbor's wife

Policy pointer:
In Table 9.1 we now factor-analyze the self-declared observance of the Western publics of Ten Commandments. In statistical terms, we can observe that factor 1 and factor 2 – the two types of non-observance are highly intercorrelated with each other, but that factor 1 has the highest factor loading with the erosion of property rights, and factor 2 has the highest loading with the non-observance of the Command regarding the Sainthood of God's name. Table 9.1 in its different components reveals the shattering reality that *"Europe is a club of the erosion of property rights"* and a *"club of the erosion the Sainthood of God's name"* above anything else, and that this phenomenon is strongest in France and weakest in Ireland, the United States and Malta. Religious observance in the United States of America is indeed different from the rest of Western democracies.

Interestingly enough, honoring parents is the command which is least explained by the factor analytical model:

Table 9.1. The Ten Commandments and their support among global publics today – *World Values Survey* data (promax factor analysis)

	initial	extraction
1st commandment applies for myself [does not apply] I am the Lord thy God. Thou shalt have no other gods before me	1,000	0,761
2nd commandment applies for myself [does not apply] Thou shalt not take the name of the Lord thy God in vain	1,000	0,766
3rd commandment applies for myself [does not apply] Remember the Sabbath day, to keep it holy	1,000	0,728
4th commandment applies for myself [does not apply] Honor thy father and thy mother	1,000	0,430
5th commandment applies for myself [does not apply] Thou shalt not kill	1,000	0,641
6th commandment applies for myself [does not apply] Thou shalt not commit adultery	1,000	0,584

Table 9.1. (Continued)

	initial	extraction
7th commandment applies for myself [does not apply] Thou shalt not steal	1,000	0,741
8th commandment applies for myself [does not apply] Thou shalt not bear false witness against thy neighbor	1,000	0,655
9th commandment applies for myself [does not apply] Thou shalt not covet your neighbor's house	1,000	0,663
10th commandment applies for myself [does not apply] Thou shalt not covet your neighbor's wife	1,000	0,663

Our two factors explain 66.3% of the total variance.

		initial *Eigen values*	
	Total	% of variance	cumulated percentage
1,000	5,172	51,723	51,723
2,000	1,459	14,591	66,315

The two resulting factors can be explained as

- the erosion of property rights
- erosion of the sainthood of God's name

If highest numerical values differ from the variable name, the highest values of the indicator for the multivariate analysis are mentioned in brackets	**erosion of property rights**	**erosion of the sainthood of God's name**
1st commandment applies for myself [does not apply] I am the Lord thy God. Thou shalt have no other gods before me	0,439	0,872
2nd commandment applies for myself [does not apply] Thou shalt not take the name of the Lord thy God in vain	0,459	**0,875**
3rd commandment applies for myself [does not apply] Remember the Sabbath day, to keep it holy	0,373	0,850
4th commandment applies for myself [does not apply] Honor thy father and thy mother	0,635	0,465
5th commandment applies for myself [does not apply] Thou shalt not kill	0,780	0,242
6th commandment applies for myself [does not apply] Thou shalt not commit adultery	0,730	0,566
7th commandment applies for myself [does not apply] Thou shalt not steal	**0,851**	0,323
8th commandment applies for myself [does not apply] Thou shalt not bear false witness against thy neighbor	0,809	0,387
9th commandment applies for myself [does not apply] Thou shalt not covet your neighbor's house	0,800	0,539
10th commandment applies for myself [does not apply] Thou shalt not covet your neighbor's wife	0,806	0,506

The correlation between the two factors is very high, especially if we consider that our data are survey results data from none the less than n = 14054 global representative citizens with available data:

	erosion of property rights	**erosion of the sainthood of God's name**
erosion of property rights	1.000	0.508
erosion of the sainthood of God's name	**0.508**	1.000

Graph 9.5 shows the factor scores of our factor analytical investigation and the decay of the observance of the Ten Commandments in the West. France, Sweden and the Netherlands are leading the field, while Malta, the United States and Ireland were the three countries where the two phenomena were least observed:

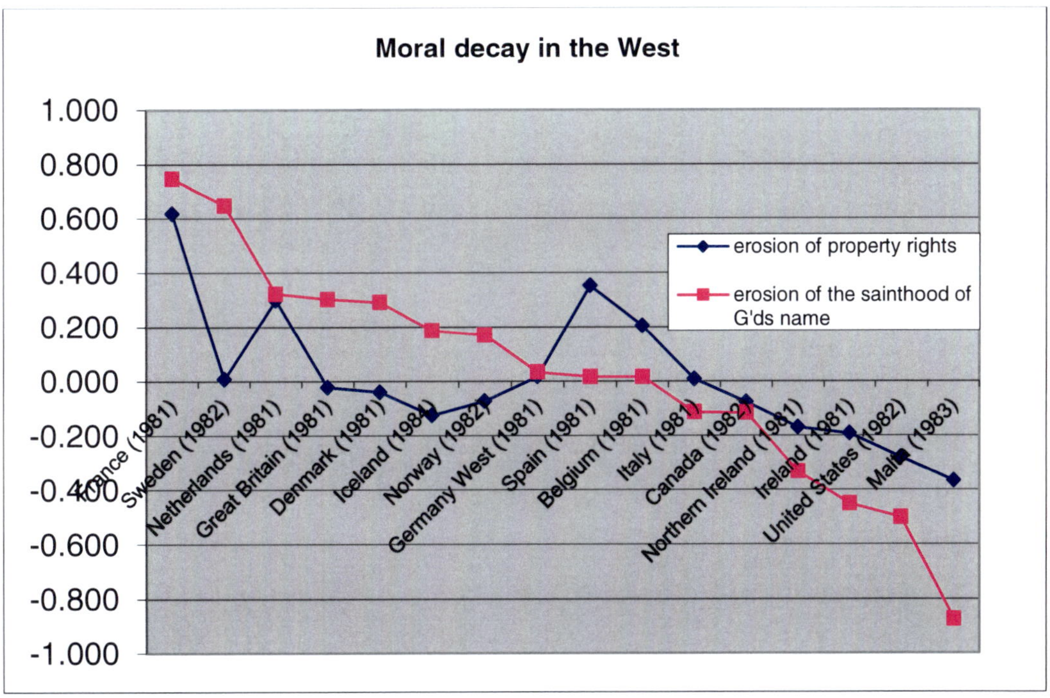

Graph 9.5. The decay of the observance of the Ten Commandments in the West.

Chapter 10

THE IMPLOSION OF "SELF-EXPRESSION" IN THE WEST AND FURTHER DOUBTS ABOUT THE CORRECTNESS OF INGLEHART'S THEORY IN THE LIGHT OF THE NEW DIRECT MEASUREMENTS OF SELF-EXPRESSION VALUES FROM THE *WORLD VALUES SURVEY*, 2010-2014

The idea that self-expression values in the West are imploding, while in other regions of the world they are rising is a major turn-around in existing global value research. An explanation for this could be that the crisis nowadays in the West brought about this change. This Chapter will document this tendency extensively. We take the Inglehartian theory at its face value, and just perform some simple operations, based on elementary algebra and simple bi-variate statistics to show that the implosion indeed happened. Any social science researcher and graduate student around the globe can again perform these calculations. The question is not whether or not Inglehart's theory is correctly explaining the world, but only which consequences can be drawn from it in a time perspective.

But we also highlight the fact that the latest wave of *World Values Survey* data, wave 6, 2010-2014, released in May 2014, contains an item which directly asked 74,044 respondents in 52 countries whether they think that self-expression is an important value for child education. The correlation between these data and Inglehart's self-expression index is negative and the R^2 between the two variables is almost 20%.

Box 10.1. The overall message of this Chapter

Data emerging from the *World Values Survey* in the first decade of the 2000s seem to suggest that the precariousness, which more and more characterizes the economies of Western countries, leads toward an implosion of what Inglehart and his sociological school of thought interpreted as *"self-expression values"*.

Our analysis of the time series element in the *World Values Survey* data shows that indeed global value change seems to correspond to various ups and downs. To this end, we calculated from the original WVS website Inglehart's own data (http://www.

worldvaluessurvey.org/wvs/articles/folder_published/article_base_54) which countries – in descending order – had very high increases or decreases in non-traditional values over the preceding WVS Survey. The very idea that self-expression values in the West are imploding, while in other regions of the world they are rising, is a major turn-around in existing global value research, and a possible indicator for the extent of the crisis that exists nowadays in the West.

The world, described by Inglehart and Baker, 2000, where in advanced industrial societies people pay large sums of money and travel long distances to experience exotic cultures now longer seems to exist for the "1.000 Euro" generation born after 1975, which experiences more and more job insecurity and hardly finds full-time tenured work opportunities. No wonder then that *"self-expression"* is dramatically declining in the West.

We also highlight the fact that the latest wave of *World Values Survey* data, wave 6, from 2010 - 2014, released in May 2014 contains an item which directly asked 74,044 respondents in 52 countries whether they think that self-expression is an important value for child education. The correlation between these data and Inglehart's self-expression index is negative and the R^2 between the two variables is almost 20%.

A look at the time series element in the *World Values Survey* data shows that indeed, global value change seems to correspond to various ups and downs. To this end, we calculated from the original WVS website Inglehart's own data (http: //www. worldvaluessurvey. org/wvs/articles/folder_published/article_base_54) which countries – in descending order – had very high increases or decreases in non-traditional values over the preceding WVS Survey.

Let us recall that self-expression values are defined by:

- Respondent gives priority to self-expression and quality-of-life (measured by the four-item materialist/*post-materialist values index,* documented in the WVS data) .59
- Respondent describes self as very *happy* .58
- Respondent *has signed and would sign a petition* .59
- *Homosexuality is always justifiable* .54
- You should be *trusting people* .44

10.1. The Implosion

Inglehart and Welzel, 2010a seem to indicate awareness of the possibility of a real implosion of value development across time:

> "Modernization theory needs to be revised for several reasons. First, modernization is not linear, moving indefinitely in the same direction. Industrialization leads to one major process of change, bringing bureaucratization, hierarchy, centralization of authority, secularization, and a shift from traditional to secular-rational values. But the postindustrial phase of modernization brings increasing emphasis on individual autonomy and self-expression values, which erode the legitimacy of authoritarian regimes and make democracy increasingly likely to emerge. The process is not deterministic; a given country's leaders and nation-specific events also matter. Moreover, modernization's changes are not irreversible. *Economic collapse can reverse them,*

as happened during the Great Depression in Germany, Italy, Japan, and Spain—and during the 1990s in most Soviet successor states." (Inglehart and Welzel, 2010a)

Similarly, Inglehart and Welzel, 2009 highlight:

> Thus, other things being equal, high levels of economic development tend to make people more tolerant and trusting bringing more emphasis on self-expression and participation in decision-making. This process is not deterministic, and any forecasts can only be probabilistic, since economic factors are not the only influence; a given country's leaders and nation-specific events also shape what happens. Moreover, modernization's changes are not irreversible. Severe economic collapse can reverse them, as happened during the Great Depression in Germany, Italy, Japan, and Spain and during the 1990s in most of the Soviet successor states. *Similarly, if the current economic crisis becomes a twenty-first-century Great Depression, we could face a struggle against renewed xenophobia and authoritarianism. (Inglehart and Welzel, 2009)*

First, our Table 10.1 will explain our methodology to measure *the trend towards secular values.* Our measurement will provide us with a clear-cut and simple trend measure of secular values over time. Table 10.2 will explain our methodology to measure the *trend towards self-expression values;* we equally hope to provide our readers with a simple and at the same time convincing measure of dynamic change along this Inglehartian category. Graph 10.1 and Table 10.3 will explain our methodology to measure the *resilience of the trend towards self-expression values;* and Graph 10.2 and Table 10.4 will explain our methodology to explain the principal component "economic growth" for the years 2008 to 2011 by showing the clear-cut *relationship between the resilience of global value change towards self-expression values and economic growth during the current crisis period.* Our methodology thus is based on simple first differences (Tables 10.1, Tables 10.2), and then standard linear OLS bi-variate regression analysis (Table 10.3 and Table 10.4).

So, to start with, Table 10.1 shows that there is an unabated tendency towards secularization and away from traditional religion in Canada, the United States, Puerto Rico, Australia, Venezuela, New Zealand, Israel, the UK, and Sweden. The "comeback" of religion above all had to be observed in former communist countries, like Estonia, Russia, Belarus, the Ukraine, Latvia, Lithuania, Moldova, Bulgaria, Armenia, and Serbia.

Table 10.2 is again based on simple differences. Jordan, Zimbabwe, Pakistan, Turkey, Bangladesh, Nigeria, Tanzania, Algeria, Georgia, and the Ukraine are the "superstars" according to Inglehart's self-expression values; while at the same time, self-expression has severely contracted in Sweden, the Netherlands, New Zealand, and in other leading Western democracies.

Graph 10.2 is the bi-variate scatterplot and linear regression analysis between the trend values from Table 10.1 on the x-axis and the trend values from Table 10.2 on the y-axis. The trend values between the two variables and their residuals can be seen in Graph 10.1 and also in Table 10.3: Table 10.3 clearly defines our concept of the resilience of the trend towards self-expression. Again so-called Third World countries or developing countries are in the lead, while the countries of the West are imploding.

Graph 10.2 describes the bi-variate scatterplot and linear regression analysis between the resilience of the trend towards self-expression values on the x-axis and the principle components of world economic growth during the crisis years 2008-2011 on the y-axis. Table

10.4 uses these concepts to highlight then the dynamics of global growth in the contemporary period. The Table again shows the devastating extent of the current economic crisis in the West and the upsurge of developing countries – several Muslim countries among them.

Let us thus start with the presentation of the time trend for Inglehart's *"secular values"*. The twenty most dramatic inroads against traditional religion had to be registered in: Canada; United States; Puerto Rico; Australia; Venezuela; New Zealand; Israel; Great Britain; Sweden; Tanzania, United Republic Of; Dominican Republic; Argentina; Nigeria; Peru; South Africa; Brazil; Chile; Mexico; Norway; and Uruguay.

The *comebacks of religion and the receding of the "secular" tendencies* presented themselves in the following countries (in descending order):

Estonia; Russian Federation; Belarus; Ukraine; Latvia; Lithuania; Moldova; Bulgaria; Armenia; Taiwan Province of China; Serbia and Montenegro; Japan; Albania; South Korea; Romania; Azerbaijan; Georgia; Macedonia, Republic of; Slovenia; Hungary; Croatia; Kyrgyzstan; Czech Republic; Bosnia and Herzegovina; Slovakia; Germany; Indonesia:

Table 10.1. Value change – secular values, deduced from the original data from the calculations by Inglehart and associates

	Traditional/Secular Rational Values WVS 1-4/2006	Traditional/Secular Rational Values Fifth Wave	Value change Secular values
Canada	-0,042	1,328	1,370
United States	-0,346	0,955	1,301
Puerto Rico	-0,894	0,267	1,161
Australia	0,075	1,066	0,990
Venezuela	-0,822	0,119	0,941
New Zealand	0,408	1,217	0,809
Israel	0,147	0,948	0,801
Great Britain	0,321	1,121	0,800
Sweden	1,234	1,966	0,732
Tanzania, United Republic Of	-0,937	-0,290	0,647
Dominican Republic	-0,444	0,152	0,596
Argentina	-0,247	0,346	0,593
Nigeria	-0,816	-0,230	0,586
Peru	-0,489	0,090	0,579
South Africa	-0,506	0,062	0,568
Brazil	-0,452	0,107	0,559
Chile	-0,431	0,106	0,537
Mexico	-0,267	0,261	0,528
Norway	0,704	1,199	0,495
Uruguay	0,020	0,499	0,479
Philippines	-0,450	0,013	0,463
Spain	0,239	0,701	0,462
Egypt	-0,713	-0,252	0,461
Netherlands	0,780	1,217	0,437
Algeria	-0,734	-0,316	0,418

	Traditional/Secular Rational Values WVS 1-4/2006	Traditional/Secular Rational Values Fifth Wave	Value change Secular values
Uganda	-0,586	-0,214	0,371
Jordan	-0,882	-0,524	0,358
Zimbabwe	-0,772	-0,419	0,353
Poland	-0,275	0,058	0,334
Pakistan	-0,763	-0,438	0,325
Singapore	-0,278	0,038	0,316
France	0,730	1,039	0,309
Bangladesh	-0,671	-0,369	0,302
Switzerland	0,706	0,962	0,256
Turkey	-0,360	-0,136	0,224
Viet Nam	-0,125	0,061	0,186
India	-0,080	0,090	0,171
Finland	0,688	0,848	0,160
Italy	0,235	0,284	0,049
Indonesia	-0,339	-0,343	-0,005
Germany	0,812	0,796	-0,016
Slovakia	0,516	0,448	-0,068
Bosnia and Herzegovina	0,226	0,038	-0,188
Czech Republic	1,093	0,893	-0,200
Kyrgyzstan	0,032	-0,227	-0,259
Croatia	0,543	0,282	-0,261
Hungary	0,429	0,153	-0,276
Slovenia	0,617	0,332	-0,286
Macedonia, Republic of	0,304	0,008	-0,296
Georgia	-0,042	-0,354	-0,312
Azerbaijan	0,018	-0,304	-0,322
Romania	0,172	-0,208	-0,379
South Korea	0,591	0,187	-0,404
Albania	0,329	-0,100	-0,428
Japan	1,152	0,700	-0,452
Serbia and Montenegro	0,503	0,028	-0,475
Taiwan Province of China	0,422	-0,138	-0,560
Armenia	0,390	-0,190	-0,580
Bulgaria	0,766	0,110	-0,656
Moldova	0,444	-0,223	-0,666
Lithuania	0,494	-0,228	-0,722
Latvia	0,902	0,157	-0,745
Ukraine	0,658	-0,199	-0,857
Belarus	0,714	-0,162	-0,877
Russian Federation	0,852	-0,166	-1,019
Estonia	1,011	-0,099	-1,109

We now apply the same simple first difference methodology to the dynamics of "self-expression". The twenty superstars of self-expression development according to the Inglehart theory are: Jordan; Zimbabwe; Pakistan; Turkey; Bangladesh; Nigeria; Tanzania, United Republic Of; Algeria; Georgia; Ukraine; Moldova; Romania; Egypt; Uganda; Poland; Azerbaijan; Venezuela; Russian Federation; Macedonia, Republic of; and Kyrgyzstan.

The following countries witnessed an implosion of their self-expression processes: Sweden; Netherlands; New Zealand; Switzerland; Australia; Finland; Norway; Japan; Great Britain; Canada; Germany; United States; Czech Republic; France; South Korea; Spain; and Israel:

Table 10.2. Value change - selfexpression, deduced from the original data from the calculations by Inglehart and associates

	Survival/Self-Expression Values WVS 1-4/2006	Survival/Self-Expression Values Fifth Wave	Value change self-expression
Jordan	-0,621	0,620	1,241
Zimbabwe	-0,765	0,398	1,163
Pakistan	-0,669	0,335	1,004
Turkey	-0,443	0,530	0,973
Bangladesh	-0,426	0,545	0,971
Nigeria	-0,276	0,624	0,900
Tanzania, United Republic Of	-0,125	0,774	0,899
Algeria	-0,441	0,426	0,867
Georgia	-0,664	0,174	0,838
Ukraine	-0,947	-0,132	0,816
Moldova	-0,993	-0,249	0,744
Romania	-0,780	-0,045	0,735
Egypt	-0,175	0,553	0,728
Uganda	-0,300	0,416	0,716
Poland	-0,353	0,329	0,683
Azerbaijan	-0,644	0,015	0,659
Venezuela	0,243	0,877	0,633
Russian Federation	-0,912	-0,292	0,621
Macedonia, Republic of	-0,600	0,009	0,610
Kyrgyzstan	-0,555	0,035	0,590
Brazil	-0,124	0,452	0,576
Philippines	-0,071	0,491	0,562
South Africa	-0,054	0,483	0,537
Armenia	-0,678	-0,141	0,537
Serbia and Montenegro	-0,662	-0,142	0,520
Peru	-0,061	0,453	0,514
Puerto Rico	0,431	0,938	0,507
Chile	-0,065	0,427	0,492
Belarus	-0,819	-0,328	0,491
Lithuania	-0,645	-0,158	0,487

	Survival/Self-Expression Values WVS 1-4/2006	Survival/Self-Expression Values Fifth Wave	Value change self-expression
Dominican Republic	0,113	0,535	0,422
Bulgaria	-0,701	-0,279	0,421
Hungary	-0,518	-0,103	0,415
Singapore	-0,158	0,226	0,384
Taiwan Province of China	-0,663	-0,298	0,364
Estonia	-0,737	-0,384	0,353
Indonesia	-0,332	-0,016	0,316
Argentina	0,052	0,355	0,303
Albania	-0,525	-0,225	0,301
Mexico	0,235	0,493	0,257
India	-0,101	0,151	0,252
Bosnia and Herzegovina	-0,288	-0,062	0,226
Latvia	-0,600	-0,390	0,209
Slovenia	-0,220	-0,045	0,175
Uruguay	0,185	0,287	0,102
Viet Nam	0,124	0,224	0,100
Slovakia	-0,184	-0,088	0,096
Italy	0,165	0,236	0,071
Croatia	-0,173	-0,104	0,069
Israel	0,111	0,111	-0,001
Spain	0,057	0,010	-0,046
South Korea	-0,041	-0,331	-0,290
France	0,259	-0,120	-0,378
Czech Republic	0,086	-0,300	-0,386
United States	0,951	0,530	-0,421
Germany	0,298	-0,137	-0,434
Canada	0,841	0,389	-0,452
Great Britain	0,661	0,193	-0,468
Japan	0,206	-0,328	-0,534
Norway	0,536	-0,110	-0,645
Finland	0,610	-0,075	-0,686
Australia	1,025	0,329	-0,695
Switzerland	0,732	-0,033	-0,766
New Zealand	1,130	0,242	-0,888
Netherlands	0,794	-0,098	-0,893
Sweden	0,902	-0,326	-1,228

From these materials, it is now very easy to deduce an analysis of the resilience of the trend towards self-expression, based on the residuals from Graph 10.1: resilient are those political cultures, which are above the regression line, and not resilient are those political cultures, which are below the regression line. The x-axis is constituted by the value change in the direction of secular values, while the y-axis is constituted by the trend towards self-expression values.

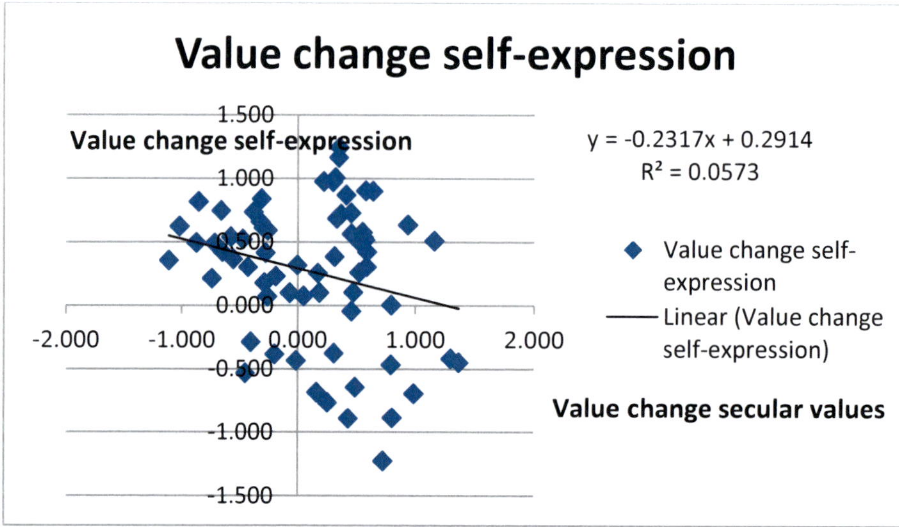

Graph 10.1. Calculating the resilience of the trend towards self-expression: value changes towards secularism (x-axis) and towards self-expression (y-axis), deduced from the original data from the calculations by Inglehart and associates.

Table 10.3 analyses the results from the exercise undertaken, and explained in Graph 10.1.

Table 10.3. Resilience of the trend towards self-expression values, deduced from the original data from the calculations by Inglehart and associates

	Value change secular values	Value change self-expression	Trend value	Residual = resilience of trend towards self-expression
Jordan	0,358	1,241	0,208	1,032
Zimbabwe	0,353	1,163	0,210	0,953
Pakistan	0,325	1,004	0,216	0,788
Tanzania, United Republic Of	0,647	0,899	0,141	0,758
Bangladesh	0,302	0,971	0,221	0,750
Nigeria	0,586	0,900	0,156	0,744
Turkey	0,224	0,973	0,240	0,734
Algeria	0,418	0,867	0,195	0,672
Venezuela	0,941	0,633	0,073	0,560
Egypt	0,461	0,728	0,185	0,543
Uganda	0,371	0,716	0,205	0,511
Puerto Rico	1,161	0,507	0,022	0,485
Georgia	-0,312	0,838	0,364	0,474
Poland	0,334	0,683	0,214	0,468
Brazil	0,559	0,576	0,162	0,414
Philippines	0,463	0,562	0,184	0,378
South Africa	0,568	0,537	0,160	0,377
Peru	0,579	0,514	0,157	0,356
Romania	-0,379	0,735	0,379	0,356
Ukraine	-0,857	0,816	0,490	0,326
Chile	0,537	0,492	0,167	0,325
Moldova	-0,666	0,744	0,446	0,298

	Value change secular values	Value change self-expression	Trend value	Residual = resilience of trend towards self-expression
Azerbaijan	-0,322	0,659	0,366	0,293
Dominican Republic	0,596	0,422	0,153	0,269
Macedonia, Republic of	-0,296	0,610	0,360	0,250
Kyrgyzstan	-0,259	0,590	0,351	0,238
Singapore	0,316	0,384	0,218	0,166
Argentina	0,593	0,303	0,154	0,149
Serbia and Montenegro	-0,475	0,520	0,401	0,119
Armenia	-0,580	0,537	0,426	0,111
Russian Federation	-1,019	0,621	0,527	0,093
Mexico	0,528	0,257	0,169	0,088
Hungary	-0,276	0,415	0,355	0,059
Lithuania	-0,722	0,487	0,459	0,029
Indonesia	-0,005	0,316	0,293	0,023
India	0,171	0,252	0,252	0,000
Belarus	-0,877	0,491	0,495	-0,003
Bulgaria	-0,656	0,421	0,443	-0,022
Taiwan Province of China	-0,560	0,364	0,421	-0,057
Uruguay	0,479	0,102	0,180	-0,078
Albania	-0,428	0,301	0,391	-0,090
Israel	0,801	-0,001	0,106	-0,106
Bosnia and Herzegovina	-0,188	0,226	0,335	-0,109
Viet Nam	0,186	0,100	0,248	-0,149
Slovenia	-0,286	0,175	0,358	-0,183
Estonia	-1,109	0,353	0,548	-0,195
Italy	0,049	0,071	0,280	-0,209
Slovakia	-0,068	0,096	0,307	-0,211
Spain	0,462	-0,046	0,184	-0,231
Latvia	-0,745	0,209	0,464	-0,255
Croatia	-0,261	0,069	0,352	-0,283
United States	1,301	-0,421	-0,010	-0,411
Canada	1,370	-0,452	-0,026	-0,426
Great Britain	0,800	-0,468	0,106	-0,574
France	0,309	-0,378	0,220	-0,598
South Korea	-0,404	-0,290	0,385	-0,675
Czech Republic	-0,200	-0,386	0,338	-0,724
Germany	-0,016	-0,434	0,295	-0,729
Australia	0,990	-0,695	0,062	-0,757
Norway	0,495	-0,645	0,177	-0,822
Japan	-0,452	-0,534	0,396	-0,930
Finland	0,160	-0,686	0,254	-0,940
New Zealand	0,809	-0,888	0,104	-0,992
Switzerland	0,256	-0,766	0,232	-0,998
Netherlands	0,437	-0,893	0,190	-1,083
Sweden	0,732	-1,228	0,122	-1,350

Among the twenty countries of our globe with a strong resilience of the self-expression tendencies there are a very large percentage of Muslim countries (i.e., members of the Organization of Islamic Cooperation) among them. Let us think for an instance Inglehart's theory to its end: the twenty superstars of a resilient trend towards self-expression are, according to the Inglehartian theory: Jordan; Zimbabwe; Pakistan; Tanzania, United Republic Of; Bangladesh; Nigeria; Turkey; Algeria; Venezuela; Egypt; Uganda; Puerto Rico; Georgia; Poland; Brazil; Philippines; South Africa; Peru; Romania; and the Ukraine.

The most notable implosions or slow developments of self-expression – independent from the secularization process – had to be noted by contrast in the following twenty countries, which are all western democracies, i.e., members of the European Union or the OECD: Sweden; Netherlands; Switzerland; New Zealand; Finland; Japan; Norway; Australia; Germany; Czech Republic; South Korea; France; Great Britain; Canada; United States; Croatia; Latvia; Spain; Slovakia; and Italy.

In Graph 10.2 we will now highlight the close, observable relationship between the resilience of the trend towards self-expression and the principle component of economic growth rates calculated from IMF data during the crisis years from 2008 to 2011 (see http://economics.uni-corvinus.hu/index.php?id=47854). The resilience of self-expression is explaining more than 1/5 of economic growth in the world system.

Muslim countries are among the trend leaders in both directions, i.e., the resilience of self-expression, and economic growth during the crisis years.

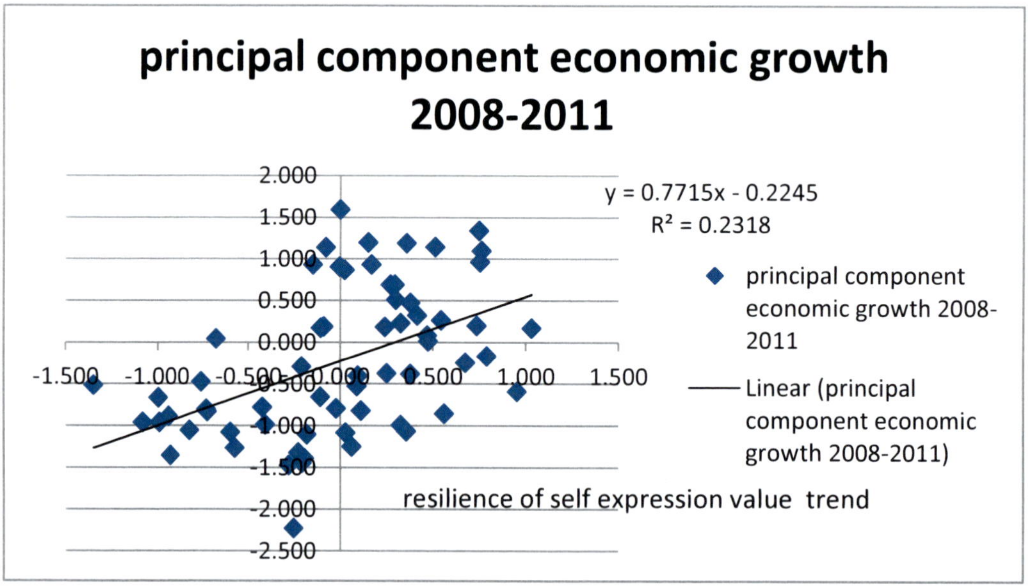

Graph 10.2. The resilience of the self-expression value trend and current economic growth, deduced from the original data from the calculations by Inglehart and associates and from IMF data.

Table 10.4 now highlights these trends and residuals; it is to be seen that member countries of the OIC – commonly called the "Muslim world", are both among the positive and negative residuals from Graph 10.2.

Table 10.4. The resilience of the trend towards self-expression and economic growth, deduced from the original data from the calculations by Inglehart and associates and from IMF data

	Resilience of self-expression	Principal component economic growth 2008-2011	Trend value principle component economic growth	Residual (growth driven by other factors beyond value change)
India	0,000	1,590	-0,225	1,815
Uruguay	-0,078	1,138	-0,285	1,423
Argentina	0,149	1,195	-0,109	1,305
Viet Nam	-0,149	0,930	-0,339	1,269
Peru	0,356	1,187	0,050	1,137
Belarus	-0,003	0,902	-0,227	1,129
Indonesia	0,023	0,867	-0,207	1,073
Singapore	0,166	0,931	-0,096	1,028
Nigeria	0,744	1,335	0,350	0,986
Uganda	0,511	1,139	0,169	0,970
South Korea	-0,675	0,044	-0,745	0,789
Sweden	-1,350	-0,514	-1,266	0,751
Tanzania, United Republic Of	0,758	1,096	0,360	0,736
Dominican Republic	0,269	0,691	-0,017	0,708
Azerbaijan	0,293	0,688	0,001	0,686
Bangladesh	0,750	0,962	0,354	0,609
Moldova	0,298	0,509	0,005	0,504
Albania	-0,090	0,188	-0,294	0,482
Israel	-0,106	0,174	-0,307	0,480
Philippines	0,378	0,473	0,067	0,406
Australia	-0,757	-0,472	-0,809	0,337
Switzerland	-0,998	-0,663	-0,994	0,331
Brazil	0,414	0,323	0,095	0,228
Kyrgyzstan	0,238	0,184	-0,041	0,225
Chile	0,325	0,230	0,026	0,204
Netherlands	-1,083	-0,959	-1,060	0,101
Slovakia	-0,211	-0,295	-0,387	0,092
Egypt	0,543	0,263	0,194	0,068
Finland	-0,940	-0,887	-0,950	0,062
New Zealand	-0,992	-0,959	-0,990	0,031
Germany	-0,729	-0,802	-0,787	-0,015
Czech Republic	-0,724	-0,831	-0,783	-0,048
Poland	0,468	0,082	0,137	-0,054
Georgia	0,474	0,013	0,141	-0,128
Turkey	0,734	0,202	0,342	-0,140
Norway	-0,822	-1,055	-0,859	-0,196
Canada	-0,426	-0,781	-0,553	-0,228
Russian Federation	0,093	-0,406	-0,153	-0,253
Macedonia, Republic of	0,250	-0,373	-0,032	-0,341

Table 10.4. (Continued)

Bosnia and Herzegovina	-0,109	-0,655	-0,308	-0,347
Mexico	0,088	-0,525	-0,156	-0,369
France	-0,598	-1,078	-0,686	-0,393
Jordan	1,032	0,169	0,572	-0,403
Japan	-0,930	-1,352	-0,942	-0,410
United States	-0,411	-0,985	-0,542	-0,443
South Africa	0,377	-0,381	0,067	-0,448
Algeria	0,672	-0,243	0,294	-0,538
Pakistan	0,788	-0,168	0,383	-0,551
Bulgaria	-0,022	-0,795	-0,242	-0,553
United Kingdom	-0,574	-1,266	-0,668	-0,599
Armenia	0,111	-0,822	-0,139	-0,683
Slovenia	-0,183	-1,105	-0,366	-0,739
Lithuania	0,029	-1,085	-0,202	-0,883
Spain	-0,231	-1,322	-0,402	-0,919
Estonia	-0,195	-1,391	-0,375	-1,015
Ukraine	0,326	-0,995	0,027	-1,022
Croatia	-0,283	-1,470	-0,443	-1,028
Italy	-0,209	-1,446	-0,386	-1,060
Venezuela	0,560	-0,856	0,208	-1,063
Hungary	0,059	-1,248	-0,179	-1,070
Zimbabwe	0,953	-0,590	0,511	-1,101
Romania	0,356	-1,066	0,050	-1,117
Latvia	-0,255	-2,229	-0,421	-1,808

Policy pointer:

Among the twenty countries of our globe with a strong resilience of the self-expression tendencies there are a very large percentage of Muslim countries (i.e., members of the Organization of Islamic Cooperation) among them. Let us think for an instance Inglehart's theory to its end: among the twenty superstars of a resilient trend towards self-expression are, according to the Inglehartian theory: Jordan; Pakistan; Bangladesh; Nigeria; Turkey; Algeria; Egypt; and Uganda.

The most notable implosions or slow developments of self-expression – independent from the secularization process – had to be noted by contrast in the western democracies.

The resilience of self-expression is explaining more than 1/5 of economic growth in the world system. Muslim countries are among the trend leaders in both directions, i.e., the resilience of self-expression, and economic growth during the crisis years.

Our Choropleth maps in this Chapter underline our contentions. A pure Inglehartian world values analysis would have to come to the conclusion that the value basis of Western society is eroding.

So while the methodology of the two approaches – Inglehart's and our own – is different, the same conclusions can be drawn from it. Map 10.5 is the final verdict in the debate about these issues: being based on Table 10.3, it just shows how powerful the Tsunami of the crisis of 2008 hit the world of the North Atlantic arena, and how the geography of global development is changing to the detriment of the hitherto dominant West.

We invite our readers to have a renewed look at Map 4.15 in Chapter 4 and at the left, first column of Table 10.1. This would be the world view, which Inglehart's theory would suggest, very much in the spirit of classical modernization theory. The Western countries, especially the Protestant countries, are seen as islands of modernity and their self-expression culture as a robust precondition of *"effective democracy"*. The data, emerging from the first four waves of the *World Values Survey,* provided some large-scale evidence to maintain such a theory. However, with all the extensions of the *World Values Survey* project over the last decades, both in terms of geography as well as the completeness of the data, also on such phenomena as the shadow economy, the world of Map 4.15 somehow receded into the memory about a world order, which no longer exists and which was severely shattered in its foundations by the *tsunami* of the global economic crisis of 2008. As we tried to show in this book, it was also shattered by the long shadows of the internal corrosion, which social decay and the loss of values brought about long before the 2008 crisis hit the North Atlantic arena. The very last wave of the *World Values Survey* project, released in May 2014, contains for the first time an item for the direct measurement of *"self-expression":* not how advanced quantitative sociology tries to impute its values from the knowledge about different other variables, but what ordinary citizens of our globe understood it to be. In presenting these data, we close this Chapter on the dynamics of self-expression values with a note of caution. Some of the countries on top of the list might still sound familiar from the research, presented by Inglehart over decades, while other countries hitherto were considered to be laggards in self-expression development. Nowadays Estonia; Azerbaijan; South Korea; Slovenia; Poland; Cyprus; Germany; Uruguay; Turkey; Sweden; Taiwan; Chile; Australia; Japan; Ukraine; New Zealand; Russia; Colombia; Belarus; and Morocco are now the global leaders in "self-expression", while the United States, for example, is only rank 39 of the 52 nations, and Russia is rank 17 of global, directly measured "self-expression". Not only the top 20 are ahead of the United States, but also Kuwait; Romania; Kazakhstan; the Occupied Palestinian Territories; Jordan; Ghana; Lebanon; Netherlands; Peru; Nigeria; Libya; Egypt; Zimbabwe; Armenia; Qatar; Mexico; Tunisia; and Algeria.

10.2. The Choropleth Maps of the Implosion Process

Our Choropleth maps underline our contentions. A pure Inglehartian world values analysis would have to come to the conclusion that the value basis of Western society is eroding.

So while the methodology of the two approaches – Inglehart's and our own – is different, the same conclusions can be drawn from it. Map 10.5 is the final verdict then in the debate about these issues: being based on Table 10.3, it just shows how powerful the Tsunami of the crisis of 2008 hit the world of the North Atlantic arena, and how the geography of global development is changing to the detriment of the hitherto dominant West.

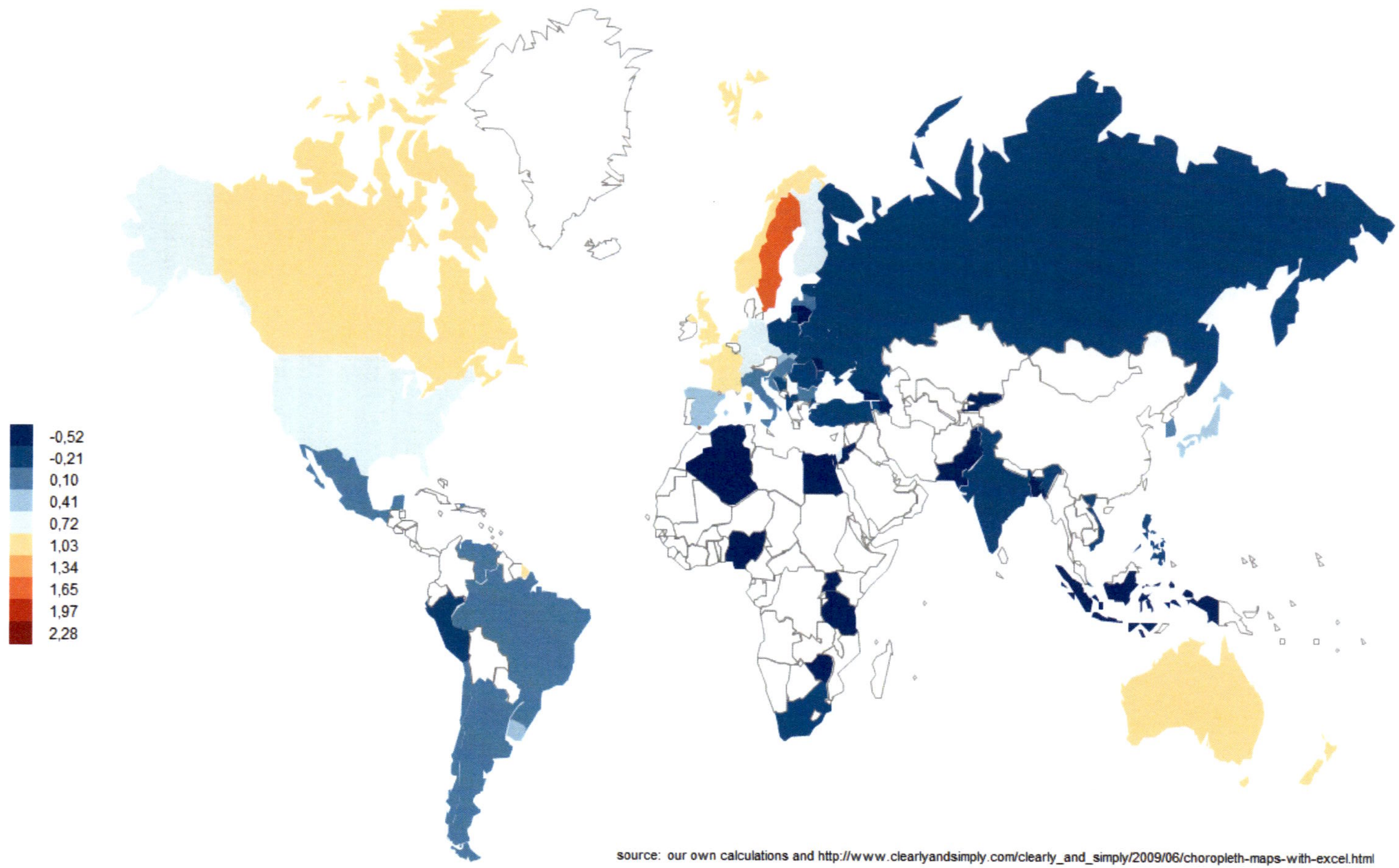

Map 10.1. Inglehart Secular Values – most recent data, deduced from the original data from the calculations by Inglehart and associates.

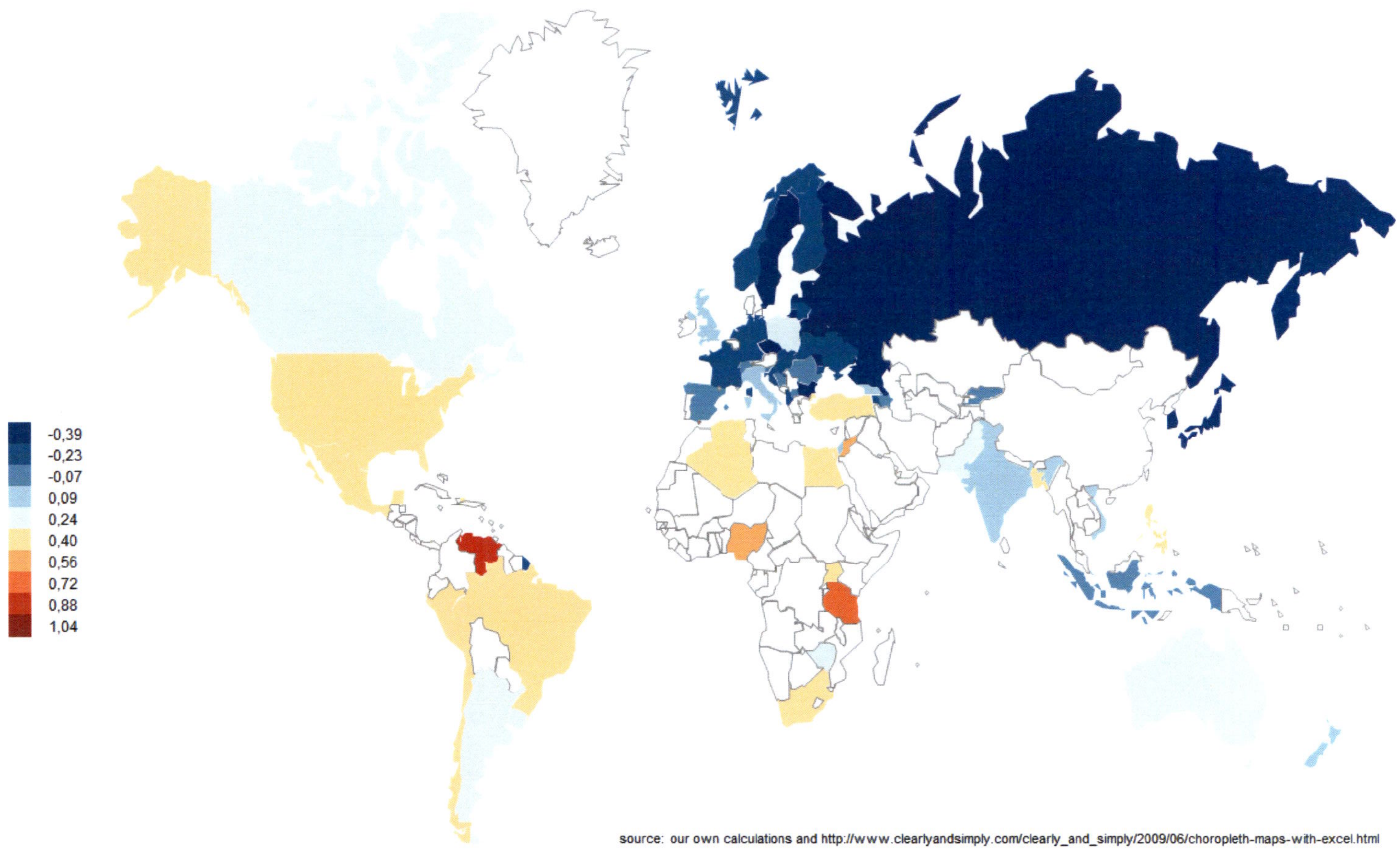

Map 10.2. Inglehart Self-Expression Values - most recent data, deduced from the original data from the calculations by Inglehart and associates.

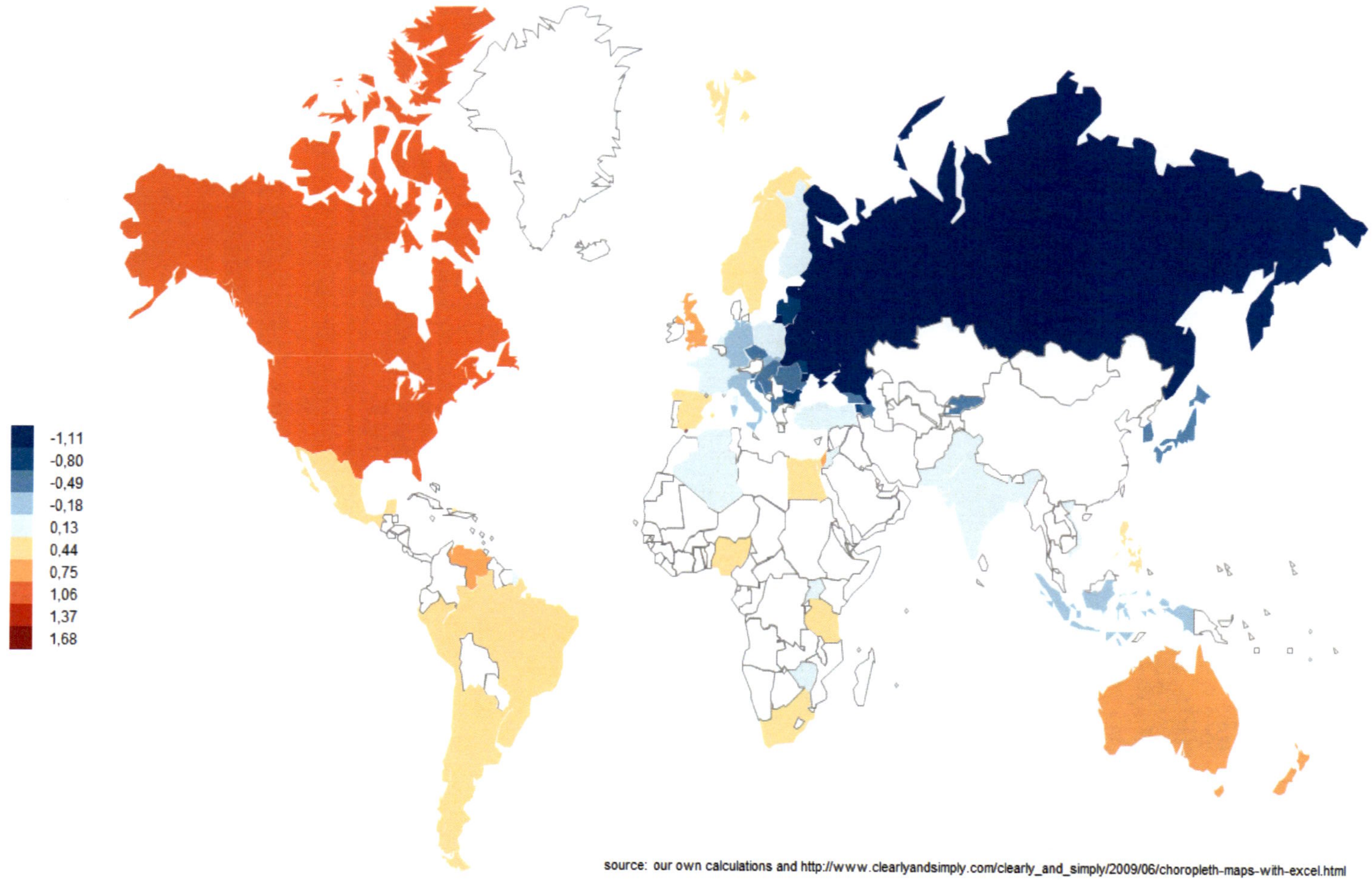

Map 10.3. DYN Secular values, deduced from the original data from the calculations by Inglehart and associates.

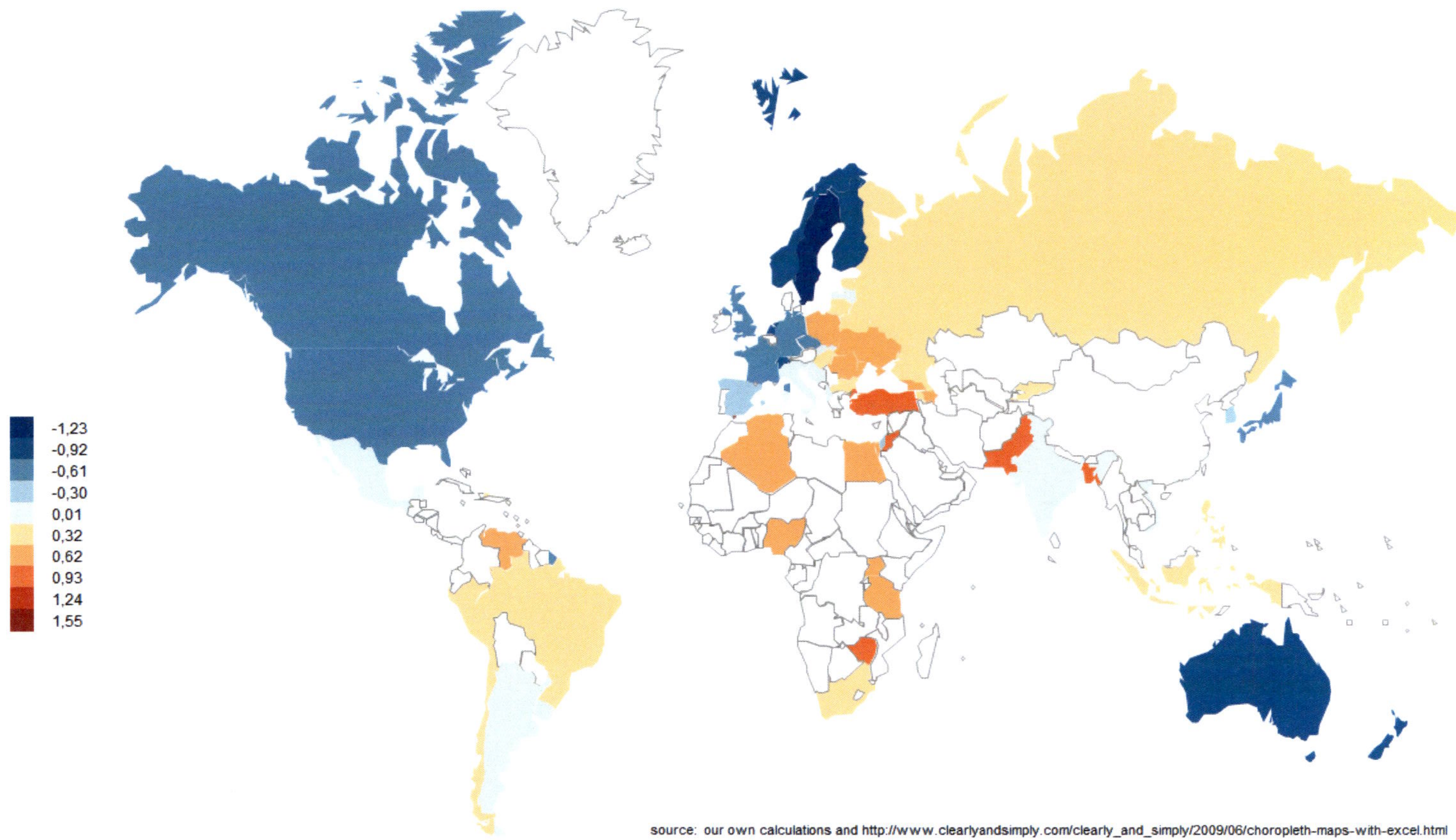

Map 10.4. DYN self-expression values, deduced from the original data from the calculations by Inglehart and associates.

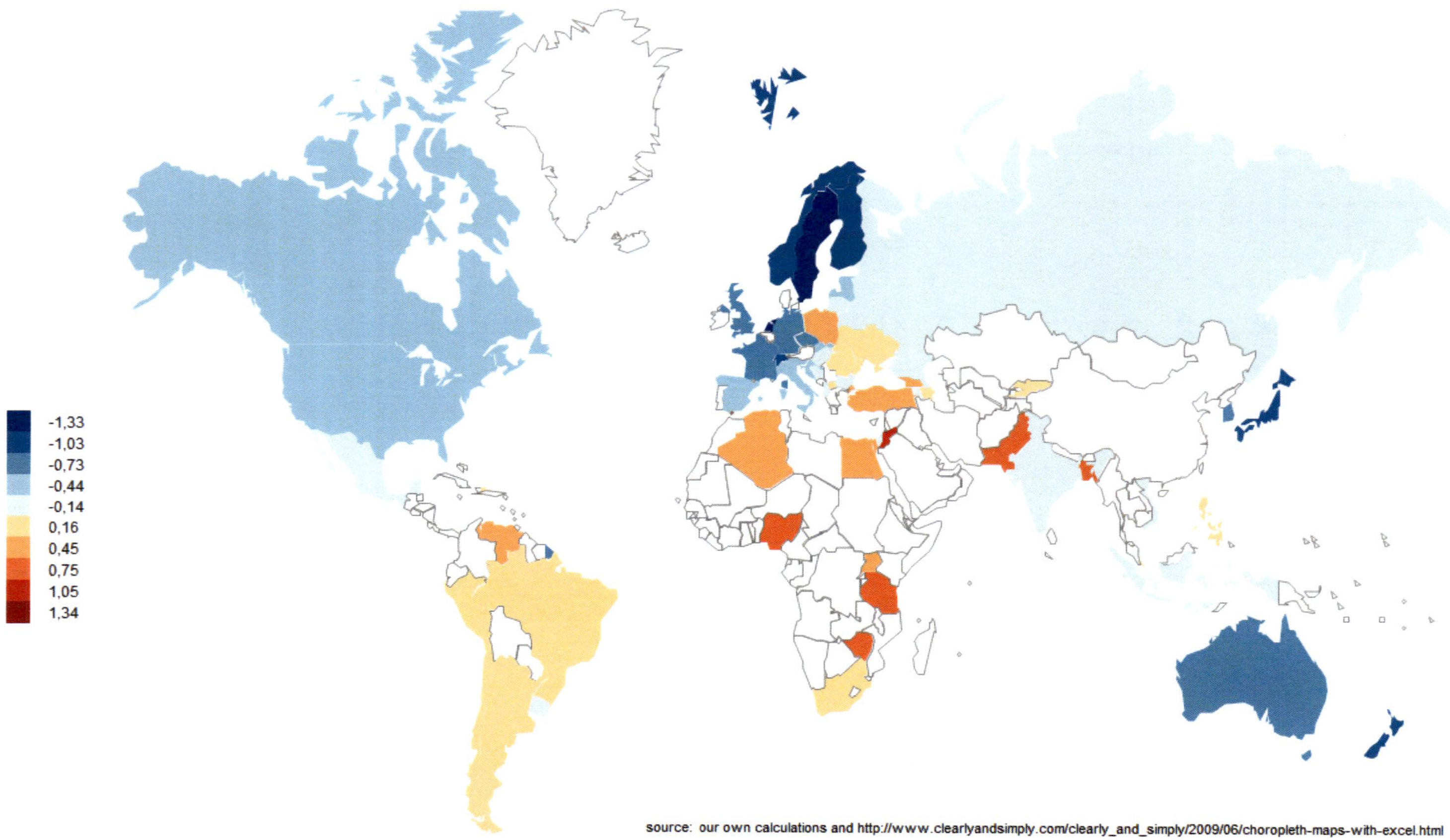

Map 10.5. Resilience of self-expression values, deduced from the original data from the calculations by Inglehart and associates.

10.3. THE DIRECT MEASUREMENT OF SELF-EXPRESSION VALUES IN THE *WORLD VALUES SURVEY* WAVE 6: 2010-2014 WITH THE ITEM "*IMPORTANT CHILD QUALITIES: SELF-EXPRESSION*" AND THE CONSEQUENCES FOR INGLEHART'S THEORY

We now invite our readers to have a renewed look at Map 4.15 in Chapter 4 and at the left, first column of Table 10.1. This would be the world view, which Inglehart's theory would suggest, very much in the spirit of classical modernization theory. The Western countries, especially the Protestant countries, are seen as islands of modernity and their self-expression culture as a robust precondition of *"effective democracy"*. The data, emerging from the first four waves of the *World Values Survey,* provided some large-scale evidence to maintain such a theory. However, with all the extensions of the *World Values Survey* project over the last decades, both in terms of geography as well as the completeness of the data, also on such phenomena as the shadow economy, the world of Map 4.15 somehow receded into the memory about a world order, which no longer exists and which was severely shattered in its foundations by the *tsunami* of the global economic crisis of 2008. As we tried to show in this book, it was also shattered by the long shadows of the internal corrosion, which social decay and the loss of values brought about long before the 2008 crisis hit the North Atlantic arena. The very last wave of the *World Values Survey* project, released in May 2014, contains for the first time an item for the direct measurement of "self-expression": not how advanced quantitative sociology tries to impute its values from the knowledge about different other variables, but what ordinary citizens of our globe understood it to be. In presenting these data in the following Table 10.5, and Maps 10.7 and 10.8, we close this Chapter on the dynamics of self-expression values with a note of caution. Some of the countries on top of the list might still sound familiar from the research, presented by Inglehart over decades, culminating in his article in 2000 (Inglehart/Baker) for the *"American Sociological Review",* while other countries hitherto were considered to be laggards in self-expression development. According to the "online data analysis" version of the newly presented *World Values Survey* website, nowadays Estonia; Azerbaijan; South Korea; Slovenia; Poland; Cyprus; Germany; Uruguay; Turkey; Sweden; Taiwan; Chile; Australia; Japan; Ukraine; New Zealand; Russia; Colombia; Belarus; and Morocco are now the global leaders in *"self-expression"* values, while the United States, for example, is only rank 39 of the 52 nations, and Russia is rank 17. Not only the top 20 are ahead of the United States, but also Kuwait; Romania; Kazakhstan; the Occupied Palestinian Territories; Jordan; Ghana; Lebanon; Netherlands; Peru; Nigeria; Libya; Egypt; Zimbabwe; Armenia; Qatar; Mexico; Tunisia; and Algeria.

Graph 10.3 portrays the relationship between the response rates from Table 10.5 and the middle column of Table 10.2. Data are available from the following countries: Armenia; Australia; Azerbaijan; Belarus; Chile; Egypt; Estonia; Germany; Japan; Jordan; Kyrgyzstan; Mexico; Netherlands; New Zealand; Nigeria; Pakistan; Peru; Philippines; Poland; Romania; Russia; Singapore; Slovenia; South Korea; Spain; Sweden; Turkey; Ukraine; United States; Uruguay; and Zimbabwe.

The correlation between the two variables, as is to be easily seen from Graph 10.3, is negative, and the $R^{\wedge 2}$ is 19.1%.

Table 10.5. ***World Values Survey*** **Wave 6: 2010-2014: Important child qualities: Self-expression**

	Important child qualities: Self-expression
Estonia	**62,90**
Azerbaijan	53,60
South Korea	**49,90**
Slovenia	**48,80**
Poland	**41,00**
Cyprus	**39,30**
Germany	**39,20**
Uruguay	38,60
Turkey	37,90
Sweden	**37,60**
Taiwan	37,50
Chile	**36,40**
Australia	**36,30**
Japan	**33,60**
Ukraine	32,30
New Zealand	**32,10**
Russia	32,10
Colombia	31,50
Belarus	29,80
Morocco	29,70
Kuwait	29,40
Romania	**29,00**
Kazakhstan	28,60
The Occupied Palestinian Territories	25,30
Jordan	24,60
Ghana	23,20
Lebanon	23,10
Netherlands	**23,00**
Peru	22,90
Nigeria	22,80
Libya	22,50
Egypt	20,60
Zimbabwe	20,60
Armenia	19,70
Qatar	19,60
Mexico	**18,60**
Tunisia	18,30
Algeria	17,80
United States	**17,60**
Pakistan	16,70
Trinidad and Tobago	15,30
Singapore	13,70
Kyrgyzstan	13,50
Ecuador	12,60
Iraq	12,60
Philippines	11,50
Malaysia	10,70
China	10,50
Rwanda	8,70
Spain	**8,30**
Uzbekistan	8,30
Yemen	6,50

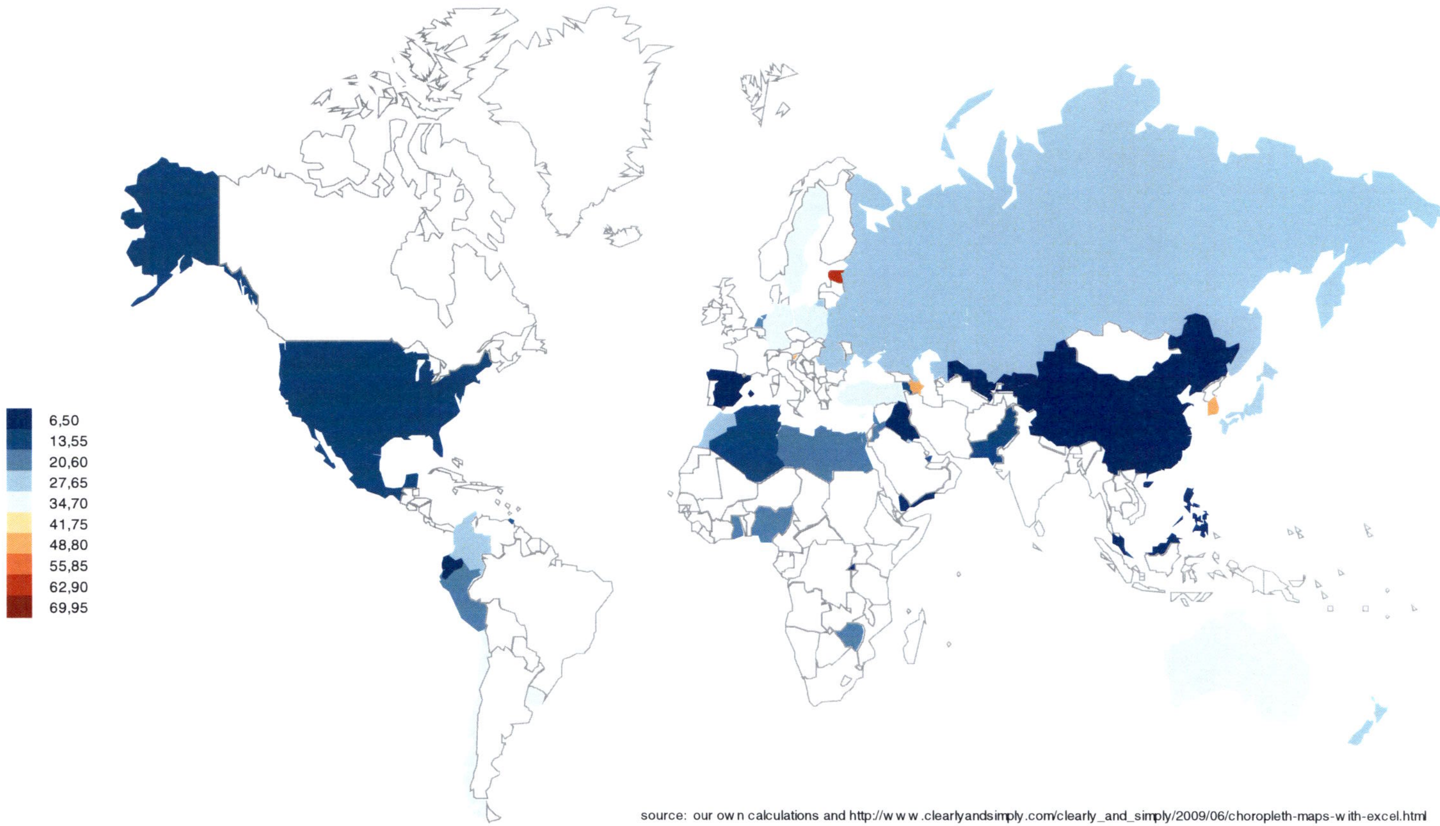

Map 10.7. The direct measurement of self-expression values in the *World Values Survey,* Wave 2010-2014 – the values from Table 10.5.

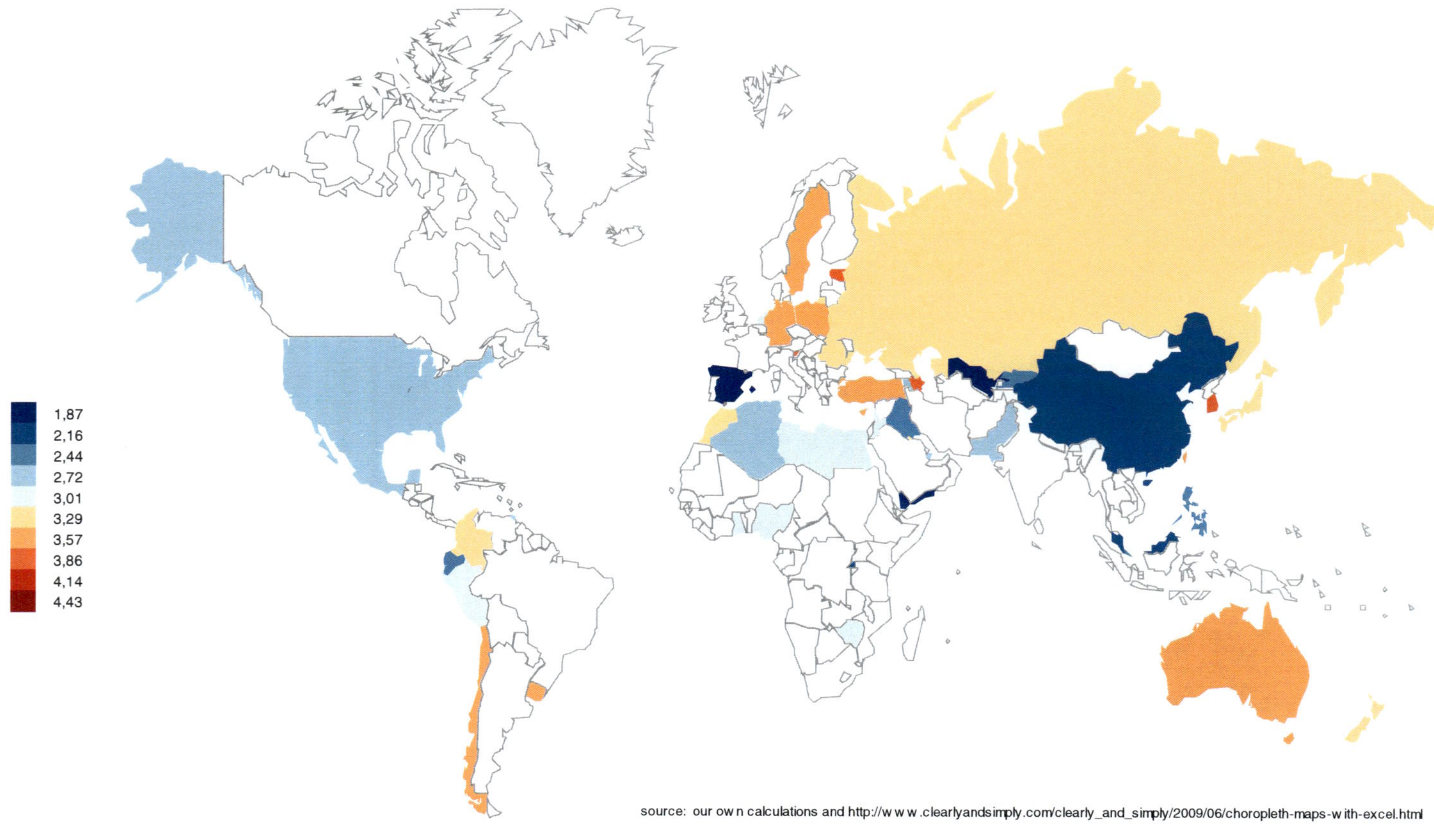

Map 10.8. The direct measurement of self-expression values in the *World Values Survey*, Wave 2010-2014 – the natural logarithm of the values from Table 10.5.

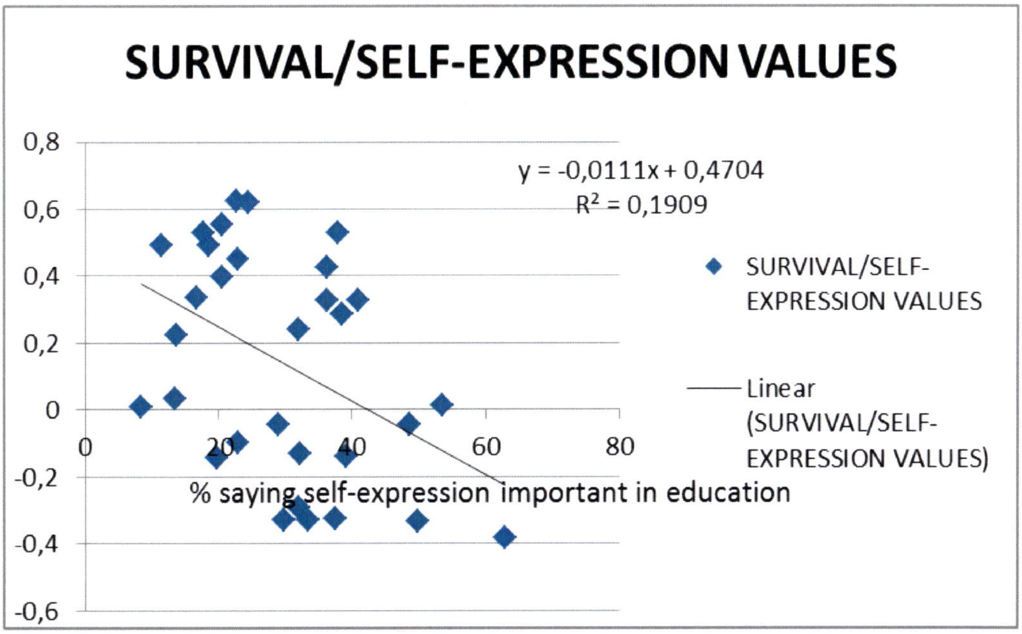

Graph 10.3. *World Values Survey* Wave 6: 2010-2014: Important child qualities: Self-expression and the Inglehart self-expression scores (latest data, see Table 10.2.

Maps 10.7 and 10.8 portray the Choropleth maps, to be derived from Table 10.5. In the concluding Chapter 12 we show that both of Inglehart's main aggregates – traditional/secular and survival/self-expression values - can well be measured by the factors derived in this work. Maps 10.7 and 10.8 contain an important message in the context of the debates of Chapter 10.1 and Chapter 10.2: self-expression values have shifted geographically around the globe, and various Western countries are in a real crisis.

Chapter 11

What the Importance Given to God by Humans Does to Society (The Way Society Benefits from the Belief in God)

Box 11.1. The overall message of this Chapter

If religions survived throughout human history, it is not by chance. Religion actually plays a more a positive role in society than acknowledged by some social scientists like Comte, Spencer, and Marx. We find religion not only at the basis of human values, but also at the end of the human strive for happiness. That is why religion, arts, science, laws, and good government are more complementary in society than contradictory. It is noteworthy that in classical Islamic thought, medieval scholars (like al-Kindi, al-Farabi, Ibn Sina, Ibn Rushd, *et cetera)* have always attempted to find the common points between philosophy (science, reason) and religion. As most of them also were deeply involved with scientific experiments and theories, their attempts at reconciling Reason and Religion, Science and God, were probably the first of the kind in human history. They were precursors even of the Enlightenment that occurred in the eighteenth-century Europe. Most of the points later debated by Enlightenment in Europe have been discussed at length in classic Islamic thought, by Mu'tazilites, Ash'arites, Ikhwan al-Safa, and several renowned philosophers.

If we don't have laws, we don't have society; if we don't have society, we don't have civilization. The law is what distinguishes humans from animals (Ibn Khaldun). It lays the foundations of human civilization.

Ibn Khaldun thought that *"social laws can only be discovered by gathering a large number of facts and observing comparable cases,"* - a task that is presently performed by our modern social sciences. Ibn Khaldun showed that *"Both past and present events can supply such facts. Social laws operate the same way in the same kinds of structure."* (Martindale, 1960: 132)

Taking into consideration the broad sociological and philosophical reflections developed in this Chapter, we also look into the discernable patterns of connection between the importance of God to one's life and social behavior.

- Instead of a clear-cut linear connection between the expressed depth of a respondent's attachment to God in one's life and patterns of social behavior or opinions, following from it, there is in most cases an U-shaped or inverted U-shaped curve. This is reminiscent of so many other U-shaped phenomena encountered in the social sciences. In 1955, economics Nobel laureate Simon Kuznets discovered that income inequality first rises and later on diminishes with rising income per capita level of a country. Correspondingly, we can discern for both global humanity as well as for the global *Umma* the following patterns to hold for both global humanity as well as the global *Umma*:

- The importance of God to one's life and the acceptancy of taking a bribe resemble an inverted U-shape, with highest levels of a disposition towards illegal behavior at middle levels of self-declared religiosity
- The importance of God to one's life and the acceptancy of cheating on taxes again resemble an inverted U-shape, with highest levels of a disposition towards illegal behavior at middle levels of self-declared religiosity
- The importance of God to one's life and the rejection of the idea that democracy is better again resemble an inverted U-shape at the world level. For Muslim respondents, the pattern resembles rather a "W", with highest levels of the rejection of the idea that democracy is better at self-declared middle and high levels of religiosity
- The importance of God to one's life and the opinion that it's fairly bad to have a democracy again resemble an inverted U-shape, with highest levels of a disposition towards rejecting democracy at middle levels of self-declared religiosity.
- The importance of God to one's life and the opinion that competition is harmful resembles an inverted U-shape, with highest levels of a disposition against competition at middle levels of self-declared religiosity
- The importance of God to one's life and life satisfaction weakly resembles a W-shape, with highest levels of life satisfaction for the global sample at very low, moderately high and very high levels of self-declared religiosity, and with highest levels of life satisfaction for the global *Umma* at moderately low, medium level and very high levels of self-declared religiosity.

Table 11.1 contains still more data on what faith does to society. The two leading "Hayek" variables in the *World Values Survey,* competition good or harmful, and private or government ownership of the means of production are not related at all with the importance of God to one's personal life, and the correlations are below +-.100.

If religions survived throughout human history, it is not by chance. Religion actually plays in society more a positive role than acknowledged by some social scientists like Auguste Comte, Spencer, and Marx. We find religion not only at the basis of human values, but also at the end of human strive for happiness. That is why religion, arts, science, laws, and good government are more complementary in society than contradictory. That is also what this chapter will show.

11.1. Durkheim and Beyond

Any social system is organized through a corpus of rules commonly called Law, revealing how one is supposed to behave in order to live up to the values of one's culture. A legal system should include all kinds of laws specifying, for example, that people cannot kill, hurt, steal, or threaten one another. These laws are linked to the values, which are shared beliefs about the goals toward which humans should strive. Values are at the core of moral doctrines first provided – in human history - by religious systems.

Jeremy Bentham put it this way: *"what is a government and its legal system for?"* He answered that the object of the law was not, as Locke had said, to protect the life, liberty, and property of the individual; indeed, the object of government was not really to deal with the individual, *per se*. Nor was its object, as Rousseau said, to express the general will. Instead, in Bentham's view, the object of government and its laws was to supply the population at large with its greatest possible happiness. (Bronowski and Mazlish, 1960: 435)

Sociologists have always stressed that while religion fills some psychological needs, it has an important social function as well. Thus, it is easier for people to believe in a religion, with a specific deity and specific rituals, than to acknowledge that "society" has power over their lives (Smelser, 1984; Durkheim, 1964, 1986). More than reflecting society, religion, according to Durkheim, also reinforces it by focusing people's attention and hopes on shared beliefs and objects of worship. (Smelser, 1984: 278) Durkheim defined religion this way:

> When a certain number of sacred things sustain relations of co-ordination or subordination with each other in such a way as to form a system having a certain unity, but which is not comprised within any other system of the same sort, the totality of these beliefs and their corresponding rites-constitutes a religion. (Durkheim, 1964: 41)

Such a definition results from Durkheim's reflection on the characteristics that distinguish all religions. In his eyes, they all share the same *unifying function* especially obvious in the practice of rituals, which actually make the distinction between the sacred and the profane. As he put it:

> The real characteristic of the religious phenomena is that they always suppose a bipartite division of the whole universe, known and knowable, into two classes which embrace all that exists, but which radically exclude each other. Sacred things are those which the interdictions protect and isolate; profane things, those to which these interdictions are applied and which must remain at a distance from the first. Religious beliefs are the representations which express the nature of sacred things and the relations which they sustain, either with each other or with profane things. Finally, rites are the rules of conduct which prescribe how a man should comport himself in the presence of these sacred objects. (Durkheim, 1964: 40-41)

The Enlightenment intellectuals were not necessarily atheists, as some people believe. They were opposed to traditional religion, but were not ready to give up religion. We see this clearly in the Deist movement, which *"numbered among its adherents such persons as Voltaire, the Encyclopedists, Hume, the Earl of Shaftesbury, Benjamin Franklin, Thomas Jefferson, and numerous other intellectuals in France, England, and North America."* (Martindale, 1960: 31) The common points on which they agreed were:

(1) the attempt to establish religion on the basis of reason rather than authority;
(2) the rejection of tradition except insofar as it was "reasonable"; the restriction of the sphere occupied in religion by "revelation" and "miracles";
(3) a critical attitude regarding religious dogmas difficult to justify rationally (such as the doctrine of Trinity);[58]
(4) the belief that there was a set of universal religious notions implanted in the minds of all men;
(5) the notion that God does not continually interfere in the natural processes of the world, but permits the natural laws to operate once he has set them in motion. (Martindale, 31)

Most of these points have been discussed at length in classic Islamic thought, by Mu'tazilites, Ash'arites, Ikhwan al-Safa, and several renowned philosophers.

Durkheim has agreed with William James, who, in *The Varieties of Religious Experience* (1902), had argued that religious beliefs rest upon real experiences whose demonstrative value, though different, is in no way inferior to that of scientific experiments. Talcott Parsons adopts the same approach when he criticizes the positivists who, since Auguste Comte and Spencer, *"believed that religion arose from the prescientific conceptions of men about the empirical facts of their own nature and their environment."* (Parsons, 1968: 4) In Comte and Spencer's view (later in Marx's opinion as well), religion would be *"the product of ignorance and error. Religious ideas would, with the progress of knowledge, be replaced by science. This was only a phase of a much wider deification of science."* (Parsons, 1968: 4) The existentialists plunged their readers into a black hole when they pretended emptying the universe of any sense related to God. Thus, Man would be alone in a universe that literally crushes and sinks Man into a vacuum of anxiety. What is the value of life if it is governed by accident or, as Albert Camus suggested: the absurd? Like in the ancient myth, Sisyphus (otherwise, Man) would roll a huge boulder to the top of a steep hill; each time the boulder neared the top it rolled back down and Sisyphus was forced to start again.

On the moral level, Dostoyevsky better underlined the dilemma, when he wondered about the value of life without God. His answer is known: *"If God doesn't exist, everything is permitted."* Indeed, no other author reflected in his works this idea pushing it to the extremity better than Marquis de Sade. Without the belief in God, all crimes are permitted. There are no morals; it is just a question of point of view.

However, we assuredly know that society and civilization are not possible without morals. Although S. Freud followed the positivists (whose trend was prevailing at his time) in infantilizing religion, linking it to a pre-scientific stage of the human behavior, he recognized all the same in one of his latest books (Civilization and Its Discontents, 1930) that the real tension between individual and society consists in the opposition between the individual's

[58] For a comprehensive survey of Christian theology, including Liberation Theology, on the subject, see *The Cambridge Companion to the Trinity*. Cambridge Companions to Religion, Editor: Peter C. Phan, Cambridge University Press, August 2011. Paperback, isbn: 9780521701136. In the interesting Ph.D. thesis "TRANSFORMING REALITIES: CHRISTIAN DISCIPLESHIP IN THE SOTERIOLOGY OF IGNACIO ELLACURÍA" submitted to the Graduate School of the University of Notre Dame in Partial Fulfillment of the Requirements for the Degree of Doctor of Philosophy by Michael Edward Lee, Graduate Program in Theology, Notre Dame, Indiana, August 2005, we read about the thought of Ignacio Ellacuria, who was assassinated by death squads on November 16th, 1989 in El Salvador, in this context. Interested readers are referred to this thesis, freely available on Google Scholar.

instinctive quest for total freedom and the society's instinctive need for conformity and repression. In other words, if there is nothing to restrain people in their quest for total freedom, the result would be the opposite of happiness, for society and civilization are not imaginable outside the Law. When he raises the question of the purpose of life, he links it to men's behavior which shows the purpose and the intention of their lives.

> What do they demand of life and wish to achieve in it? The answer to this can hardly be in doubt. They strive after happiness; they want to become happy and to remain so. (Freud, 1962: 23)

After enumerating several paths which may lead to human happiness, Freud admits that none of them is certain. Even religion, in his opinion, *"cannot keep its promise"*. When we ask: why? He gives the following answer:

> If the believer finally sees himself obliged to speak of 'God's inscrutable decrees', he is admitting that all that is left to him as a last possible consolation and source of pleasure in his suffering is an unconditional submission. And if he is prepared for that, he could probably have spared himself the detour he has made. (Freud, 1962: 32)

Thus, Freud understood very well the appeasing role of religion and its importance in giving a sense of purpose to human life and to society as well. But apparently he did not understand the sense of the "detour", which is implied of many human experiences (Ghazali, Bunyan, H. Hess, *et cetera)*, leading people from a loss of meaning to a kind of spiritual regeneration through religion. The "unconditional submission" by the way, is included in the Arabic word: *Islam*, although all religions require it from their followers.

Oddly enough, both William James and Durkheim adopted an approach that reminds us of the position of the classical Muslim rationalists (such as Ibn Sina/Avicenna and Ibn Rushd/ Averroes) toward religion. Both believed that science and religion are more including than excluding each other.

Policy pointer:

An approach assuming that science could not be reconciled with religion, Durkheim argued, does not correspond to the religious believer's own account of the nature of his experience, which is less one of thought than of action: *"The believer who has communicated with his God,"* Durkheim observed, *"is not merely a man who sees new truths of which the unbeliever is ignorant; he is a man who is stronger. He feels within him more force, either to endure the trials of existence, or to conquer them."*

The mere ideas of the individual are clearly insufficient to this purpose; and, in this sociological version of the principle "outside the Church there is no salvation" (*"extra ecclesiam nulla salus")*, Durkheim thus insisted that it is the repeated acts of the cult which give rise to *"impressions of joy, of interior peace, of serenity, of enthusiasm which are, for the believer, an experimental proof of his beliefs."* (Jones, 1986: 115-155)

11.2. The Law

If we don't have laws, we don't have society; if we don't have society, we don't have civilization. The law is what distinguishes humans from animals (Ibn Khaldun). It lays the foundations of human civilization.

The Christian tradition reflected this fact of life through the *Ten Commandments* (Thou shalt not kill; Thou shalt not covet thy neighbor's wife; Honor thy father and thy mother...*et cetera)* and the teachings of Jesus:

> Do not think that I have come to abolish the Law or the Prophets; I have not come to abolish them but to fulfill them. For truly I tell you, until heaven and earth disappear, not the smallest letter, not the least stroke of a pen, will by any means disappear from the Law until everything is accomplished. Therefore anyone who sets aside one of the least of these commands and teaches others accordingly will be called least in the kingdom of heaven, but whoever practices and teaches these commands will be called great in the kingdom of heaven (Matt 5: 17-19).

Similarly, the *Umma* bound by Islam since the time of the Prophet Muhammad (S), has been founded on the concept of 'rule of law.' The *Umma* here has a religious connotation that should distinguish it from the *Umma* as a political entity of modern times. In Arabic, the term *Umma*, is commonly used in the sense of *nation or community*. Obviously, if we use *Umma* in the political sense, we can hardly say that in the Islamic world it is firmly founded on the concept of *'rule of law'*. The modern Muslim-majority nations have not been really so far a model regarding the rule of law. We associate the rule of law with freedom, because it is only in countries where citizens choose freely their rulers and/or where the rulers respect the rights of their fellow-citizens that the rule of law is credible.

A simple look at the map of freedom in the world since 1988 (http://www.freedomhouse.org/report/freedom-world/freedom-world-2013) would convince any skeptic that none of the Muslim-majority nations has obtained the status of "free country", according to the criteria of *Freedom House*. The only exception is Indonesia (only since 2006). Indonesia is now the world's third most populous democracy, the world's largest archipelagic state, and the world's largest Muslim-majority nation (https: //www.cia.gov/library/publications/the-world-factbook/geos/id.html).

If we consider *Freedom House* criteria, it is only in non- Muslim nations (mostly in Europe, North America, Australia, South Africa, and a few Asian nations), that one can find Muslim communities enjoying freedom. As it happens, those nations are democracies. We designate those Muslims as: global Muslims. How about the rest of Muslim populations? From Afghanistan to Morocco, one is well forced to admit that the rule of law is not exactly the best characteristic of these societies.

Many Islamist authors think that this is due to the ignorance of Islamic laws *(Shari'ah)*. However, if we come back to the concept of *Umma* with its religious connotation, we will find that it acquires its full meaning with the *Shari'ah,* which is the Islamic law influencing the legal code in most Muslim countries. A movement to allow *shari'ah* to govern personal status law, a set of regulations that pertain to marriage, divorce, inheritance, and custody, is even expanding into the West.

The controversy about Shari'ah is not limited to the Western countries, though. In Muslim countries too, it stirs a fervent debate between "literalists" (justifying punishments such as amputation and stoning, as well as uneven treatment of women in inheritance, dress, and independence), and progressive-liberal intellectuals.

Furthermore, the *Shari'ah*-led countries, claim that the criteria used by *Freedom House* and other alike organizations become unfair when applied to them, since the sources of the laws in the West and in the concerned Islamic countries are different. Values are therefore different in time and space. And *"science achieves its full explanatory power only by becoming value-free and independent of our preferences."* (Martindale, 1960: 379)

Nevertheless, beyond the debate over the universality and the relativity of moral values, religions all the same shape the mind of the individual through their moral values, their rites and code of behavior. Even Marx recognizes in *The German Ideology (*1845) that "*Men are the producers of their conceptions, ideas, et cetera.*", although "*The phantoms formed in the human brain are also, necessarily, sublimates of their material life-process, which is empirically verifiable and bound to material premises. Morality, religion, metaphysics, all the rest of ideology and their corresponding forms of consciousness, thus no longer retain the semblance of independence. They have no history, no development; but men, developing their material production and their material intercourse, alter, along with this their real existence, their thinking and the products of their thinking. Life is not determined by consciousness, but consciousness by life.*"

If we admit that everything is thus interconnected and that religions, moral values, and people's opinion are interdependent with their material life and influence one another, we should admit that comparing the behavior of religious and non-religious citizens in society, and their attitudes toward the same topics, is a real test for any hypothesis about the positive or negative influence of religion in society. Such issue still needs to be further explored by social scientists.

So far, and in the absence of comparative data especially focused on this issue, we don't have an objective answer concerning the importance of God in the life of the society or even the individual. Concerning values such as: democracy, freedom, justice, welfare, happiness, etc…individuals have different answers, according to their creeds. The assumption that those for whom God is very important are very much tilting toward these values does not exclude the opposite. The data collected from different sources confirm that religious people value democracy, freedom, justice, welfare, and feel happy with their way of life. This does not entail that atheists do not consider the same values very important in their life. Actually, to be good is not the exclusiveness of the believers, atheists would argue. However, although we do not think that *"religion, family, state, law, morality, science, art, et cetera,* ***are only*** *[our emphasis]* particular *modes of production, and fall under its general law"*, (Marx, *Economic and Philosophic Manuscripts* of 1844), we do think that they still are modes of production, plus something else, something particular that places Religion for example – and to some extent morality - above production relations. As Max Weber (1922/1964) suggested, different social classes are attracted to different kinds of religious beliefs, the upper classes adopting religious beliefs that justify their privileges, the lower middle and working classes embracing beliefs that stress future, rather than present rewards, *et cetera.*

For atheists, some religious laws are immoral regarding the values of freedom, equality, human rights, *et cetera.* In thousands of years, they argue, religion did nothing to improve the life of millions of followers: the slave remained slave although he (she) was a believer. The same holds for the classes of poor and deprived laborers. The same holds for sexually-exploited and cloistered women. The same holds for labor-exploited children. *Crimes, murders, rapes, unjust imprisonment have been performed throughout centuries by people claiming that they are good believers against other good believers. As we are writing down these lines, right now, Muslims are massacring other Muslims while millions of other Muslims are watching their brethren being massacred by other brethren on the land of Islam. This also has happened between Christians across Europe, as it happened between people from different religions. In modern times, it has become the custom for any terrorist, any criminal, and any dictator to invoke religion in order to justify awful crimes.* One must just remember the victims of two world wars, not to mention the civil wars from Spain to Ireland. This is just for the 20th Century. But if we consider the history of mankind, we should wonder how many wars have been waged in the name of religious creeds?

However, we can hardly say that this is the domain of the sacred (which defines religion, according to Durkheim). Very often, political conflicts lead to their justification by religious movements. What is at fault here is not the religion, but rather its hijacking by some social groups for their own interests. At this point, religion becomes ideology, and Marx is probably right in referring to it as *"opium,"* insofar as it is used to exploit and enslave people at the service of some groups (whether they are a social class, the founders or the guardians of an order, the rulers, or some militant or terrorist organizations).

We should indeed distinguish between:

(1) Religion as a system of beliefs and rites *"by which a group of people interprets and responds to what they feel is supernatural and sacred"* (Johnstone, 1975: 20; Smelser, 1984: 269); and
(2) Religion as an ideology, which reflects the transformation of the sacred in the context of social practice and production.

In this context, religion becomes subjected to social conflict.

As we know, *"the chief usefulness of the concept of ideology concerns the critique of domination."* (Giddens, 1979: 187) This was Marx concern indeed. In *The German ideology*, he says:

> The entire body of German philosophical criticism from Strauss to Stirner is confined to criticism of religious conceptions. The critics started from real religion and actual theology. What religious consciousness and a religious conception really meant was determined variously as they went along. Their advance consisted in subsuming the allegedly dominant metaphysical, political, juridical, moral and other conceptions under the class of religious or theological conceptions; and similarly in pronouncing political, juridical, moral consciousness as religious or theological, and the political, juridical, moral man – "man" in the last resort – as religious. The dominance of religion was taken for granted. Gradually every dominant relationship was pronounced a religious relationship and transformed into a cult, a cult of law, a cult of the State, et cetera. On all sides it was only a question of dogmas and belief in dogmas. (Marx, 1845)

Policy pointer:

According to Ibn Khaldun, nothing is more powerful than religious propaganda at the service of a political power. "The reason for this" as he explains, "is that religious coloring does away with mutual jealousy and envy among people who share in a group feeling, and causes concentration upon the truth. When people (who have a religious coloring) come to have the (right) insight into their affairs, nothing can withstand them, because their outlook is one and their object one of common accord. They are willing to die for (their objectives)." (Ch. III: 5)

When *"structures of signification are mobilised to legitimate the sectional interests of hegemonic groups"* (Giddens, 1979: 188) as has been the case for the use of religion so many times in history, we are then facing a conflict of interests, which is far away from the domain of the sacred by which religions are defined.

Thus, the Muslim nation is taken between hammer and anvil. On the one hand, corrupt despots violating human rights and their own laws and governing large countries according to their whims; on the other hand, opponents using Islam as a tool to mobilize people, organize resistance and sedition. In such a battle, where moderate voices are lost, the message from the luminous liberal and progressive side of the Islamic patrimony, does not reach the masses; and if it does, it is distorted and represented as "westernized" and even non-Islamic.

Now, there is all the same another interesting trend concerning the effect of religion on society. It concerns the notion of the *"right"* way to live. Thus, Johnstone (1975) has listed five basic elements common to belief systems regarded as religions. First, the group of believers that all religions imply. Second, the sacred domain. Third, specific creeds defining human nature, nature, society, and the supernatural. Fourth, specific rituals which are patterns of behavior toward the sacred. And finally the precepts or norms supposed to guide the behavior.

Most surveys based on the gathering of quantitative data from opinion polls are concerned with this last point related to the "right" way to live that is embedded in the attitudes and the values.

Opinion polls, we know since Bourdieu (1980), are just a tool to measure the state of the mind at a determined moment for a determined object. They are not the scientific truth, but just a glimpse into the general mood at a precise moment. If we rely only on opinion polls to make a scientific assessment of any situation, then we are facing a serious shortcoming. Opinions change. They are like the stock exchange values. They rise, they fall, and they are in eternal fluctuation following the activity on the markets. Can we build anything solid and durable only on the basis of quantitative data collected from opinion polls? Even with a projection in the future, our forecasting would be limited in time. What we know today may change tomorrow. Science is subject to change (Kuhn T. 1962) and such is society, values, and morality as well (Taylor, 1992).

The only domain that seems to be immutable is the sacred. However, our knowledge of it has always stopped at the threshold. We do not scrutinize the sacred because science is interested only in the phenomena that may be subjected to experimentation and empirical observation. God is beyond this domain. However, though we don't control God, God (or rather the belief in God) is still exerting an unmatchable influence over society and civilization. This sociological and historical fact is observable, maybe even measurable with improved accuracy in the future. Individual and group attitudes reflect and reveal it. We can only induce some social laws from these attitudes.

Whatever we may emphasize in this context has already been observed by some great minds centuries ago. Thus, Ibn Khaldun thought that *"social laws can only be discovered by gathering a large number of facts and observing comparable cases,"* - a task that is presently performed by our modern social sciences. Ibn Khaldun showed that *"Both past and present events can supply such facts. Social laws operate the same way in the same kinds of structure."* (Martindale, 1960: 132) In Ibn Khaldūn's view, *"a state can be established only by conflict in which victory goes to the more cohesive and compact group. Similarly, a new religion can establish itself only by strife, but it will succeed only if it enlists the help of a powerful social solidarity ['Asabiyya']. Once established, a religion can reinforce social solidarity (...) by centering emotion and thought in a common purpose. Religion is the most powerful cohesive force in a large sedentary people."* (Martindale, 1960: 133) In Ibn Khaldūn's own words (Ch. III: 4): If *"Dynasties of wide power and large royal authority have their origin in religion based either on prophecy or on truthful propaganda, this is because royal authority results from superiority. Superiority results from group feeling. Only by God's help in establishing His religion do individual desires come together in agreement to press their claims, and hearts become united."*

Policy pointer:

Thus, we can never stress enough the depth, religions reach in our societies regardless of the kind of religion (Islam, Judaism, Buddhism, Christianity, etc…) and the kind of society (European, Asian, Middle-Eastern, etc…). Nor can we underestimate the influence religion may exert in shaping the mind and attitudes of millions of people, or in creating group solidarity (*Asabiyya,* according to Ibn Khaldun), or in developing militant movements and political blocks, or in determining the outcome of elections (in democratic societies, the USA is a good example) and other social processes, or in the outcome of power conflicts.

Taking into consideration what we have stated above, we will now look into the discernable patterns of connection between the importance of God to one's life and social behavior.

The empirical and mathematical analysis of results of opinion surveys reveals many interesting and often underlying dimensions of cultures and constituencies. Is it really the case that, for example, people who stress that God is highly important for them are less inclined to take a free ride on public transport? What is called in the language of the Noble Quran the *"religions of the book"* they all make exactly many references to people, who hypocritically profess to be religious, but in their daily lives and practices do not keep these rules. And is it so that the importance, given to God, has, say, the same effects in New York City, in Beijing, in Sao Paulo, Mumbai or Karachi?

- Graph 11.1 to Graph 11.6 show that instead of a clear-cut linear connection between the expressed depth of a respondent's attachment to God in one's life, and patterns of social behavior or opinions, there is in most cases a U-shaped or inverted U-shaped curve emerging. Correspondingly, we can discern for both global humanity as well as the global *Umma* the following very interesting and perhaps also very surprising patterns to hold:
- Graph 11.1: The importance of God to one's life and the acceptancy of taking a bribe resemble an inverted U-shape, with highest levels of a disposition towards illegal behavior at middle levels of self-declared religiosity
- Graph 11.2: The importance of God to one's life and the acceptancy of cheating on taxes again resemble an inverted U-shape, with highest levels of a disposition towards illegal behavior at middle levels of self-declared religiosity
- Graph 11.3: The importance of God to one's life and the rejection of the idea that democracy is better again resemble an inverted U-shape at the world level, with highest levels of a disposition towards rejecting the idea that democracy is better to be found at middle levels of self-declared religiosity. For Muslim respondents, the pattern resembles rather a "W", with highest levels of the rejection of the idea that democracy is better at self-declared middle and high levels of religiosity
- Graph 11.4: The importance of God to one's life and the opinion that it's fairly bad to have a democracy again resemble an inverted U-shape, with highest levels of a disposition towards rejecting democracy at middle levels of self-declared religiosity
- Graph 11.5: The importance of God to one's life and the opinion that competition is harmful resembles an inverted U-shape, with highest levels of a disposition against competition at middle levels of self-declared religiosity
- Graph 11.6: The importance of God to one's life and life satisfaction – weakly resembling a W-shape, with highest levels of life satisfaction for the global sample at very low, moderately high and very high levels of self-declared religiosity, and with highest levels of life satisfaction for the global *Umma* at moderately low, medium level and very high levels of self-declared religiosity.

Policy pointer:

From the viewpoint of the neo-liberal, mainstream contemporary interpretation of democracy and the acceptancy of the competitive market economy, the biggest problems seem to arise not with the strongly secularist segments of global society and the deeply religious segments of global society, but with the *"middle of the road"* segments who attach only a smaller importance of God to their own personal life, i.e., they are religious, but only by tradition.

Our detailed graphs are the following:

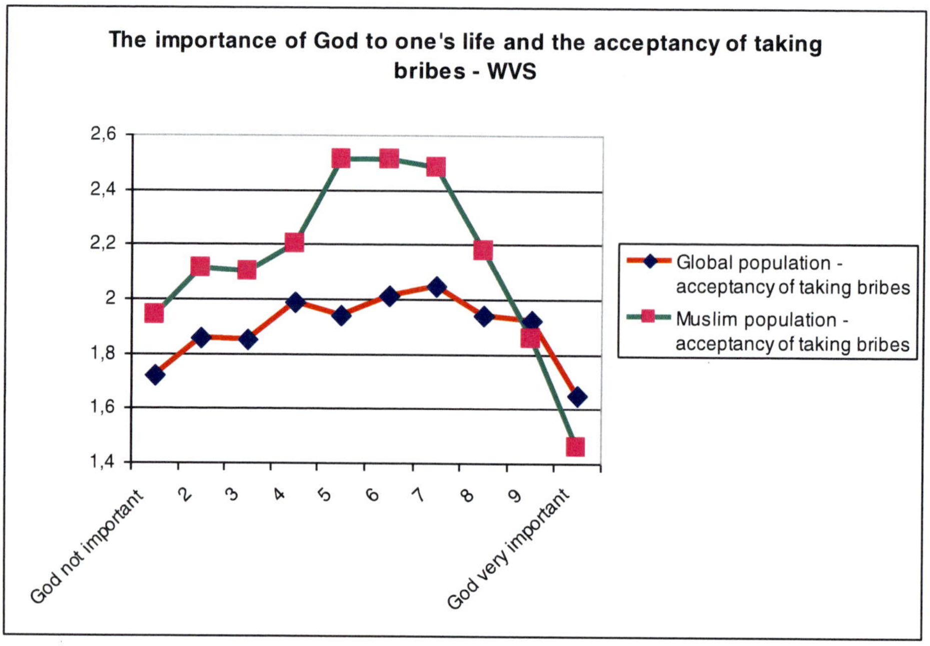

Graph 11.1. The importance of God to one's life and the acceptancy of taking a bribe – global results and global Muslim results.

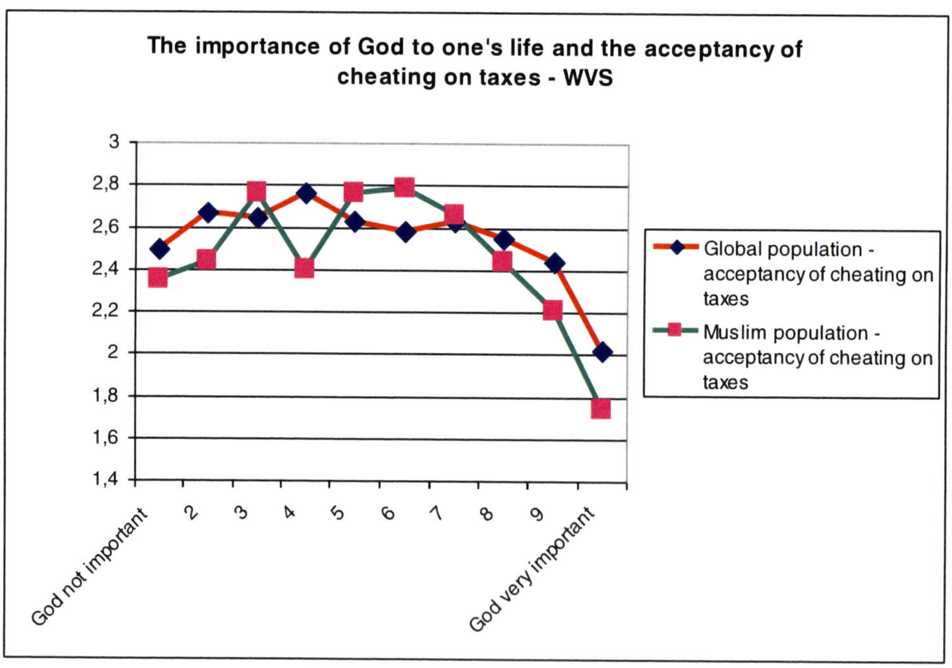

Graph 11.2. The importance of God to one's life and the acceptancy of cheating on taxes – global results and global Muslim results.

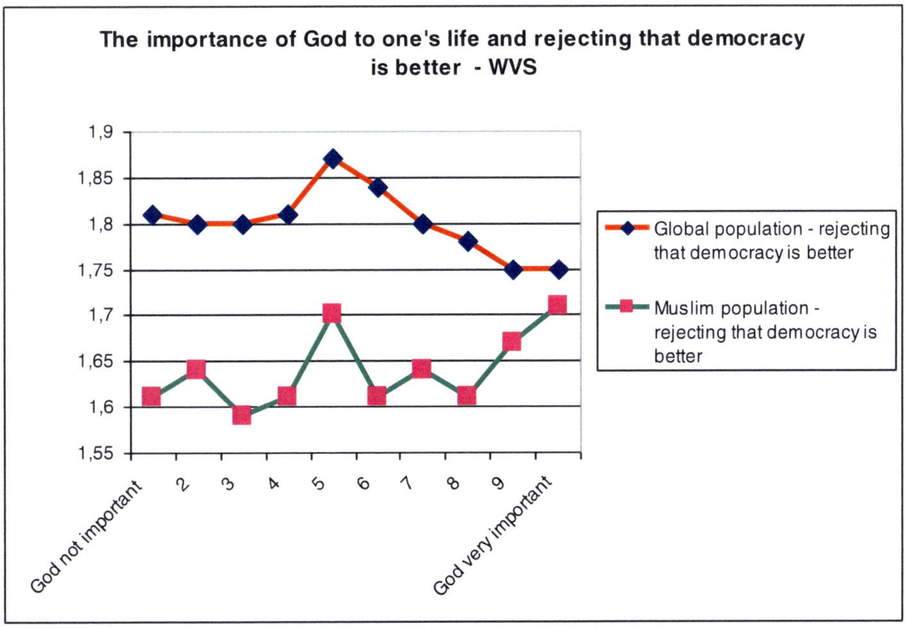

Graph 11.3. The importance of God to one's life and the rejection of the idea that democracy is better – global results and global Muslim results.

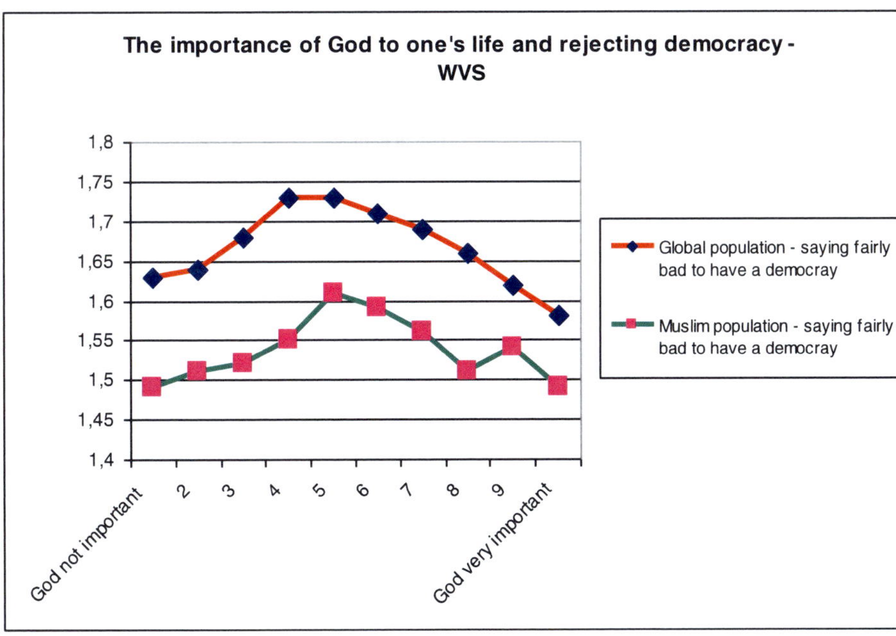

Graph 11.4. The importance of God to one's life and the opinion that it's fairly bad to have a democracy – global results and global Muslim results.

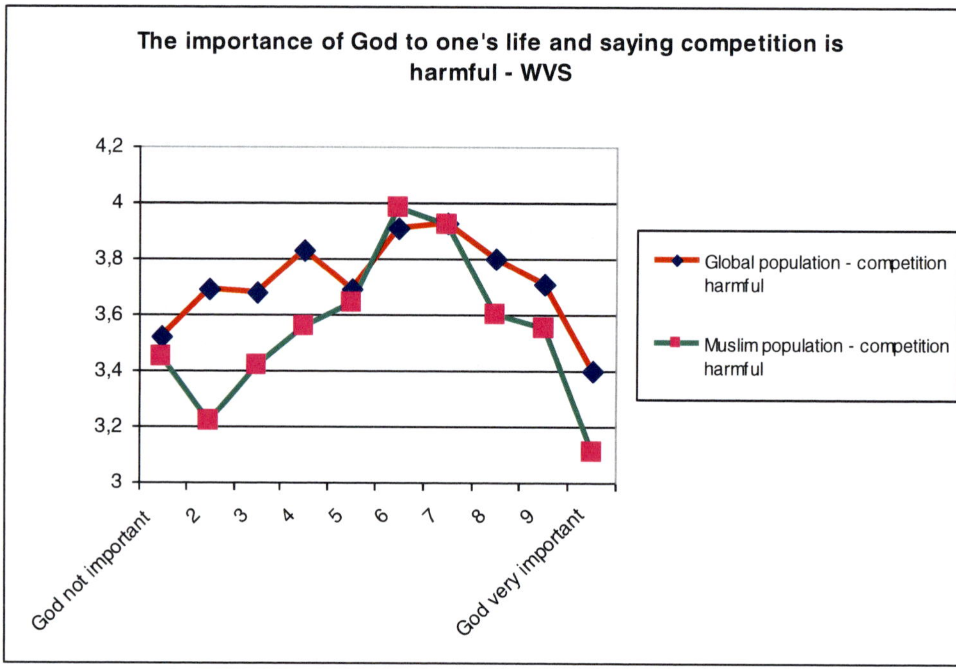

Graph 11.5. The importance of God to one's life and the opinion that competition is harmful – global results and global Muslim results.

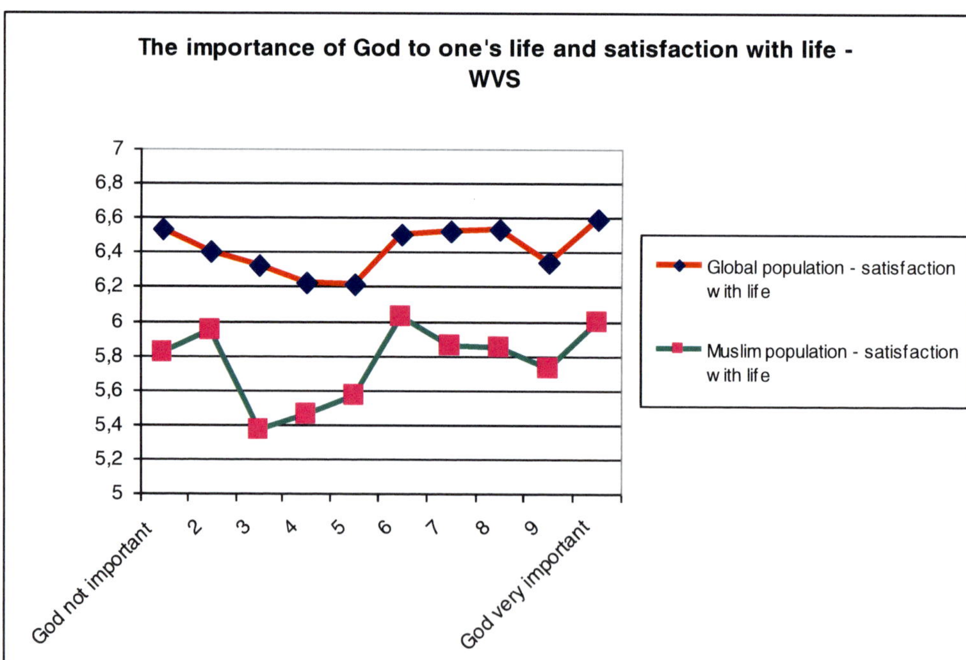

Graph 11.6. The importance of God to one's life and life satisfaction – global results and global Muslim results.

Table 11.1 contains still more data on what faith – expressed by the assigned importance to God in one's life – does to society. We distinguish here between the effects on the level of the global sample and on the level of global Orthodoxy, Protestantism, Hinduism, Muslims, and Catholicism. The size of the statistical linear effects is rather small and rather seems to suggest that, again, U-shaped patterns seem to be at work. In the following, we show the effects of the importance an individual assigns to God for the individual social and ethical convictions as documented in the *World Values Survey* for the different major global denominations with sufficient numbers of respondents to be included in Table 11.1. For most major denominations and for the global publics as a whole (believers and non-believers alike, people with and without denominations), the effects of the importance assigned to God in one's life seem to concern mainly attitudes about religious affairs in general. While on a global level, the acceptancy of non-believing political candidates is determined to around 18% by the importance assigned to God in one's life; this percentage is around 16% for Orthodox publics, 22.1% for Protestant publics, only 1.3% for Hindu publics, only 5% for Muslim publics, and only 8.5% for Roman Catholic global publics. It's important to remember that on a global level, 82% of the variance of the variable: *Politicians who don´t believe in God are fit for public office*[59] is NOT determined by the importance an individual assigns to God in one's life. People "distant" or "near" to God in their personal life will have a preference for either secular or religious politicians. Orthodox and Protestant publics predominantly live in societies with a higher degree of secularism, while Hindu, Muslim and Roman Catholic publics also are heavily present in countries with a lower degree of secularism. Differences in attitudes are hardly determined in such societies by *individual religious attitudes*; rather, *secular and non-secular individuals* in these countries hold their attitudes regarding the election of public officials independent from the strength of their own personal religious convictions.

Our materials confirm the hypothesis that individual faith has its most direct consequences for individual morality and family values. In quantitative terms, the strongest effects for all denominations are on the acceptancy or rejection of abortion, while even for the acceptancy of homosexuality and prostitution the percentages of explained variances are already rather low.

Our empirical materials are indeed full of other surprises as well. The much-debated issue of *"Islamophobia",* measured here simply by the rejection of Muslim neighbors, is not determined at all by the extent of religiosity, as being measured by the importance, an individual in Orthodox, Protestant, Hindu, Muslim, Catholic or global cultures assigns to God in one's life. The same happens with the rejection of neighbors who are homosexuals, immigrants or foreign workers, neighbors of a different race, or people who suffer from aids. The only exception, interestingly enough, is Protestantism, where believers seem to be rather more conservative concerning such attitudes than the rest of society, and where we are confronted with a significant effect of religiosity on homophobia (rejection of neighbors because they are homosexuals), or AIDSophobia (rejection of neighbors because they suffer from AIDS). These correlations also have an effect on the global relationships between the variables. Religious Protestants also tend to be more sexist in their attitude concerning University studies of female children, and they also favor income inequality. Religious Protestants, on the other hand, still correspond to the old stereotype of Max Weber's sociology because they form the only religious denomination where there is a significant correlation

[59] The original wording of the question in the WVS was reverse. For reasons of comparability, we reversed the variable.

between the importance of hard work and the importance of obedience on the one hand and the importance of God to one's personal life on the other hand. These correlations also have an effect on the global relationships between the variables. Interestingly enough, there is also a significant negative correlation between the importance, assigned to God by an individual, and the value of independence in child education in the civilization of Protestantism and – somewhat weaker, and still significant – also in Roman Catholicism.

The two leading "Hayek" variables in the *World Values Survey,* competition good or harmful, and private or government ownership of the means of production are not related at all with the importance of God to one's personal life, and the correlations are below +-.100. Religious Protestants tend to distrust the police more than secular Protestants, but else, the variables of trust in institutions and social capital exhibit little surprising results. Religious Roman Catholics tend to have a higher trust in the Armed Forces, and religious Roman Catholics and Muslims tend to have a higher trust in NATO, reflecting especially the situation in the Southern Mediterranean countries from Portugal to Turkey. The correlation between the importance assigned to God in one's personal life and the political self-classification of an individual on a left-wing to right-wing scale, ranging from 1, left, to 10, right, is below +.100 for Protestantism and Islam, and is below +-.200 for Orthodoxy, Hinduism and Catholicism, and is also below +-.200 on a global level. Since explained percentages of variance are nothing but squared correlation coefficients, we can say that explained variances are not bigger than 2.8%.

Policy pointer:

It's important to remember that on a global level, 82% of the variance of the variable: *Politicians who don´t believe in God are fit for public office* is NOT determined by the importance an individual assigns to God in one's life. People "distant" or "near" to God in their personal life will have a preference for either secular or religious politicians. Our materials confirm the hypothesis that individual faith has its most direct consequences for individual morality and family values. In quantitative terms, the strongest effects for all denominations are on the acceptancy or rejection of abortion, while even for the acceptancy of homosexuality and prostitution; the percentages of explained variances are already rather low.

The two leading "Hayek" variables in the *World Values Survey,* competition good or harmful, and private or government ownership of the means of production are not related at all with the importance of God to one's personal life, and the correlations are below +-.100.

Global publics: 1% or more of the following variables are explained by the importance given to God

	correlation	R^2
	Globe	Globe
No confidence: Churches	-0,497	24,701
Important child qualities: religious faith	0,438	19,184
Politicians who don't believe in God are fit for public office	-0,425	18,063
Justifiable: abortion	-0,381	14,516

	correlation	R^2
	Globe	Globe
Justifiable: euthanasia	-0,331	10,956
Justifiable: divorce	-0,259	6,708
Justifiable: suicide	-0,251	6,300
Justifiable: homosexuality	-0,246	6,052
Justifiable: prostitution	-0,211	4,452
Important child qualities: obedience	0,174	3,028
Important child qualities: independence	-0,132	1,742
[Right wing] self-positioning in political scale (1 - left ... 10 right)	0,125	1,563
Important child qualities: determination and perseverance	-0,113	1,277
Jobs scarce: Men should have more right to a job than women (reject)	-0,110	1,210
Rejecting neighbors: Homosexuals	0,107	1,145

Orthodox global publics: 1% or more of the following variables are explained by the importance given to God

	correlation	R^2
	Orthodoxy	Orthodoxy
No confidence: Churches	-0,407	16,565
Politicians who don´t believe in God are fit for public office	-0,399	15,920
Important child qualities: religious faith	0,334	11,156
Justifiable: abortion	-0,228	5,198
Justifiable: euthanasia	-0,172	2,958
Justifiable: divorce	-0,157	2,465
[Right wing] self-positioning in political scale (1 - left ... 10 right)	0,135	1,823
Justifiable: suicide	-0,125	1,563
Important child qualities: determination and perseverance	-0,113	1,277

Protestant global publics: 1% or more of the following variables are explained by the importance given to God

	correlation	R^2
	Protestantism	Protestantism
Justifiable: abortion	-0,492	24,206
Important child qualities: religious faith	0,481	23,136
No confidence: Churches	-0,478	22,848
Politicians who don´t believe in God are fit for public office	-0,470	22,090

Table. (Continued)

Justifiable: homosexuality	-0,416	17,306
Justifiable: divorce	-0,404	16,322
Justifiable: euthanasia	-0,394	15,524
Justifiable: prostitution	-0,311	9,672
Justifiable: suicide	-0,290	8,410
Rejecting neighbors: Homosexuals	0,271	7,344
Important child qualities: hard work	0,226	5,108
Important child qualities: obedience	0,222	4,928
Important child qualities: independence	-0,206	4,244
Important child qualities: imagination	-0,177	3,133
Important child qualities: feeling of responsibility	-0,163	2,657
Rejecting neighbors: People who have AIDS	0,145	2,103
Justifiable: cheating on taxes	-0,122	1,488
No confidence: The Police	0,117	1,369
Rejecting sexist position: University is more important for a boy than for a girl	-0,115	1,323
Income equality	0,101	1,020

Hindu global publics: 1% or more of the following variables are explained by the importance given to God

	correlation	R^2
	Hinduism	Hinduism
No confidence: Churches	-0,219	4,796
Important child qualities: religious faith	0,186	3,460
[Right wing] self-positioning in political scale (1 - left ... 10 right)	0,163	2,657
Justifiable: abortion	-0,117	1,369
Politicians who don´t believe in God are fit for public office	-0,114	1,300
Justifiable: euthanasia	-0,103	1,061

Muslim global publics: 1% or more of the following variables are explained by the importance given to God

	correlation	R^2
	Muslim denomination	Muslim denomination
No confidence: Churches	-0,242	5,856
Important child qualities: religious faith	0,240	5,760
Justifiable: abortion	-0,238	5,664
Politicians who don´t believe in God are fit for public office	-0,224	5,018
Justifiable: euthanasia	-0,189	3,572
Justifiable: prostitution	-0,171	2,924
Justifiable: homosexuality	-0,168	2,822

	correlation	R^2
	Muslim denomination	Muslim denomination
Justifiable: suicide	-0,156	2,434
Justifiable: someone accepting a bribe	-0,145	2,103
No confidence: NATO	-0,132	1,742
Justifiable: cheating on taxes	-0,128	1,638
Justifiable: divorce	-0,116	1,346
Jobs scarce: Men should have more right to a job than women (reject)	-0,104	1,082
Justifiable: claiming government benefits even if one is not entitled to them	-0,102	1,040

Roman Catholic global publics: 1% or more of the following variables are explained by the importance given to God

	correlation	R^2
	Catholicism	Catholicism
No confidence: Churches	-0,436	19,010
Justifiable: abortion	-0,353	12,461
Important child qualities: religious faith	0,323	10,433
Politicians who don´t believe in God are fit for public office	-0,291	8,468
Justifiable: euthanasia	-0,236	5,570
Justifiable: suicide	-0,226	5,108
Justifiable: divorce	-0,201	4,040
Justifiable: prostitution	-0,191	3,648
Justifiable: homosexuality	-0,187	3,497
[Right wing] self-positioning in political scale (1 - left ... 10 right)	0,160	2,560
Important child qualities: independence	-0,132	1,742
Important child qualities: obedience	0,118	1,392
No confidence: Armed Forces	-0,106	1,124
No confidence: NATO	-0,105	1,103

Table 11.1. What faith does to society: the Pearson Bravais correlations of the importance given to God and human attitudes and behavior according to the *World Values Survey*, broken down by different denominations

	Orthodoxy	**Protes-tantism**	**Hinduism**	**Muslim denomi-nation**	**Catho-licism**	**Globe**
[Right wing] self-positioning in political scale (1 - left ... 10 right)	**0,135**	0,065	**0,163**	0,098	**0,160**	**0,125**
Competition good or harmful	-0,003	-0,029	-0,046	-0,073	-0,038	-0,045
Government ownership vs. private ownership of business (scale: 1=1 Private ownership of business should be increased; 10=Government ownership of business should be increased)	0,021	0,059	-0,044	0,086	0,075	0,041

Table 11.1. (Continued)

	Orthodoxy	Protes-tantism	Hinduism	Muslim denomi-nation	Catho-licism	Globe
Important child qualities: determination and perseverance	**-0,113**	-0,073	-0,020	-0,058	-0,093	**-0,113**
Important child qualities: feeling of responsibility	-0,018	**-0,163**	-0,028	0,007	-0,011	-0,064
Important child qualities: hard work	-0,013	**0,226**	-0,027	-0,037	-0,040	0,012
Important child qualities: imagination	0,023	**-0,177**	-0,061	-0,057	-0,066	-0,084
Important child qualities: independence	-0,061	**-0,206**	-0,013	-0,050	**-0,132**	**-0,132**
Important child qualities: obedience	0,031	**0,222**	0,026	0,058	**0,118**	**0,174**
Important child qualities: religious faith	**0,334**	**0,481**	**0,186**	**0,240**	**0,323**	**0,438**
Important child qualities: thrift, and saving money and things	0,006	-0,042	0,019	-0,043	-0,025	-0,054
Important child qualities: tolerance and respect for other people	0,028	-0,092	0,055	0,015	0,020	-0,004
Important child qualities: unselfishness	-0,025	0,022	-0,073	0,010	0,037	0,012
Income equality	0,027	**0,101**	0,064	0,073	0,067	0,065
Jobs scarce: Men should have more right to a job than women (reject)	-0,024	-0,085	0,004	**-0,104**	-0,028	**-0,110**
Justifiable: abortion	**-0,228**	**-0,492**	**-0,117**	**-0,238**	**-0,353**	**-0,381**
Justifiable: avoiding a fare on public transport	-0,070	-0,038	-0,064	-0,093	-0,015	-0,054
Justifiable: cheating on taxes	-0,075	**-0,122**	-0,049	**-0,128**	-0,089	-0,093
Justifiable: claiming government benefits even if one is not entitled to them	-0,039	0,015	-0,049	**-0,102**	-0,009	-0,018
Justifiable: divorce	**-0,157**	**-0,404**	-0,082	**-0,116**	**-0,201**	**-0,259**
Justifiable: euthanasia	**-0,172**	**-0,394**	**-0,103**	**-0,189**	**-0,236**	**-0,331**
Justifiable: homosexuality	-0,081	**-0,416**	-0,018	**-0,168**	**-0,187**	**-0,246**
Justifiable: prostitution	-0,098	**-0,311**	-0,084	**-0,171**	**-0,191**	**-0,211**
Justifiable: someone accepting a bribe	-0,034	-0,009	-0,052	**-0,145**	-0,041	-0,039
Justifiable: suicide	**-0,125**	**-0,290**	-0,068	**-0,156**	**-0,226**	**-0,251**
No confidence: Armed Forces	-0,039	-0,021	-0,012	-0,077	**-0,106**	-0,088
No confidence: Churches	**-0,407**	**-0,478**	**-0,219**	**-0,242**	**-0,436**	**-0,497**
No confidence: Labor Unions	-0,035	0,030	0,007	-0,037	-0,003	-0,001
No confidence: Major Companies	-0,026	-0,077	-0,041	-0,058	-0,087	-0,088
No confidence: NATO	-0,075	-0,037	-0,054	**-0,132**	**-0,105**	-0,068
No confidence: Parliament	-0,013	-0,005	-0,043	-0,081	0,032	-0,030
No confidence: Television	-0,025	-0,045	-0,058	-0,068	-0,048	-0,057
No confidence: The Civil Services	-0,024	-0,041	-0,057	-0,081	0,006	-0,042
No confidence: The Environmental Protection Movement	-0,011	-0,043	-0,011	-0,068	-0,094	-0,041
No confidence: The Government	0,001	-0,064	-0,050	-0,066	-0,069	-0,080

	Orthodoxy	**Protes-tantism**	**Hinduism**	**Muslim denomi-nation**	**Catho-licism**	**Globe**
No confidence: The Police	-0,024	**0,117**	-0,003	-0,062	-0,006	-0,007
No confidence: The Political Parties	-0,017	-0,036	-0,016	-0,027	0,012	-0,005
No confidence: The Press	-0,032	-0,090	0,012	-0,074	-0,038	-0,061
No confidence: The Women´s Movement	-0,051	-0,093	-0,027	-0,046	-0,078	-0,039
Politicians who don´t believe in God are fit for public office	**-0,399**	**-0,470**	**-0,114**	**-0,224**	**-0,291**	**-0,425**
Rejecting neighbors: Homosexuals	0,044	**0,271**	-0,038	0,021	0,075	**0,107**
Rejecting neighbors: Immigrants/foreign workers	0,028	0,090	0,002	0,066	-0,016	0,050
Rejecting neighbors: Muslims	0,053	0,013	0,019	-0,086	-0,009	-0,015
Rejecting neighbors: People of a different race	0,039	0,083	0,008	0,062	-0,012	0,049
Rejecting neighbors: People who have AIDS	-0,019	**0,145**	-0,066	0,019	0,018	0,050
Rejecting sexist position: University is more important for a boy than for a girl	0,038	**-0,115**	0,078	-0,022	-0,021	-0,057

Chapter 12

HOFSTEDE, SCHWARTZ, INGLEHART AND THE FUTURE OF WORLD DEVELOPMENT

In many European countries, bitter debates on human values take place, be it on matters of sexuality, family policy, religion, or – increasingly – on the possible accession of Turkey to the European Union. Similar debates are to be witnessed in many other Western countries outside the European Union. For us as empirical social scientists, the main questions arising out of such debates are for example: what are the empirical relationships between acceptancy of prostitution and acceptancy of homosexuality on a global scale? Is there a positive, a non-existent, or a negative correlation between the two?

Box 12.1. The overall message of this Chapter

This Chapter is intended to provide new and additional insights into our rethinking of the *World Values Survey* data and analysis. For one, we would like to present a still more conclusive proof of the interrelationship between the different types of permissiveness and the weight these factors have in relationship to the other variables contained in the *World Values Survey* data. Based on our analysis of the complete available data based on 28 items from the *World Values Survey* from 70 countries of the world, including the OIC member countries Albania; Azerbaijan; Bangladesh; Bosnia and Herzegovina; Burkina Faso; Indonesia; Jordan; Kyrgyzstan; Mali; Nigeria; Turkey; and Uganda we attempt to show the interrelationships between permissiveness, the shadow economy, educational values, and other socio-political variables, like fundamental positions on the market economy and democracy.

In the second part of this Chapter we will then look at the summary relationships between our own new factors and those presented by Hofstede, Schwartz and Inglehart. We will come to the conclusion that our factors sufficiently correlate with the factors presented by Hofstede, Schwartz and Inglehart, without implying what we perceive to be their shortcomings. In Table 12.6 and 12.7 we provide our readers with clear-cut Pearson-Bravais correlation coefficients between the data presented by Hofstede, Schwartz and Inglehart and the factor scores from our own analytical dimensions.

Current debates in Europe on issues of human values have acquired an almost quasi-religious character, with each side claiming that it is in the sole possession of truth, while in reality the complex relationships between value patterns are often overlooked. Tables 12.7 and 12.8 pick out the two issues which were at the heart of the recent debates in France – the acceptancy of homosexuality, and the acceptancy of prostitution. Looking at the partial correlations between the acceptancy of homosexuality and 20 other key-indicators of trust and behavior, we find that there are significant and higher correlations between the unconditional acceptancy of homosexuality with accepting a bribe, avoiding a fare on public transport, cheating on taxes, and the generalized loss of trust in basic institutions of democratic society.

Hard core feminism indicators and economic freedom indicators are far more influential than value change indicators in bringing about effective democracy even at low levels of development. Inglehart's "self-expression" scale achieves about the same results as Hofstede's "power distance" and our own "authoritarian personality" scale. The absence of militarism and of an oil exporting economy has also a beneficial effect on "effective democracy". So, while feminism and a free economy generally affect effective democracy in a positive fashion, we have to emphasize as well that Maastricht-type of savings programs and also very high inward migration rates are not a driver, but a bottleneck of effective democracy.

As the manuscript to this book was about to be finished, the new data of the *World Values Survey*, 2010-2014 were released, containing an enormous wealth of new data, including on the Muslim world. We have chosen to concentrate on two phenomena, which received a large attention on the pages of this book – tolerance and democracy.

In Table 12.12 we calculate a simple new UNDP Human Development Index type of Index of Tolerance, minimizing the rejection of neighbors along the following criteria:

- People who speak a different language
- People of a different religion
- Immigrants/foreign workers
- People of a different race

According to these data the most tolerant nation on earth is Uruguay, followed by Sweden; New Zealand; Spain; Trinidad and Tobago; Poland; Rwanda; Colombia; Chile and Australia.

The good news for the growing forces of moderation and democracy in the Arab world and the Muslim world in general is that Uzbekistan, Morocco and Kazakhstan are nowadays ahead of Germany; and Pakistan, Qatar and Tunisia are more tolerant than the EU-member country Romania. The bad news is the poor performance of such Muslim countries as Turkey, and the very poor performance of the countries on the bottom of Table 12.12.

The following Chapter is intended to provide new and additional insights into our rethinking of the *World Values Survey* data and analysis. For one, we would like to present a still more conclusive proof of the interrelationship between the different types of permissiveness and the weight these factors have in relationship to the other variables, contained in the *World Values Survey* data. Based on our analysis of the complete available data based on 28 items from the *World Values Survey* from 70 countries of the world, including the OIC member countries Albania; Azerbaijan; Bangladesh; Bosnia and Herzegovina; Burkina Faso; Indonesia; Jordan; Kyrgyzstan; Mali; Nigeria; Turkey; and Uganda we attempt to show the interrelationships between permissiveness, the shadow economy, educational values, and other socio-political variables, like fundamental positions on the market economy and democracy. Much of what we will be presenting will be a counterpoint to the hitherto dominant interpretations offered by Professor Ronald Inglehart, who seems to suggest that for a society to have a an effective democracy, it needs to have very pronounced self-expression values, including the acceptancy of homosexuality.

In the second part of this Chapter, we will then look at the summary relationships between our own new factors and those presented by Hofstede and Inglehart. We will come to the conclusion that our factors sufficiently correlate with the factors presented by Hofstede and Inglehart, without implying what we perceive to be their shortcomings.

12.1. One, Two, Three Types of Permissiveness and Their Effects on Value Systems

We have maintained throughout this work that Inglehart overlooks the negative side-effects of secular and self-expression values. We have shown in the previous chapters that this especially relevant for the issue of the shadow economy.

To re-iterate our point, we should not underestimate the role which self-expression values, associated by Inglehart and associates with participation in decision making, political activism, environmental protection, gender equality, and growing tolerance of ethnic minorities, foreigners, gays and lesbians play for their theory of democracy, which rests, according to them on a culture of trust and tolerance in which people place high value on individual freedom and self-expression and have activist political orientations:

> "The shift from survival to self-expression values is linked with the rise of postindustrial society. It reflects a cultural shift that occurs when younger generations emerge who have grown up taking survival for granted. Survival values give top priority to economic and physical security and conformist social norms. *Self-expression values give high priority to freedom of expression, participation in decision making, political activism, environmental protection, gender equality, and growing tolerance of ethnic minorities, foreigners, gays and lesbians. Growing emphasis on these values produces a culture of trust and tolerance in which people place high value on individual freedom and self-expression and have activist political orientations. These attributes are crucial to democracy – and thus explain how economic growth, which takes societies from agrarian to industrial and then from industrial to post-industrial, leads to democratization.* The unprecedented economic growth of the past 50 years has meant that an increasing share of the world's population has grown up taking survival for granted. Time series data from the values surveys indicates that mass priorities have shifted from an overwhelming emphasis on economic and physical security toward an emphasis on subjective well-being, selfexpression, participation in decision-making and a relatively trusting and tolerant outlook. *Both dimensions are closely linked with economic development: the value systems of high-income countries differ dramatically from those of low-income countries.* Every nation that the World Bank defines as high income ranks relatively high on both dimensions -- with a strong emphasis on both self-expression and secular-rational values. All of the low-income and lowermiddle income societies rank relatively low on both dimensions. The

> upper-middle income countries fall somewhere in between. To a remarkable degree, the values and beliefs of a given society reflect its level of economic development—just as modernization theory implies." (Inglehart and Welzel, 2009)

In the following, we will now extend the argument by using Schumpeter's hypothesis (Schumpeter, 1950) about the generalized loss of family values as the possible trigger of the overall decline of the capitalist order.

To arrive at an empirical and theoretical synthesis of our analysis, we will attempt to show the relationship between the decay of traditional family values, the loss of religious values and the rise of the shadow economic mentality. To this end, we will perform a promax factor analysis for a very large number of countries with complete data on

- Acceptance or rejection of income inequality
- Democratic values
- Economic and social values, rejection of the shadow economy, government benefits fraud, bribery and corruption
- Family values and sexual norms
- Opinions on markets and state intervention
- Religious values
- Values guiding the education of children – independence; hard work; feeling of responsibility; imagination; tolerance and respect for other people; thrift, saving money and things; determination and perseverance; religious faith; unselfishness; obedience

Are the convictions of religious human beings incompatible with modernization and *"effective democracy"*, and are religiously motivated persons, sharing traditional religious values, the antithesis to the values of *"selfexpression"*, which implies a low importance given to religion in society, a low importance of obedience in child education, a weak sense of national pride, a disconnection of one's life from the pride of one's parents, the full acceptancy of divorce and abortion, no desire for limits on selling foreign goods, and little respect for authority in society? As Inglehart and associates seem to suggest? Or are Barro and Schumpeter right in their conviction that the belief in eternal judgment brings about an improvement of the societal ethics of contractual obligations (Barro) or that traditional family values are important for the existence of the capitalist enterprise and economic growth (Schumpeter) which threatens to be substituted by a socialist bureaucracy?

The countries with complete data for these questions are a fair sample of the available data from the *World Values Survey* project

1. Albania
2. Andorra
3. Argentina
4. Armenia
5. Australia
6. Azerbaijan
7. Bangladesh
8. Belarus

9. Bosnia and Herzegovina
10. Brazil
11. Bulgaria
12. Burkina Faso
13. Canada
14. Chile
15. Croatia
16. Cyprus
17. Czech Republic
18. Dominican Republic
19. Estonia
20. Ethiopia
21. Finland
22. Georgia
23. Germany
24. Ghana
25. Guatemala
26. Hong Kong
27. Hungary
28. India
29. Indonesia
30. Italy
31. Japan
32. Jordan
33. Kyrgyzstan
34. Latvia
35. Lithuania
36. Macedonia
37. Mali
38. Mexico
39. Moldova
40. New Zealand
41. Nigeria
42. Norway
43. Peru
44. Philippines
45. Poland
46. Puerto Rico
47. Romania
48. Russian Federation
49. Serbia
50. Serbia and Montenegro
51. Slovakia
52. Slovenia
53. South Africa
54. South Korea

55. Spain
56. Sweden
57. Switzerland
58. Taiwan
59. Tanzania
60. Thailand
61. Trinidad and Tobago
62. Turkey
63. Uganda
64. Ukraine
65. United States
66. Uruguay
67. Venezuela
68. Viet Nam
69. Zambia
70. Zimbabwe

The questions chosen from the WVS are the following:

1. How important is God in your life
2. How often do you attend religious services
3. Justifiable: claiming government benefits even if one is not entitled to them
4. Justifiable: avoiding a fare on public transport
5. Justifiable: cheating on taxes
6. Justifiable: someone accepting a bribe
7. Justifiable: homosexuality
8. Justifiable: prostitution
9. Justifiable: abortion
10. Justifiable: divorce
11. Justifiable: euthanasia
12. Justifiable: suicide
13. Important child qualities: independence
14. Important child qualities: hard work
15. Important child qualities: feeling of responsibility
16. Important child qualities: imagination
17. Important child qualities: tolerance and respect for other people
18. Important child qualities: thrift, and saving money and things
19. Important child qualities: determination and perseverance
20. Important child qualities: religious faith
21. Important child qualities: unselfishness
22. Important child qualities: obedience
23. Self-positioning in political scale (left to right)
24. Income equality (rejection or acceptance)
25. Private vs. state ownership of business
26. Competition good or harmful
27. Political system: Having the army rule (reject)

28. Political system: Having a democratic political system (reject)

Table 12.1 shows the percentages of the total variance explained by our investigation for each variable of our model.

Table 12.1. Promax factor analysis, based on the latest edition of the combined *World Values Surveys* on religion, on what is justifiable, and on democracy (quantitative analysis of *"Permissiveness"*)

	initial	extraction
How important is God in your life	1,000	0,642
How often do you attend religious services	1,000	0,617
Justifiable: claiming government benefits even if one is not entitled to them	1,000	0,532
Justifiable: avoiding a fare on public transport	1,000	0,615
Justifiable: cheating on taxes	1,000	0,642
Justifiable: someone accepting a bribe	1,000	0,550
Justifiable: homosexuality	1,000	0,641
Justifiable: prostitution	1,000	0,610
Justifiable: abortion	1,000	0,655
Justifiable: divorce	1,000	0,601
Justifiable: euthanasia	1,000	0,527
Justifiable: suicide	1,000	0,491
Important child qualities: independence	1,000	0,501
Important child qualities: hard work	1,000	0,639
Important child qualities: feeling of responsibility	1,000	0,560
Important child qualities: imagination	1,000	0,468
Important child qualities: tolerance and respect for other people	1,000	0,491
Important child qualities: thrift, and saving money and things	1,000	0,516
Important child qualities: determination and perseverance	1,000	0,430
Important child qualities: religious faith	1,000	0,561
Important child qualities: unselfishness	1,000	0,573
Important child qualities: obedience	1,000	0,433
Self-positioning in political scale	1,000	0,578
Income equality	1,000	0,601
Private vs. state ownership of business	1,000	0,632
Competition good or harmful	1,000	0,546
Political system: Having the army rule	1,000	0,381
Political system: Having a democratic political system	1,000	0,293

The nine factors to be extracted from the data for more than 90.000 representative respondents in 70 countries are the following:

- moral (sexual) permissiveness ('Permissiveness 1')
- acceptancy of the shadow economy('Permissiveness 2')
- distance from religion ('Permissiveness 3')
- educational values: independence and imagination
- distance to market economy values
- education values: responsibility and tolerance
- educational values: determination and perseverance and being against saving
- right wing acceptance of inequality
- educational values: favoring unselfishness, rejecting hard work

Contrary to Inglehart's expectations about a positive role of the low importance given to religion in society, and divorce and abortion being fully accepted, it emerges that the two factors of permissiveness (permissive family values and the loss of hard-core Max Weberian economic values) are closely interrelated with one another and with the loss of religious values, and that the three Barro-type and Schumpeterian *"permissiveness factors"* together already explain some 30,4% of the total variance of individual opinions under investigation on a global scale.

Table 12.2. The *Eigen values* for our analysis of latest edition of the combined *World Values Surveys* on religion, on what is justifiable, and on democracy (quantitative analysis of *"Permissiveness")*

		initial *Eigen values*	
	Total	% of variance	cumulated percentage
1,000	4,506	16,094	16,094
2,000	2,525	9,017	25,110
3,000	1,476	5,273	30,383
4,000	1,301	4,646	35,030
5,000	1,224	4,372	39,402
6,000	1,120	3,999	43,401
7,000	1,107	3,953	47,354
8,000	1,065	3,802	51,156
9,000	1,002	3,578	54,734

Table 12.3 shows the factor loadings for each of the variables analyzed here. The variables with a high importance for "effective democracy", i.e., tolerance and respect for other people, rejection or acceptance of corruption, and the assessment of democracy as such and vis-à-vis military rule, are highlighted in our Table 12.3. *Nowhere there is a notable negative or positive factor analytical loading of beyond 0. 333 (>10% of variance explained) confirming that religious people are antidemocratic, right-wing, and pro or anti-market.*

In addition, the structure of the factor loadings even suggests the following:

a) distance from religion is even a motive to reject a democratic political system
b) moral/sexual permissiveness goes hand in hand with economic and social decay

	moral (sexual) permissiveness ('Permissiveness 1')	**distance from religion ('Permissiveness 3')**
Political system: Having a democratic political system (reject)	-0,099	**0,125**
Competition good or harmful (harmful)	0,038	0,004
Private vs. state ownership of business (state)	-0,064	-0,027
Justifiable: avoiding a fare on public transport	**0,213**	0,056
Justifiable: cheating on taxes	**0,239**	0,088
Justifiable: claiming government benefits even if one is not entitled to them	**0,104**	0,001
Justifiable: someone accepting a bribe	**0,218**	-0,043

Table 12.4 shows the all-important correlations between the promax factors, extracted from the correlation matrix between the variables of our model. Table 12.5 and Maps 12.1 to 12.9 show the country values for our analysis ("factor scores") as well as the cascades of moral and social decay in the Western countries and also the evidence for the Muslim countries with available data. Graph 12.5 finally summarizes the pessimistic research findings, which rather support the views of Barro and Schumpeter against the logic, proposed by Inglehart.

Policy pointer:

Contrary to Inglehart's expectations about a positive role of the low importance given to religion in society, and divorce and abortion being fully accepted, it emerges that the two factors of permissiveness (permissive family values and the loss of hard-core Max Weberian economic values) are closely interrelated with one another and with the loss of religious values.

Nowhere there is a notable negative or positive factor analytical loading of beyond 0. 333 (>10% of variance explained) confirming that religious people are antidemocratic, right-wing, and pro or anti-market.

In addition, the structure of the factor loadings even suggests the following:

a) distance from religion is even a motive to reject a democratic political system
b) moral/sexual permissiveness goes hand in hand with economic and social decay

For the analysis of the choropleth maps of global values, it's again necessary, as in previous Chapters, to analyze in more detail the factor definitions (loadings which are higher than or are equal to 0.5) as well as the countries with the highest and lowest factor scores each.

Table 12.3. Multivariate model, based on 91649 global representative respondents with complete data from 70 countries and territories, based on the *World Values Survey* from the combined latest edition [2012] of global opinion investigation, integrating the fifth wave of investigations, 2005-2008. Is (*'Permissiveness'*) really a precondition of democracy? – Some continued doubts about implications of the Inglehart School

If highest numerical values differ from the variable name, the highest values of the indicator for the multivariate analysis are mentioned in brackets	moral (sexual) permissiveness ('Permissiveness 1')	acceptancy of the shadow economy('Permissiveness 2')	distance from religion ('Permiss-iveness 3')	educational values: independence and imagination	distance to market economy values	education values: responsibility and tolerance	educational values: determination and perseverance and being against saving	right wing acceptance of inequality	educational values: favoring unselfishness, rejecting hard work
Competition good or harmful (harmful)	0,038	0,134	0,004	0,080	0,719	0,006	-0,138	-0,100	0,138
How important is God in your life	-0,380	-0,019	-0,789	-0,149	0,071	-0,122	-0,071	0,130	0,036
How often do you attend religious services (never)	0,352	-0,003	0,770	0,075	-0,037	0,159	0,063	-0,115	-0,011
Important child qualities: determination and perseverance	0,028	0,039	0,304	0,160	-0,065	0,041	0,523	0,074	-0,058
Important child qualities: feeling of responsibility	0,095	-0,032	0,157	0,138	-0,038	0,730	-0,006	-0,007	-0,013
Important child qualities: hard work	-0,145	0,021	0,010	-0,259	-0,013	-0,346	-0,011	0,117	-0,675
Important child qualities: imagination	0,165	-0,025	0,062	0,610	0,015	-0,076	0,071	-0,152	0,085
Important child qualities: independence	0,105	0,010	0,164	0,695	-0,067	0,176	0,061	0,058	-0,044
Important child qualities: obedience	-0,150	0,014	-0,286	-0,437	0,122	-0,459	-0,177	0,007	0,251
Important child qualities: religious faith	-0,255	-0,031	-0,743	-0,196	0,032	-0,171	-0,105	0,064	0,052
Important child qualities: thrift, and saving money and things	-0,116	0,019	0,083	-0,004	-0,009	-0,080	-0,650	0,066	-0,146
Important child qualities: tolerance and respect for other people	0,134	-0,118	-0,045	-0,298	-0,004	0,532	0,274	-0,125	0,058
Important child qualities: unselfishness	0,019	-0,013	0,031	-0,180	-0,011	-0,218	0,094	0,003	0,658

If highest numerical values differ from the variable name, the highest values of the indicator for the multivariate analysis are mentioned in brackets	moral (sexual) permissiveness (‘Permissiveness 1’)	acceptancy of the shadow economy(‘Permissiveness 2’)	distance from religion (‘Permiss-iveness 3’)	educational values: independence and imagination	distance to market economy values	education values: responsibility and tolerance	educational values: determination and perseverance and being against saving	right wing acceptance of inequality	educational values: favoring unselfishness, rejecting hard work
Income equality (large differences needed)	-0,025	-0,002	-0,032	-0,086	-0,002	-0,026	0,186	0,711	-0,185
Justifiable: abortion	0,788	0,146	0,436	0,110	-0,107	0,139	0,157	-0,120	-0,054
Justifiable: avoiding a fare on public transport	0,213	0,774	0,056	0,005	0,123	-0,033	-0,028	-0,007	0,016
Justifiable: cheating on taxes	0,239	0,783	0,088	0,013	0,042	-0,047	-0,015	0,015	-0,033
Justifiable: claiming government benefits even if one is not entitled to them	0,104	0,725	0,001	0,027	0,181	-0,056	-0,118	0,018	0,073
Justifiable: divorce	0,769	0,094	0,331	0,082	-0,102	0,174	0,184	-0,132	0,002
Justifiable: euthanasia	0,692	0,149	0,414	0,067	-0,095	0,140	0,089	-0,046	-0,008
Justifiable: homosexuality	0,779	0,127	0,219	0,198	0,000	0,226	0,189	-0,182	0,102
Justifiable: prostitution	0,759	0,283	0,167	0,134	0,007	0,084	0,098	-0,102	0,040
Justifiable: someone accepting a bribe	0,218	0,728	-0,043	0,021	0,115	-0,108	-0,118	0,071	0,022
Justifiable: suicide	0,668	0,295	0,207	0,132	0,043	0,047	0,010	-0,070	0,018
Political system: Having a democratic political system (reject)	-0,099	0,232	0,125	-0,038	0,300	-0,044	-0,395	0,035	0,197
Political system: Having the army rule (very bad)	0,253	-0,231	0,173	-0,030	-0,342	0,214	0,436	-0,288	-0,195
Private vs. state ownership of business (state)	-0,064	0,073	-0,027	-0,099	0,762	-0,059	-0,027	0,049	-0,064
Self-positioning in political scale (right wing)	-0,145	0,008	-0,114	0,051	-0,058	-0,033	-0,220	0,717	0,116

Table 12.4. Promax factor analysis, based on the latest edition of the combined *World Values Surveys* on religion, on what is justifiable, and democracy (quantitative analysis of *"Permissiveness"*) - The inter-correlations between the promax factors

	correlation matrix of the components								
	moral (sexual) permissiveness (*'Permissiveness 1'*)	acceptancy of the shadow economy (*'Permissiveness 2'*)	distance from religion (*'Permissiveness 3'*)	educational values: independence and imagination	distance to market economy values	education values: responsibilit y and tolerance	educational values: determination and perseverance and being against saving	right wing acceptance of inequality	educational values: favoring unselfishness, rejecting hard work
moral (sexual) permissiveness ('Permissiveness 1')	1,000	0,199	0,348	0,130	-0,082	0,194	0,203	-0,183	0,013
acceptancy of the shadow economy ('Permissiveness 2')	**0,199**	1,000	0,040	0,050	0,195	-0,102	-0,156	0,073	0,051
distance from religion (*'Permissiveness 3'*)	**0,348**	0,040	1,000	0,178	-0,074	0,154	0,101	-0,070	-0,098
educational values: independence and imagination	**0,130**	0,050	**0,178**	1,000	-0,031	0,168	-0,029	-0,011	0,009
distance to market economy values	-0,082	**0,195**	-0,074	-0,031	1,000	-0,084	-0,194	0,014	0,138
education values: responsibility and tolerance	**0,194**	**-0,102**	**0,154**	**0,168**	-0,084	1,000	0,170	-0,100	-0,031
educational values: determination and perseverance and being against saving	**0,203**	**-0,156**	**0,101**	-0,029	**-0,194**	**0,170**	1,000	-0,103	-0,146
right wing acceptance of inequality	**-0,183**	0,073	-0,070	-0,011	0,014	**-0,100**	**-0,103**	1,000	-0,055
educational values: favoring unselfishness, rejecting hard work	0,013	0,051	-0,098	0,009	**0,138**	-0,031	**-0,146**	-0,055	1,000

n = 91649,000 global representative citizens with available data

Table 12.5. Promax factor analysis, based on the latest edition of the combined *World Values Surveys* on religion, on what is justifiable, and democracy (quantitative analysis of *"Permissiveness"*) - The data [country means from Promax factor analysis] for new maps of global values from 70 countries and territories, based on the *World Values Survey* from the combined latest edition [2012] of global opinion investigation, integrating the fifth wave of investigations, 2005-2008 [Member countries of the Organization of Islamic Cooperation are highlighted in green color]

Country/region	**moral (sexual) permissiveness (*'Permissiveness 1'*)**	**acceptancy of the shadow economy (*'Permissiveness 2'*)**	**distance from religion (*'Permissiv-eness 3'*)**	**educational values: independence and imagination**	**distance to market economy values**	**education values: responsibility and tolerance**	**educational values: determination and perseverance and being against saving**	**right wing acceptance of inequality**	**educational values: favoring unselfishness, rejecting hard work**
Albania	-0,313	0,058	0,077	-0,143	-0,381	-0,276	-0,217	-0,035	-0,185
Andorra	1,525	-0,134	0,705	0,004	0,119	0,490	0,409	-0,477	-0,185
Argentina	0,221	-0,141	-0,200	0,001	0,266	0,261	0,174	-0,206	-0,078
Armenia	-0,160	0,528	0,567	-0,227	0,320	-0,279	0,097	0,282	-0,516
Australia	0,732	-0,361	0,497	0,056	-0,306	0,187	0,350	-0,293	0,310
Azerbaijan	-0,518	0,266	0,204	0,105	-0,098	-0,002	-0,110	-0,070	-0,695
Bangladesh	-1,017	-0,677	-0,819	0,598	-0,228	-0,058	-0,116	0,864	-0,363
Belarus	0,098	0,541	0,873	-0,529	0,074	-0,153	-0,098	0,184	-0,672
Bosnia and Herzegovina	-0,287	-0,312	0,107	-0,086	-0,306	-0,107	0,091	-0,186	-0,029
Brazil	-0,204	0,555	-0,652	-0,381	0,297	-0,210	-0,264	-0,010	0,358
Bulgaria	0,282	-0,075	0,881	-0,120	-0,052	-0,287	0,095	0,072	-0,346
Burkina Faso	-0,435	0,061	-0,930	-0,667	-0,002	-0,888	-0,004	0,044	-0,260
Canada	0,515	-0,293	0,120	0,117	-0,251	0,241	0,381	-0,276	0,211
Chile	-0,087	0,200	-0,297	0,093	0,603	0,270	-0,163	-0,436	0,646
Croatia	0,589	0,624	0,382	0,193	-0,734	0,073	0,383	-0,315	-0,342
Cyprus	0,046	-0,165	-0,147	-0,140	0,011	0,007	0,083	-0,317	0,175
Czech Republic	0,940	0,227	1,054	-0,476	-0,179	-0,006	0,048	0,080	-0,464
Dominican Republic	-0,132	-0,307	-0,680	-0,323	0,125	0,289	0,386	0,507	-0,287
Estonia	0,154	0,025	1,102	-0,271	-0,159	0,062	0,266	-0,070	-0,794
Ethiopia	-0,771	-0,294	-0,654	0,764	-0,137	-0,622	-0,316	0,440	0,006
Finland	0,610	-0,155	0,585	0,390	-0,080	0,707	0,122	-0,409	0,525
Georgia	-0,536	-0,042	-0,079	-0,181	-0,115	0,030	0,167	0,444	-0,690
Germany	0,709	-0,205	0,784	0,766	-0,085	0,691	0,145	-0,732	-0,128

Table 12.5. (Continued)

Country/region	**moral (sexual) permissiveness** (*'Permissiveness 1'*)	**acceptancy of the shadow economy** (*'Permissiveness 2'*)	**distance from religion** (*'Permissiv-eness 3'*)	**educational values: independence and imagination**	**distance to market economy values**	**education values: responsibility and tolerance**	**educational values: determination and perseverance and being against saving**	**right wing acceptance of inequality**	**educational values: favoring unselfishness, rejecting hard work**
Ghana	-0,556	-0,222	-1,048	-0,565	0,211	-0,722	0,540	0,586	-0,349
Guatemala	-0,210	0,519	-0,861	-0,319	0,510	-0,104	-0,335	0,328	0,564
Hong Kong	0,019	-0,042	0,842	0,128	0,165	-0,490	-0,440	-0,092	0,184
Hungary	0,380	0,450	0,803	-0,065	-0,274	0,361	0,114	-0,554	0,136
India	-0,359	-0,085	-0,269	-0,062	-0,104	-0,498	-0,261	-0,175	-0,105
Indonesia	-0,946	-0,250	-1,047	0,359	0,521	-0,083	-0,570	0,980	0,276
Italy	-0,002	-0,370	-0,059	0,083	-0,032	0,483	0,411	-0,217	0,120
Japan	0,305	-0,397	0,846	0,594	-0,154	0,534	0,222	-0,033	0,327
Jordan	-1,029	-0,102	-1,192	-0,053	0,231	-0,505	-0,137	0,280	0,694
Kyrgyzstan	-0,628	0,192	0,238	0,165	0,232	-0,113	-0,266	0,140	-0,290
Latvia	0,370	0,471	0,842	-0,400	-0,129	0,165	0,117	0,072	-0,731
Lithuania	-0,184	0,234	0,475	0,017	-0,122	-0,022	-0,063	0,018	-0,475
Macedonia	-0,417	-0,082	0,240	0,154	-0,441	0,373	0,018	-0,232	0,479
Mali	-0,005	0,500	-0,822	-0,286	0,300	-0,716	-0,253	0,358	-0,148
Mexico	-0,169	0,469	-0,363	0,076	0,294	-0,083	-0,361	0,065	0,689
Moldova	-0,252	0,530	0,139	-0,151	0,360	-0,058	-0,132	0,122	-0,539
New Zealand	0,752	-0,411	0,603	0,152	-0,395	0,194	0,444	-0,270	0,131
Nigeria	-0,707	-0,193	-1,083	-0,805	0,025	-0,951	0,439	0,154	-0,460
Norway	0,799	-0,328	0,672	0,850	-0,201	0,769	0,372	-0,401	0,275
Peru	-0,172	0,012	-0,570	-0,397	0,145	0,072	0,090	0,098	-0,147
Philippines	-0,315	0,713	-0,807	0,070	0,580	-0,392	-0,500	0,466	0,201
Poland	-0,114	-0,004	-0,498	-0,063	0,765	0,426	-0,463	0,199	0,403
Puerto Rico	-0,365	-0,325	-0,896	-0,438	-0,012	0,320	0,076	0,128	0,492
Romania	-0,174	-0,149	-0,247	0,089	-0,467	0,097	-0,049	0,035	-0,455
Russian Federation	-0,093	0,386	1,186	-0,650	0,467	-0,092	-0,279	0,177	-0,580
Serbia	0,393	1,353	0,119	0,132	0,312	-0,205	-0,129	0,216	-0,018
Serbia and Montenegro	-0,060	-0,171	0,611	-0,168	-0,400	-0,137	0,335	-0,147	-0,396
Slovakia	0,544	0,379	0,166	-0,469	0,128	0,027	-0,018	-0,199	-0,519
Slovenia	0,621	0,191	0,662	0,255	-0,277	0,376	0,173	-0,395	0,228
South Africa	-0,274	0,009	-0,654	-0,206	-0,043	-0,222	-0,061	-0,042	-0,049

Country/region	**moral (sexual) permissiveness (*'Permissiveness 1'*)**	**acceptancy of the shadow economy (*'Permissiveness 2'*)**	**distance from religion (*'Permissiv-eness 3'*)**	**educational values: independence and imagination**	**distance to market economy values**	**education values: responsibility and tolerance**	**educational values: determination and perseverance and being against saving**	**right wing acceptance of inequality**	**educational values: favoring unselfishness, rejecting hard work**
South Korea	-0,088	-0,051	0,555	0,613	0,167	0,241	-0,529	0,161	-0,567
Spain	0,833	-0,177	0,409	-0,242	0,240	0,092	0,229	-0,477	-0,213
Sweden	1,309	-0,185	0,895	0,624	-0,320	0,852	0,248	-0,270	0,344
Switzerland	1,003	-0,255	0,364	0,418	-0,349	0,641	0,370	-0,632	0,170
Taiwan	-0,112	-0,165	0,745	0,279	0,073	0,487	-0,489	0,382	-0,010
Tanzania	-0,786	-0,517	-0,977	-0,035	-0,249	-0,783	0,121	-0,044	0,076
Thailand	-0,384	0,613	-0,369	0,254	0,763	-0,285	-0,540	0,520	0,574
Trinidad and Tobago	-0,372	0,056	-0,804	-0,551	0,050	-0,196	0,019	0,284	-0,129
Turkey	-0,629	-0,438	-0,290	-0,143	0,408	-0,192	-0,229	0,066	-0,111
Uganda	-0,695	0,272	-0,839	-0,574	-0,538	-0,701	0,339	0,206	-0,225
Ukraine	-0,061	0,604	0,566	-0,311	0,248	-0,219	-0,053	0,221	-0,469
United States	0,237	-0,287	-0,307	-0,012	-0,464	0,114	0,220	-0,121	0.088
Uruguay	0,429	-0,288	0,285	0,138	0,286	0,288	0,179	-0,342	0.749
Venezuela	-0,403	-0,057	-0,455	-0,227	0,133	0,075	-0,291	0,145	0,550
Viet Nam	-0,791	-0,233	0,853	0,196	0,186	-0,340	-0,657	1,352	0,066
Zambia	0,006	0,641	-0,768	-0,204	0,320	-0,647	0,125	0,423	-0,156
Zimbabwe	-0,910	-0,501	-0,958	-0,854	-0,363	-0,610	0,454	-0,436	-0,301

Factor definitions:

- moral (sexual) permissiveness ('Permissiveness 1')

Definition:
Justifiable: abortion
Justifiable: homosexuality
Justifiable: divorce
Justifiable: prostitution
Justifiable: euthanasia
Justifiable: suicide

Highest values:
Andorra
Sweden
Switzerland
Czech Republic
Spain

Lowest values:
Jordan
Bangladesh
Indonesia
Zimbabwe
Viet Nam

- acceptancy of the shadow economy('Permissiveness 2')

Definition:

Justifiable: cheating on taxes
Justifiable: avoiding a fare on public transport
Justifiable: someone accepting a bribe
Justifiable: claiming government benefits even if one is not entitled to them

Highest values:
Serbia
Philippines
Zambia
Croatia
Thailand

Lowest values:
Bangladesh
Tanzania
Zimbabwe
Turkey
New Zealand

- **distance from religion ('Permissiveness 3')**

Definition:

How often do you attend religious services (never?)
negative loading: How important is God in your life
negative loading: important child qualities: religious faith

Highest values:
Russian Federation
Estonia
Czech Republic
Sweden
Bulgaria

Lowest values:
Jordan
Nigeria
Ghana
Indonesia
Tanzania

- **educational values: independence and imagination**

Definition:

Important child qualities: independence
Important child qualities: imagination

Highest values:
Norway
Germany
Ethiopia
Sweden
South Korea

Lowest values:
Zimbabwe
Nigeria
Burkina Faso
Russian Federation
Uganda

- **distance to market economy values**

Definition:
Favoring private or state ownership of business (state ownership)
Competition good or harmful (harmful)

Highest values:
Poland
Thailand
Chile
Philippines
Indonesia

Lowest values:
Croatia
Uganda
Romania
United States
Macedonia

- **education values: responsibility and tolerance**

Definition:
Important child qualities: feeling of responsibility
Important child qualities: tolerance and respect for other people
Negative loading: obedience (-0,459)

Highest values:
Sweden
Norway
Finland
Germany
Switzerland

Lowest values:
Nigeria
Burkina Faso
Tanzania

Ghana
Mali

- **educational values: determination and perseverance and being against saving**

Definition:
Important child qualities: determination and perseverance
Negative loading: Important child qualities: thrift, and saving money and things

Highest values:
Ghana
Zimbabwe
New Zealand
Nigeria
Italy

Lowest values:
Viet Nam
Indonesia
Thailand
South Korea
Philippines

- **right wing acceptance of inequality**

Definition:
Self-positioning in political scale (right wing)
Income equality (large differences needed)

Highest values:
Viet Nam
Indonesia
Bangladesh
Ghana
Thailand

Lowest values:
Germany
Switzerland
Hungary
Spain
Andorra

- **educational values: favoring unselfishness, rejecting hard work**

Definition:
Important child qualities: unselfishness
Negative loading: important child qualities: hard work

Highest values:
Uruguay
Jordan
Mexico
Chile
Thailand

Lowest values:
Estonia
Latvia
Azerbaijan
Georgia
Belarus

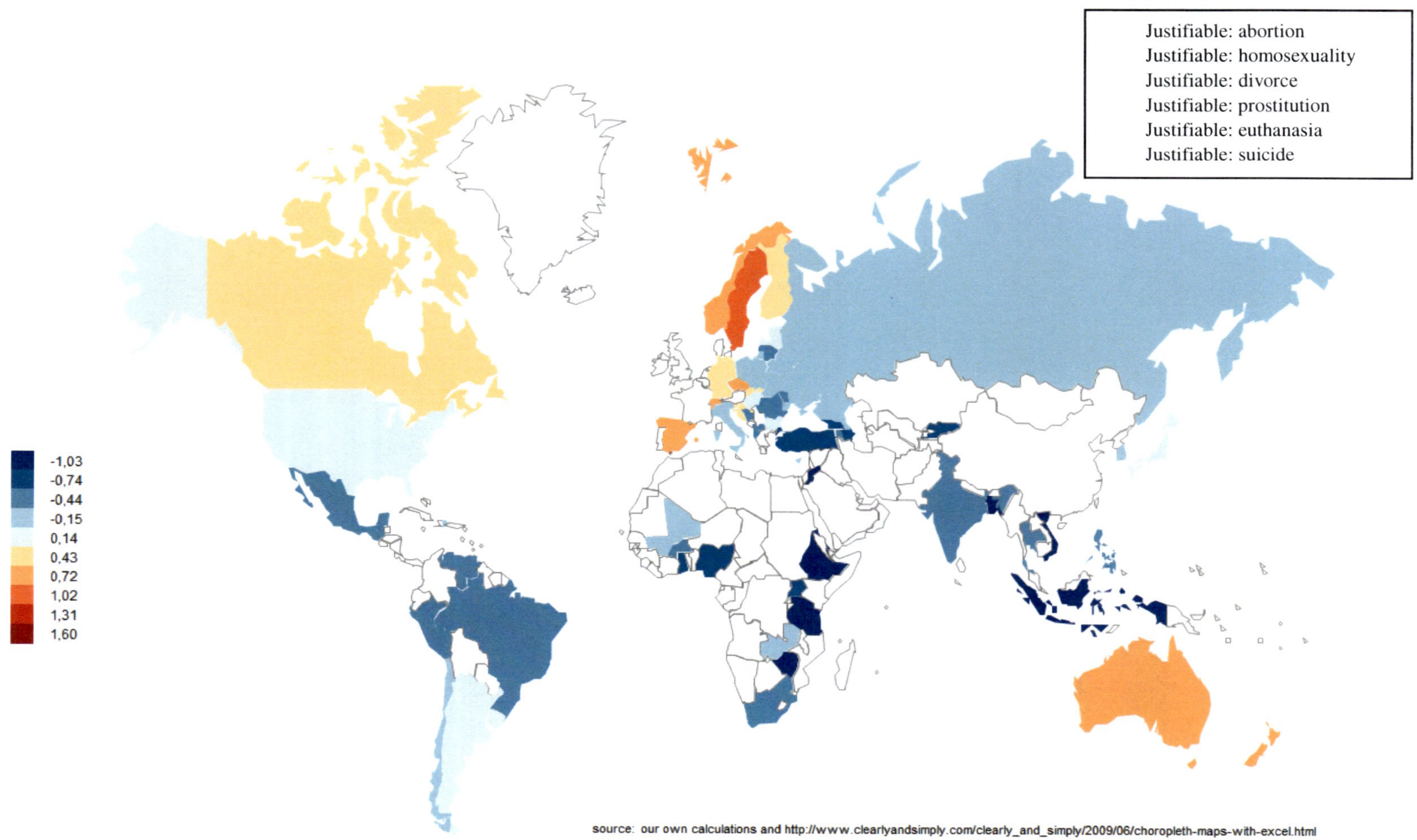

Map 12.1. Moral (sexual) permissiveness (*'Permissiveness 1'*).

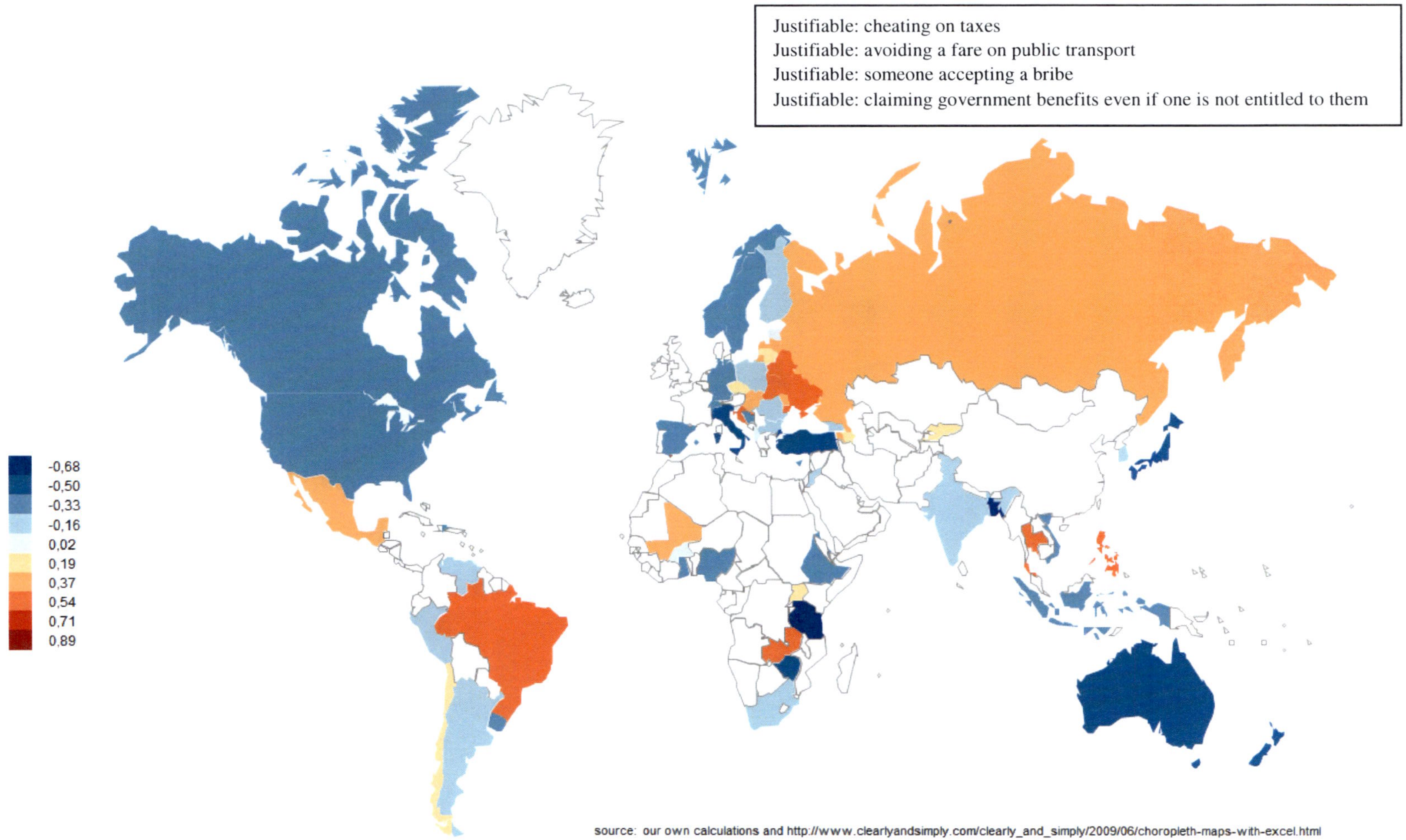

Map 12.2. Acceptancy of the shadow economy *('Permissiveness 2')*.

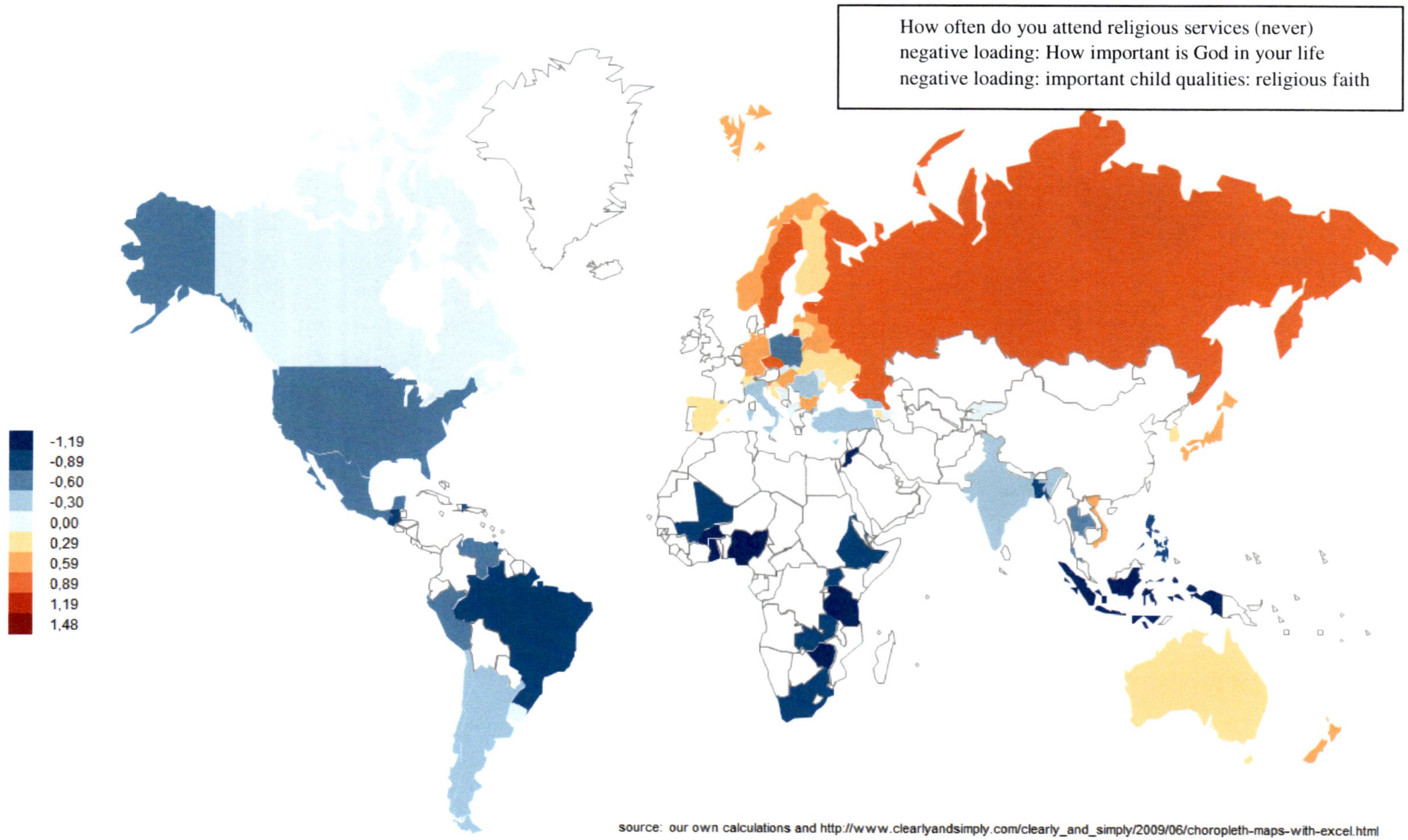

Map 12.3. Distance from religion (*'Permissiveness 3'*).

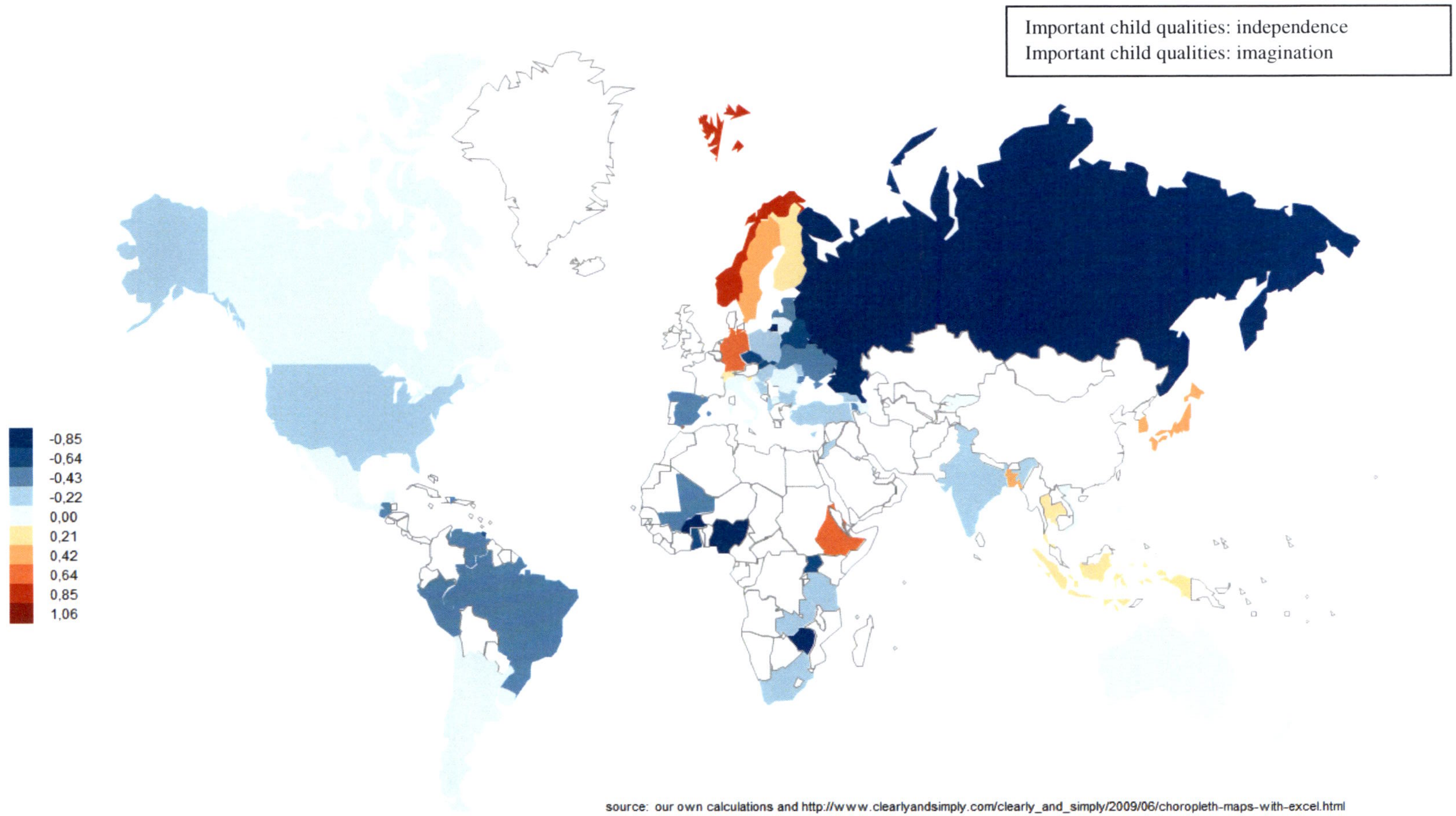

Map 12.4. Educational values: independence and imagination.

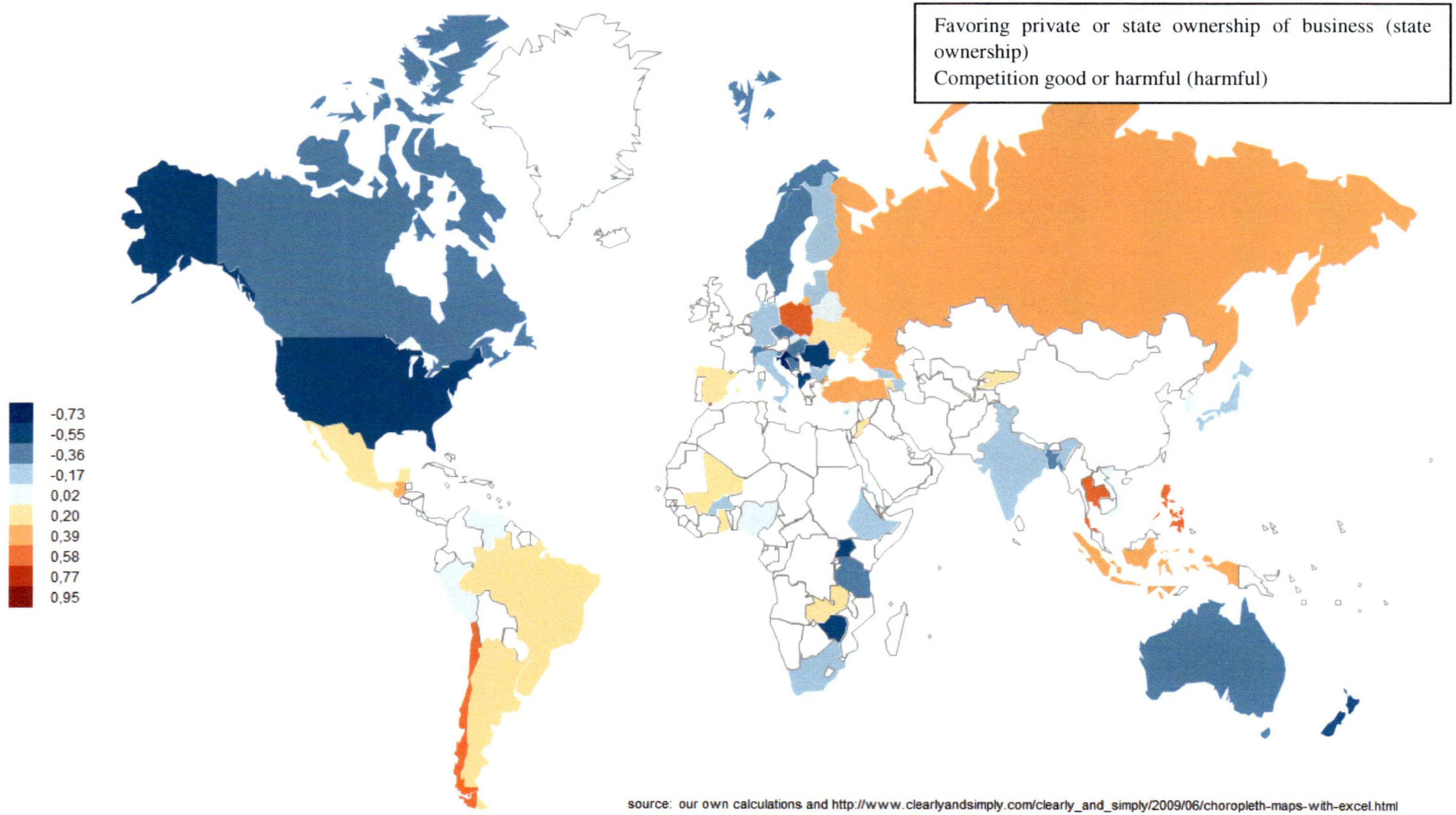

Map 12.5. Distance to market economy values.

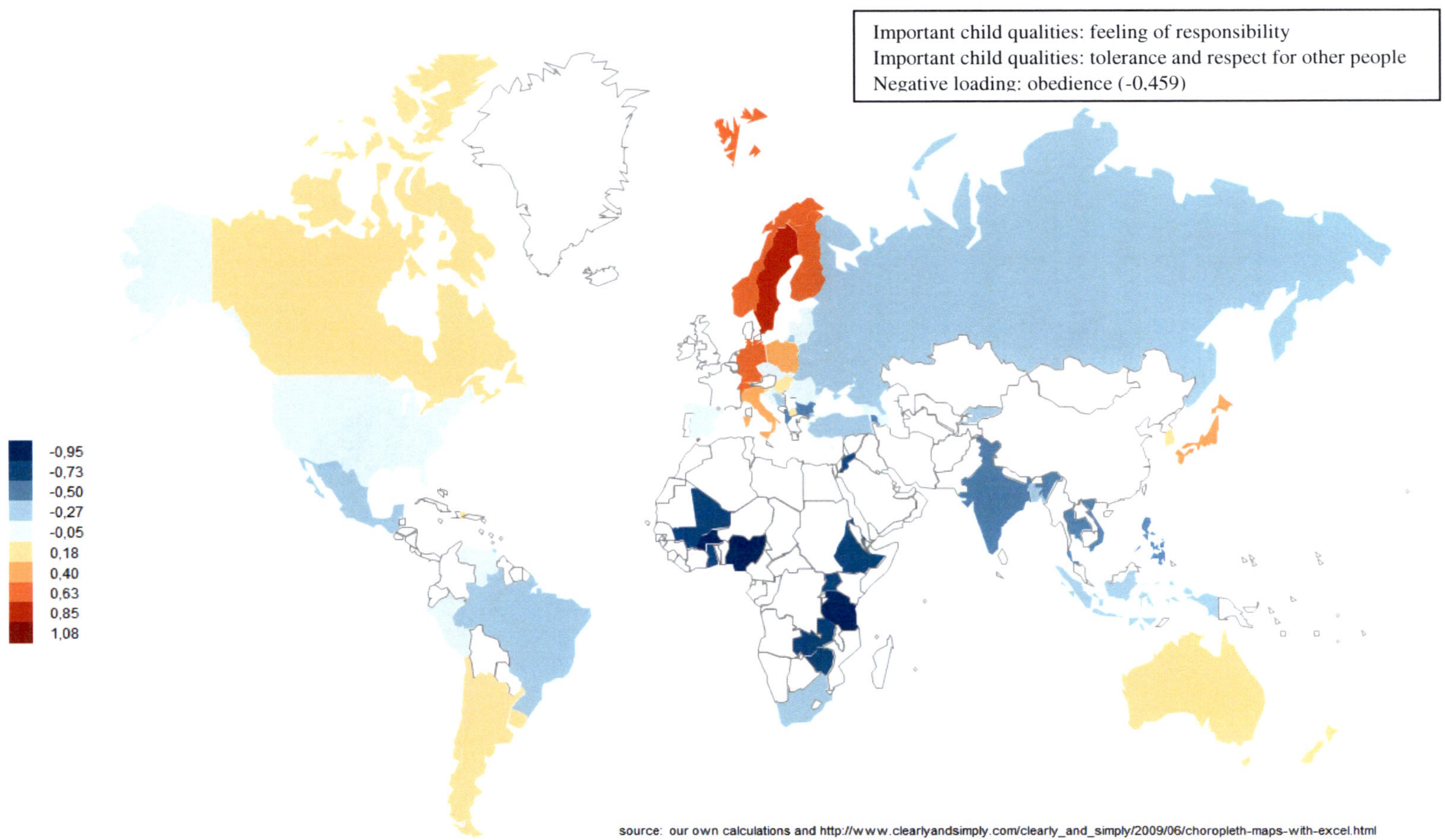

Map 12.6. Educational values: responsibility and tolerance.

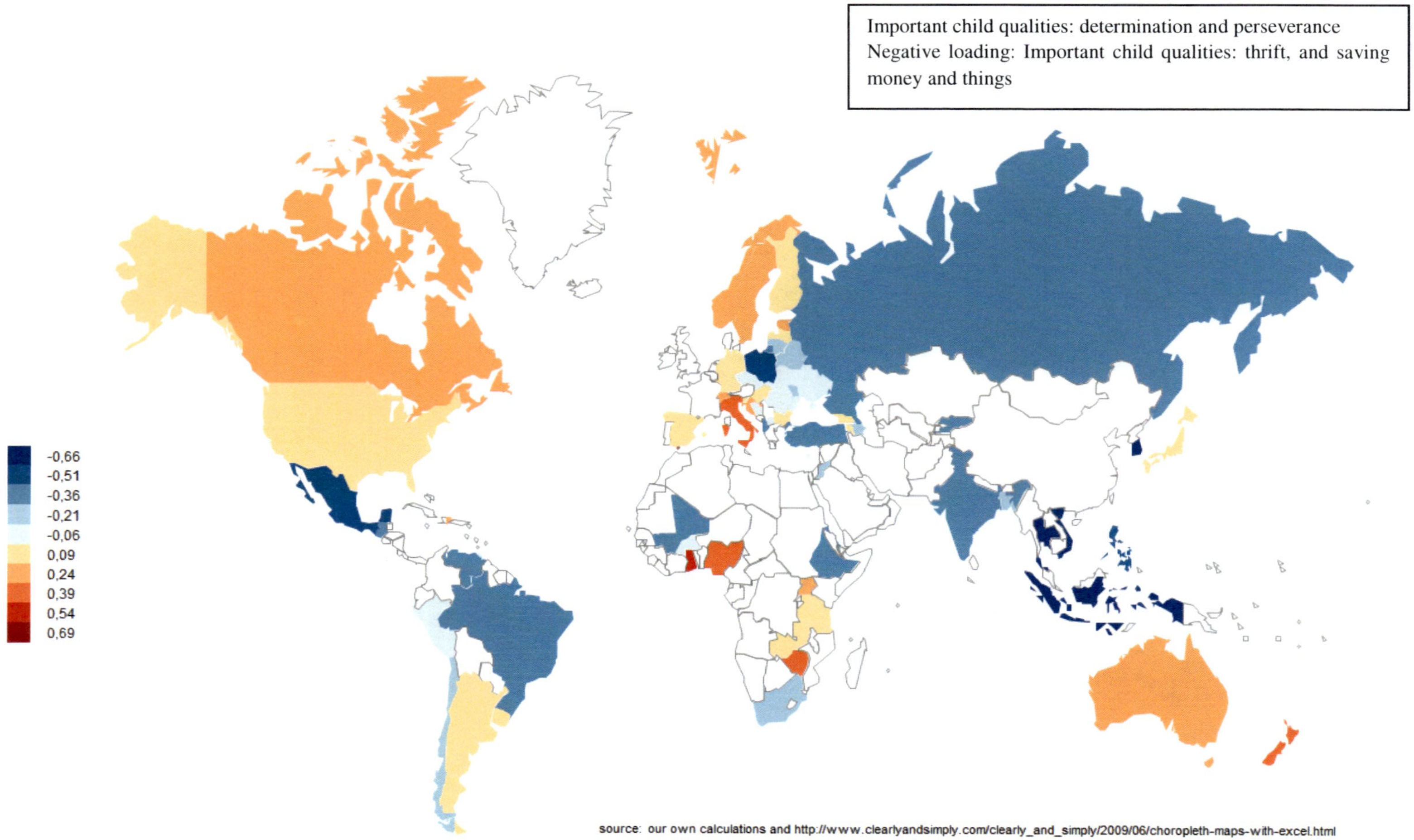

Map 12.7. Educational values: determination and perseverance and being against saving.

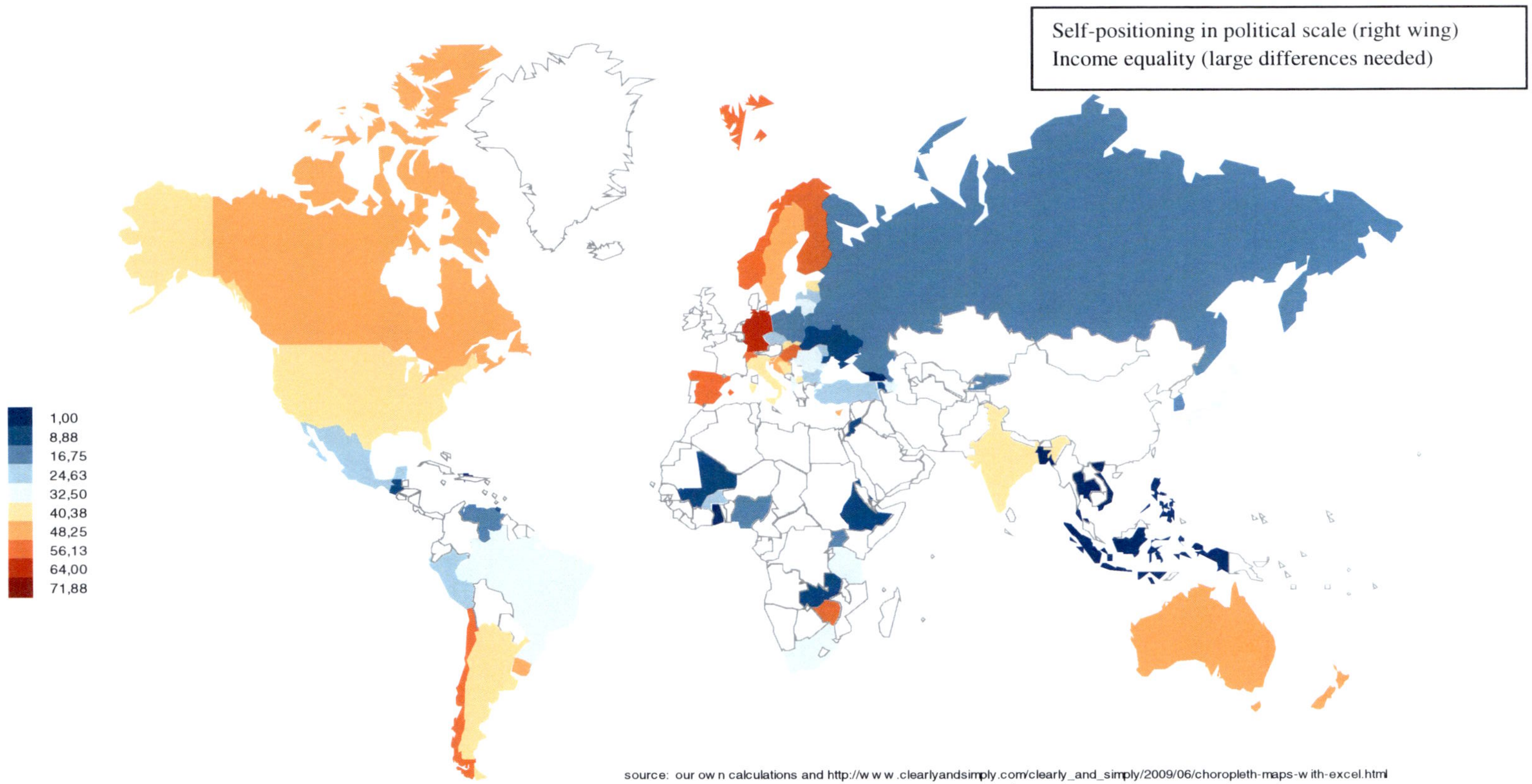

Map 12.8. Right wing acceptance of inequality (global rank, right wing acceptance of inequality).[60] High right-wing acceptance of inequality: marked in blue; low right-wing acceptance of inequality: marked in red.

---

[60] The indicator did not produce meaningful maps with the original values, so we transformed it to a ranking.

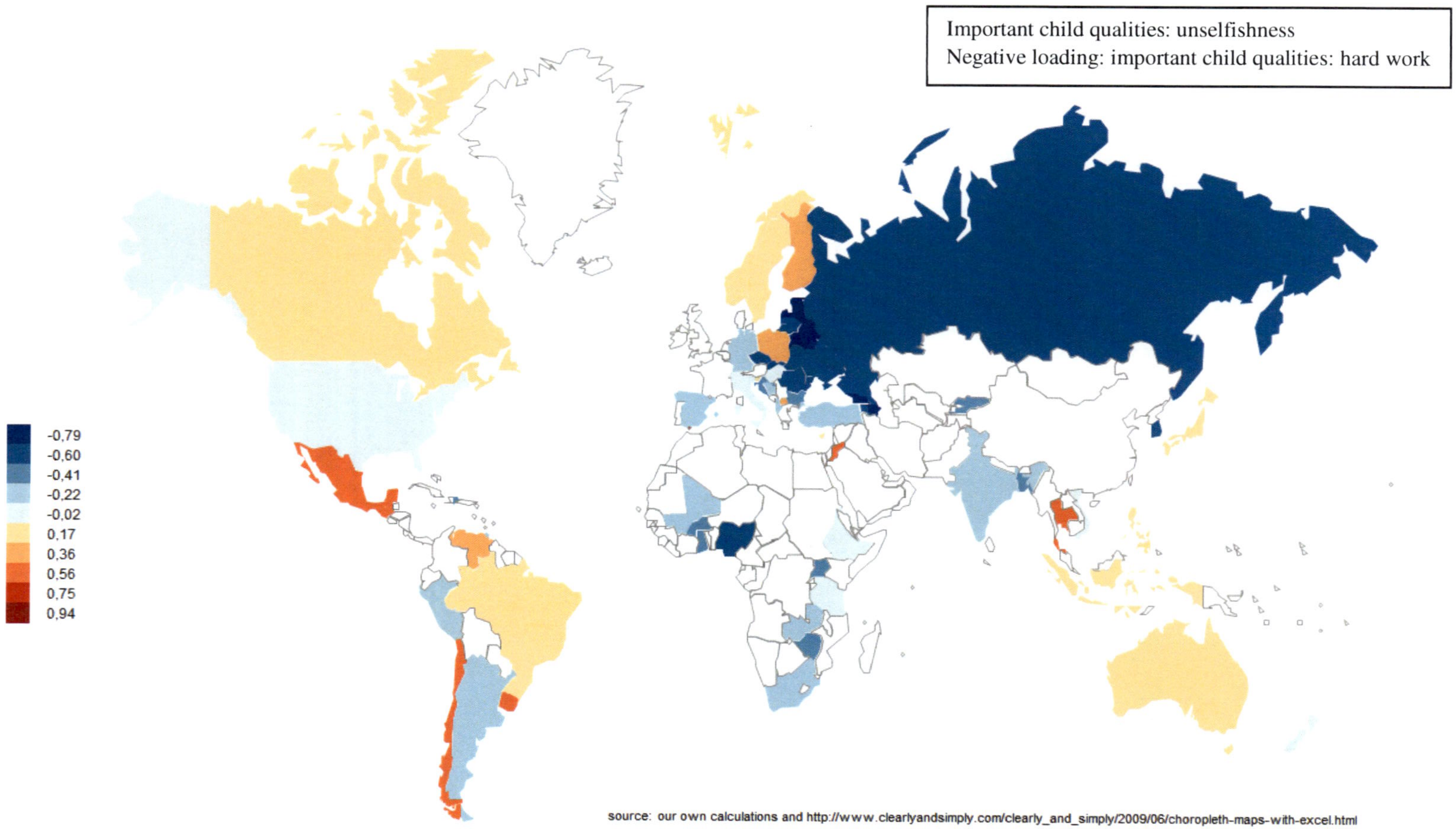

Map 12.9. Educational values: favoring unselfishness, rejecting hard work.

12.2. HOFSTEDE, SCHWARTZ, INGLEHART AND OUR TYPOLOGIES

We have now almost reached the end of our journey. With lots of data, shown in Tables, Graphs and Maps, we have tried to present our picture of the analyses, as they emerge from the data of the *World Values Survey,* and the *Arab Opinion Index.* In Table 12.6 we provide our readers with clear-cut Pearson-Bravais correlation coefficients between the data presented by Hofstede and Inglehart and the factor scores from our own analytical dimensions, presented in this work. Table 12.7 shows the Pearson-Bravais correlations between the Schwartz dimensions and our results.

In many ways, we can show that Hofstede's Power Distance, Individualism versus Collectivism, Long-Term Orientation, and Indulgence versus Restraint very well correspond to our own factor analyses. The same happens with Inglehart's main dimensions, traditional versus secular, and survival versus self-expression. In all cases, however, we could avoid some of the problematic assumptions still inherent in the research by Hofstede and Inglehart.

Our factors show correlations of +-.500 and more with those mentioned dimensions of Hofstede's and Inglehart's research. Hofstede's other factors – Masculinity versus Femininity, and Uncertainty Avoidance, found no confirmation of a correlation of +-.500 or more with our own factors. But this does not necessarily imply that Hofstede's research is falsified. It might also very well be that

1) Relationships between phenomena changed over time or
2) Hofstede's samples (employees of the transnational corporation IBM) exhibit other tendencies and correlations as the populations at large.

In the following, we list the comparisons of the research results by Hofstede and Inglehart and our own with more detail. The data for this exercise are freely available from https://www.academia.edu/7504018/Choropleth_maps_of_global_values_and_background_variables).[61]

Our own factors which confirmed the theoretical work by Hofstede and Inglehart:

- traditional religion
- the 'ego' company (rejection of obedience + unselfishness)
- feminism analysis: distance to traditionalist forms of religion
- feminism analysis: feminism
- re-analysis: moral (sexual) permissiveness ('Permissiveness 1')
- re-analysis: distance from religion ('Permissiveness 3')
- re-analysis: education values: responsibility and tolerance

[61] Readers are reminded here that the mentioned EXCEL files, downloadable from https://www.academia.edu/7504018/Choropleth_maps_of_global_values_and_background_variables serve a dual purpose: easy data documentation for standard variables, used in this work, and the drawing of choropleth maps. For that very reason, data for Hongkong and Macau, which are territories belonging to the People's Republic of China, could not be used further. Likewise, we should draw our reader's attention that available data for Taiwan, Province of China, were used in the choropleth maps.

- re-analysis: educational values: favoring unselfishness, rejecting hard work

Our own factors which confirmed the theoretical work by Hofstede only:

- authoritarian character
- tolerance and respect + postmaterialism
- re-analysis: educational values: determination and perseverance and being against saving
- re-analysis: right wing acceptance of inequality

Our own factors which confirmed the theoretical work by Inglehart only:

- economic permissiveness

Our own factors which confirmed earlier theories by a correlation of less than +-0.50:

- Hofstede: Masculinity versus Femininity
- Hofstede: Uncertainty Avoidance Index

Our own factors which were absent from earlier theoretical attempts:

- racism
- higher education of the younger generation (education gap between the generations)
- distrust of the army and the press
- [predominantly] female rejection of the market economy and democracy
- re-analysis: acceptancy of the shadow economy ('Permissiveness 2')
- re-analysis: educational values: independence and imagination
- re-analysis: distance to market economy values

In the following two Tables, we evaluate the relationship of the Schwartz factors with the other aggregates, presented in this book. Again, the data for this exercise are freely available from https://www.academia.edu/7504018/Choropleth_maps_of_global_values_and_background_variables).[62] The Schwartz values (raw country scores) were listed among others in Ralston, Egri, Reynaud et al., 2011.

The Schwartz's factors *Affective Autonomy* and *Harmony* do not achieve any correlations which have more than 25% of variance in common with our own factors, and in addition, the following dimensions from our own research are untapped, it seems, by the Schwartz' factors (to judge from the less than 25% of variance they have in common with the Schwartz factors):

[62] See note before.

Chapter 5: racism
Chapter 5: higher education of the younger generation (education gap between the generations)
Chapter 5: distrust of the army and the press
Chapter 5: female rejection of the market economy and democracy
Chapter 12 re-analysis: shadow economy mentality ('Permissiveness 2')
Chapter 12 re-analysis: educational values: independence and imagination
Chapter 12 re-analysis: education values: unselfishness and being against hard work
Chapter 6: feminism analysis: Generation gap in educational level

In addition, the following factors from Hofstede and Inglehart are untapped; it seems, by Schwartz's theories (again to judge from the less than 25% of variance they have in common with the Schwartz factors)

Hofstede: Masculinity versus Femininity
Hofstede: Uncertainty Avoidance Index
Inglehart: Self-Expression Values (WVS 1-4, 2006)

Policy pointer:

We provide our readers with clear-cut Pearson-Bravais correlation coefficients between the data presented by Hofstede and Inglehart and the factor scores from our own analytical dimensions, presented in this work on the bases of promax factor analysis with individual data from up to more than 80 countries. In many ways, we can show that Hofstede's Power Distance, Individualism versus Collectivism, Long-Term Orientation, and Indulgence versus Restraint very well correspond to our own factor analyses. The same happens with Inglehart's main dimensions, traditional versus secular, and survival versus self-expression. In all cases, however, we could avoid some of the problematic assumptions, still inherent in the research by Hofstede and Inglehart. Our factors show correlations of +-.500 and more with those mentioned dimensions of Hofstede's and Inglehart's research. Hofstede's other factors – Masculinity versus Femininity, and Uncertainty Avoidance, found no confirmation of a correlation of +-.500 or more with our own factors.

We also evaluate the relationship of the Schwartz factors with the other aggregates, presented in this book. The Schwartz factor "Embeddedness" has the highest correlation with the absence of sexual permissiveness (permissiveness 1 – Chapter 12 of this work). Hierarchy is most closely correlated with the absence of feminism (see Chapter 6 of this work). Mastery is closely related to the absence of secular values from Inglehart's theory; and Egalitarianism also has almost 50% of variance with that Inglehartian factor (absence of secular values) in common. Affective Autonomy and Harmony are only weakly connected to the other global value aggregates, presented in this book. Intellectual Autonomy is somehow related to Hofstede's Indulgence Factor, but the other correlations are relatively small. Schwartz' Egalitarianism is strongest in religious societies; it negatively correlates with Hofstede's long-term orientation and our "ego factor". Interestingly enough, Muslim population shares and OIC membership present high correlations with the Schwartz factors "Embeddedness", "Hierarchy" and "Mastery".

The Schwartz factor "Embeddedness" has the highest correlation with the absence of sexual permissiveness (permissiveness 1 – Chapter 12 of this work). Hierarchy is most closely correlated with the absence of feminism (see Chapter 6 of this work). Mastery is closely related to the absence of secular values from Inglehart's theory; and Egalitarianism also has almost 50% of variance with that Inglehartian factor (absence of secular values) in common. Affective Autonomy and Harmony are only weakly connected to the other global value aggregates, presented in this book. Intellectual Autonomy is somehow related to Hofstede's Indulgence Factor, but the other correlations are relatively small. Schwartz' Egalitarianism is strongest in religious societies; it negatively correlates with Hofstede's long-term orientation and our "ego factor".

Table 12.8 now shows the correlations of the country scores from Schwartz' work with standard socio-economic indicators. Interestingly enough, Muslim population shares and OIC membership present high correlations with the Schwartz factors "Embeddedness", "Hierarchy" and "Mastery". For the other correlations of the Schwartz factors, we refer our readers to Table 12.8.

12.3. On Sexual Behavior Values and Effective Democracy

In these concluding remarks we will also visit current debates in France and in Europe. These days, we read stories in the leading world newspapers like the one below which tell us a lot about the conflicts about global values in countries like Europe today. *Is prostitution justifiable? Is homosexuality justifiable? Are both justifiable? Is one justifiable, the other is not? Are both not justifiable?* The French Socialists seem to think that one is not, and the other is. President Hollande and his administration put considerable political energy into legalizing homosexual marriages and prohibiting prostitution. But global citizens hold another view, and there is *a high positive correlation of 0.632 between the two items in the "World Values Survey", based on 218877 individuals from around the globe.* For a correlation at the individual level from such a large global survey, the coefficient is very high indeed i.e., people in favor of the complete acceptability of homosexuality will also be in favor of the complete acceptability of prostitution and vice versa. But empirical *World Values Survey* research has many other surprises in store for all ideological camps, and empirical *World Values Survey* research more often than not contradicts ideologies, left and right, religiously motivated and secularist alike.

The Turkish newspaper *Hürriyet* brought the recent debate to a point, when it said:

> "Hürriyet November/22/2013: French lawmakers will next week consider a bill that punishes the clients of prostitutes and has sparked fierce debate in a country with a long history of liberal attitudes to sex. The bill, which the government says is aimed at preventing violence against women, has come under fire from celebrities like Catherine Deneuve and Charles Aznavour, leading intellectuals and prostitutes themselves. Spearheaded by Women's Rights Minister Najat Vallaud-Belkacem, the law will fine clients of prostitutes 1,500 euros ($2,025) for a first offense and double that for repeat offenders. Prostitution itself is legal in France but soliciting, pimping and minors selling sex are prohibited. The government estimates about

20,000 prostitutes operate in France. […] About 60 people, including Deneuve, Aznavour and former culture minister Jack Lang released an open letter this month opposing the bill and calling for "a real debate" on prostitution "without ideological prejudice." Surprising some, among the most vocal opponents of the bill has been the philosopher and feminist Elisabeth Badinter, who has pleaded that "the state has no place legislating on individual sexual activity." Badinter said she saw no direct link between male sexuality and violence against women, accusing some of having "a stereotypical view that is very negative and moralistic and which I reject." Her remarks drew a sharp rebuke from Health Minister Marisol Touraine. […] It comes after a deeply divisive debate on another contentious sexuality issue -- the legalisation of gay marriage. President Francois Hollande's Socialist government legalised homosexual marriages and adoptions earlier this year, but only after tens of thousands took to the streets in a series of protests to denounce the move. Europe is also in the midst of a wider debate over prostitution, which is legal in several countries on the continent."
http://www.hurriyetdailynews.com/fierce-debate-as-france-looks-to-punish-prostitutes-clients.aspx?pageID=238&nID=58376&NewsCatID=351

The emotionalized and very bitter debate in France took place without any real reference to the empirical realities of global values "on the ground". France, like many other European countries, seems to be disconnected from empirical global *World Values Survey* science research. The freely available archive of all the leading French language scholarly journals, *Persee,* lists just 17 articles mentioning the word *"World Values Survey"* at all.[63] Not a single article of these 17 articles deals with the problem of homosexuality, which was the issue of a real ideological warfare in the country recently. So we can say that French language social science hardly works at all with the *World Values Survey.*

Current debates in Europe on issues of human values have acquired an almost quasi-religious character, with each side claiming that it is in the sole possession of truth, while in reality the complex relationships between value patterns are often overlooked. Tables 12.9 and 12.10 pick out the two issues which were at the heart of the recent debates in France – *the acceptancy of homosexuality, and the acceptancy of prostitution.* Looking at the partial correlations between the acceptancy of homosexuality and 20 other *World Values Survey* key-indicators of trust and behavior, we find that apart from the high correlations with acceptancy rates for such phenomena as divorce, abortion, and large-scale non-confidence in Churches, which mainstream secular and left of the middle socio-political currents in Europe will perceive as largely unproblematic, *there are also significant and higher correlations between the unconditional acceptancy of homosexuality with accepting a bribe, avoiding a fare on public transport, cheating on taxes, and the generalized loss of trust in basic institutions of democratic society.* We used *World Values Survey* data WVS, version *"wvs1981_2008_v20090914. sav",* IBM-SPSS 21 software, and partial correlation (keeping constant income and education level). Our correlations are based on the WVS data for all the representative citizens of our globe with complete available data.

[63] http://www.persee.fr/web/revues/home/prescript/revues/alph?_Prescripts_Search_tabs1= advanced&_ Prescripts_ Search_oai=false&_Prescripts_Search_revueDisplay=advanced

Table 12.6. Hofstede, Inglehart and the factors from our work – Pearson Bravais correlations of the country results

	Hofstede: Power Distance	**Hofstede: Individualism vs. Collectivism**	**Hofstede: Masculinity vs.Femininity**	**Hofstede: Uncertainty Avoidance Index**	**Hofstede: Long-Term Orientation**	**Hofstede: Indulgence vs. Restraint**	**Inglehart: Traditional/Secular Values**	**Inglehart: Survival vs. Self-Expression Values**
economic permissiveness	0,304	-0,138	0,064	0,162	0,091	-0,051	0,272	**-0,715**
traditional religion	0,447	-0,439	0,288	0,082	**-0,546**	0,179	**-0,911**	0,110
racism	0,426	-0,491	0,087	-0,247	0,189	-0,401	-0,362	-0,020
higher education of the younger generation (education gap between the generations)	-0,086	-0,038	-0,210	-0,019	0,047	0,119	-0,310	-0,087
distrust of the army and the press	-0,435	0,332	-0,043	0,371	-0,058	0,246	0,424	-0,405
authoritarian character	**0,567**	-0,365	0,200	0,355	0,086	-0,374	-0,238	-0,297
tolerance and respect + postmaterialism	**-0,686**	0,597	-0,071	-0,079	-0,196	**0,530**	0,293	-0,247
the 'ego' company (rejection of obedience + unselfishness)	-0,067	0,122	-0,196	-0,032	**0,796**	**-0,600**	**0,640**	**0,700**
[predominantly] female rejection of the market economy and democracy	0,094	-0,238	-0,065	0,461	-0,039	0,191	0,146	-0,140
feminism analysis: distance to traditionalist forms of religion	-0,424	0,526	-0,159	-0,086	**0,523**	-0,057	**0,916**	0,300
feminism analysis: feminism	0,349	-0,064	0,247	-0,013	**0,513**	0,446	**0,531**	-0,399
re-analysis: moral (sexual) permissiveness ('Permissiveness 1')	**-0,560**	0,650	-0,076	-0,028	0,130	0,180	**0,705**	**0,561**
re-analysis: acceptancy of the shadow economy ('Permissiveness 2')	0,386	-0,210	-0,038	0,150	0,045	-0,133	0,236	-0,439
re-analysis: distance from religion ('Permissiveness 3')	-0,383	0,404	-0,222	-0,029	**0,578**	-0,304	**0,913**	0,065
re-analysis: educational values: independence and imagination	-0,326	0,081	-0,152	-0,218	0,130	0,104	0,277	0,456
re-analysis: distance to market economy values	0,431	-0,480	0,025	0,212	-0,182	0,006	-0,229	-0,230
re-analysis: education values: responsibility and tolerance	**-0,594**	**0,522**	-0,104	0,154	0,095	0,260	**0,555**	0,476
re-analysis: educational values: determination and perseverance and being against saving	**-0,631**	**0,654**	-0,033	0,054	-0,020	0,171	0,185	0,330

Table 12.6. (Continued)

	Hofstede: Power Distance	Hofstede: Individualism vs. Collectivism	Hofstede: Masculinity versus Femininity	Hofstede: Uncertainty Avoidance Index	Hofstede: Long-Term Orientation	Hofstede: Indulgence versus Restraint	Inglehart: Traditional/Secular Values	Inglehart: Survival vs. Self-Expression Values
re-analysis: right wing acceptance of inequality	0,470	**-0,564**	-0,042	-0,204	-0,025	-0,213	-0,321	-0,324
re-analysis: educational values: favoring unselfishness, rejecting hard work	-0,110	-0,035	0,097	-0,039	**-0,595**	**0,665**	-0,280	**0,549**
	Hofstede: Power Distance	**Hofstede: Individualism vs. Collectivism**	**Hofstede: Masculinity versus Femininity**	**Hofstede: Uncertainty Avoidance Index**	**Hofstede: Long-Term Orientation**	**Hofstede: Indulgence versus Restraint**	**Inglehart: Traditional/Secular Values**	**Inglehart: Survival vs. Self-Expression Values**

Table 12.7. The correlations of the country results of Schwartz with the country results by Hofstede, Inglehart and the factors from our work

	Embeddedness	Hierarchy	Mastery	Affective Autonomy	Intellectual Autonomy	Egalitarianism	Harmony
Chapter 5: traditional religion	**0,567**	0,370	0,403	0,167	0,453	**0,598**	0,263
Chapter 5: racism	0,373	0,370	0,111	-0,391	-0,364	-0,193	-0,108
Chapter 5: higher education of the younger generation (education gap between the generations)	0,185	0,156	0,379	0,344	0,444	0,208	0,214
Chapter 5: distrust of the army and the press	-0,333	-0,469	-0,249	0,214	0,382	0,033	-0,073
Chapter 5: authoritarian character	**0,518**	0,325	0,299	-0,058	-0,016	0,015	0,350
Chapter 5: tolerance and respect + postmaterialism	**-0,525**	-0,556	-0,284	0,145	0,228	0,100	-0,137
Chapter 5: the 'ego' company (rejection of obedience + unselfishness)	-0,341	-0,180	-0,331	-0,210	-0,400	**-0,640**	-0,395
Chapter 5: female rejection of the market economy and democracy	0,162	-0,070	-0,133	-0,148	0,094	0,113	-0,044
Chapter 12 re-analysis: moral (sexual) permissiveness ('Permissiveness 1')	**-0,727**	**-0,642**	-0,462	0,171	-0,018	-0,230	-0,148

Table 12.7. (Continued)

	Embeddedness	Hierarchy	Mastery	Affective Autonomy	Intellectual Autonomy	Egalitarianism	Harmony
Chapter 12 re-analysis: shadow economy mentality ('Permissiveness 2')	0,093	-0,261	-0,145	-0,281	-0,133	-0,156	0,044
Chapter 12 re-analysis: distance from religion ('Permissiveness 3')	**-0,516**	-0,337	-0,372	-0,191	-0,486	**-0,641**	-0,249
Chapter 12 re-analysis: educational values: independence and imagination	-0,299	-0,187	-0,202	-0,089	-0,096	-0,092	-0,340
Chapter 12 re-analysis: distance to market economy values	0,497	**0,557**	0,122	-0,219	-0,039	0,172	0,023
Chapter 12 re-analysis: education values: responsibility and tolerance	**-0,518**	**-0,533**	-0,370	0,028	0,210	-0,052	-0,154
Chapter 12 re-analysis: educational values: determination and perseverance and being against saving	**-0,557**	**-0,598**	-0,230	0,220	0,267	-0,007	-0,016
Chapter 12 re-analysis: right wing acceptance of inequality	0,452	**0,545**	0,188	-0,206	-0,394	-0,110	-0,068
Chapter 12 re-analysis: education values: unselfishness and being against hard work	0,160	0,054	0,165	0,100	0,274	0,490	0,161
Chapter 6: feminism analysis: distance to traditionalist forms of religion	**-0,702**	**-0,616**	**-0,542**	-0,082	-0,283	**-0,557**	-0,390
Chapter 6: feminism analysis: Generation gap in educational level	-0,181	-0,115	-0,042	0,191	0,116	-0,191	-0,252
Chapter 6: feminism analysis: feminism	**-0,677**	**-0,749**	**-0,516**	0,145	0,207	-0,101	-0,392
Hofstede: Power Distance	**0,570**	**0,569**	0,365	-0,216	-0,119	-0,002	0,258
Hofstede: Individualism vs. Collectivism	**-0,567**	**-0,532**	-0,328	0,251	0,136	-0,078	-0,147
Hofstede: Masculinity versus Femininity	-0,187	-0,065	-0,103	0,028	0,065	0,052	-0,046
Hofstede: Uncertainty Avoidance Index	0,050	-0,238	-0,080	-0,091	0,325	0,213	0,362
Hofstede: Long-Term Orientation	-0,378	-0,218	-0,400	-0,322	-0,488	**-0,685**	-0,379
Hofstede: Indulgence versus Restraint	0,062	-0,006	0,203	0,492	**0,501**	**0,586**	0,210
Inglehart: Secular Values (WVS 1-4, 2006)	**-0,705**	**-0,619**	**-0,635**	-0,187	-0,351	**-0,709**	-0,447
Inglehart: Self-Expression Values (WVS 1-4, 2006)	-0,412	-0,243	-0,154	0,437	0,207	0,154	-0,223

Table 12.8. The correlations of the country results of Schwartz with country aggregate and value research data

	Embeddedness	Hierarchy	Mastery	Affective Autonomy	Intellectual Autonomy	Egalitarianism	Harmony
Membership in the Islamic Conference	**0,677**	**0,613**	**0,537**	0,098	0,089	0,321	0,480
Muslim population share per total population	**0,678**	**0,613**	**0,526**	0,074	0,059	0,298	0,459
% women in government, all levels	-0,301	-0,346	-0,116	0,291	0,148	0,041	-0,123
2000 Economic Freedom Score	-0,261	-0,124	-0,145	0,239	0,065	-0,061	-0,412
ecological footprint (g ha /cap)	-0,327	-0,237	-0,206	0,288	0,026	-0,182	-0,251
ESI-Index Environment Sustainability Index (Yale Columbia)	-0,369	-0,482	-0,313	0,134	0,219	-0,016	-0,103
Happy life years	-0,382	-0,260	-0,276	0,288	0,130	0,050	-0,149
Happy Planet Index, HPI	0,198	0,206	0,119	-0,028	0,087	0,302	0,260
homicide rate	0,130	0,123	0,142	0,082	0,215	0,170	-0,009
Immigration - Share of population 2005 (%)	0,003	0,169	0,105	0,287	0,081	-0,098	-0,162
Life Satisfaction (0-10)	-0,311	-0,193	-0,220	0,338	0,143	0,098	-0,097
WVS: mean acceptance competition	0,032	-0,063	-0,133	-0,006	0,254	0,187	-0,002
WVS: mean acceptance gov benefits fraud	0,235	0,116	0,083	0,015	0,263	0,205	0,149
WVS: mean left-right position	0,433	0,440	0,297	-0,134	-0,204	0,148	0,238
Shadow economy as % of the GDP	0,456	0,325	0,265	-0,153	0,083	0,149	0,254
Global tolerance index	**-0,595**	**-0,609**	-0,404	0,329	0,250	-0,076	-0,361
Human development index (HDI) value 2004	**-0,610**	**-0,563**	-0,498	0,080	0,015	-0,218	-0,386
ln GDP per capita	**-0,594**	**-0,552**	-0,492	0,149	0,036	-0,224	-0,386
Middle class share (share of quintiles 2-4), average 60-96	-0,484	-0,379	-0,456	-0,097	-0,432	**-0,553**	-0,377
principal component growth 2008-2011, final version	0,399	**0,596**	0,403	-0,024	-0,067	0,136	0,115
quintile share income difference between richest and poorest 20%	0,224	0,227	0,269	0,169	0,389	***0,437***	0,130
social security expenditure per GDP average 1990s (ILO)	**-0,626**	**-0,728**	-0,494	0,079	0,145	-0,115	-0,172
Total crimes (per capita)	-0,473	**-0,576**	-0,251	0,220	-0,006	-0,040	-0,238
unemployment rate	0,283	0,009	0,182	-0,037	0,169	0,356	0,287
Value Development	-0,359	-0,128	-0,049	0,213	0,077	0,093	-0,085
World Economic Forum Global Competitiveness	**-0,549**	-0,337	-0,387	0,186	-0,153	-0,315	**-0,521**

Table 12.9. Was Schumpeter right after all? The loss of capitalist family values as measured by social acceptancy rates of homosexuality and their correlates from the *World Values Survey*

	Control variables: Income level & Education level (recoded)		
	Justifiable: homosexuality		
	partial correlation	**error p**	**df. (degrees of freedom)**
Justifiable: prostitution	0,628	0,000	43318
Justifiable: divorce	0,475	0,000	43318
Justifiable: abortion	0,474	0,000	43318
Justifiable: suicide	0,419	0,000	43318
Justifiable: euthanasia	0,397	0,000	42476
No confidence: Churches	0,217	0,000	43318
Justifiable: drinking alcohol	0,197	0,000	14292
Justifiable: someone accepting a bribe	0,188	0,000	43318
Justifiable: avoiding a fare on public transport	0,167	0,000	43318
Justifiable: cheating on taxes	0,164	0,000	43318
No confidence: Armed Forces	0,123	0,000	43318
Justifiable: claiming government benefits	0,105	0,000	42795
No confidence: The Government	0,086	0,000	43318
No confidence: The Civil Services	0,081	0,000	43318
No confidence: Parliament	0,078	0,000	43318
No confidence: The Press	0,076	0,000	43318
No confidence: Television	0,072	0,000	43318
No confidence: Major Companies	0,060	0,000	42066
No confidence: The Political Parties	0,058	0,000	42473
No confidence: Labor Unions	0,017	0,001	39781

Whatever way, the French Socialists overlooked the fact that the global citizens saying homosexuality is always justified will also have a high tendency to say as well that prostitution is always justified. The correlation between the two phenomena, independent from income and education, is 0.628, i.e., 39.4% of the variance is in common. Considering that these are results from individual interviews across the globe and across cultures, this correlation again is really extraordinary. So if there is a majority of citizens voting "yes" in a hypothetical global referendum on legalizing prostitution and gay marriages, the same constituencies would be in favor or against the motion.

Table 12.10. Social acceptancy rates of prostitution and their correlates from the *World Values Survey*

	Control variables: Income level & Education level (recoded)		
	Justifiable: prostitution		
	partial correlation	**error p**	**df. (degrees of freedom)**
Justifiable: homosexuality	0,628	0,000	43318
Justifiable: abortion	0,491	0,000	43566
Justifiable: suicide	0,454	0,000	43566
Justifiable: divorce	0,434	0,000	43566
Justifiable: euthanasia	0,385	0,000	42476
Justifiable: someone accepting a bribe	0,299	0,000	43566
Justifiable: cheating on taxes	0,272	0,000	43553
Justifiable: avoiding a fare on public transport	0,231	0,000	43566
Justifiable: drinking alcohol	0,214	0,000	14325
No confidence: Churches	0,188	0,000	43566
Justifiable: claiming government benefits	0,162	0,000	42795
No confidence: Armed Forces	0,113	0,000	43566
No confidence: The Government	0,093	0,000	43566
No confidence: The Civil Services	0,092	0,000	43566
No confidence: Parliament	0,085	0,000	43323
No confidence: The Press	0,069	0,000	43566
No confidence: Television	0,066	0,000	43566
No confidence: The Political Parties	0,062	0,000	42473
No confidence: Major Companies	0,056	0,000	42066
No confidence: Labor Unions	0,043	0,000	39781
No confidence: The Police	0,040	0,000	43566
No confidence: The United Nations	0,018	0,000	41746
No confidence: The Women´s Movement	0,018	0,000	40824

Beyond this specifically European point, a larger issue emerges, which really goes to the heart of the whole argument, i.e., *whether changing sexual norms away from the standards of the father/mother/child family, prescribed by the major world religions, should be important for democracy.* Inglehart and Welzel (2009) are very explicit about this, making it very clear that for them it is a vital issue for the future of democracy:

> "[…] Modernization does not automatically lead to democracy. Rather, it, in the long run, brings social and cultural changes that make democratization increasingly probable. Simply attaining a high level of per capita GDP does not produce democracy: if it did, Kuwait and the United Arab Emirates would have become model democracies. (These countries have not gone through the modernization process described above.) But the emergence of postindustrial society brings certain social and cultural changes that are specifically conducive to democratization. Knowledge societies cannot function effectively without highly educated publics that have become increasingly accustomed to thinking for themselves. Furthermore, rising levels of economic security bring a growing emphasis on a syndrome of self-expression

values -- one that gives high priority to free choice and motivates political action. Beyond a certain point, accordingly, it becomes difficult to avoid democratization, because repressing mass demands for more open societies becomes increasingly costly and detrimental to economic effectiveness. Thus, in its advanced stages, modernization brings social and cultural changes that make the emergence and flourishing of democratic institutions increasingly likely.

The core idea of modernization theory is that economic and technological development bring a *coherent set of social, cultural, and political changes.* A large body of empirical evidence supports this idea. *Economic development is, indeed, strongly linked to pervasive shifts in people's beliefs and motivations, and these shifts in turn change the role of religion, job motivations, human fertility rates, gender roles, and sexual norms. And they also bring growing mass demands for democratic institutions and for more responsive behavior on the part of elites. These changes together make democracy increasingly likely to emerge, while also making war less acceptable to publics."* (Inglehart and Welzel, 2009).

At one point in their essay, they maintain:

"The shift from traditional to secular-rational values is linked to the shift from agrarian to industrial societies. Traditional societies emphasize religion, respect for and obedience to authority, and national pride. These characteristics change as societies become more secular and rational.

The shift from survival to self-expression values is linked to the rise of postindustrial societies. It reflects a cultural shift that occurs when younger geerations emerge that have grown up taking survival for granted. Survival values give top priority to economic and physical security and conformist social norms. *Self-expression values give high priority to freedom of expression, participation in decision-making, political activism, environmental protection, gender equality, and tolerance of ethnic minorities, foreigners, and gays and lesbians. A growing emphasis on these latter values engenders a culture of trust and tolerance in which people cherish individual freedom and self-expression and have activist political orientations. These attributes are crucial to democracy -- and thus explain how economic growth, which takes societies from agrarian to industrial and then from industrial to postindustrial, leads to democratization.* The unprecedented economic growth of the past 50 years has meant that an increasing share of the world's population has grown up taking survival for granted. Time-series data from the values surveys indicate that mass priorities have shifted from an overwhelming emphasis on economic and physical security to an emphasis on subjective well-being, self-expression, participation in decision-making, and a relatively trusting and tolerant outlook." (Inglehart and Welzel, 2009)

This shift away from traditional family values now happens in advanced Western democracies at an unprecedented scale. The replacement of *"traditional gender roles"* by concepts like *"parent 1"* and *"parent 2"* in official documents has already begun in several industrialized western democracies, including the United States of America:

"The words "mother" and "father" will be removed from U.S. passport applications and replaced with gender neutral terminology, the State Department says. "The words in the old form were 'mother' and 'father,'" said Brenda Sprague, deputy assistant Secretary of State for Passport Services. *"They are now 'parent one' and 'parent two.'"* A statement on the State Department website noted: "These improvements are being made to provide a gender neutral description of a child's parents and in recognition of different types of families." The statement didn't note if it was for child applications only. The State Department said the new passport applications, not yet available to the public, will be available online soon. Sprague said the decision to remove the traditional parenting names was not an act of political correctness. "We

> find that with changes in medical science and reproductive technology that we are confronting situations now that we would not have anticipated 10 or 15 years ago," she said. Gay rights groups are applauding the decision. "Changing the term mother and father to the more global term of parent allows many different types of families to be able to go and apply for a passport for their child without feeling like the government doesn't recognize their family," said Jennifer Chrisler, executive director of Family Equality Council. Her organization lobbied the government for several years to remove the words from passport applications. "Our government needs to recognize that the family structure is changing," Chrisler said. "The best thing that we can do is support people who are raising kids in loving, stable families." (http://www.foxnews.com/politics/2011/01/07/passport-applications-soon-gender-neutral/)

Major mainstream religious denominations like the Roman Catholic Church now began to join the debate on *"parent 1" and "parent 2"* by voicing their absolute discontent:

> "ROME, September 13, 2013 (LifeSiteNews.com) – *Replacing terms like mother and father with "Parent 1 and Parent 2" is not an innocuous defense of "equality" but the beginning of a complete re-write of Italian society, a prominent Italian cardinal said this week. Changing the categories of family life and genealogy could send "a real earthquake" through the social structure.* The expression, that has already been proposed for all official documents related to public housing and kindergarten placements in the city of Venice, will make "the category of fatherhood and motherhood disappear, to be replaced by the generic category of 'parenting,'" the Cardinal Archbishop of Bologna, Carlo Caffarra, said. Speaking in a keynote address at a conference at the Teatro Manzoni in Bologna this week, Cardinal Caffarra identified the largest question facing Italian society: "Is marriage a reality at the disposal of the men, or does it have a 'hard core' that is not open to human preference?" (http://www.lifesitenews.com/news/cardinal-replacing-mom-and-dad-with-parent-1-and-2-will-send-an-earthquake)

From the perspective of developing countries, striving for stable institutions of democracy, one really has to question whether at the end of the day the political decision makers should start re-writing the identity cards and passports of their citizens and start hosting *"love parades"* in the name of *"tolerance for gays and lesbians",* said to be so important for achieving effective democracy? Should they follow the lead of the Parliamentary Assembly of the *"Council of Europe"* (http://assembly.coe.int/ASP/Doc/XrefDocDetails_ E.asp?FileID= 12456) and adopt new legislation of their own also to eliminate the words *"father"* and *"mother"* from official documents and can they hope then that their political system would become *"an effective democracy"* by adopting such measures?

Let us return for a moment to the Choropleth maps 3.11 and 3.12 of this work and the data on "effective democracy". As Alexander, Inglehart and Welzel, 2011, explain the index of effective democracy combines two processes: *civil liberties ratings* and the *political rights ratings,* which are averaged to obtain an *overall index of democratic rights.* This component index is transformed into a 0-to-100 range. The dimension of effectiveness in the index is provided by the use of the World Bank's *Rule of Law index.*

Table 12.11. The drivers of *"effective democracy"* irrespective of development level (partial correlations, Kuznets curve = development level constant; non-linear formulation)

	Pearson correlation with effective democracy	**R^2**
% women in government, all levels	0,484	23,383
WEF closing of global gender gap overall score 2009	0,461	21,278
Investment Freedom	0,449	20,144
closing political gender gap	0,449	20,119
WEF Global Gender Gap Report Political Empowerment 2009	0,449	20,119
Dummy for oil exporting	**-0,447**	**19,949**
WEF Gender Gap Report Economic Participation and Opportunity 2009	0,439	19,298
% women in government, sub ministerial level	0,436	19,041
UNDP gender empowerment index	0,433	18,773
military personnel rate ln (MPR+1)	**-0,419**	**17,579**
Hofstede: Power Distance	-0,419	17,573
Inglehart: Self-Expression	0,419	17,557
Hofstede: Individualism vs. Collectivism	0,409	16,702
Heshmati/Karoui/Tausch: authoritarian character	-0,402	16,131
Protestant population share, 2000	0,398	15,855
Heshmati/Karoui/Tausch: feminism analysis: feminism	0,398	15,811
average surplus-deficit 2000-2007	-0,387	15,002
military expenditures per GDP	**-0,373**	**13,915**
% women in government, ministerial level	0,373	13,898
Heshmati/Karoui/Tausch: tolerance and respect + postmaterialism	0,372	13,826
Immigration - Share of population 2005 (%)	**-0,371**	**13,800**
Ratio of estimated female to male earned income	0,371	13,734
Tausch: Global tolerance index	0,368	13,578
Property Rights	0,348	12,088
Financial Freedom	0,346	11,996
Kearney technological globalization	0,346	11,993
Heshmati/Karoui/Tausch: re-analysis: right wing acceptance of inequality	-0,328	10,747
Heshmati/Karoui/Tausch: Value Development Index	0,326	10,650

Source: Our own calculations from https://uibk.academia.edu/ArnoTausch/Documentation-for-books-and-articles - Analysis of the shadow economy.

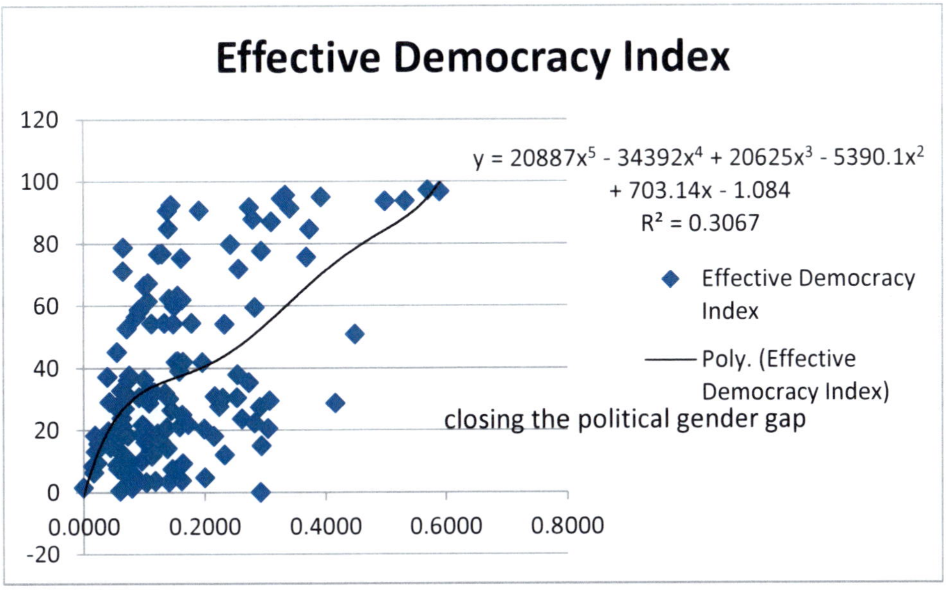

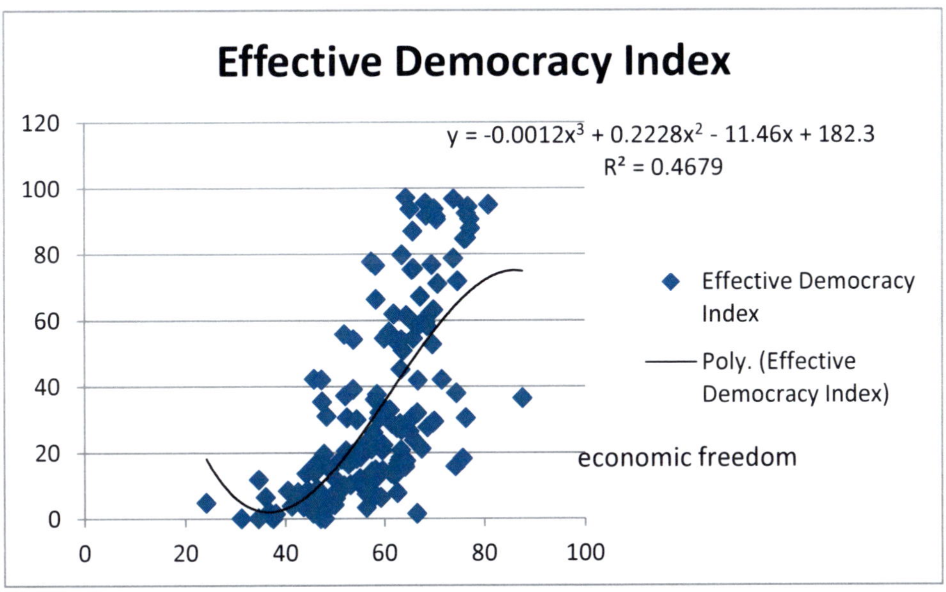

Graph 12.1. The drivers of "effective democracy".

If we keep development levels – as in Map 3.12 of this work – constant, it emerges that hard core, economic and social "bread and butter" feminism indicators and economic freedom indicators are far more influential than value change indicators in bringing about effective democracy even at low levels of development. Inglehart's *"self-expression"* scale achieves about the same results as Hofstede's "power distance" and our own "authoritarian personality" scale. The absence of militarism and of an oil exporting economy has also a beneficial effect on *"effective democracy"*. So, while feminism and a free economy generally affect effective democracy in a positive fashion, we have to emphasize as well that Maastricht-type of savings programs and also very high inward migration rates are not a driver, but a bottleneck of effective democracy.

Policy pointer:

Is prostitution justifiable? Is homosexuality justifiable? The French socialists, it seems, for example seem to think that one is not, and the other is. President Hollande and his administration put considerable political energy into legalizing homosexual marriages and prohibiting prostitution. But global citizens hold another view, and there is a high positive correlation of 0.632 between the two items in the *World Values Survey*, based on 218877 individuals from around the globe.

The emotionalized and very bitter debate in France took place without any real reference to the empirical realities of global values "on the ground". France, like many other European countries, seems to be disconnected from global social science research. The freely available archive of all the leading French language scholarly journals, *Persee,* lists just 17 articles mentioning the word "*World Values Survey*" at all. Not a single article of these 17 articles deals with the problem of homosexuality, which was the issue of a real ideological warfare in the country recently.

Current debates in Europe on issues of human values have acquired an almost quasi-religious character, with each side claiming that it is in the sole possession of truth, while in reality the complex relationships between value patterns are often overlooked. There are significant and higher correlations between the unconditional acceptancy of homosexuality with such phenomena as accepting a bribe, avoiding a fare on public transport, cheating on taxes, and the generalized loss of trust in basic institutions of democratic society.

Whatever way, the French socialists overlooked the fact that the global citizens, saying homosexuality is always justified, will have a high tendency to say as well that prostitution is always justified. The correlation between the two phenomena, independent from income and education, is 0.628, i.e., 39.4% of the variance in common. Considering that these are results from individual interviews across the globe and across cultures, this correlation is really extraordinary. So if there is a majority of citizens voting "yes" in a hypothetical global referendum on legalizing prostitution and gay marriages, the same constituencies would be in favor or against the motion.

Graph 12.1 highlights the message of our book on the drivers of "effective democracy": a sound gender political agenda, ending the political discrimination of women, and economic freedom will be conducive to "effective democracy".

Graph 12.1 highlights the politically, socially and ethically robust and globally applicable message of our book on the drivers of *"effective democracy":* a sound gender political agenda, ending the political discrimination of women, and economic freedom will be conducive to *"effective democracy"*. Nevertheless the path towards *"effective democracy"* will be one of ups and downs, and especially in developing countries, there will be also certain limits for a too rapid economic liberalization.

12.4. A GLANCE AT THE *WORLD VALUES SURVEY* WAVE, 6, 2010-2014

As the manuscript to this book was about to be finished, the new data of the *World Values Survey,* 2010-2014 were released, containing yet another enormous wealth of new data, including on the Muslim world. We have chosen to concentrate on two phenomena, which received a large attention on the pages of this book – tolerance and democracy. *World Values Survey* data for the 6th wave are now available for Algeria; Armenia; Australia; Azerbaijan; Belarus; Chile; China; Colombia; Cyprus; Ecuador; Estonia; Germany; Ghana; Iraq; Japan; Jordan; Kazakhstan; Kuwait; Kyrgyzstan; Lebanon; Libya; Malaysia; Mexico; Morocco; Netherlands; New Zealand; Nigeria; Pakistan; Peru; Philippines; Poland; Qatar; Romania; Russia; Rwanda; Singapore; Slovenia; South Korea; Spain; Sweden; Taiwan; The Occupied Palestinian Territories; Trinidad and Tobago; Tunisia; Turkey; Ukraine; United States; Uruguay; Uzbekistan; Yemen; and Zimbabwe.

In Table 12.12 we calculate a simple UNDP Human Development Index type of Index of Tolerance, minimizing the rejection of neighbors with the following characteristics:

- People who speak a different language
- People of a different religion
- Immigrants/foreign workers
- People of a different race

Table 12.12. Xenophobia rates in the world system, 2010-2014: Rejecting different types of neighbors

	Rejection rate in %	Rejection rate in %	Rejection rate in %	Rejection rate in %	Index of tolerance (0-1)
	People who speak a different language	People of a different religion	Immigrants/foreign workers	People of a different race	Overall tolerance index
Uruguay	2,50	2,90	1,70	1,60	0,993
Sweden	3,20	3,70	3,50	2,80	0,971
New Zealand	5,50	1,40	5,90	2,90	0,955
Spain	3,10	3,10	7,50	4,80	0,949
Trinidad and Tobago	6,70	2,90	6,80	1,50	0,943
Poland	3,20	4,60	7,20	5,50	0,939
Rwanda	7,70	4,80	5,40	3,50	0,924
Colombia	7,50	6,50	4,70	3,20	0,922
Chile	5,40	5,80	7,60	5,60	0,917
Australia	9,90	4,00	10,50	5,00	0,885
Zimbabwe	6,90	7,10	13,70	8,10	0,863
Slovenia	5,00	8,50	13,60	10,90	0,858
China	6,60	9,20	12,20	10,50	0,852
United States	12,90	3,40	13,60	5,60	0,851

Table 12.12. (Continued)

	Rejection rate in %	Rejection rate in %	Rejection rate in %	Rejection rate in %	Index of tolerance (0-1)
	People who speak a different language	People of a different religion	Immigrants/foreign workers	People of a different race	Overall tolerance index
Taiwan	7,00	4,50	20,20	8,40	0,845
Peru	10,90	12,30	10,70	9,30	0,820
Netherlands	14,50	2,70	19,60	8,20	0,806
Uzbekistan	7,10	15,10	12,30	14,00	0,806
Mexico	14,10	15,60	11,60	10,20	0,776
Morocco	11,30	18,90	10,80	13,80	0,767
Kazakhstan	11,30	12,10	27,80	11,20	0,736
Ukraine	11,60	15,00	19,30	16,90	0,733
Germany	13,40	14,10	21,40	14,80	0,725
Singapore	10,20	10,80	35,80	12,60	0,709
Pakistan	12,00	23,80	20,90	15,70	0,689
Qatar	9,80	11,60	46,00	8,80	0,681
Tunisia	12,90	29,70	18,80	16,90	0,659
Romania	16,70	19,10	21,30	23,50	0,642
Russia	18,90	14,30	32,20	17,20	0,629
Ghana	21,90	22,50	20,20	19,90	0,612
Philippines	30,70	16,30	14,10	21,60	0,599
Belarus	17,60	18,00	33,40	23,10	0,590
Cyprus	14,40	20,10	35,50	24,80	0,586
Nigeria	23,70	26,50	20,10	20,90	0,577
Estonia	16,50	20,70	37,50	25,40	0,558
Algeria	18,10	42,70	28,00	19,80	0,513
Japan	19,90	32,60	36,30	22,30	0,499
Armenia	16,20	56,60	18,40	31,80	0,451
Kyrgyzstan	25,90	34,10	30,70	28,10	0,450
Jordan	27,40	28,70	37,20	27,20	0,441
Iraq	27,80	28,20	39,00	27,70	0,430
Kuwait	28,90	*29,00*[64]	37,20	28,10	0,425
Turkey	30,00	36,80	30,50	35,80	0,377
Ecuador	34,20	32,10	34,60	34,50	0,358
Lebanon	32,40	33,80	40,40	36,30	0,330
South Korea	33,30	31,10	44,20	34,10	0,329
Yemen	31,30	53,30	24,20	34,00	0,329
Malaysia	29,20	30,30	59,70	31,30	0,306
Azerbaijan	26,30	34,80	40,60	58,10	0,269
The Occupied Palestinian Territories	39,20	50,30	39,90	44,00	0,177
Libya	39,30	54,10	59,00	55,10	0,028

[64] To be able to work with the full number of countries with otherwise available data, we had to impute the missing observation for Kuwait for the variable *"rejecting neighbors: of a different religion"* from the closely correlating variable: *"rejecting people of a different race"*. Since the two variables correlate with each other to the tune of +0.87 (explained R^2 = 75.21%), we used the linear OLS regression function

*y (religio-phobia) = 0.9644 * racism + 1.9079*

from Table 12.12 to arrive at the imputed value of 29.0% religio-phobic people in Kuwait. This is the only instance in this book that we use such an imputation.

According to the data the most tolerant nation on earth now is Uruguay, followed by Sweden; New Zealand; Spain; Trinidad and Tobago; Poland; Rwanda; Colombia; Chile and Australia.

Uzbekistan, Morocco and Kazakhstan are nowadays ahead of Germany; and Pakistan, Qatar and Tunisia are more tolerant than the EU-member country Romania. Muslim countries such as Turkey (which is still ahead of the OECD-member country South Korea) have still a poor performance.

Table 12.13 and Maps 12.9 to 12.12 list the *World Values Survey* results for the average importance given by the global publics to democracy and the standard deviation of this indicator. Where the standard deviation is low, opinions on democracy – either way – are undivided, while high standard deviations indicate that the publics are – often bitterly – divided on the issue of democracy.

Countries with an above than average importance assigned to democracy and very high internal divisions on this issue are Tunisia; Mexico; Romania; Armenia and Yemen. While there is a general consensus that democracy is important, there are important dissenting voices. Nostalgia for past more authoritarian patterns of government can go hand in hand with economic discontent with present conditions. Countries with an above than average importance assigned to democracy, and very low internal divisions on this issue are the *Netherlands; Egypt; Sweden; Turkey; and Cyprus. For anyone, attempting to turn back the clocks of history in such countries could result to be a very costly error. The recent introduction of internet censorship in Turkey would be just one example showing the relevance of this hypothesis.*

Countries with a below than average importance assigned to democracy, but very high internal divisions on the issue are *Libya; Philippines; Qatar; the Occupied Palestinian Territories; and Russia.* In these countries and territories, *debates on the issue of democracy will surge, one way or the other. While the average importance assigned to democracy is still lower than the world average, the divisions on the issue are already very high, and unforeseen events could trigger a popular movement for more participation and democracy.* Finally, countries with a below than average importance assigned to democracy, and very low internal divisions on this issue are Singapore; Rwanda; South Korea; Estonia; and Lebanon. One might expect that the current stagnation in the democratic development of the given country will continue: publics don't assign a great importance to democracy, and they are hardly divided on this issue.

Policy pointer:

As the manuscript to this book was about to be finished, the new data of the *World Values Survey*, 2010-2014 were released, containing an enormous wealth of new data, including on the Muslim world. In Table 12.12 we calculate a simple UNDP Human Development Index type of Index of Tolerance, minimizing the rejection of neighbors with the following characteristics:

- People who speak a different language
- People of a different religion
- Immigrants/foreign workers
- People of a different race

According to these *World Values Survey* data, the most tolerant nation on earth now is Uruguay, followed by Sweden; New Zealand; Spain; Trinidad and Tobago; Poland; Rwanda; Colombia; Chile and Australia. The good news for the growing forces of moderation and democracy in the Arab world and the Muslim world in general is that Uzbekistan, Morocco and Kazakhstan are nowadays ahead of Germany. Pakistan, Qatar and Tunisia is more tolerant than the EU-member country Romania. The bad news is the poor performance of such Muslim countries as Turkey (which is still ahead of the OECD-member country South Korea).

Table 12.13 and our maps also have another, more immediate and direct implication: the dire state of the support of democracy in many Western countries, currently hit by the economic crisis and austerity packages, and the surge of democracy in the Muslim world and the Arab world in particular. That Egypt is ahead of Germany, Uzbekistan ahead of the EU-members Poland and Spain, and a number of other Arab and Muslim countries in general ahead of the United States; and Qatar ahead of the EU-member Estonia with justification should be celebrated by the Arab and Muslim readership of this book. Such comparison could be continued endlessly, and it is time for our readers now to lay down this book and start to click:

http://www.worldvaluessurvey.org/WVSOnline.jsp

Table 12.13. The importance of democracy

	Importance of democracy (1-10)	**standard deviation: importance of democracy**
Sweden	9,29	1,53
Cyprus	9,15	1,60
Uruguay	9,00	1,61
Egypt	8,95	1,51
Germany	8,94	1,70
Taiwan	8,91	1,61
Netherlands	8,87	1,40
Australia	8,83	1,90
Uzbekistan	8,79	1,77
Zimbabwe	8,77	1,90
New Zealand	8,73	1,94
Poland	8,70	1,91
Spain	8,64	1,61
Kazakhstan	8,62	1,81
Armenia	8,62	2,13
Malaysia	8,61	1,63
Trinidad and Tobago	8,61	2,09
Turkey	8,57	1,58
Romania	**8,57**	**2,25**
Mexico	**8,54**	**2,28**
Chile	**8,53**	**1,95**
Ecuador	8,51	1,82
Morocco	8,49	2,09
Yemen	8,49	2,13
Tunisia	8,46	2,38
China	8,43	1,59
Ghana	8,43	1,84
United States	**8,41**	**2,05**
Peru	8,39	2,08
Qatar	8,34	2,49
South Korea	**8,31**	**1,82**
Estonia	**8,30**	**1,93**
Jordan	8,29	2,07
Japan	**8,27**	**2,00**
Kuwait	8,17	2,37
Colombia	8,13	2,24
Libya	8,13	2,68
Algeria	8,12	2,32
Slovenia	**8,11**	**2,17**
Azerbaijan	8,08	2,02
Philippines	**8,04**	**2,56**
Lebanon	8,01	1,96
Iraq	7,95	2,13
Kyrgyzstan	7,89	2,20
Nigeria	7,86	1,99
Pakistan	7,85	2,07
Ukraine	7,83	2,35
The Occupied Palestinian Territories	7,78	2,44
Rwanda	7,69	1,78
Singapore	7,65	1,58
Belarus	7,48	2,24
Russia	7,42	2,42

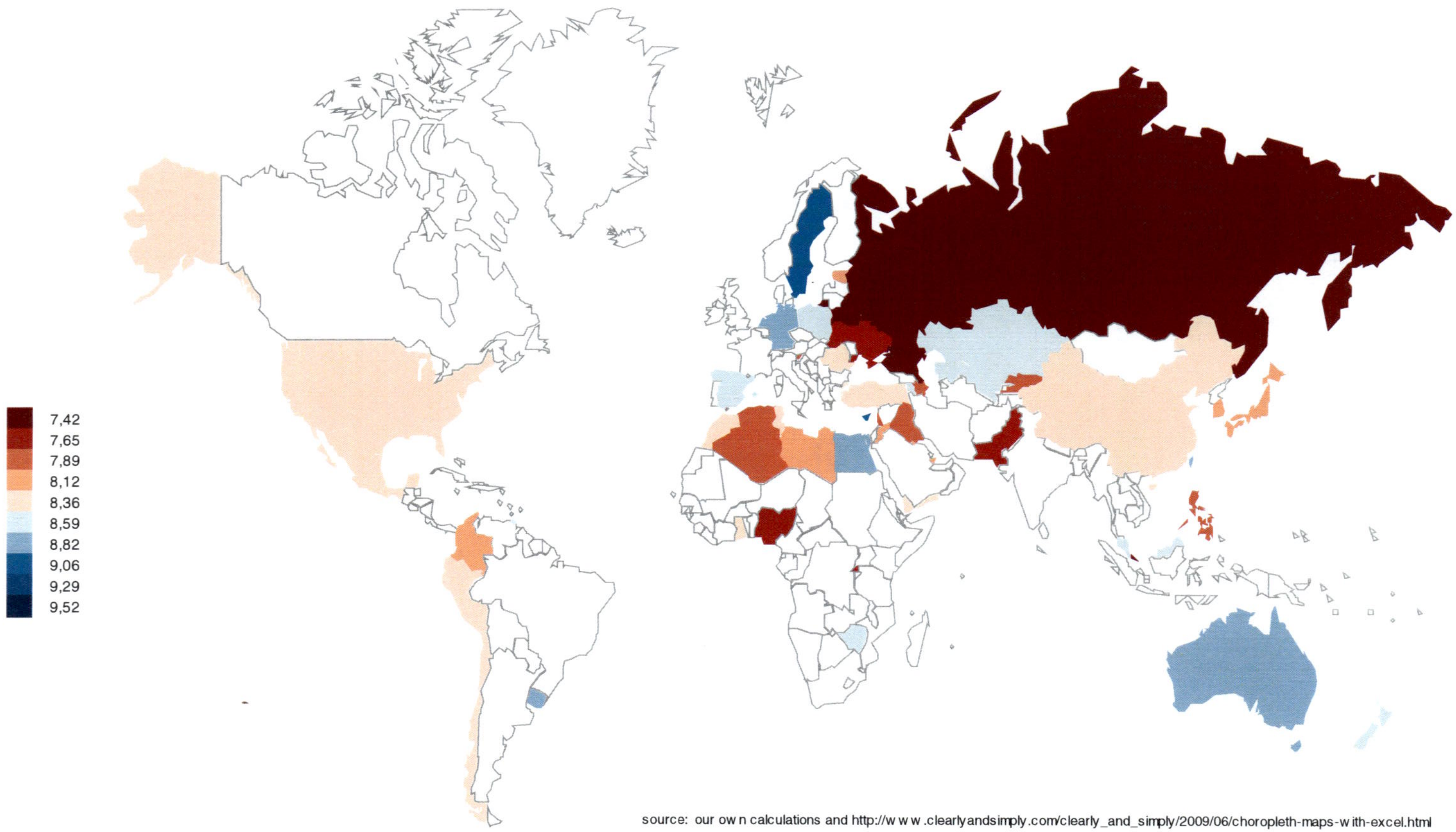

Map 12.9. The importance of democracy (1-10) – *World Values Survey,* Wave 6, 2010-2014.

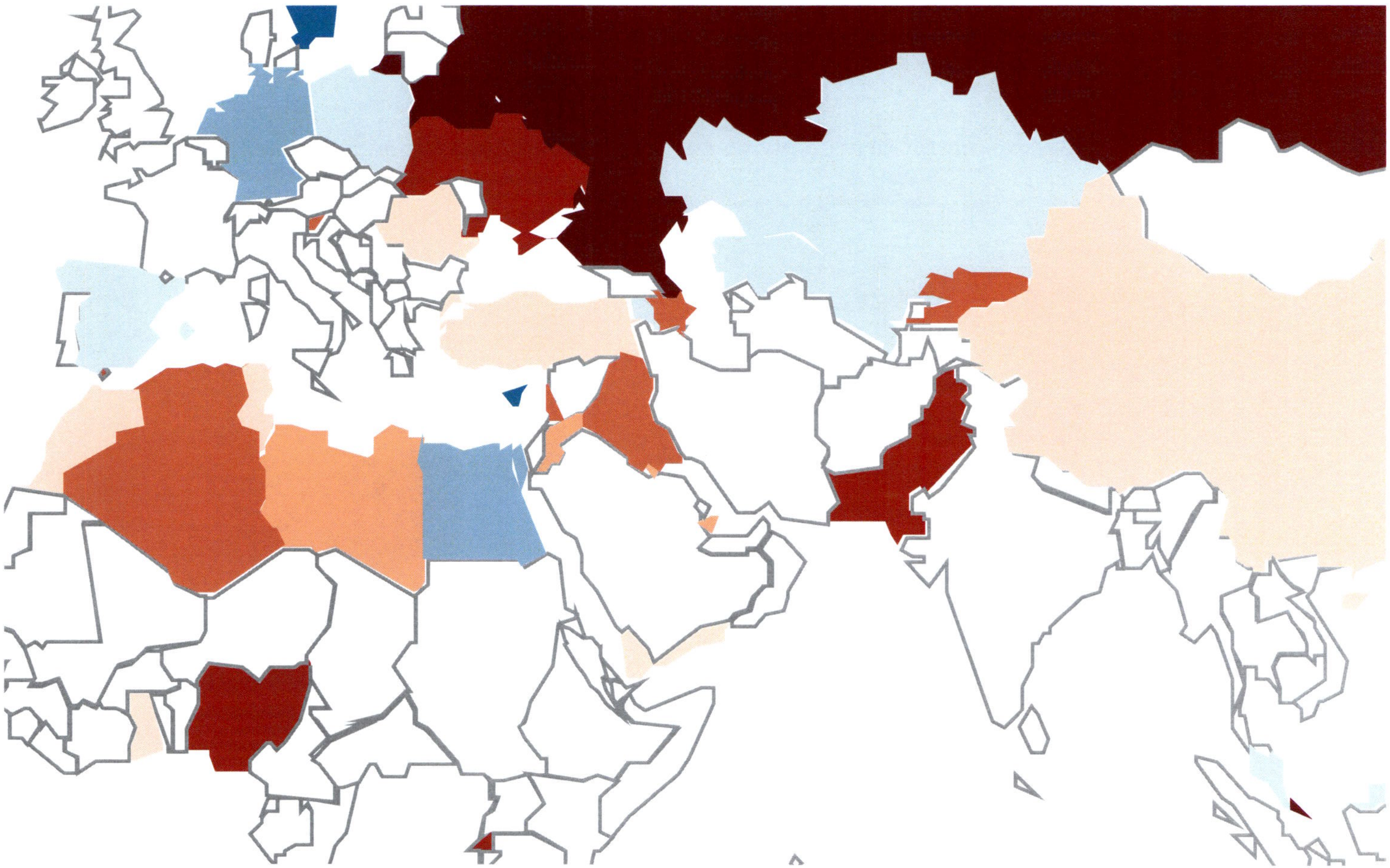

Map 12.10. The importance of democracy (1-10) – *World Values Survey,* Wave 6, 2010-2014 – close-up for the Middle-East region.

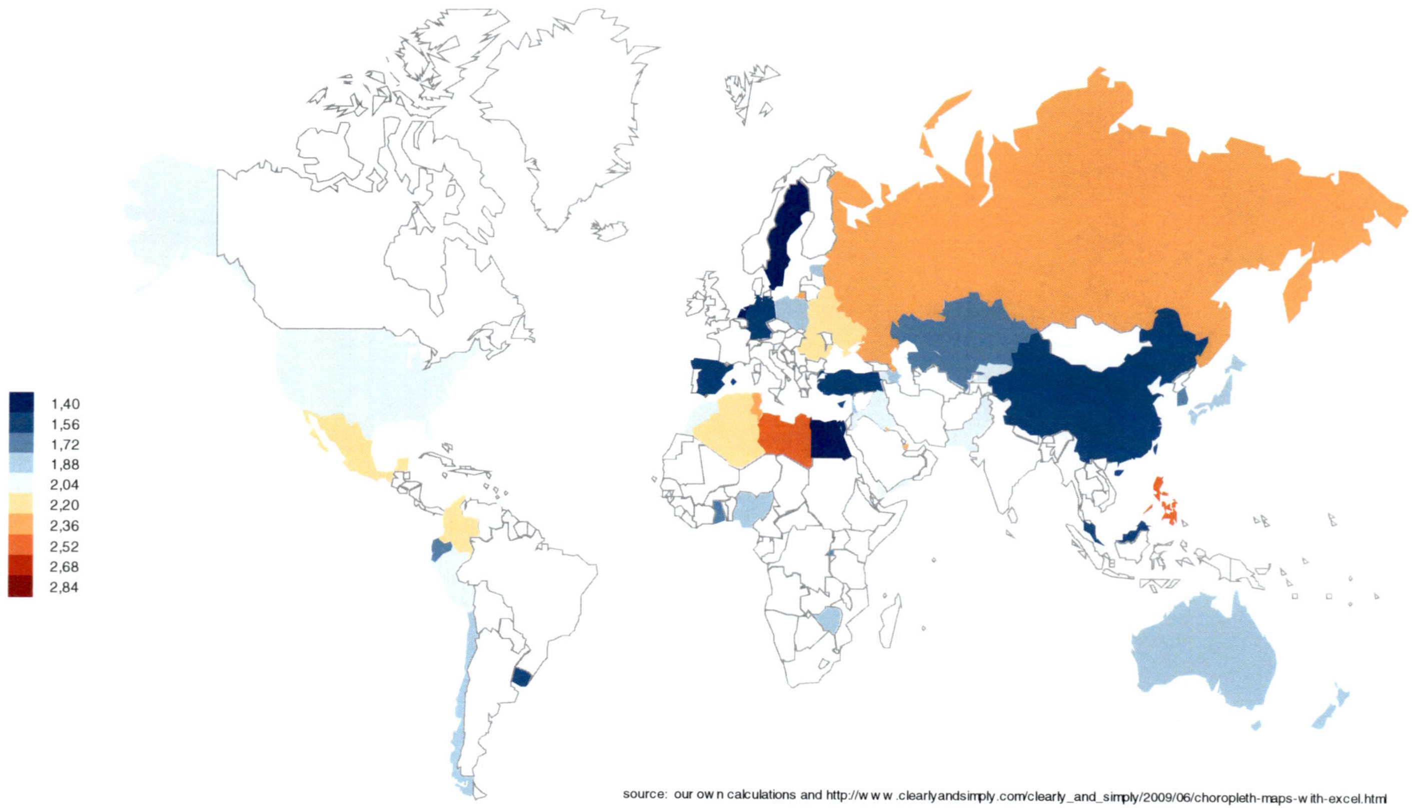

Map 12.11. Where the value of democracy is weakly or strongly contested: standard deviations of the importance, assigned by publics to democracy (1-10) – *World Values Survey,* Wave 6, 2010-2014.

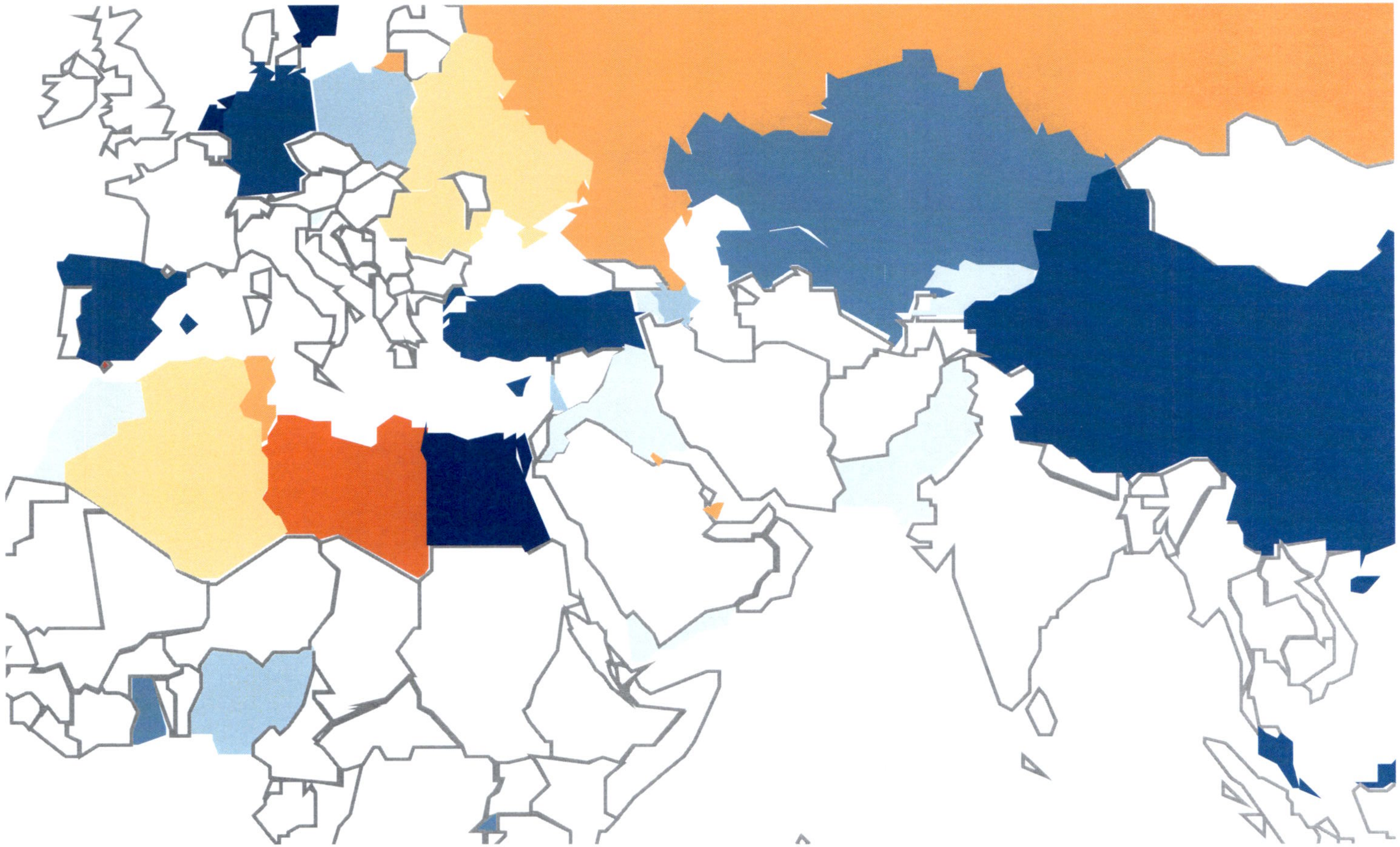

Map 12.12. Where the value of democracy is weakly or strongly contested: standard deviations of the importance, assigned by publics to democracy (1-10) – *World Values Survey,* Wave 6, 2010-2014 - – close-up for the Middle-East region.

Chapter 13

64 THEMES FOR STUDENT AND COURSE PARTICIPANTS' ESSAYS

After having worked with this book, readers can check their progress in dealing with issues of global values by preparing 10-20 minute statements (i.e., approximately 1000-2000 words) on the following issues:

1. Please shortly describe the four main approaches in the social sciences today on global values and global value change. Where are their common points and where are their differences? (Hofstede, Schwartz, Inglehart, Barro/Alesina)
2. In what sense one can speak about a crisis of the West?
3. Ibn Khaldūn (1332-1406) and the role of values in civilization
4. The Khaldūn/Buddenbrooks cycle
5. Robert Barro and the study of human values
6. Barro and the cycles of convergence and divergence
7. The decline of Western countries, the analysis of global values and the world system position of the Arab and the Muslim world
8. The *World Values Survey* project and data from the Muslim world
9. The Composite Index Methodology
10. Non-parametric Indices
11. Parametric Indices
12. Varimax and Promax Rotation Procedures
13. Correlation Relationships
14. Partial Correlation Analysis
15. Simple, Multiple and Stepwise Regression Analysis
16. Survey analysis
17. Aggregate data
18. Sampling, sampling error and margins
19. Shadow economic activities and their measurement
20. Shadow economic activities in industrialized and transition economies
21. Tax morale, trust, institutional quality and shadow economy
22. Shadow economy, corruption, Information and Communication Technologies (ICT) and environment
23. Debt, default risk, financial development and size of informal economy
24. Economic Freedom, Corruption and Economic Growth

25. Corruption, economic freedom and development
26. Gains and well-being from economic freedom
27. Causality between economic freedom and economic growth
28. Effective Democracy
29. Democracy, natural resources and aid
30. Political culture, press freedom and outcome of economic policy
31. Political democracy, economic democracy and degrowth policies
32. Democracy, state capacity and economic growth
33. Geert Hofstede
34. Shalom Schwartz and Eldad Davidov
35. Inglehart's theory
36. The Re-analysis of Inglehart's data by Tausch/Heshmati/Karoui
37. The Kuznets curves of human values along the paths of human development
38. The factor analytical model
39. The new choropleth maps of global human values
40. Conclusions from the country factor scores
41. Adorno's F-scale and global economic growth in the post-2008 crisis era
42. The trajectories of global value change along the paths of human development
43. The performance of Muslim countries along these trajectories
44. The discovery of a Global Value Development Index and regional value differences at the sub-national level
45. Muslim feminism – the view of a quantitative social scientist
46. Women in Islam
47. The choropleth maps of global feminism
48. The regional geography of Muslim feminism
49. Some sociological maps guiding towards an analysis of the "Arab Spring"
50. The Arab Opinion Index
51. Islamism and Salafism
52. What the "Arab masses" really think and what the Arab-Muslim attitudes towards the West really are – the evidence from the Arab Opinion Index
53. Some new evidence on the basis of the *World Values Survey* project
54. How Muslims perceive the problem of knowledge and morals
55. Values and human security are also related in classic and modern Islamic thought
56. The U-shaped recovery of Muslim societies since 1800
57. The erosion of the respect for property rights in the "Christian Club"
58. The implosion of "self-expression" in the West and further doubts about the correctness of Inglehart's theory in the light of the new direct measurements of self-expression values from the *World Values Survey*, 2010-2014
59. The direct measurement of self-expression values in the *World Values Survey* Wave 6: 2010-2014 with the item "Important child qualities: Self-expression" and the consequences for Inglehart's theory
60. What the importance given to God by humans does to society
61. One, two, three types of permissiveness and their effects on value systems
62. Hofstede, Schwartz, Inglehart and our typologies
63. On sexual behavior values and effective democracy
64. A glance at the *World Values Survey* wave, 6, 2010-2014

LITERATURE AND SUGGESTED FURTHER READINGS

Abadir K. and Talmain G. 2002. Aggregation, persistence and volatility in a macro model. *Review of Economic Studies*. 69 (4): 749-779.

Abbt Ch. und Schoeller D. (Ed.) 2009. *Im Zeichen der Religion. Gewalt und Friedfertigkeit in Christentum und Islam.* Frankfurt/Main: Campus Verlag.

Abdollahian M, Coan T., Oh H., and Yesilada B. 2008. Dynamics of Cultural Change: The Human Development Perspective. *World Values Research,* 1 (4): 94-119. Available at: http://www.worldvaluessurvey.org/the_paperseries.

Abdullah M. 2012. The Moral Economy of Classical Islam: A Fiqhiconomic Model, by El-Shaikh Salah: A Review. *The Muslim World,* Vol. 98, 116, Oxford, Blackwell Publishing Lmt. SSRN Papers: http://papers.ssrn.com/sol3/papers.cfm?abstract_id=2217932.

Abdullah M. S. and Khoury A. Th. 1984. *Mohammed für Christen. Eine Herausforderung. Freiburg,* Basel, Wien: Herder.

Abrahamson M. and Carter V. J. 1986. Tolerance, Urbanism and Region. *American Sociological Review,* Vol. 51, No. 2, Apr., 1986. pp. 287-294.

Acemoglu D. 2012. The World Our Grandchildren will Inherit: The Rights Revolution and Beyond. *Working Paper 12-, Massachusetts Institute of Technology, Department of Economics, Working Paper Series.*

Acevedo G. A. 2008. *Islamic Fatalism and the Clash of Civilizations: An Appraisal of a Contentious and Dubious Theory. Social Forces,* 86 (4): 1711-52.

Adams C. C. 1929. Muhammad 'Abduh the Reformer. *The Moslem World,* Vol. XIX. No. 1 (January 1929), 266: 264-273.

Adams, C. C. 2010. *Islam and Modernism in Egypt: A Study of the Modern Reform Movement Inaugurated by Muhammad 'Abduh.* Kuala Lumpur: Islamic Book Trust.

Adamson P. and Taylor R. C. (Eds.) 2005. The Cambridge companion to Arabic philosophy. Cambridge University Press.

Adorno, Th. W. et al. 1950. *The Authoritarian personality. Studies in prejudice.* New York: Harper. Social studies series (American Jewish Committee), publication no. 3.

Agénor P. R. 2003. *Does Globalization Hurt the Poor?,* World Bank, available at http: //elibrary. worldbank. org/doi/book/10. 1596/1813-9450-2922.

Ahmad N. A., Abdi O. S., Ahmad F. I. 2012. Muhammad Abduh's Contributions To Modernity. *Asian Journal Of Management Sciences And Education,* ISSN: 2186-845X ISSN: 2186-8441 Print Vol. 1. No. 1, April. Retrievable at: http://www.ajmse.leena-luna.co.jp/AJMSEPDFs/Vol.1(1)/AJMSE2012(1.1-07).pdf.

Ahmuda H., Alvaredo F. and A. Canavese A. 2008. The monetary method to measure the shadow economy: The forgotten problem of the initial conditions. *Economics Letters,* 101, 97-99.

Aiginger K. 2009. *A Comparison of the Current Crisis with the Great Depression as Regards their Depth and the Policy Responses.* Lecture at the NERO-Meeting on "The Causes and Consequences of the Financial Crisis", OECD, Paris, 21 September 2009, WIFO-Lecture, No. 105, Vienna: Austrian Institute for Economic Research (URL: available at: http://karl. aiginger.wifo.ac.at/index. php?id=19.

Ajami F. 2012. The Arab Spring at One. *Foreign Affairs* (March/April 2012): retrievable at http: //www. foreignaffairs. com/articles/137053/fouad-ajami/the-arab-spring-at-one.

Ajrouch K and Moaddel M. 2007. Social Structure versus Perception: A Cross-National Comparison of the Determinants of Self-Rated Health in Egypt, Iran, Jordan, and the US. in '*Values and Perceptions of the Islamic and Middle Eastern Publics: Findings from Values Survey'.* Mansoor Moaddel, ((Ed.)) New York: Palgrave Macmillan. Pp. 181-208.

Akamatsu K. 1961. A Theory of Unbalanced Growth in the World Economy. *Weltwirtschaftliches Archiv – Review of World Economics.* 86 (2): 196-217.

Akcomak S. and ter Weel B. 2009. Social capital, innovation and growth: Evidence from Europe. *European Economic Review,* 53 (5): 544-567.

Albano R. and Barbera F. 2010. Social Capital, Welfare State, and Political Legitimacy. *American Behavioral Scientist,* Jan 2010; vol. 53, 677 - 690.

Al-Daffa A. A. (Ali Abdallah) 1977. *The Muslim contribution to mathematics.* Atlantic Highlands, N.J.: Humanities Press.

Aleksynska M. 2007. Quantitative Assessment of Immigrants' Civic Activities: Exploring the European Social Survey. In: *Highly Active Immigrants,* Frankfurt a M.: Peter Lang: 59-76.

Aleksynska M. 2011. Civic Participation of Immigrants in Europe: Assimilation, Origin and Destination Country Effects. *European Journal of Political Economy,* 27 (3): 566-585.

Aleksynska M. and Chiswick B. 2011. *Religiosity and Migration: Travel into One's Self versus Travel Across* Cultures. IZA: Forschungsinstitut zur Zukunft der Arbeit, available at http://ideas. repec.org/p/iza/izadps/dp5724.html.

Alesina A. 2013. Women, Fertility and the Rise in Modern Capitalism. *Science,* 342(6157): 427–428.

Alesina A. and Angeletos G. M. 2005. Fairness and Redistribution. *American Economic Review,* 95(4), 960–980.

Alesina A. and Fuchs-Schündeln N. 2007. Goodbye Lenin (or Not?): The Effect of Communism on People. *American Economic Review,* 97(4): 1507–1528.

Alesina A. and Giuliano P. 2010. The Power of the Family. *Journal of Economic Growth,* 15, 93–125.

Alesina A. and Giuliano P. 2011. Family Ties and Political Participation, *Journal of the European Economic Association,* 9(5): 817–839.

Alesina A. and Guiliano P. 2013. Culture and Institutions. *NBER Working Paper Series, Working Paper 19750,* available at http://scholar.harvard.edu/alesina/publications

Alesina A., Cozzi, G., and Mantovan N. 2012. The Evolution of Ideology, Fairness, and Redistribution. *The Economic Journal,* 122 (565), 1244–1261.

Alesina A., Di Tella R., and MacCulloch R. 2004. Inequality and Happiness: Are Europeans and Americans Different? *Journal of Public Economics* 2004, 88(9–10) 2009–2042.

Alexander A. C., R. Inglehart R. F. and Welzel C. 2011. *Measuring Effective Democracy: A Defense. World Values Research,* 4 (1): 1-40. Available at: http://www. worldvaluessurvey.org/the_paperseries and file:///C:/Users/ta1/Downloads/WVR_04_01_Alexander_et_al.pdf.

Alexander C. A. and Welzel C. 2011. Islam and Patriarchy: How Robust Is Muslim Support for Patriarchal Values? *World Values Research,* 4 (2): 41-72. Available at: http: //www. worldvaluessurvey. org/the_paperseries

Al-Ghazali M. (Mohamed) 1978. *Khuluq al-Muslim [Muslim's Ethics].* Cairo: Dar al-Rayyan litturath.

Al-Hajjawi M. A. (Musa ibn Ahmed) 2011. *Sharh Manzumat al-Adab al shar'yya [Interpretation of legal behavior system].* Saudi Arabia: Dar Ibn al-Jawzi.

Aliriza B. and Aras B. 2012. *U. S. -Turkish Relations. A review at the beginning of the third decade of the post–cold war era.* SAM Center for Strategic Research, Republic of Turkey, Ministry of Foreign Affairs and CSIS, Center for Strategic and International Studies, Washington (November 2012), SAM Reports (November 2012) retrievable at http: //sam. gov. tr/u-s-turkish-relations-a-review-at-the-beginning-of-the-third-decade-of-the-post-cold-war-era/.

Al-Kadhi M. A. (Misha'al ibn Abdullah) 1995. *What did Jesus really say?* Preface, pp. 8-13: available at: http://www.islambasics.com/view.php?authID=16.

Alm J., Martinez-Vazque J., and Torgler B. 2006. Russian attitudes toward paying taxes - Before, during, and after the transition. *International Journal of Social Economics,* 33 (12), pp. 832-857.

Al-Mawdudi A. A. (Abul Aala) 1986. Islamic Way of Life, The Islamic Foundation, 1986. Retrievable at: http://www.islambasics.com/index.php?act=download&BID=70.

Al-Mutavkkil A, Heshmati A. and Hwang J-S. 2009. Development of Telecommunication and Broadcasting Infrastructure Indices at the Global Level. *Telecommunication Policy* 33 (3-4), 176-179.

Al-Najjar A. H. A. (Abdel Hadi Ali) 1990. *Al-Islam wal-Iqtisad [Islam and Economy].* Kuwait: Alam al-Maarifa.

Altmann J., Heshmati A. and Othman B. T. 2013. *Evaluation of E-Infrastructure Deployment in OECD and MENA Countries.* Available at http://ideas. repec.org/p/snv/dp2009/ 2013102.html.

Amaney J and Nooruddin I. 2010. The Democratic Utility of Trust: A Cross-National Analysis. *The Journal of Politics,* 72 (1): 45-59.

Amaney J. 2007a. *Barriers to Democracy.* Princeton: Princeton University Press.

Amaney J. 2007b. When Is Social Trust a Desirable Outcomes?: Examining Levels of Trust in the Arab World. *Comparative Political Studies,* 40 (11): 1328-1349.

Amin S. 1976. *Unequal Development: An Essay on the Social Formations of Peripheral Capitalism* New York: Monthly Review Press.

Amin S. 1989. *Eurocentrism. Translated by Russell Moore.* New York: Monthly Review Press.

Amin S. 1994. *Re-reading the postwar period: an intellectual itinerary.* Translated by Michael Wolfers. New York: Monthly Review Press.

Andersen R. and Fetner T. 2008a. Cohort Differences in Tolerance of Homosexuality: Attitudinal Change in Canada and the United States, 1981-2000. *The Public Opinion Quarterly,* 72 (2): 311-330.

Andersen R. and Fetner T. 2008b. Economic Inequality and Intolerance: Attitudes toward Homosexuality in 35 Democracies. *American Journal of Political Science*, Vol. 52 (4): 942-958.

Andersen T. M. and Herbertsson T. T. 2003. *Measuring Globalization.* IZA Discussion Paper 2003: 817. Bonn: IZA. Available at: http: //papers. ssrn. com/sol3/papers. cfm?abstract_id=434540.

Apergis N., Dincer O. C. and Payne J. E. 2012. Live free or bribe: On the causal dynamics between economic freedom and corruption in U. S. states. *European Journal of Political Economy,* 28, 215-226.

Arafa M. A. (Mohamed A.) 2012 Corruption and Bribery in Islamic Law: Are Islamic Ideals Being Met in Practice? *Annual Survey of International & Comparative Law,* Vol. 18: Iss. 1, Article 9, pp. 171-242. Available at: http://digitalcommons.law.ggu.edu/annlsurvey/vol18/iss1/9.

Aras B. 2000 and Caha O. 2000. Fethullah Gulen and His Liberal 'Turkish Islam' Movement. *MERIA Journal* (Gloria Center, Israel), Volume 04, Number 04 (Dec 2000), available at http: //www. gloria-center. org/2000/12/aras-and-caha-2000-12-04/.

Archibugi D. and Coco A. 2004. A New Indicator of Technological Capabilities for Developed and Developing Countries (ArCo), *World Development* 32 (4), 629–654.

Archibugi D., Denni M. and Filippetti A. 2009. The technological capabilities of nations: The state of the art of synthetic indicators. *Technological Forecasting & Social Change*, 76, 917–931.

Arda, Berna, and Chaim, Vardit Rispler, Editors (2012), *Islam and Bioethics,* Ankara, retrievable from: http://kitaplar.ankara.edu.tr/dosyalar/pdf/846.pdf

Arikan G. 2011. *Economic Individualism and Government Spending. World Values Research,* 4 (3): 73-96. Available at: http: //www. worldvaluessurvey. org/the_paperseries.

Armstrong K. 1992. *Muhammad: a biography of the prophet.* San Francisco, Calif.: Harper SanFrancisco.

Armstrong K. 1993. *A history of God: the 4000-year quest of Judaism,* Christianity, and Islam. New York: A. A. Knopf: Distributed by Random House.

Armstrong K. 2000. *Islam: a short history.* New York: Modern Library.

Armstrong K. 2006. *Muhammad: a prophet for our time.* New York: Atlas Books/ HarperCollins.

Arrighi G. 1989. *The Developmentalist Illusion: A Reconceptualization of the Semiperiphery. Paper,* presented at the Thirteenth Annual Political Economy of the World System Conference, University of Illinois at Urbana - Champaign, April 28 - 30.

Arrighi G. 1991. *World Income Inequalities and the Future of Socialism.* Fernand Braudel Centre, State University of New York at Binghamton.

Arrighi G. 1995. *The Long 20th Century.* Money, Power, and the Origins of Our Times. London, New York: Verso.

Arrighi G. and Silver, B. J. 1984. *Labor Movements and Capital Migration: The United States and Western Europe in World - Historical Perspective.* In: Labor in the Capitalist World - Economy (Bergquist Ch. (Ed.)): 183-216, Beverly Hills: Sage.

Arrighi G. et al. 1991. *The Rise of East Asia. One Miracle or Many?* State University of New York at Binghamton: Fernand Braudel Centre.

Arrighi G. et al. 1996a. *Beyond Western Hegemonies.* State University of New York at Binghamton: Fernand Braudel Center.

Arrighi G. et al. 1996b. *Modelling Zones of the World-Economy: A Polynomial Regression Analysis (1964-1994).* State University of New York at Binghamton: Fernand Braudel Center.

Arrighi G. et al. 1996c. *The Rise of East Asia in World Historical Perspective.* State University of New York at Binghamton: Fernand Braudel Center.

Arrighi G., Silver B. J. and Brewer B. D. 2003. Industrial convergence, globalization, and the persistence of the North-South divide. *Studies in Comparative International Development.* 38 (1): 3-31.

Arzheimer K. 2008. Protest, Neo-Liberalism or Anti-Immigrant Sentiment: What Motivates the Voters of the Extreme Right in Western Europe? *Zeitschrift für Vergleichende Politikwissenschaft/Comparative Governance,* 2 (2): 173-197.

Arzheimer K. and Carter E. 2009. Christian Religiosity and Voting for West European Radical Right Parties. *West European Politics,* 32 (5): 985-1011.

Asiedu E. and Lien D. (2011. Democracy, foreign direct investment and natural resources, *Journal of International Economics,* 84, 99-111.

Attinà F. 2003. 'Organisation, Competition and Change of the International System.' *International interactions.* 16 (4): 317-333.

Attinà F. 2005. *'State aggregation in defence pacts: systematic explanations.'* Jean Monnet Working Papers in Comparative and International Politics, *Jean Monnet Centre EuroMed, Department of Political Studies,* University of Catania, URL: http://www.fscpo.unict.it/EuroMed/jmwp56.pdf.

Au K. Y. 2000. Intra-cultural variation as another construct of international management: A study based on secondary data of 42 countries. *Journal of International Management* 6 (3), pp. 217-238.

Awan A. N. 2008. Antecedents of Islamic Political Radicalism among Muslim Communities in Europe. PS: *Political Science and Politics,* 41 (1): 13-17.

Aydin H. et al. 2003. *'Euro-Islam'. Das neue Islamverständnis der Muslime in der Migration.* Stiftung Zentrum für Türkeistudien, Institut an der Universität Duisburg-Essen. Available at: http: //www.renner-institut.at/download/texte/euroisla.pdf.

Azman-Saini W. N. W., Baharumshah A. Z. and Law S. H. 2010. Foreign direct investment, economic freedom and economic growth, *Economic Modelling,* 27, 1079-1089.

Babetskii I.; Komarek L.; and Komarkova Z. 2007. Financial integration of stock markets among new EU member states and the euro area. Finance a Uver. *Czech Journal of Economics and Finance.* 57 (7-8): 341-362.

Badawi J. 1971. The Status of Women in Islam, *Al-Ittihad,* Vol. 8, No. 2, Sha'ban 1391/Sept.

Baglioni A. S. and Cherubini U. 2010. *Marking-to-Market Government Guarantees to Financial Systems: An Empirical Analysis of Europe* (November 10, 2010). URL: Available at SSRN: http: //ssrn. com/abstract=1715405 or http: //dx. doi. org/10. 2139/ssrn. 1715405).

Bail C. 2008. The Configuration of Symbolic Boundaries against Immigrants in Europe. *American Sociological Review,* 73 (1): 37-59.

Balic S. 2001. *Islam für Europa: neue Perspektiven einer alten Religion.* Köln and Vienna: Böhlau.

Barakat H. 1993. *The Arab world: society, culture, and state.* Univeristy of California Press.

Barber B. 1992. Jihad vs. McWorld. The Atlantic Online. http: //www. theatlantic. com/politics/foreign/barberf. htm.

Bardakoglu A. 2006. *Religion and Society.* New Perspectives from Turkey. Ankara: P. P. R. A.

Bariun F. (Fawzia) 1991. Social Change as Seen by Malik Bennabi. *The American Journal of Islamic Social Sciences,* Vol.8, N°1: pp.189-194.

Barnell A. C. 1997. *A Dictionary of Econometrics,* Edward Elgar: Cheltenham.

Barrett D. B. 1982. *World Christian encyclopedia: a comparative study of churches and religions in the modern world,* AD 1900-2000. Oxford, London and New York: Oxford University Press.

Barro R. J and Hwang J. 2007. Religious Conversion in 40 Countries. *NBER Working Paper Series Working Paper 13689.*

Barro R. J. 1991. Economic Growth in a Cross-Section of Countries. *Quarterly Journal of Economics.* 106 (2): 407-443.

Barro R. J. 1998. Determinants of Economic Growth: A Cross-country Empirical Study. *Lionel Robbins Lectures.* Cambridge, MA: MIT Press, first edition.

Barro R. J. 2004. Spirit of Capitalism Religion and Economic Development. *Harvard International Review,* vol. 25, no. 4, pp. 64-67.

Barro R. J. 2012. *Convergence and Modernization Revisited.* Department of Economics, Harvard University.

Barro R. J. 2013. Health and Economic Growth. Annals of Economics and Finance. 14 (2): 329-366.

Barro R. J. and McCleary R. M. 2003. Religion and Economic Growth across Countries. *American Sociological Review,* 68 (5): 760-781.

Barro R. J. and McCleary R. M. 2004. *Religion and economic growth.* The Milken Institute Review (Milken, Santa Monica, Cal., 6, 2004. 2, S. 36-45.

Barro R. J. and Sala-i-Martin X. 1992. Convergence. *Journal of Political Economy.* 100 (2): 223-251.

Barro R. J. and Sala-i-Martin, X. 2003. *Economic Growth.* Cambridge, MA: MIT Press, second edition.

Barro R. J. and Ursúa J. F. 2008. Macroeconomic Crises since 1870. Brookings Papers on Economic Activity. 2008 (1): 255-350.

Barro R. J. et al. 2010. Religious Conversion in 40 Countries. *Journal for the Scientific Study of Religion,* 49 (1): 15–36.

Barro R. J., Nakamura E., and Ursúa J. F. 2011. *Crises and Recoveries in an Empirical Model of Consumption Disasters. As yet unpublished.* Cambridge, M. A., Harvard University, Department of Economics.

Barro R. J., Sala-i-Martin X. et al. 1991. Convergence Across States and Regions. *Brookings Papers on Economic Activity.* 1991 (1): 107-182.

Bartlett J. E., Kotrlik J. W. and Higgins C. 2001. Organizational research: Determining appropriate sample size for survey research, *Information Technology, Learning, and Performance Journal,* 19 (1), 43-50.

Bartlett M. S. 1946. On the theoretical specification of sampling properties of autocorrelated time series. *Journal of Royal Statistical Society,* Series B, 8, 27–27.

Barysch K. et al. 2005. *Why Europe should embrace Turkey.* Centre for European Reform, abstract available at http: //www.cer.org.uk/.

Beck N. 1991. The Illusion of Cycles in International Relations. *International Studies Quarterly.* 35 (4): 455-476.

Beckfield J. 2006. European Integration and Income Inequality. *American Sociological Review,* 71 (6), 964-985.

Belasen A. R. and Hafer R. W. 2012. Well-being and economic freedom: evidence from the States, *Intelligence,* 40, 306-316.

Bennabi M. (Malik) 1981. *Wijhat al 'Alam al Islami, [Orientation of Islamic World].* Damascus: Dar al Fikr.

Bennabi M. (Malik) 1986. *Shurut al Nahdah, [Conditions of Renaissance].* Damascus: Dar al Fikr.

Berger P. L. 2001. *Holy war, Inc.: inside the secret world of Osama bin Laden.* New York: Free Press.

Bergesen A. J. 2007. *'The Sayyid Qutb Reader. Selected Writings on Politics, Religion, and Society.* Edited by Albert J. Bergesen.' Routledge, London and New York.

Bergesen A. J. and Han Y. 2005. New Directions for Terrorism Research. *International Journal of Comparative Sociology,* 46: 133-151.

Berman P. 2003. The Philosopher of Islamic Terror. *New York Times Magazine,* March 23, 2003, freely available at http: //www. nytimes. com/2003/03/23/magazine/the-philosopher-of-islamic-terror. html?pagewanted=all&src=pm.

Berry B. J. L.; Kim H.; and Baker E. S. 2001. Low-frequency waves of inflation and economic growth: Digital spectral analysis. *Technological Forecasting and Social Change.* 68 (1): 63-73.

Biard J. and Roshdi R. 1996. *Descartes et le Moyen Âge: actes du colloque organisé à la Sorbonne du 4 au 7 juin 1996 par le Centre d'histoire des sciences et des philosophies arabes et médiévales (URA 1085, CNRS/ÉPHÉ) à l'occasion du quatrième centenaire de la naissance de Descartes.* Paris: J. Vrin, Études de philosophie médiévale 75.

Biswas A. K., Farzanegan M. R. and Thum M. 2012. Pollution shadow economy and corruption-theory and evidence, *Ecological Economics,* 75, 114-125.

Bjørnskov C. 2007. *Determinants of Generalized Trust: A Cross-Country Comparison.* Public Choice, (30 (1/2): 1-21.

Bjørnskov C. 2010. Do elites benefit from democracy and foreign aid in developing countries?, *Journal of Development Economics,* 92, 115-124.

Blackburn K., Bose N. and Capasso S. 2012. Tax evasion, the underground economy and financial development. *Journal of Economic Behavior & Organization,* 83, 243-253.

Blalock H. M. 1972. Social statistics. New York: McGraw-Hill.

Blaydes L and Linzer D. 2008. The Political Economy of Women's Support for Fundamentalist Islam. *World Politics,* 60 (4): 576-609.

Blinkert B. 2010. Unsicherheitsbefindlichkeit als "sozialer Tatbestand". Kriminalitätsfurcht und die Wahrnehmung von Sicherheit und Unsicherheit in Europa. *Monatsschrift für Kriminologie und Strafrechtsreform,* 93 (2): 106-125.

Bloomfield P. 1976. *Fourier analysis of time series.* New York: John Wiley and Sons.

BM. I. SIAK 2006. *Perspektiven und Herausforderungen in der Integration muslimischer MitbürgerInnen in Österreich.* Federal Ministry of the Interior, Republic of Austria, available at: http: //www. bmi. gv. at/downloadarea/asyl_fremdenwesen/Perspektiven_ Herausforderungen. pdf.

Boccara, P. 2008. *Transformations et crise du capitalisme mondialisé: quelle alternative? Paris*: Pantin: Le Temps des crises.

Boff L. 2005. *Global civilization: challenges to society and to Christianity. London*; Oakville: Equinox Publishing.

Boli J. and Lechner, F. J. 2002. Globalization and World Culture. In *International Encyclopedia of the Social and Behavioral Sciences.* Amsterdam: Elsevier.

Boli J. and Thomas G. 1999. Constructing World Culture. Stanford: Stanford University Press.

Bolt J. and van Zanden J. L. 2013. *The First Update of the Maddison Project; Re-Estimating Growth Before 1820. Maddison Project Working Paper 4.* URL: http: //www. ggdc. net/maddison/maddison-project/home. htm.

Bonini, A. N. 2008. Cross-national variation in individual life satisfaction: Effects of national wealth, human development, and environmental conditions. *Social Indicators Research,* 87 (2), pp. 223-236.

Boomgaarden H. and Freire A. 2009. Religion and Euroscepticism: Direct, Indirect or No Effects?. *West European Politics,* 32 (6): 1240-1265.

Bornschier V. 1996. *Western society in transition.* New Brunswick, N. J.: Transaction Publishers.

Bornschier V. 2000. Trust and Tolerance – Enabling Social Capital Formation for Modern Economic Growth and Societal Change. *Paper prepared for the conference at the University of Technology,* Sidney, Australia, December 14-16, 2000, on 'Organizing Knowledge Economies and Societies', organized by APROS.

Bornschier V. and Chase-Dunn Ch. 1998. *The Future of Global Conflict.* London: Sage.

Bornschier V. and Chase-Dunn, Ch. K 1985. *Transnational Corporations and Underdevelopment* N. Y., N. Y.: Praeger.

Boroujerdi M. 1997. Iranian Islam and the Faustian Bargain of Western Modernity. J*ournal of Peace Research,* Vol. 34, No. 1 (Feb., 1997), pp. 1-5.

Bourdieu P. 1980. Questions de sociologie. Paris: *Les Éditions de Minuit, coll.* « Documents ».

Bowen G. L. 2011. Has outreach to the Muslim World by the Obama Administration had an Impact on Muslim Attitudes towards Terrorists and Terrorism? *MERIA Journal Volume* 15, Number 01 (Mar 2011) http: //www. gloria-center. org/2011/08/has-outreach-to-the-muslim-world-by-the-obama-adminstration-had-an-impact-on-muslim-attitudes-toward-terrorists-and-terrorism-2/.

Bowker J. (Ed.) 2000. *The concise Oxford dictionary of world religions.* Oxford, London and New York: Oxford University Press.

Box G. E. P., and Jenkins G. M. 1976. Time series analysis: *Forecasting and control,* Rev. ed. San Francisco: Holden-Day.

Braizat al F. 2002. Muslims and Democracy: An Empirical Critique of Fukuyama's Culturalist Approach. *International Journal of Comparative Sociology,* 43 (3-5): 269-99.

Brammer S., Williams G. and Zinkin J. 2007. Religion and Attitudes to Corporate Social Responsibility in a Large Cross-Country Sample. *Journal of Business Ethics,* 71 (3): 229-243.

Brites R. and Torres A. 2007. European attitudes and values: The perspective of gender in a transverse analysis. *Portuguese Journal of Social Science,* 5 (3): 179-214.

Browers M. 2006. *Democracy and Civil Society in Arab Political Thought.* Syracuse University Press: 30.

Bsteh A. 1996. *Peace for humanity: principles, problems, and perspectives of the future as seen by Muslims and Christians.* New Delhi: Vikas Pub. House.

Bsteh A. and Anawati G. C. 1978. *Der Gott des Christentums und des Islams.* Mödling: Verl. St. Gabriel.

Bsteh A. and Dupré W. 2007. *Christian faith in dialogue with Islam. Lectures, questions, interventions.* Mödling St. Gabriel Publ.

Bsteh A. and Khoury A. Th. 1994. *Der Islam als Anfrage an christliche Theologie und Philosophie.* Mödling: Verlag St. Gabriel.

Butterworth Ch. and Blake A. K. 1994. *The Introduction of Arabic philosophy into Europe.* Leiden: E. J. Brill, Studien und Texte zur Geistesgeschichte des Mittelalters / herausgegeben von Albert Zimmermann Bd. 39.

Cabral M. V. 2007. Class effects and societal effects: Elite and working class attitudes towards political citizenship in European comparative perspective. *Portuguese Journal of Social Science,* 5 (3): 159-178.

Cao L. and Stack, S. 2010. Exploring terra incognita: Family values and prostitution acceptance in China. *Journal of Criminal Justice,* 38 (4), pp. 531-537.

Cavalcanti T. V., Parente S. L. and Zhao R. 2007. Religion in macroeconomics: a quantitative analysis of Weber's thesis. *Economic Theory,* 32: 105–123.

Ceobanu A. M. 2010. Usual suspects? Public views about immigrants' impact on crime in European countries. *International Journal of Comparative Sociology,* 52 (1-2): 114-131.

Chambers R. L. and Skinner C. J. (2003), Editors, Analysis of Survey Data, Wiley.

Chandler J. 2006. *Cross National Survey Research on Islamic Publics: Recent Findings. Department of Political Science,* City University of New York, Graduate School & University Center.

Chandy L. and Gertz G. 2011. *With Little Notice, Globalization Reduced Poverty.* YaleGlobal, 5 July 2011, URL: http: //yaleglobal. yale. edu/content/little-notice-globalization-reduced-poverty.

Chaudhuri K., Schneider F. and Chattopadhyay S. 2006. The size and development of the shadow economy-an empirical investigation from states of India, *Journal of Development Economics,* 80, 428-443.

Cheng P. S. 2011. *Radical love: an introduction to queer theology.* New York: Seabury Books.

Child D. 2006. *Essentials of factor analysis.* New York: Continuum.

Chomsky N. 2001. 9–11. New York: Seven Stories Press.

Choudhury M. A. 1997. *Money in Islam: A Study in Islamic Political Economy.* London and New York: Routledge.

Chu S. H. and Freund S. 1996. Volatility estimation for stock index options: A GARCH approach. *Quarterly Review of Economics and Finance.* 36 (4): 431-450.

Ciftci S. 2010. Modernization, Islam, or Social Capital: What Explains Attitudes Toward Democracy in the Muslim World? *Comparative Political Studies,* 43 (11): 1442-1470.

Citrin J. and Sides J. M. 2008. Immigration and the Imagined Community in Europe and the United States. *Political Studies,* 56 (1): 33-56.

Clark T. E. and West K. D. 2006. Using out-of-sample mean squared prediction errors to test the martingale difference hypothesis. *Journal of Econometrics.* 135 (1-2): 155-186.

Clesse A. and Tashan S. 2004. *Turkey and the European Union: 2004 and beyond.* Amsterdam: Dutch University Press.

Coenders M. Gijsberts M. and Scheepers P. 2004. Resistance to the presence of immigrants and refugees in 22 countries. In: *Nationalism and exclusion of migrants: cross-national comparisons,* Aldershot: Ashgate: 97-120.

Coenders M. Lubbers M. and Scheepers P. 2005. *Majorities' attitudes towards minorities, findings from the Eurobarometer and the European Social Survey.* Vienna: European Fundamental Rights Agency, available at: https: //fra. europa. eu/fraWebsite/ attachments/Report-1. pdf.

Collard F. 1999. Spectral and persistence properties of cyclical growth. *Journal of Economic Dynamics and Control.* 23 (3): 463-488.

Compton R. A., Giedeman D. C. and Hoover G. A. 2011. Panel evidence on economic freedom and growth in the United States. *European Journal of Political Economy,* 27, 423-435.

Comte A. 1896, posthumously. *The positive philosophy of Auguste Comte;* Auguste Comte; Harriet Martineau; Frederic Harrison. London: G. Bell & sons.

Connor P. 2010. Contexts of immigrant receptivity and immigrant religious outcomes: the case of Muslims in Western Europe. *Ethnic and Racial Studies,* 33 (3): 376-403.

Cook R. D. 1977. Detection of influential observations in linear regression. *Technometrics,* 19, 15–18.

Cooper J. 2006. Of BRICs and brains: Comparing Russia with China, India, and other populous emerging economies. *Eurasian Geography and Economics.* 47 (3): 255-284.

Crane T. 2013. *The Routledge Encyclopedia of Philosophy,* online. Available at http: //www. rep. routledge. com/LOGIN?sessionid=8b89f7b449f95dffc942351dcce7 fb40& authstatuscode=400.

Crepaz M. M. L. and Damron R. 2008. Constructing Tolerance: How the Welfare State Shapes Attitudes About Immigrants. *Comparative Political Studies,* 42 (3): 437-463.

Cronin A. K. 2003. Behind the Curve: Globalization and International Terrorism. *International Security,* 27: 30-58.

Crowley P. M. 2010. *Long cycles in growth: explorations using new frequency domain techniques with US data.* Bank of Finland Research Discussion Papers, 6, 2010. URL: http: //patrickmcrowley. com/Research/workingpapers. htm.

Cryer J. D. 1986. Time series analysis. Boston, Mass.: Duxbury Press.

Curtis J. E., Baer D. E. and Grabb E. G., 2001. Nations of Joiners: Explaining Voluntary Association Membership in Democratic Societies. *American Sociological Review*, 66 (6): 783-805.

D'Hernoncourt D. and Meon P. G. 2012. The not so dark side of trust: Does trust increase the size of the shadow economy?, *Journal of Economic Behavior & Organization,* 81, 97-121.

Dalacoura K. 2012. The 2011 uprisings in the Arab Middle East: political change and geopolitical implications. *International Affairs,* 88: 1. The Royal Institute of International Affairs.

Dard D., Dustmann C., and Preston I. 2005. *Understanding Attitudes to Immigration: The Migration and Minority module of the first European Social Survey.* Centre for Research and Analysis of Migration. Available at: http: //ideas. repec. org/p/crm/wpaper/0503. html.

Das J., DiRienzo C., and Tiemann T. 2008. A Global Tolerance Index. *Competitiveness Review: An International Business Journal.* Vol. 18 No 3, 2008. pp. 192-205.

Davidov E. 2008. A Cross-Country and Cross-Time Comparison of the Human Values Measurements with the Second Round of the European Social Survey. *Survey Research Methods,* 2 (1): 33-46.

Davidov E. 2010. Testing for comparability of human values across countries and time with the third round of the European Social Survey. *International Journal of Comparative Sociology,* 51 (3): 171-191.

Davidov E., Meuleman B., Billiet J., and Schmidt P. 2008. Values and Support for Immigration: A Cross-Country Comparison. *European Sociological Review,* 24 (5): 583-599.

Davidov E., Schmidt P. and Billiet J. 2011. *Cross-cultural analysis: methods and applications.* New York: Routledge.

Davidov E., Schmidt P., and Schwartz S. H. 2008. *Bringing Values Back In: The Adequacy of the European Social Survey to Measure Values in 20 countries.* Public Opinion Quarterly, 72 (3): 420-445.

Davidson R. and MacKinnon J. M. 1993. *Estimation and Inference in Econometrics,* Oxford University Press: Oxford.

Davis D. W., Dowley, K. M., and Silver B. D. 1999. Postmaterialism in world societies: Is it really a value dimension? *American Journal of Political Science* 43 (3), pp. 935-962.

Davis N. J. and Robinson R. 2006. The Egalitarian Face of Islamic Orthodoxy: Support for Islamic Law and Economic Justice in Seven Muslim-Majority Nations. *American Sociological Review,* 71 (2): 167-90.

De Grauwe P. and Ji Y. 2012. Mispricing of Sovereign Risk and Macroeconomic Stability in the Eurozone. *JCMS – Journal of Common Market Studies.* 50 (6): 866-880.

De Groot B. and Franses P. H. 2008. Stability through cycles. *Technological Forecasting and Social Change.* 75 (3): 301-311.

De Rooij E. A. 2011. *Patterns of Immigrant Political Participation: Explaining Differences in Types of Political Participation between Immigrants and the Majority Population in Western Europe. European Sociological Review,* 27 (1) first published online February 17, 2011 doi: 10. 1093/esr/jcr010.

de Soysa I. and Nordas R. 2006. *Islam's Bloody Innards? Religion and Political Terror,* 1980 – 2000. Conference Paper, International Peace Research Institute, Oslo (PRIO. available at: http: //www. prio. no/files/file48349_desoysa_nordas_wg3meeting. pdf.

Decancq K. and Lugo M. A. 2009. *Setting Weights in Multidimensional Indices of Well-being and Deprivation,* Oxford Poverty & Human Development Initiative (OPHI) Working Paper No. 18.

Dekker P. 2009. *Strategic Europe.* Sociaal en Cultureel Planbureau. Available at: http://www.scp.nl/english/Publications/Publications_by_year/Publications_2009/Strategic_Europe.

Dempster A. P. 1969. *Elements of Continuous Multivariate Analysis.* Reading, MA: Addison-Wesley.

Deutsch K. W. 1961. Social Mobilization and Political Development. *The American Political Science Review,* 55 (3): 493-514.

DeVellis R. F. 2003. *Scale development: Theory and applications,* Thousand Oaks, CA: Sage.

Devezas T. C. (Ed.) 2006. *Kondratieff waves, warfare and world security: [proceedings of the NATO Advanced Research Workshop on The Influence of Chance Events and Socioeconomic Long Waves in the New Arena of Asymmetric Warfare, Covilhã, Portugal, 14 - 18 February 2005].* Amsterdam: IOS Press.

Devezas T. C. 2010. Crises, depressions, and expansions: Global analysis and secular trends. *Technological Forecasting and Social Change.* 77 (5): 739-761.

Devezas T. C. 2012. The Recent Crisis under the Light of the Long Wave Theory. In: Kondratieff Waves. *Dimensions and Prospects at the Dawn of the 21st Century* (Grinin L. E., Devezas T. C., and Korotayev A. V. (Eds.)), 138-175, Volgograd, Russia: Uchitel.

Devezas T. C. and Corredine J. T. 2001. The biological determinants of long-wave behavior in socioeconomic growth and development. *Technological Forecasting and Social Change.* 68 (1): 1-57.

Diebold F. X., Kilian L. and Nerlove M. 2013. Time series analysis. *The New Palgrave Dictionary of Economics.* Second Edition. (Eds. Steven N. Durlauf and Lawrence E. Blume. Basingstoke, U. K.: Palgrave Macmillan, 2008. The New Palgrave Dictionary of Economics Online. Palgrave Macmillan. 04 June 2013 <http: //www. dictionaryofeconomics. com/article?id=pde2008_T000064> doi: 10. 1057/ 9780230226203. 1707.

Dien J., Beal D. J. and Berg P. 2005. Optimizing principal components analysis of event-related potentials: Matrix type, factor loading weighting, extraction, and rotations, *Clinical Neurophysiology,* 16, 1808-1825.

Diez-Nicolas J. 2003. Two Contradictory Hypotheses on Globalization; Social Convergence of Civilization Differentiation and Clash. In *'Human Values and Social Change: Findings from the Values Surveys'.* (Inglehart R. F. ((Ed.)) Leiden, The Netherlands: Brill: 465-493.

Diez-Nicolas J. 2007. Value Systems of Elites and Publics in the Mediterranean: Convergence and Divergence. In *'Values and Perceptions of the Islamic and Middle Eastern Publics'* (Moaddel M. (Ed.)): 47 - 70, Houndmills, Basingstoke, Hampshire: Palgrave Macmillan.

Diez-Nicolas J. 2009. *Two Contradictory Hypotheses on Globalization: Societal Convergence or Civilizational Differentiation? World Values Research,* 2 (4): 78-105. Available at: http: //www. worldvaluessurvey. org/the_paperseries.

Diez-Nicolas J. 2010. *Cultural Differences on Values about Conflict, War, and Peace. World Values Research,* 3 (1): 1-20. Available at: http: //www. worldvaluessurvey. org/the_paperseries.

Dittmann J. and Scheuer A. 2008. *Lebensbedingungen und Wohlbefinden in Europa. In: Datenreport 2008.* Ein Sozialbericht der Bundesrepublik Deutschland, Bonn: Bundeszentrale für Politische Bildung: 435-441.

Dittmar R. D; Gavin W. T; and Kydland F. E. 2005. Inflation persistence and flexible prices. *International Economic Review.* 46 (1): 245-261.

Dittrich M. 2006. Muslims in Europe: addressing the challenges of radicalisation. *European Policy Centre in strategic partnership with the King Baudouin Foundation and the Comagnia di San Paolo,* Brussels, available at http: //www. theepc. be/TEWN/pdf/ 602431467_EPC%20Working%20Paper%2023%20Muslims%20in%20Europe. pdf.

Dobewall H. and Strack M. 2013. Relationship of Inglehart's and Schwartz's value dimensions revisited. *International Union of Psychological Science.* Vol. 49, Issue 3: DOI: 10.1002/ijop.12004.

Donno D. and Russett B. 2004. Islam, Authoritarianism, and Female Empowerment: What Are the Linkages? *World Politics,* vol. 56, no. 4, pp. 582-607.

Dostoyevsky F. M. 1976, posthumously. *The Brothers Karamazov: the Constance Garnett translation revised by Ralph E. Matlaw: backgrounds and sources, essays in criticism.* New York: Norton, 1976.

Dreher A. 2006. Does Globalization Affect Growth? Evidence from a new Index of Globalization, Applied Economics, 38 (10), 1091-1110.

Dreher A., Gaston N. and Martens P. 2008. *Measuring Globalisation: Gauging its Consequences.* New York: Springer.

Dronkers J. and Fleischmann F. 2010. *The Educational Attainment of Second Generation Immigrants from different Countries of Origin in the EU member-states.* In: Quality and Inequality of Education. Cross-National Perspectives, Dordrecht. Heidelberg London New York: Springer Verlag: 163-204.

Duch R. M. and Taylor, M. A. 1993. Postmaterialism and the Economic Condition. *American Journal of Political Science,* 37 (3): 747-779.

Durant, W. 1950, *The Age of Faith, Simon and Shuster,* New York. Chapter VIII; p.174. See also: The Foundation for Science, Technology and Civilisation (FSTC) Fri 16 September, 2005: http://www.muslimheritage.com/topics/default.cfm?articleID=544

Durkheim E. 1964, posthumously. *The Elementary Forms of the Religious Life,* Tr. Joseph Ward Swain. London: George Allen & Unwin LTD.

Durkheim E. 1986, posthumously. *Durkheim on politics and the state.* Émile Durkheim; Anthony Giddens. Stanford, Calif.: Stanford University Press.

Durlauf St. N. et al. 2008. Are any growth theories robust? *The Economic Journal,* 118 (1), 329–346.

Dziuban C. D. and Shirkey E. C. 1974. When is a correlation matrix appropriate for factor analysis?. *Psychological Bulletin,* 81, 358–361.

Eilat Y. and Zinnes C. 2002. The Shadow Economy in Transition Countries: Friend or Foe? A Policy Perspective, *World Development,* 30 (7), 1233-1254.

Elgin C. 2013. Internet usage and the shadow economy Evidence from panel data, *Economic Systems,* 37 (1), 111-121.

Elgin C. and Uras B. R. 2012. Public debt, sovereign default risk and shadow economy, *Journal of Financial Stability, forthcoming.*

Elkins Z. and Sides J. 2007. Can Institutions Build Unity in Multiethnic States? *The American Political Science Review,* 101 (4): 693-708.

Ellacuria I. SJ and Sobrino J. SJ. 1993. *Mysterium liberationis: fundamental concepts of liberation theology.* Maryknoll, N.Y.: Orbis Books.

Ellacuria I. SJ. 1976. *Freedom made flesh: the mission of Christ and His church.* Maryknoll, N.Y.: Orbis Books.

Ellacuria I. SJ. 2013. *Ignacio Ellacuria: essays on history, liberation, and salvation.* By Ignacio Ellacuría; Michael Edward Lee. Maryknoll, New York.: Orbis Books.

Elliott M. and Hayward R. D. 2009. Religion and Life Satisfaction Worldwide: The Role of Government Regulation. *Sociology of Religion,* 70 (3): 285-310.

Enders W. and Sandler T. 2002. Patterns of Transnational Terrorism, 1970–1999: Alternative Time-Series Estimates. *International Studies Quarterly,* 46: 145-165.

Erber G. 2013. *The Austerity Paradox: I See Austerity Everywhere, But Not in the Statistics* (February 28, 2013). URL: Available at SSRN: http: //ssrn. com/abstract=2226319 or http://dx. doi.org/10. 2139/ssrn. 2226319.

Erdenir B. 2006. *The Future of Europe: Islamophobia?* Secretariat General for EU Affairs (EUSG) of Turkey, available at: http://www.turkishpolicy.com/ default.asp?show=fall_ 2005_erdenir (Turkish Policy Quarterly, 2006).

Erlinghagen M. 2007a. *Beschäftigungsunsicherheit in Europa.* WSI-Mitteilungen, 60 (3): 123-130.

Erlinghagen M. 2007b. Self-Perceived Job Insecurity and Social Context. A Multi-Level Analysis of 17 European Countries. *European Sociological Review,* 24 (2): 183-197.

Erlinghagen M. 2008. Informelle Arbeit in alternden Gesellschaften. Möglichkeiten und Grenzen des ‚produktiven Alterns'. *Schmollers Jahrbuch,* 128 (2): 237-259.

Erlinghagen M. and Knuth M. 2010. Unemployment as an Institutional Construct? Structural Differences in Non-Employment in Selected European Countries and the United States. *Journal of Social Policy,* 39 (1): 71-94.

Erlinghagen M. and Zink L. 2008. Arbeitslos oder erwerbsunfähig? Unterschiedliche Formen der Nicht-Erwerbstätigkeit in Europa und den USA. *Kölner Zeitschrift für Soziologie und Sozialpsychologie,* 60 (3): 579-608.

Ervasti H. 2004. Attitudes Towards Foreign-born Settlers: Finland in a Comparative Perspective. *Yearbook of Population Research in Finland,* 40: 25-44.

Ervasti H. and Venetoklis T. 2006. Unemployment and Subjective Well-being: Does Money Make a Difference. *Government Institute for Economic Research.*

Ervasti H., Fridberg T., Hjerm M., and Ringdal K. 2008. *Nordic Social Attitudes in a European Perspective.* Cheltenham: Edward Elgar Publishing.

Esmer Y, and Pettersson T. 2007. *Measuring and Mapping Cultures: 25 Years of Comparative Values Surveys.* (Yilmaz Esmer and Thorlief Pettersson, (Ed.)) Leiden and Boston: Brill.

Esmer Y. 2003. Is There an Islamic Civilization? In *'Human Values and Social Change: Findings from the Values Surveys'.* (Inglehart R. F. (Ed.)) Leiden, The Netherlands: Brill: 265-98.

Esposito J. L. 1995. *The Oxford encyclopedia of the modern Islamic world.* Oxford, London and New York: Oxford University Press.

Esposito, John L. 1984. *Islam and Politics.* New York: Syracuse University Press.

Etzioni A. 1968. *The active society: a theory of societal and political processes.* London, Collier-Macmillan; New York, Free Press.

Etzioni A. 1996. *The new golden rule: community and morality in a democratic society.* New York: Basic Books.

European Social Survey 2006. *Data materials.* Freely available from: http: //www. europeansocialsurvey. org/.

Fahrenberg J., and Steiner J. M. 2004. *Adorno und die autoritäre persönlichkeit* | [Adorno and the Authoritarian Personality]. Koelner Zeitschrift fur Soziologie und Sozialpsychologie, 56 (1), pp. 127-152+211.

Fedáková D., Stangl A., and Veira A. 2008. Determinants of female labour supply in Europe: Evidence from the ESS round two data. *Sotsiologicheski Problemi:* 113-130.

Feld L. P. and Kirchgässner G. 2000. Direct democracy, political culture, and the outcome of economic policy: A report on the Swiss experience, *European Journal of Political Economy,* 16, 287-306.

Ferguson N. 2004. *Economics, religion and the decline of Europe.* IEA, Institute of Economic Affairs, Economic Affairs, 2004. Oxford: Blackwell Publishing.

Fetzer J. S. 2011a. *Luxembourg as an Immigration Success Story: The Grand Duchy in Pan-European Perspective.* Lanham, Maryland, USA: Lexington Books.

Fetzer J. S. 2011b. *The Evolution of Public Attitudes toward Immigration in Europe and the United States,* 2000-2010. European University Institute.

Finch H. 2006. Comparison of the Performance of Varimax and Promax Rotations: Factor Structure Recovery for Dichotomous Items, *Journal of Educational Measurement,* 43 (1), 39-52.

Finch H. 2006. Comparison of the Performance of Varimax and Promax Rotations: Factor Structure Recovery for Dichotomous Items. *Journal of Educational Measurement,* 43 (1): 39–52.

Fischer-Schreiber I. (Ed.) 1989. *The Encyclopedia of Eastern philosophy and religion: Buddhism,* Hinduism, Taoism, Zen. Fischer-Schreiber, Ingrid; Schuhmacher, Stephan; Woerner, Gert. Vienna: Shambhala.

Fish S. 2002. Islam and Authoritarianism. *World Politics,* 55 (1): 4-37.

Flam H. 2004. *Turkey and the EU: Politics and Economics of Accession. Munich, CESifo Economic Studies,* 50 (1): 171 – 210, abstract available at: http: //www. cesifo. de/cesifo/newsletter/turkey. htm.

Fleischmann F. and Dronkers J. 2007. *The Effects of Social and Labour Market Policies of EU-countries on the Socio-Economic Integration of First and Second Generation Immigrants from Different Countries of Origin.* European University Institute, Firenze, European University Institute RSCAS Working Papers, 2007/19, available at: http: //ideas. repec. org/p/rsc/rsceui/2007-19. html.

Fleischmann F. and Dronkers J. 2010. Unemployment among immigrants in European labour markets: an analysis of origin and destination effects. *Work, Employment and Society,* 24 (2): 337-354.

Fleischmann F. and Phalet K. 2012. Integration and religiosity among the Turkish second generation in Europe: a comparative analysis across four capital cities. *Ethnic and Racial Studies,* 35 (2): 320-341.

Flere S. 1991. Explaining ethnic antagonism in Yugoslavia. *European Sociological Review,* 7 (3), pp. 183-193.

Forrester J. W. 1977. *Growth Cycles. Economist.* 125 (4): 525-543.

Frank A. G. 1969. *Capitalism and underdevelopment in Latin America; historical studies of Chile and Brazil.* New York: Monthly Review Press.

Frank A. G. 1998. ReOrient: *Global Economy in the Asian Age.* Ewing, U. S. A.: University of California Press.

Freese J. 2004. Risk Preferences and Gender Differences in Religiousness: Evidence from the World ValuesSurvey. *Review of Religious Research,* 46 (1): 88-91.

Freitag M. and Kirchner A. 2011. Social capital and unemployment. A macro-quantitative analysis of the European regions. *Political Studies,* 59 (2): 389-410.

Freitag, M. 2003. Beyond Tocqueville: The Origins of Social Capital in Switzerland. *European Sociological Review,* 19 (2), pp. 217-232+i.

Freud S. 1962, posthumously. *Civilization and its Discontents.* New York, W. W. Norton & Company, Inc.: 23; 32.

Frie K. G., Eikemo T. A., Knesebeck O. v. d. 2009. Education and self-reported health care seeking behaviour in European welfare regimes: results from the European Social Survey. *International Journal of Public Health,* 55 (3): 217-220.

Friedman Th. 1999. *The Lexus and the Olive Tree.* New York: Anchor Press.

Fukuyama F. 2001. Social Capital, Civil Society and Development. *Third World Quarterly,* 22 (1): 7-20.

Fuller W. A. 1976. *Introduction to statistical time series.* New York: John Wiley and Sons.

Furia P. and Lucas R. 2006. Determinants of Arab Public Opinion on Foreign Relations. *International Studies Quarterly,* Vol. 50, No. 3, pp. 585–605.

Furia P. and Lucas R. 2008. Arab Muslim Attitudes Toward the West: Cultural, Social, and Political Explanations. *International Interactions,* 34: 186–207.

Gal A. and Davidov E. 2011. Can we Rate Public Support for Democracy in a Comparable Way? Cross-National Equivalence of Democratic Attitudes in the World Value Survey. *Social Indicators Research,* 104 (2): 271-286.

Gallego A. 2007. Unequal Political Participation in Europe. *International Journal of Sociology,* 37 (4): 10-25.

Gargatz D. A. and G. W. Hill 1996. Analysis of Survey. *Preventive Veterinary Medicine,* 28, 225-237.

Gehring K. 2013. Who benefits from economic freedom? Unraveling the effect of economic freedom on subjective well-being. *World Development,* 50, 74-90.

Genzel R. and Cesarsky C. J. 2000. Extragalactic results from the infrared space observatory. *Annual Review of Astronomy and Astrophysics.* 38: 761–814.

Giacolone R. A., Jurkiewicz C. L. and Deckop J. R. 2008. On ethics and social responsibility: the impact of materialism, postmaterialism, and hope. *Human Relations*, Apr 2008; vol. 61, 483 - 514.

Gibson J. 2006. Enigmas of Intolerance: Fifty Years after Stouffer's Communism, Conformity, and Civil Liberties. *Perspectives on Politics,* 4 (1): 21-34.

Gibson J. L. and Gouws, A. 2001. Making Tolerance Judgments: The Effects of Context, Local and National. *The Journal of Politics,* 63 (4): 1067-1090.

Giddens A. 1979. *Central Problems in Social Theory.* Berkeley, and Los Angeles: University of California Press.

Gijsberts M. and Dagevos J. 2010. At home in the Netherlands. Trends in integration of non-Western migrants. *Annual Report on Integration* 2009.. Den Haag: Sociaal en Cultureel Planbureau.

Gilley B. 2006. The Determinants of State Legitimacy: Results for 72 Countries. *International Political Science Review / Revue internationale de science politique,* Vol. 27 (1): 47-71.

Girès J. 2011. La mobilité sociale intergénérationnelle en Belgique. Rôle de l'origine sociale dans les destinées sociales et scolaires et dimension sexuée de la fluidité sociale. *Recherches sociologiques et anthropologiques,* 42 (1): 3-23.

Goermez M. 2012. 'Religion and Secularism in the Modern World: A Turkish Perspective.' *SAM Papers, SAM Center for Strategic Research, Republic of Turkey, Ministry of Foreign Affairs* (2, March), 2012: retrievable at http: //sam. gov. tr/religion-and-secularism-in-the-modern-world-a-turkish-perspective/.

Goodwin J. 2006. A Theory of Categorical Terrorism. *Social Forces,* 84 (4): 2027-2046.

Gore C. 2000. The rise and fall of the Washington consensus as a paradigm for developing countries. *World Development.* 28 (5): 789-804.

Gorodnichenko Y. and Roland G. 2011. Which Dimensions of Culture Matter for Long-Run Growth? *American Economic Review,* 101(3): 492-98.

Gorodzeisky A. 2010. Who are the Europeans that Europeans prefer? Economic conditions and exclusionary views toward European immigrants. *International Journal of Comparative Sociology,* 52 (1): 100-113.

Gorodzeisky A. and Semyonov M. 2009. Terms of exclusion: public views towards admission and allocation of rights to immigrants in European countries. *Ethnic and Racial Studies,* 32 (3): 401-423.

Gorsuch R. L. 1983. *Factor analysis.* Hillsdale, N. J.: Erlbaum.

Gourio F. 2012. Disaster Risk and Business Cycles. *American Economic Review.* 102 (6): 2734-66.

Grabb E. G., 1979. Working-Class Authoritarianism and Tolerance of Outgroups: A Reassessment. *The Public Opinion Quarterly,* 43 (1): 36-47.

Gran P. 2007. An attempt to get beyond Huntington's polarity of thought in the case of western democracy and Islam. A rethinking of modern Egypt and Italy. *Oriente Moderno, Nuova serie,* 87 (2): 407-430.

Granger C. W. J. 1981. Some properties of time-series data and their use in econometric model specification. *Journal of Econometrics.* 16 (1): 121-130.

Grant R. and Short J. R., eds 2002. *Globalization and the Margins.* Basinstoke and New York: Palgrave Macmillan.

Green E. G., T. 2007. Guarding the gates of Europe: A typological analysis of immigration attitudes across 21 countries. *International Journal of Psychology,* 42 (6): 365-379.

Green E. G., T. 2009. Who can enter? A Multilevel Analysis on Public Support for Immigration Criteria across 20 European Countries. *Group Processes & Intergroup Relations,* 12 (1): 41-60.

Green E. G., T., Fasel N. and Sarrasin O. 2010. The More the Merrier? The Effects of Type of Cultural Diversity on Exclusionary Immigration Attitudes in Switzerland. *International Journal of Conflict and Violence,* 4 (2): 177-190.

Greene W. H. 2008. *Econometric analysis,* Prentice Hall, 6th Edition.

Grier R. 1997. The Effect of Religion on Economic Development: A Cross National Study of 63 Former Colonies. *Kyklos,* 50 (1): 47-62.

Grinin L. E.; Devezas T. C. and Korotayev A. V. 2012. Kondratieff's Mystery. In: *Kondratieff Waves. Dimensions and Prospects at the Dawn of the 21st Century* (Grinin L. E., Devezas T. C., and Korotayev A. V. (Eds.)), 5-22, Volgograd, Russia: Uchitel.

Grupp A. and Mogee M. E. 2004. Indicators for national science and technology policy: how robust are composite indicators?, *Research Policy,* 33, 1373-1384.

Guérin D. and Pelletier R. 2000. Cultural Nationalism and Political Tolerance in Advanced Industrial Societies: The Basque country and Catalonia. *Nationalism and Ethnic Politics,* 6 (4): 1-22.

Guerin D., Petry F., et al. 2004. Tolerance, Protest and Democratic Transition: Survey Evidence from 13 Post-communist Countries. *European Journal of Political Research,* 43 (3): 371-395.

Gueveli A. and Platt L. 2011. Understanding the Religious Behaviour of Muslims in the Netherlands and the UK. Sociology. *The Journal of the British Sociological Association,* 45, 6: 1008-1027.

Guger A., Marterbauer M. and Walterskirchen E. 2004. *Growth Policy in the Spirit of Steindl and Kalecki. WIFO Working papers,* 240, 2004, available at: http: //ewald. walterskirchen.wifo.ac.at/.

Guger A., Marterbauer M. and Walterskirchen E. 2006a. Growth Policy in the Spirit of Steindl and Kalecki. *Metroeconomica,* 57 (3), 428–442.

Guger A., Marterbauer M. and Walterskirchen E. 2006b. Zur Aktualität der Politischen Ökonomie von Josef Steindl. *Kurswechsel,* 4/2006, 18-26, available at http: //ewald. walterskirchen. wifo. ac. at/.

Guiso L., Sapienza P. and Zingales L. 2006. Does Culture Affect Economic Outcomes? *The Journal of Economic Perspectives, 20* (2): 23-48.

Gujarati D. 1999. *Essentials of Econometrics,* 2nd Edition. Boston: McGraw Hill.

Gurr T. R. 1970. *Why Men Rebel.* Princeton: Princeton University Press.

Habib S. 2010. *Islam and homosexuality.* Santa Barbara, Calif.: Praeger.

Hadenius A., and Teorell J. 2005. Cultural and economic prerequisites of democracy: Reassessing recent evidence. *Studies in Comparative International Development.* 39 (4), pp. 87-106.

Hadjimichalis C. 2011. Uneven geographical development and socio-spatial justice and solidarity: European regions after the 2009 financial crisis. *European Urban and Regional Studies.* 18 (3): 254-274.

Hainmueller J. and Hiscox M. J. 2007. Educated Preferences: Explaining Attitudes Toward Immigration in Europe. *International Organization,* 61 (2): 399-442.

Haller M. 2002. Theory and method in the comparative analysis of values. Critique and Alternative to Inglehart. *European Sociological Review,* 18 (2): 139-158.

Haller M. and Hadler M. 2006. How Social Relations and Structures Can Produce Happiness and Unhappiness: An InternationalComparative Analysis. *Social Indicators Research,* 75 (2): 169-216.

Hambleton R. K., Swaminathan H., and Rogers H. J. 1991. *Fundamentals of item response theory,* Newbury Park, CA: Sage.

Hanahan D. and Weinberg R. A. 2000. The hallmarks of cancer. *Cell.* 100 (1): 57-70.

Hanahan D. and Weinberg R. A. 2011. Hallmarks of cancer: the next generation. *Cell. Mar* 4;144 (5): 646-74.

Harell A. 2008. The Micro-story of Multiculturalism: Diverse Social Networks and the Socialization of Tolerance. *PHD Thesis.* McGill University Canada. 2008. 250 p.

Harman H. H. 1976. *Modern Factor Analysis,* 3rd ed. Chicago: University of Chicago Press.

Haustein H. D. and Neuwirth E. 1982. Long Waves in World Industrial Production, Energy Consumption, Innovations, Inventions, and Patents, and their Identification by Spectral Analysis. *Technological Forecasting and Social Change.* 22 (1): 53-89.

Hayes B. C., McAllister I., and Studlar D. T. 2000. Gender, postmaterialism, and feminism in comparative perspective. *International Political Science Review,* 21 (4), pp. 425-439.

Heath A., Fisher S., and Smith S. 2005. The Globalization of Public Opinion Research. *Annual review of political science* (Palo Alto, Calif. Print), 8: 297-333.

Heeringa S. G. 2007. Probability Sampling and the Scientific Survey Method for Population Studies: Application to Survey Research in Islamic Countries. In *'Values and Perceptions of the Islamic and Middle Eastern Publics: Findings from Values Survey'.* (Mansoor Moaddel, (Ed.)) New York: Palgrave Macmillan: 323-338.

Henzel Ch. 2005. *The Origins of al Qaeda's Ideology: Implications for US Strategy. Parameters,* 35, 1, (2005): 69-80, retrievable at: http://www.carlisle.army. mil/USAWC/parameters/.

Herda D. 2010. How Many Immigrants? Foreign-Born Population Innumeracy in Europe. *Public Opinion Quarterly,* 74 (4): 674-695.

Herreros F. and Criado H. 2008a. Social Trust, Social Capital and Perceptions of Immigration. *Political Studies,* 57 (2): 337-355.
Herreros F. and Criado H. 2008b. The State and the Development of Social Trust. *International Political Science Review,* 29 (1): 53-71.
Heshmati A. 2006a. Measurement of a Multidimensional Index of Globalization, *Global Economy Journal,* 6 (2), Article 1.
Heshmati A. 2006b. *The Relationship between Income Inequality Poverty, and Globalization, in The Impact of Globalization on the World's Poor* (M. Nissanke and E. Thorbecke, (Eds.)), Basinsgtoke and New York: Palgrave Macmillan, pp. 59-93.
Heshmati A. 2007. *Global Trends in Income Inequality.* Hauppauge, New York: Nova Science Publishers.
Heshmati A. and Lee S. 2010. The Relationship between Globalization, Economic Growth and Income Inequality. *Journal of Globalization Studies,* 1 (2), 87-117.
Heshmati A. and Oh J-E. 2006. Alternative Composite Lisbon Development Strategy Indices: A Comparison of EU, USA, Japan and Korea. *The European Journal of Comparative Economics,* 3 (2), 133-170.
Heshmati A. and Tausch A. 2007. Eds. *Roadmap to Bangalore? Globalization, the EUs Lisbon Process and the Structures of Global Inequality.* Hauppauge: Nova Science Publishers.
Heshmati A., Tausch A. and Bajalan C. 2008. Measurement and Analysis of Child Well-Being in Middle and High Income Countries. *European Journal of Comparative Economics,* 5 (2), 227-286.
Hinkelammert F. 1983. The economic roots of idolatry: entrepreneurial metaphysics. In *The idols of death and the God of life: a theology.* B. E. Campbell, B. Shepard, Trans.; (Richard, P., et al., Eds.) pp. 165-193. Maryknoll, NY: Orbis Books.
Hinkelammert F. 1986. *The Ideological Weapons of Death: A Theological Critique of Capitalism, trans.* by Phillip Berryman with a foreword by Cornel West. Maryknoll, N. Y.: Orbis, 1986.
Hironaka A. 2005. *Neverending Wars: The International Community, Weak States, and the Perpetuation of Civil War.* Cambridge: Harvard University Press.
Hirschle J. 2011. *The Affluent Society and its Religious Consequences: An Empirical Investigation of 20 European Countries. Socio-Economic Review,* 9 (2): 261-285.
Hjerm M. 2007. Do Numbers Really Count? Group Threat Theory Revisited. *Journal of ethnic and migration studies,* 33 (8): 1253-1275.
Hjerm M. 2009. Anti-Immigrant Attitudes and Cross-Municipal Variation in the Proportion of Immigrants. *Acta Sociologica,* 52 (1): 47-62.
Hjerm M. and Berg L. 2010. National Identity and Political Trust. *Perspectives on European Politics and Society,* 11 (4): 390-407.
Hjerm M. and Nagayoshi K. 2011. The composition of the minority population as a threat: Can real economic and cultural threats explain xenophobia? *International Sociology,* 26 (6): 815-843.
Hodson R., Sekulic, D., and Massey, G. 1994. National Tolerance in the Former Yugoslavia. *The American Journal of Sociology,* 99 (6): 1534-1558.
Hofstede G. 2001. *Culture's consequences: comparing values, behaviors, institutions, and organizations across nations.* Thousand Oaks, Calif.: Sage Publications.
Hofstede G. and Minkov M. 2010. Long- versus short-term orientation: new perspectives. *Asia Pacific Business Review, 16* (4): 493–504.

Hofstede G., Hofstede G. J. and Minkov, M. 2010. *Cultures and Organizations: Software of the Mind.* Revised and expanded 3rd Edition. New York: McGraw-Hill.

Holcombe R. G. 2013. South Korea's economic future: Industrial policy or economic democracy?, *Journal of Economic Behavior & Organization,* 88, 3-13.

Hooghe M. and Reeskens T. 2007. Are cross-national surveys the best way to study the extreme-right vote in Europe?. *Patterns of Prejudice,* 41 (2): 177-196.

Hooghe M., Reeskens T., and Stolle D. 2007. Diversity, Multiculturalism and Social Cohesion: Trust and Ethnocentrism in European Societies. In: Belonging? *Diversity, Recognition and Shared Citizenship in Canada,* Montreal: Institute for Research on Public Policy: 387-410.

Hooghe M., Reeskens T., Stolle D., and Trappers A. 2008. Ethnic Diversity and Generalized Trust in Europe. A Cross-National Multilevel Study. *Comparative Political Studies,* 42 (2): 198-223.

Hotelling H. 1933. Analysis of a Complex of Statistical Variables into Principal Components. *Journal of Educational Psychology.* 24, 417-441 and 24, 498-520.

Howard M. M. 2002. Postcommunist civil society in comparative perspective. *Demokratizatsiya,* 10 (3), pp. 285-305.

Hudson Institute (Vidino L.) 2006. *Aims and Methods of Europe's Muslim Brotherhood. Current Trends in Islamist Ideology* 4, available at: http: //www. futureofmuslimworld. com/research/pubID. 55/pub_detail. asp.

Hughes K. 2004. *The Political Dynamics of Turkish Accession to the EU: A European Success Story or the EU's Most Contested Enlargement?* Swedish Institute for European Policy Studies, SIEPS, 9, available at: http: //www. lse. ac. uk/collections/europeanInstitute/ pdfs/Turkey-EU%20report%20Sieps%20final. pdf.

Huntington S. P. 1993. The Clash of Civilizations? *Foreign Affairs,* 72 (3): 22 - 49.

Huntington S. P. 1996. *The clash of civilizations and the remaking of world order.* New York: Simon & Schuster.

Husson M. and Louça F. 2012. Late Capitalism and Neo-Liberalism – A perspective on the Current Phase of the Long Wave of Capitalist Development. In: *Kondratieff Waves. Dimensions and Prospects at the Dawn of the 21st Century* (Grinin L. E., Devezas T. C., and Korotayev A. V. (Eds.)), 176-187, Volgograd, Russia: Uchitel.

Hutchison M., and Gibler D. 2007. Political Tolerance and Territorial Threat: A Cross National Study. *Journal of Politics,* 69 (1): 128-142.

IBM. 2011. *IBM SPSS Statistics 20 Algorithms.* Armonk, New York. (URL: http: //www-01. ibm. com/support/docview. wss?uid=swg27021213#en).

Ibn Kathir, *Stories of the Prophets,* Translated from Arabic by Muhammad Mustapha Geme'ah, Ryadh, KSA: Al-Azhar Darussalam (N.D.)

Ibn Khaldoun (1332-1406), 2005. *The Muqaddimah: an introduction to history. Ibn Khaldun, translated and introduced by Franz Rosenthal;* abridged and edited by N. J. Dawood, with a new introduction by Bruce B. Lawrence. Princeton, N. J.: Princeton University Press.

Ibn Taymiyyah T. D. A. (Taqy al-Din Ahmad) 1999. *Al Ubudiyya [Servitude].* Al-Isma'iliyya: Dar al-Asala.

IBS Research Consultancy, European Round Table of Industrialists 2005. *Turkey – A New Corporate World for Europe.* Brussels: ERT; available at: http: //www. ert. be/pb/pbb/enbb_frame. htm.

Iglič H. 2010. Voluntary Associations and Tolerance: An Ambiguous Relationship. *American Behavioral Scientist,* 53 (5): 717-736.

Indjikian R. and Siegel D. S. 2005. The impact of investment in IT on economic performance: implications for developing countries, *World Development,* 33, 681-700.

Inglehart R. F. 1970. The New Europeans: Inward or Outward-Looking? *International Organization,* 24 (1): 129-139.

Inglehart R. F. 1971. The Silent Revolution in Europe: Intergenerational Change in Post-Industrial Societies. *American Political Science Review,* 65 (December): 991–1017.

Inglehart R. F. 1977. *The Silent Revolution: Changing Values and Political Styles in Advanced Industrial Society.* Princeton, NJ: Princeton University Press.

Inglehart R. F. 1990. *Culture shift in advanced industrial societies.* Princeton, N. J.: Princeton University Press.

Inglehart R. F. 1997. *Modernization and Postmodernization: Cultural, Economic and Political Change in 43 Societies.* Princeton, NJ: Princeton University Press.

Inglehart R. F. 2000. *Globalization and Postmodern Values.* The Washington Quarterly, 23 (1): 215-228.

Inglehart R. F. 2002. Islam, Gender, Culture, and Democracy. *International Journal of Comparative Sociology,* 43 (2-5): 224-28.

Inglehart R. F. 2003a. How Solid Is Mass Support for Democracy: And How Can We Measure It? PS: *Political Science and Politics,* 36 (1): 51-57.

Inglehart R. F. 2003b. *Human Values and Social Change.* (Inglehart R. F. (Ed.)) Leiden, The Netherlands: Brill.

Inglehart R. F. 2003c. *Islam, Gender, Culture, and Democracy.* (Ronald Inglehart Ed). Willowdale, ON, Canada: de Sitter.

Inglehart R. F. 2007. The Worldviews of Islamic Publics in Global Perspective. In *'Values and Perceptions of the Islamic and Middle Eastern Publics: Findings from Values Survey'.* (Mansoor Moaddel, (Ed.)) New York: Palgrave Macmillan: 25-46.

Inglehart R. F. and Baker, W. E. 2000. Modernization, Cultural Change, and the Persistence of Traditional Values. *American Sociological Review.* 65 (1): 19-51.

Inglehart R. F. and Norris P. 2003a. Islamic culture and democracy: Testing the 'clash of civilizations' thesis. *Comparative Sociology,* 1 (3-4): 235-263.

Inglehart R. F. and Norris P. 2003b. *Rising tide: gender equality and cultural change around the world.* Cambridge, UK; New York: Cambridge University Press, 2003.

Inglehart R. F. and Norris P. 2003c. *The True Clash of Civilizations.* Foreign Policy, 135 (Mar. - Apr., 2003): 62-70.

Inglehart R. F. and Norris P. 2010. Are high levels of existential security conducive to secularization? *A response to our critics. Paper for Mid-West Political Science Association annual meeting,* Chicago, April 22, 2010.

Inglehart R. F. and Norris P. 2012. The Four Horsemen of the Apocalypse: Understanding Human Security. *Scandinavian Political Studies,* 35(1): 71-95.

Inglehart R. F. and Welzel C. 2003. Political Culture and Democracy: Analyzing Cross-Level Linkages. *Comparative Politics, 36* (1): 61-79.

Inglehart R. F. and Welzel C. 2005. *Modernization, Cultural Change and Democracy.* New York: Cambridge University Press, 2005.

Inglehart R. F. and Welzel C. 2009. *How Development Leads to Democracy. What We Know About Modernization.* Foreign Affairs, March, April (freely available at: http://www.foreignaffairs.com/articles/64821/ronald-inglehart-and-christian-welzel/how-development-leads-to-democracy).

Inglehart R. F. and Welzel C. 2010a. Changing Mass Priorities: The Link between Modernization and Democracy. *Perspectives on Politics,* 8(2): 551-567.

Inglehart R. F. and Welzel C. 2010b. Agency, Values, and Well-Being: A Human Development Model. *Social Indicators Research*, 10.1007/s11205-009-9557-z

Inglehart R. F. et al. 2004. *Human Beliefs and Values. A cross-cultural sourcebook based on the 1999-2002 values surveys.* (Ed.) by R. Inglehart, M. Basanez, J. Diez-Medrano, L. Halman and R. Luijkx. Mexico: Siglo XXI.

Inglehart R. F., and Norris P. 2000. The developmental theory of the gender gap: Women's and men's voting behavior in global perspective. *International Political Science Review,* 21 (4), pp. 441-463.

Inglehart R. F., Moaddel M., and Tessler M. 2006. Xenophobia and In-Group Solidarity in Iraq: A Natural Experiment on the Impact of Insecurity. *Perspectives on Politics,* 4 (3): 495-505.

Internet Sacred Text Archive 2013. *Internet Sacred Text Archive,* available at: http: //www. sacred-texts. com/isl/index. htm.

Islamic Philosophy Online. 2014. *Islamic Philosophy Online.* http://www.muslimphilosophy. com/ Philosophia Islamica. Webmaster: Muhammad Hozien.

Issawi Charles and Leaman Oliver, 1998, "Ibn Khaldūn, 'Abd al-Rahman (1332–1406)," in *Routledge Encyclopedia of Philosophy* (London: Routledge), vol. 4, 623–627.

Ivarsflaten E. 2007. What Unites Right-Wing Populists in Western Europe? Re-Examining Grievance Mobilization Models in Seven Successful Cases. *Comparative Political Studies,* 41 (1): 3-23.

Jabber P. 2001. Impact of the War on Terror on Certain Aspects of US Policy in the Middle East. *A Medium-Term Assessment. Prepared for the United States National Intelligence Council.* Available at: http: //www. fas. org/irp/nic/jabber_paper. htm.

Jackson D. J. and Borgatta E. F. (Editors) 1981. *Factor Analysis and Measurement in Sociological Research. A Multi-Dimensional Perspective.* London and Beverly Hills: Sage, Sage Studies in International Sociology, 21.

Jacobs A. 2006. *Reformislam.* Arbeitspapiere, Nr. 155, Sankt Augustin, 21. Sept. 2006. Herausgeber: Konrad-Adenauer-Stiftung e. V., available at: http: //www. kas. de/wf/de/33. 8230/.

Jaeger J. and Springler E. 2012. *Oekonomie der Internationalen Entwicklung.* Eine kritischer Einfuehrung in die Volkswirtschaftslehre. Wien: Mandelbaum.

Jamal A. 2006. Reassessing Support for Islam and Democracy in the Arab World? Evidence from Egypt andJordan. *World Affairs,* 169 (2): 51-63.

Jamal A. 2007. When Is Social Trust a Desirable Outcome? Examining Levels of Trust in the Arab World. *Comparative Political Studies*, Nov 2007; vol. 40: 1328 - 1349.

Jamal al-Din, Nadia, 1994, "Miskawayh (A.H. 320-421/A.D. 932-1030)" in: *"Thinkers on Education" published by Prospects: the quarterly review of comparative education* (Paris, UNESCO: International Bureau of Education), vol. 24, no. 1/2, p. 131–52.

Janmaat J. G. and Braun R. 2009. Diversity and Postmaterialism as Rival Perspectives in Accounting for Social Solidarity: Evidence from International Surveys. *International Journal of Comparative Sociology,* Feb 2009; vol. 50, 39-68.

Jen M. H., Jones K. and Johnston R. 2009. Global variations in health: Evaluating Wilkinson's income inequality hypothesis using the World Values Survey. *Social Science and Medicine,* 68, 643-653.

Johanisova N. and Wolf S. 2012. Economic democracy: A path for the future, *Futures,* 44, 562-570.

Johnson N. D. and Mislin A. 2012. How much should we trust the World Value Survey trust question?, *Economics Letters,* 116, 210-212.

Johnston J. and DiNardo J. 1997. *Econometric Methods,* 4th Edition, McGraw Hill: New York.

Johnston R. L. 1975. *Religion and Society in Interaction,* Englewood Cliffs, NJ: Prentice-Hall.

Jones R. A. 1986. *Emile Durkheim: An Introduction to Four Major Works.* Beverly Hills, CA: Sage Publications, Inc., Pp. 115-155.

Juergensmeyer M. 2001. *Terror In The Mind Of God: The Global Rise Of Religious Violence.* Berkeley: University of California Press.

Jungbauer-Gans M. and Gross C. 2007. Verteilung des sozialen Kapitals. *Eine makrosoziologische Analyse des European Social Survey 2002 und 2004.* Kölner Zeitschrift für Soziologie und Sozialpsychologie, 59 (47): 211-240.

Junttila J. 2001. Structural breaks, ARIMA model and Finnish inflation forecasts. *International Journal of Forecasting.* 17 (2): 203-230.

Kääriäinen J. 2007. Trust in the Police in 16 European Countries. A Multilevel Analysis. *European Journal of Criminology,* 4 (4): 409-435.

Kalecki M. 1968. *Theory of economic dynamics; an essay on cyclical and long-run changes in capitalist economy. New* York: Monthly Review Press.

Kalecki M. 1968. Trend and Business Cycles Reconsidered. *Economic Journal.* 78 (310): 263-276.

Kalecki M. 1971. *Selected essays on the dynamics of the capitalist economy 1933-1970.* Cambridge [Eng.]: Cambridge University Press.

Kalecki M. and Feiwel G. R. 1972. *The last phase in the transformation of capitalism.* New York: Monthly Review Press.

Kalenborn C. and Lessmann C. 2013. The impact of democracy and press freedom on corruption: Conditionality matters, *Journal of Policy Modeling*, in press.

Kalin Ibrahim, 2002, "Three Views of Science in the Islamic World", in: Ted Peters, Muzaffar Iqbal, Syed Nomanul Haq (Eds.) God, Life and the Cosmos: *Christian and Islamic Perspectives,* (Ashgate), pp. 43-75.

Kang S. M. 2002. A Sensitivity Analysis of the Korean Composite Environmental Index. *Ecological Economics,* 43, 159-174.

Karlinger F. and Freitas G. SJ. 1972. *Brasilianische Mårchen. Dusseldorf: E. Diederichs,* Serie: Mårchen der Weltliteratur.

Karoui H. 2012. Arab Spring, A Middle East Studies Book. *Middle East Studies Online Journal,* Charleston S. C. 2012: 4, 111.

Karstedt S. 2006. Democracy, Values, and Violence: Paradoxes, Tensions, and Comparative Advantages of Liberal Inclusion. *Annals of the American Academy of Political and Social Science,* 605: May 2006: 50-81.

Karyotis G. and Patrikios S. 2010. Religion, securitization and anti-immigration attitudes: The case of Greece. *Journal of Peace Research,* 47 (1): 43-57.

Katnik A. 2002. Religion, Social Class and a Political Tolerance: A Cross-national Analysis. *International Journal of Sociology,* 32 (1): 14-38.

Kaya M. K. and Cornell S. E. 2012. The big split: the differences that led Erdoğan and the Gülen movement to part ways. *Turkey Analyst* (Central Asia-Cacucasus Institute, Johns

Hopkins University), vol. 5, no. 5 (5 March 2012), retrievable at: http: //www. silkroadstudies. org/new/inside/turkey/2012/120305a. html.

Kearney A. T., Inc., *The Carnegie Endowment for International Peace.* 2002. Globalization's Last Hurrah?. Foreign Policy. January/February: 38-51.

Kearney A. T., Inc., The Carnegie Endowment for International Peace. 2003. Measuring Globalization: Who's Up, Who's Down?. *Foreign Policy.* January/February: 60-72.

Keddie, Nikki R. 1983, *An Islamic Response to Imperialism, Political and Religious Writings of Sayyid Jamal al-Din al-Afghani,* (Berkeley: University of California Press), pp. 130-174.

Kehberg J. E. 2007. Public Opinion on Immigration in Western Europe: Economics, Tolerance, and Exposure. *Comparative European Politics,* 5 (3): 264 - 281.

Kentor J. 2001. The Long Term Effects of Globalization on Income Inequality, Population Growth, and Economic Development. *Social Problems,* 48 (4), Special Issue on Globalization and Social Problems: 435-455.

Kentor J. and Boswell T. 2003. Foreign Capital Dependence and Development: A New Direction. *American Sociological Review,* 68 (2): 301-313.

Keown D. 2000. *Buddhism: a very short introduction.* Oxford, London and New York: Oxford University Press

Kestilä E. 2006. Is There Demand for Radical Right Populism in the Finnish Electorate?. *Scandinavian Political Studies,* 29 (3): 169-191.

Khan M. Z. (Muhammad Zafrullah) 2008. *Islam.* International Publications Limited, Tilford, Surrey, U.K.

Khawaja M. S. (Muhammad Saeed) 2011. Islamic Art and Its Spiritual Message. *International Journal of Humanities and Social Science,* Vol. 1 No. 2; February.

Khoury A. Th. 1980. *Toleranz im Islam.* München: Kaiser und Grünwald.

Khoury A. Th. 1991. *Was ist los in der islamischen Welt?* Die Konflikte verstehen. Freiburg, Basel, Vienna: Herder.

Khoury A. Th. 1994. *Christen unterm Halbmond: religiöse Minderheiten unter der Herrschaft des Islam.* Freiburg [im Breisgau]: Herder.

Khoury A. Th. 2001. Der *Islam und die westliche Welt: religiöse und politische Grundfragen.* Darmstadt: Primus.

Khoury A. Th. 2002. *Mit Muslimen in Frieden leben: Friedenspotentiale des Islam.* Würzburg: Echter, 2002.

Khoury A. Th. 2005. *Der Koran erschlossen und kommentiert. Patmos.*

Khoury A. Th. 2007a. *Der Koran. Guetersloh: Guetersloher Verlagshaus.*

Khoury A. Th. 2007b. *Lexikon religiöser Grundbegriffe,* Judentum - Christentum – Islam. Wiesbaden: Marix Verlag.

Khoury A. Th. 2008a. *Der Hadith. Quelle der islamischen Tradition.* Guetersloher Verlagshaus.

Khoury A. Th. 2008b. *Muhammad, Der Prophet und seine Botschaft.* Freiburg: Herder Verlag.

Khoury A. Th. et al. 2006. *Islam-Lexikon A-Z.* Freiburg: Herder.

Kieffer K. M. 1998. Orthogonal versus oblique rotation: A review of the literature regarding the pros and cons. *Paper presented at the Annual Meeting of the Mid-South Educational Research Association,* New Orleans, LA, November 1998.

Kilburn H. W. 2009. Personal Values and Public Opinion. *Social Science Quarterly,* 90 (4): 868-885.

Kim J. -O. and Mueller Ch. W. 1978a. *Introduction to Factor Analysis.* What It Is and How to Do It. London and Beverly Hills: Sage, Series: Quantitative Applications in the Social Sciences, 13.

Kim J. -O. and Mueller Ch. W. 1978b. *Factor Analysis. Statistical Methods and Practical Issues.* London and Beverly Hills: Sage, Series: Quantitative Applications in the Social Sciences, 14.

Klausen J. 2005. The Re-Politicization of Religion in Europe: The Next Ten Years. *Perspectives on Politics,* 3 (3): 554-55.

Knott K. 2000. *Hinduism: a very short introduction.* Oxford, London and New York: Oxford University Press.

Knutsen C. H. 2013. Democracy, state capacity, and economic growth, *World Development,* 43 (1), 1-18.

Koesebalaban H. 2011. Turkey and the New Middle East: Between Liberalism and Realism. Perceptions (SAM Center for Strategic Research, Republic of Turkey, Ministry of Foreign Affairs), 26 (Autumn 2011): 93-114, retrievable at http: //sam. gov. tr/category/ publications/perceptions/.

Komarkova Z. and Komarek L. 2007. Integration of the foreign exchange markets of the selected EU new member states. *Politicka Ekonomie.* 55 (3): 315-333.

Kueng H. 1986. *Christianity and the world religions: paths of dialogue with Islam, Hinduism, and Buddhism.* Hans Küng... [*et al.*]; translated by Peter Heinegg. New York: Doubleday.

Kuhn Th. S. 1962. The Structure of Scientific Revolutions (1st ed.). University of Chicago Press.

Kunitzsch P. 2004. *Stars and numbers: astronomy and mathematics in the medieval Arab and western worlds.* Burlington, Vt.: Ashgate.

Kutub M. 1984. *Islam The Misunderstood Religion.* Kuwait: International Islamic Federation.

Kuznets S. 1940. Schumpeters Business Cycles, *American Economic Review,* 30 (2), 157-169.

Kuznets S. 1976. *Modern Economic Growth: Rate, Structure and Spread.* New Haven, CT: Yale University Press.

Kvamn K. E.; Schearing L. S.; Valarie H Ziegler V. H. 1999. *Eve and Adam: Jewish, Christian, and Muslim readings on Genesis and gender.* Bloomington: Indiana University Press.

Lam P. -Y. 2006. Religion and Civic Culture: A Cross-National Study of Voluntary Association Membership. *Journal for the Scientific Study of Religion,* 45 (2): 177-193.

Leaman O. 1998. *Islamic Philosophy.* From the Routledge Encyclopedia of Philosophy. General Editor Edward Craig, Churchill College, University of Cambridge, UK. Available at: http: //www. muslimphilosophy. com/ip/rep. htm .

Lee K. 2009. Towards a Reformulation of Core/Periphery Relationship: A Critical Reappraisal of the Trimodality of the Capitalist World-Economy in the Early 21st Century. *Perspectives on Global Development and Technology,* 8 (2-3): 263-294.

Lehmann E. L. and Romano J. P. 2005. *Testing Statistical Hypotheses,* 3rd Edition. New York: Springer.

Lejour A. M. *et al.* 2004. *Assessing the economic implications of Turkish accession to the EU.* CPB Netherlands Bureau for Economic Policy Analysis, 56, 2004, available at: http: //www. cpb. nl/eng/pub/document/56/doc56. pdf .

Lewis B. 1993. *Islam and the West.* New York: Oxford University Press.

Lewis B. 2002. *What went wrong?: Western impact and Middle Eastern response.* Oxford; New York: Oxford University Press, 2002.

Lewis B. 2003. *The crisis of Islam: holy war and unholy terror.* New York: Modern Library.

Li C.-L., Zinn H. C., Chick G. E., Absher J. D., Graefe A. R., Hsu Y. -C. 2007. Segmentation of culturally diverse visitors' values in forest recreation management. *Forest Snow and Landscape Research* 81 (1-2), pp. 19-29.

Li M., Xiao F. and Zhu A. 2007. Long Waves, Institutional Changes, and Historical Trends: A Study of the Long-Term Movement of the Profit Rate in the Capitalist World Economy. *Journal of World-Systems Research,* 13 (1): 33-54. URL: http://www.jwsr.org/archive/volume-13-issue-1-2007.

Li Q. and Schaub D. 2004. Economic Globalization and Transnational Terrorism: A Pooled Time-Series Analysis. *Journal of Conflict Resolution* 48 (2): 230-258.

Liman M. W. L. and Bond M. H. 2010a. *Does Individual Secularism Promote Life Satisfaction? The Moderating Role of Societal Development.* World Values Research, 3 (3): 14-28. Available at: http: //www. worldvaluessurvey. org/the_paperseries .

Liman M. W. L. and Bond M. H. 2010b. Analyzing National Change in Citizen Secularism Across Four Time Periods in the World Values Surveys. *World Values Research,* 3 (2): 0-13. Available at: http: //www. worldvaluessurvey. org/the_paperseries .

Lippl B. 2007. Soziales Engagement und politische Partizipation in Europa. In: Sozialkapital. Grundlagen und Anwendungen. *Kölner Zeitschrift für Soziologie und Sozialpsychologie,* Sonderheft 47, Wiesbaden: VS Verlag für Sozialwissenschaften: 420-449.

Lipset S. M. 1968. *Revolution and Counterrevolution,* New York: Basic Books.

Listhaug O. and Ringdal K. 2008. Trust in political institutions. In: *Nordic social attitudes in a European perspective,* Cheltenham: Edward Elgar Publishing: 131-151.

Listhaug O. and Ringdal K. 2009. Competition between institutions in multilevel governance: Implications for political trust. In: *Religion, democratic values and political conflict. Festschrift in the honour of Torleif Petterson,* Uppsala: Uppsala University: 231-253.

Lizardo O. 2006. The effect of economic and cultural globalization on anti-U. S. transnational terrorism: 1971-2000. *Journal of World Systems Research,* 12: 149-186.

Loades A. 1990. *Feminist theology. A reader.* London: SPCK.

Lockwood B. 2004. How Robust is the Foreign Policy-Kearney Globalization Index?. *The World Economy,* 27, 507-523.

Lockwook B. and M. Redoano 2005. *The CSGR Globalization Index: An Introductory Guide.* CSGR Working Paper 155/04.

Login A. and Areas S. 2007. Public Opinion on Immigration in Western Europe: Economics, Tolerance, and Exposure. *Comparative European Politics,* 5: 264-281.

Lubbers M. and Scheepers P. 2007. Explanations of political Euro-scepticism at the individual, regional and national levels. *European Societies: The Official Journal of the European Sociological Association,* 9 (4): 643-669.

Lubbers M. and Scheepers P. 2008. Euro-scepticism and Extreme Voting Patterns in Europe. In: *Measuring meaningful data in social research,* Leuven: Acco: 71-92.

Luxemburg R. 1964. *The accumulation of capital.* New York, Monthly Review Press.

Mace D. and Mace V. 1960. *Marriage: East and West.* Dolphin Books, Doubleday and Co., Inc., N.Y., 1960.

Macintosh R. 1998. A confirmatory factor analysis of the affect balance scale in 38 nations: A research note. *Social Psychology Quarterly* 61 (1), pp. 83-91.

Macri Y. 2012. *"The End of A World"*, CILE, Essays, 01/10/2012: http://cilecenter.org/areas-of-research/economics/essays/essays-details?articleID=34

Maddison A. 1964. *Economic growth in the West: comparative experience in Europe and North America; a Twentieth Century Fund study.* N. Y. New York: Twentieth Century Fund.

Maddison A. 1969. *Economic growth in Japan and the USSR.* London: Allen and Unwin.

Maddison A. 1971. *Class structure and economic growth. India and Pakistan since the Moghuls.* London, Allen and Unwin.

Maddison A. 1982. *Phases of capitalist development.* Oxford; New York: Oxford University Press.

Maddison A. 1985. *Two crises: Latin America and Asia, 1929-38 and 1973-83.* Paris: Development Centre of the Organisation for Economic Co-operation and Development; Washington, D. C.: OECD Publications and Information Center.

Maddison A. 1988. *Comparisons of real output in manufacturing.* Washington, D. C.: International Economics Dept., World Bank.

Maddison A. 1989. *The world economy in the 20th century.* Paris: Development Centre of the Organisation for Economic Co-operation and Development; Washington, D. C.: OECD Publications and Information Center.

Maddison A. 1991. *Dynamic forces in capitalist development: a long-run comparative view.* Oxford; New York: Oxford University Press.

Maddison A. 1995. *Explaining the economic performance of nations: essays in time and space.* Aldershot, Hants, England; Brookfield, Vt., USA: E. Elgar.

Maddison A. 2003. *The world economy: historical statistics.* Paris, France: Development Centre of the Organisation for Economic Co-operation and Development.

Maddison A. 2007. *Contours of the world economy, 1-2030 AD: essays in macro-economic history.* Oxford; New York: Oxford University Press.

Mahasneh H. I. 2003. Faiths and Ecology. In: *Faith in Conservation.* By Martin Palmer with Victoria Finlay, published by the World Bank. Retrievable at: http://www.arcworld.org/faiths.asp?pageID=75

Malchow-Møller N., Munch J. R., Schroll S., and Skaksen J. R. 2009. Explaining Cross-Country Differences in Attitudes Towards Immigration in the EU-15. *Social Indicators Research,* 91 (3): 371-390.

Malchow-Møller N., Munch J. R., Schroll S., and Skaksen J. R. 2008. Attitudes towards Immigration--Perceived Consequences and Economic Self-Interest. *Economics Letters,* 100 (2): 254-257.

Mandel E. 1995. *Long waves of capitalist development. A Marxist interpretation; based on the Marshall lectures given at the University of Cambridge.* London: Verso.

Manji I. 2005. *The Trouble with Islam Today.* New York: St. Martin's Press.

Mankiw G. N., Romer D. and Weil D. N. 1992. A Contribution to the Empirics of Economic Growth. *The Quarterly Journal of Economics.* 107 (2): 407-437.

Mann Th. 1924. *Buddenbrooks, translated from the German of Thomas Mann, by H. T. Lowe-Porter.* New York: A. A. Knopf.

Mansour F. 1992. *The Arab world: nation, state, and democracy. Fawzy Mansour; with a contributory chapter by Samir Amin.* United Nations University Press Zed Books, Studies in African political economy.

Marchetti C. 1980. Society as a Learning System. Discovery, Invention and Innovation Cycles Revisited. *Technological Forecasting and Social Change.* 18 (4): 267-282.

Marchetti C. 2006. Is History Automatic and Are Wars a la Carte? The Perplexing Suggestions of a System Analysis of Historical Time Series. In: *Kondratieff waves, warfare and world security: [proceedings of the NATO Advanced Research Workshop on The Influence of Chance Events and Socioeconomic Long Waves in the New Arena of Asymmetric Warfare,* Covilhã, Portugal, 14 - 18 February 2005]. (Devezas T. C. (Ed.)), 173-179. Amsterdam: IOS Press.

Marquart-Pyatt S. and Paxton P. 2007. In Principle and in Practice: Learning Political Tolerance in Eastern and Western Europe. *Political Behavior,* 29 (1): 89-113.

Martin-Artiles A. and Molina O. 2011. Crisis, economic uncertainty and union members' attitudes towards immigration in Europe. *Transfer,* 17 (4): 453-469.

Martindale D. 1960. *The Nature and Types of Sociological Theory.* Boston: Houghton and Mifflin Company.

Martínez-Herrera E. and Moualhi D. 2007. Predispositions to discriminatory immigration policies in western Europe: An exploration of political causes. *Portuguese Journal of Social Science,* 5 (3): 215-233.

Maseland R. and Van Hoorn A. 2011. Why Muslims like democracy yet have so little of it. *Public Choice,* 147: 481–496.

Masso A. 2009. A Readiness to Accept Immigrants in Europe? Individual and Country-Level Characteristics. *Journal of Ethnic and Migration Studies,* 35 (2): 251-270.

Mastekaasa A. 1983. Post-Materialist Values and Subjective Satisfaction: Testing Ronald Inglehart's Hypotheses. *Acta Sociologica,* Jan 1983; 26, 141-159.

Masûdi 1896. *Livre de l'avertissement, 1896, trad. B. Carra de Vaux,* Paris, Imprimerie nationale.

Matsumoto D. and Yoo S. H. 2006. Toward a New Generation of Cross-Cultural Research. *Perspectives on Psychological Science,* Sep 2006 (1), 234-250.

Mau S. and Burkhardt C. 2009. Migration and welfare state solidarity in Western Europe. *Journal of European Social Policy,* 19 (3): 213-229.

Maxwell R. 2010. Evaluating Migrant Integration: Political Attitudes Across Generations in Europe. *International Migration Review,* 44 (1): 25-52.

McCleary R. M. and Barro R. J. 2006a. Religion and Economy. *The Journal of Economic Perspectives,* 20 (2): 49-72.

McCleary R. M. and Barro R. J. 2006b. Religion and Political Economy in an International Panel. *Journal for the Scientific Study of Religion,* 45 (2): 149-175.

McDonald R. P. 1997. *Test theory: A unified treatment.* Mahwah, N. J.: Erlbaum.

McLeod L. D., Swygert K. A. and Thissen D. 2001. Factor analysis for items scored in two categories. In: *Test scoring* (D. Thissen & H. Wainer (Eds.)), pp. 189–216). Mahwah, N. J.: Erlbaum.

Meloen J. D., Van Der Linden G., and De Witte H. 1996. A test of the approaches of Adorno *et al.*, Lederer and Altemeyer of authoritarianism in Belgian Flanders: A research note. *Political Psychology,* 17 (4), pp. 643-656.

Messner D. and Freitas G. SJ. 1988. *Studia lusitanica Salzburgensia; Geraldo de Freitas zum 60. Geburtstag.* Institut für Romanistik (Salzburg), Salzburg: Inst. für Romanistik [u.a.], Salzburger romanistische Schriften, 11.

Metz R. 2008. *Auf der Suche nach den Langen Wellen der Konjunktur.* Stuttgart: Franz Steiner Verlag.

Metz R. 2011. Do Kondratieff waves exist? How time series techniques can help to solve the problem. *Cliometrica.* 5 (3): 205-238.

Meuleman B. 2010a. Perceived Economic Threat and Anti-immigration Attitudes: Effects of Immigrant Group Size and Economic Conditions Revisited. In: *Cross-Cultural Analysis: Methods and Applications,* New York: Routledge Academic: 281-310.

Meuleman B. 2010b. Religious Involvement: Its Relation to Values and Social Attitudes: A Simultaneous Test of Measurement and Structural Models Across European Countries. In: *Cross-Cultural Analysis: Methods and Applications,* New York: Routledge Academic: 173-206.

Meuleman B., Davidov E., and Billiet J. 2008. Changing attitudes toward immigration in Europe, 2002-2007: a dynamic group conflict theory approach. *Social Science Research,* 38 (2): 352-365.

Meulemann H. 2006. Religiosität: Immer noch die Persistenz eines Sonderfalls. *Aus Politik und Zeitgeschichte. Beilage zur Wochenzeitung,* Berlin, Jul 24 2006, p. 30.

Meulemann H. 2009. Religiosity in Europe and in the Two Germanies: The Persistence of a Special Case – as revealed by the European Social Survey. In: *Church and Religion in Contemporary Europe: Results from Empirical and Comparative Research,* Wiesbaden: VS Verlag für Sozialwissenschaften: 35-48.

Meyer K, and Price A. M. 2008. Religion and Support for Democracy: A Cross-National Examination. *Sociological Spectrum,* 28 (5): 625-653.

Middleton R. 1976. Regional Differences in Prejudice. *American Sociological Review.* 41 (1): 94-117.

Mill J. St. 1988. *The Subjection of Women,* Hackett Publishing Co, Inc; New Ed edition.

Miller A. 2000. Going to Hell in Asia. *Review of Religious Research,* 42 (1): 5-18.

Miller A. 2002. *Subset Selection in Regression,* Second Edition, Chapman & Hall/CRC Press.

Miller A. S. and Stark R. 2002. Gender and Religiousness: Can Socialization Explanations Be Saved? *American Journal of Sociology,* 107 (6): 1399-1423.

Minkenberg M. and Perrineau P. 2007. The Radical Right in the European Elections 2004. *International Political Science Review / Revue internationale de science politique,* 28 (1): 29-55.

Minkov M. 2009. Predictors of Differences in Subjective Well-Being Across 97 Nations. *Cross-Cultural Research,* May 2009; vol. 43, 152 - 179.

Minkov M. and Hofstede G. 2011. *Cultural differences in a globalizing world.* Bingley, UK: Emerald.

Minkov M. and Hofstede G. 2013. *Cross-cultural analysis: the science and art of comparing the world's modern societies and their cultures.* Los Angeles: Sage.

Moaddel M. 1994. Political Conflict in the World Economy: A Cross-National Analysis of Modernization and World-System Theory. *American Sociological Review,* 276-303.

Moaddel M. 1996. The Social Bases and the Discursive Context of the Rise of Islamic Fundamentalism: The Cases of Iran and Syria. *Sociological Inquiry,* 330-355.

Moaddel M. 1998. Religion and Women: Islamic Modernism versus Islamic Fundamentalism. *Journal for the Scientific Study of Religion,* 37 (1), 108-130.

Moaddel M. 2001. Conditions for Ideological Production: The Origins of Islamic Modernism in India, Egypt, and Iran. *Theory and Society,* 30: 669-731.

Moaddel M. 2002a. Discursive Pluralism and the Rise of Islamic Modernism in Egypt. *Arab Studies Quarterly,* 24 (1): 1-29.

Moaddel M. 2002b. The Singularity of the Jordanian Religious Experience. *International Journal of Society, Culture, and Politics,* 15 (4): 527-68.

Moaddel M. 2002c. The Study of Islamic Culture and Politics: An Overview and Assessment. *Annual Review of Sociology,* 28: 359-86.

Moaddel M. 2003a. 2002 Keynote Address: September 11, Global Exigencies, and Sociology: A New Challenge. *Michigan Sociological Review,* 17 (3): 1-5.

Moaddel M. 2003b. Public Opinion in Islamic Countries: Survey Results. *Footnotes,* 31 (1), available at http://mevs.org/pubs/abs/1250

Moaddel M. 2003c. The Worldviews of Islamic Publics: The Cases of Egypt, Iran, and Jordan. In *'Human Values and Social Change: Findings from the Values Surveys'.* (Ronald Inglehart, (Ed.)) Leiden, The Netherlands: Brill.

Moaddel M. 2004a. Prospects for Change in Saudi Arabia. *Footnotes,* 32, 8 (November).

Moaddel M. 2004b. The future of Islam after 9/11. *Futures,* 36 (9): 961-977.

Moaddel M. 2006. The Saudi Public Speaks: Religion, Gender, and Politics. *International Journal of Middle East Studies,* 38 (1): 79-108.

Moaddel M. 2007a. Class and Labor in Iran: Did the revolution matter? *American Journal of Sociology,* 113 (3): 876-878.

Moaddel M. 2007b. What the Iraqi Study Group Missed: The Iraqi People. *Footnote,* 35 (1): 1, 4, 10.

Moaddel M. 2008. Religious Regimes and Prospects for Liberal Politics: Futures of Iran, Iraq, and Saudi-Arabia. *World Values Research,* 1 (2): 36-58. Available at: http: //www. worldvaluessurvey. org/the_paperseries .

Moaddel M. 2009. The Iranian Revolution and its Nemesis: the Rise of Liberal Values among Iranians. *Comparative Studies of South Asia, Africa, and the Middle East,* 29: 126-136.

Moaddel M. 2010. Iranians and Their Pride: Modalities of Political Sovereignty. *Lecture Notes in Computer Science,* Volume 6007/2010.

Moaddel M. and Abdel-Latif A. 2006. Events and Value Change: The Impact of September 11, 2001 on the Worldviews of Egyptians and Moroccans. *Interdisciplinary Journal of Research on Religion,* 2: 1-48.

Moaddel M. and Azadarmaki T. 2002. The Worldviews of Islamic Publics: The Cases of Egypt, Iran, and Jordan. *Comparative Sociology,* 1, 3-4: 299-319.

Moaddel M. and Karabenick S. 2008. Religious Fundamentalism among Young Muslims in Egypt and Saudi Arabia. *Social Forces,* 86 (4): 1675-1710.

Moaddel M., de Jong J, and Dagher M. 2011. Beyond Sectarianism in Iraq. *Contexts,* 10 (3): 66-67.

Moaddel M., Tessler M., and Inglehart R. F. 2008a. Foreign Occupation and National Pride. *Public Opinion Quarterly,* 72 (4): 677-705.

Moaddel M., Tessler M., and Inglehart R. F. 2008b. Saddam Hussein and the Sunni Insurgency: Findings from Values Surveys. *Political Science Quarterly,* 123 (4): 623-644.

Moors G. 2007. Testing the Internal Validity of the Inglehart Thesis by Means of a Latent Class Choice Model. *Acta Sociologica,* Jun 2007 (50): 147-160.

Morales C. E. 2014. *Entrepreneurship and postmaterialistic values. An individual-based analysis.* Federación Española de Sociología, available at: http://www.fes-web.org/uploads/files/modules/congress/10/grupos-trabajo/ponencias/438.pdf.

Morel J. SJ. 1968. *Religion in der kommunistischen Presse. Eine Inhaltsanalyse.* Innsbruck, Austria: Ungarisches Kirchensoziologisches Institut und Österreichische Kommissionsbuchhandlung in Kommission.

Morel J. SJ. 1972. *Glaube und Säkularisierung: Religion im Christentum als Problem.* Innsbruck, Wien, München: Tyrolia-Verlag.

Morel J. SJ. 1977. *Enthüllung der Ordnung: Grundbegriffe und Funktionen der Soziologie.* Innsbruck [u.a.]: Tyrolia-Verlag.

Morel J. SJ. 1980. *Führungsforschung: kritische Beiträge.* Göttingen; Toronto; Zürich: Verlag für Psychologie Hogrefe, 1980.

Morel J. SJ. 1986. *Ordnung und Freiheit: die soziologische Perspektive.* Innsbruck: Tyrolia-Verlag.

Morel J. SJ. 1997. *Soziologische Theorie: Abriss der Ansätze ihrer Hauptvertreter.* München [u.a.]: Oldenbourg.

Morel J. SJ. 1998. *Glauben hat Zukunft: persönliche Wege einer neuen Sinnfindung.* Innsbruck; Wien: Tyrolia-Verlag.

Morel J. SJ. 2003. *Radikale Kirchenreform: für eine mutige Erneuerung: Fakten und religionssoziologische Theorien über die Krise der katholischen Kirche von 1950 bis in die Gegenwart als Entscheidungsgrundlagen notwendiger Reformen.* Innsbruck: Tyrolia-Verlag.

Morselli D. and Passini S. 2012. Disobedience and support for democracy: Evidences from the World Values Survey, *The Social Science Journal*, 49, 284-294.

Mossialos E. and Lear J. 2012. Balancing economic freedom against social policy principles: EC competition law and national health systems. *Health Policy,* 106, 127-137.

Mueller J. 1988. Trends in Political Tolerance. *The Public Opinion Quarterly,* 52 (1): 1-25.

Mueller T. 2009. Religiosity and Attitudes towards the Involvement of Religious Leaders in Politics. *World Values Research,* 2 (1): 1-30. Available at: http: //www. worldvaluessurvey. org/the_paperseries .

Muenz R. F. 2006. *Population Change and the Impact of Demographic Aging: Consequences and Policy Options for Europe.* International Conference on Cultural and Political Conditions for the Reform and Modernisation of Social Models in Europe and the U. S., Vienna, May 19-20, 2006, Institute for Human Sciences, Vienna.

Müller A. *et al.* 2000. *Global Capitalism, Liberation Theology, and the Social Sciences: An Analysis of the Contradictions of Modernity at the Turn of the Millenium.* Nova Sciences: Huntington and Commack, New York.

Mūsá Khuwārizmī M. (Muḥammad ibn) 1896 (posthumously). *The algebra of Mohammad ben Musa.* Hildesheim; New York: G. Olms.

Narayan P. K., Nayaran S. and R. Smyth 2011. Does democracy facilitate economic growth or does economic growth facilitate democracy? An empirical study of Sub-Saharan Africa. *Economic Modelling,* 28, 900-910.

Navas S. 2004. La idea liberal de tolerancia en las democracias actuales. *Política y Cultura,* primavera 2004, num. 21, pp. 37-56.

Neeleman J., Halpern D., *et al.* 1997. Tolerance of Suicide, Religion and Suicide Rates: an Ecological and Individual Study in 19 Western Countries. *Psychological Medicine,* 27 (5): 1165.

Neller K. 2004. Der European Social Survey (ESS): Neue Analysemöglichkeiten für die international vergleichende empirische Sozialforschung. *Politische Vierteljahresschrift,* 45 (2): 259-261.

Neller K. and Deth J. W. v. 2006. Politisches Engagement in Europa. *Aus Politik und Zeitgeschichte,* 30: 30-38.

Netton I. R. 1999. *Al-Farabi and His School.* London and New York: Routledge. Available at Muslim Philosophy Online: http://www.muslimphilosophy.com/ip/rep/H021.htm

Neusner J. (Ed.) 2000/2004. *The encyclopaedia of Judaism. edited by Jacob Neusner, Alan J. Avery-Peck, William Scott Green.* Leiden: Brill, 2000-2004.

Nicholson A., Rose R. and Bobak M. 2009. Association between attendance at religious services and self-reported health in 22 European countries. *Social Science and Medicine,* 69 (4): 519-528.

Noland M. 2004. *Religion and economic performance.* Washington D. C.: The Peterson Institute, available at: http: //www. petersoninstitute. org/publications/wp/03-8. pdf also available at IDEAS http: //www. iie. com/publications/wp/03-8. pdf.

Noland M. 2005. *Explaining Middle Eastern Authoritarianism.* Washington D. C., Institute for International Economics, Working Paper Series 05-5 (entire. available at: http: //www. iie. com/publications/wp/wp05-5. pdf.

Noland M. and Pack H. 2004. *Islam, Globalization, and Economic Performance in the Middle East.* International Economics Policy Briefs, Washington D. C., Institute for International Economics, PB 04-4, June (entire. available at: http: //www. iie. com/publications/ pb/pb04-4. pdf).

Noorbakhsh F. 1998. The Human Development Index: Some Technical Issues and Alternative Indices. *Journal of International Development,* 10, 589-605.

Noren R. 2011. Towards a more integrated, symmetric and viable EMU. *Journal of Policy Modeling.* 33 (6): 821-830.

Norris P. and Inglehart R. F. 2003. Islamic Culture and Democracy: Testing the 'Clash of Civilizations' Thesis. In *'Human Values and Social Change: Findings from the Values Surveys'.* (Inglehart R. F. (Ed.))) Leiden, The Netherlands: Brill: 5-33.

Norris P. and Inglehart R. F. 2004. *Sacred and Secular: Reexamining the Secularization Thesis.* New York and Cambridge: Cambridge University Press.

Norris P., Inglehart R. F., and Welzel C. 2003. Gender Equality and Democracy. in *'Human Values and Social Change: Findings from the Values Surveys'.* (Inglehart R. F. (Ed.)) Leiden, The Netherlands: Brill: 321-46.

O'Connell M. 2005. Economic forces and anti-immigrant attitudes in Western Europe: a paradox in search of an explanation. *Patterns of Prejudice,* 39 (1): 60-74.

Oesch D. 2008. Explaining Workers' Support for Right-Wing Populist Parties in Western Europe: Evidence from Austria, Belgium, France, Norway, and Switzerland. *International Political Science Review,* 29 (3): 349-373.

Ogg J. 2005. Social exclusion and insecurity among older Europeans: the influence of welfare regimes. *Ageing & Society,* 25 (1): 69-90.

O'Leary De L. 1922. A*rabic thought and its place in history.* London: Kegan Paul, available at https://archive.org/details/arabicthoughtits00olea.

Olofsson A. and Oehmans S. 2007. Cosmopolitans and Locals: An Empirical Investigation of Transnationalism. *Current Sociology,* Nov 2007; 55, 877-895.

Olzak S. 2006. *The Global Dynamics of Racial and Ethnic Mobilization.* Stanford: Stanford University Press.

Oskarson M. 2007. Social Risk, Policy Dissatisfaction, and Political Alienation. A Comparision of Six European Countries. In: *The Political Sociology of the Welfare State,* Stanford: Stanford University Press: 117-148.

Oskarsson S. 2010. Generalized trust and political support: A cross-national investigation. *Acta Politica,* 45 (4): 423-443.

Pääkkönen J. 2010. Economic freedom as driver of growth in transition, *Economic Systems,* 34, 469-479.

Palmer M. and Finlay V. 2013. Faith in Conservation. New Approaches to Religion and the Environment, Revised edition, *The Alliance of Religions and Conservation.* Retrievable at: http://www.arcworld.org/downloads/Faith-in-Conservation-2013.pdf

Pape R. A. 2003. The Strategic Logic of Suicide Terrorism. *American Political Science Review,* 97 (3): 343-361.

Parboteeah K. P., Hoegl M. and Cullen J. B. 2008. Ethics and Religion: An Empirical Test of a Multidimensional Model. *Journal of Business Ethics,* 80 (2): 387-398.

Patel I. A. 1997. *Islam, the Choice of Thinking Women,* Ta Ha Publishers.

Pearson K. 1901. On Lines and Planes of Closest fit to Systems of Points in Space. *Philosophical Magazine* 6 (2), 559-72.

Peffley M. and Rohrschneider R. 2003. Democratization and Political Tolerance in Seventeen Countries: A Multi-level Model of Democratic Learning. *Political Research Quarterly,* 56 (3): 243.

Pehrson S. and Green E. G., T. 2010. Who We Are and Who Can Join Us: National Identity Content and Entry Criteria for New Immigrants. *Journal of Social Issues,* 66 (4): 695-716.

Pena Lopez J. A. and Sanchez Santos J. M. 2008. Effects of competition on religious markets: some empirical evidence. *Applied Economics Letters,* 15, 371–374.

Pereira C. Vala J. Costa-Lopes R. 2010. From prejudice to discrimination: The legitimizing role of perceived threat in discrimination against immigrants. *European Journal of Social Psychology,* 40 (7): 1231-1250.

Perez C. 1983. Structural Change and Assimilation of New Technologies in the Economic and Social Systems. *Futures.* 15 (5): 357-375.

Perman R. and Tavera C. 2005. A cross-country analysis of the Okun's Law coefficient convergence in Europe. *Applied Economics.* 37 (21): 2501-2513.

Peterson L. and Price A. M. 2009. *Protest Potential in the Middle East. Presented at the American Sociological Association Annual Meeting,* San Francisco, November.

Petrella I. 2004. *The Future of Liberation Theology: An Argument and Manifesto.* Aldershot, Hants, England; Burlington, VT: Ashgate.

Petrella I. 2005. *Latin American liberation theology: the next generation.* Maryknoll, N. Y.: Orbis Books.

Pettersson T. 2007. Muslim Immigrants in Western Europe: Persisting Value Difference or Value Adaptation. In *'Values and Perceptions of the Islamic and Middle Eastern Publics: Findings from Values Survey'.* (Mansoor Moaddel, Ed). New York: Palgrave Macmillan: 71-102.

Pettersson T. and Esmer Y. 2008. *Changing Values, Persisting Cultures; Case Studies in Value Change.* Leiden, The Netherlands: Brill.

PEW Research Center for the People and the Press. 2006. *The Great Divide: How Westerners and Muslims View Each Other. Europe's Muslims More Moderate.* PEW, Washington D. C.: http: //pewglobal. org/reports/display. php?ReportID=253 (June 22, 2006).

Pieroni L. and d'Agostino G. 2013. Corruption and the effects of economic freedom. *European Journal of Political Economy,* 29, 54-72.

Pipes D. 2002. God and Mammon: Does Poverty Cause Militant Islam? *The National Interest,* Winter 2002, available at: http://www. danielpipes.org/article/104.

Polanyi K. 1944/1957. *The Great Transformation. The political and economic origins of our time. Foreword* by Robert MacIver. Boston: Beacon Press.

Polanyi K., Lewis J. Kitchin D. 1972. *Christianity and the social revolution.* Freeport, N. Y.: Books for Libraries Press.

Pöschl J. et al. 2005. Turkey: Macroeconomic Vulnerability, Competitiveness and the Labour Market' WIIW Current Analyses and Country Profiles. *Vienna Institute for International Economic Studies',* abstract available at: http: //wiiw66. wsr. ac. at/cgi-bin/t3cgi. exe/publ/lastpubl. taf?_function=detail&BERICHTLAND_uid1=CA21.

Pratt D. 2012. *Islamic Prospects for Inter-Religious Dialogue: The Contribution of Fethullah Gülen.* Available at the official website of the Turkish Gülen movement at http://en.fgulen. com/component/search/?searchword=douglas+pratt&ordering=&searchphrase=all.

Price A. M. 2008a. Colonial History, Muslim Presence, and Gender Equity Ideology: A Cross-National Analysis. *International Journal of Sociology*, 38 (1): 81-103.

Price A. M. 2008b. *Gender, Work, and Ideology in the Middle East.* Presented at the American Sociological Association Annual Meeting, San Francisco, November 2009.

Pryor F. 2007. Culture and Economic Systems. *American Journal of Economics and Sociology,* 66 (4): 817—855.

Pryor F. L. 2010. Capitalism and freedom?, *Economic Systems,* 34, 91-104.

Puranen B., and Widenfalk O. 2007. The Rentier State: Does Rentierism Hinder Democracy? The Rentier Mentality Hypothesis Tested in Seven Middle Eastern Countries. in *'Values and Perceptions of the Islamic and Middle Eastern Publics: Findings from Values Survey'.* (Mansoor Moaddel, (Ed.)) New York: Palgrave Macmillan: 160-78.

Quaisser W. and Wood S. 2004. *EU Member Turkey? Preconditions, Consequences and Integration Alternatives.* Forost Arbeitspapier, Munich, 25, available at http://www.fak12. uni-muenchen.de/forost/forosthope/fo_pub_AP25_abstract.htm.

Quenouville M. H. 1949. Approximate tests of correlation in time series. *Journal of the Royal Statistical Society,* Series B, 11, 68–68.

Radford-Ruether R. 1983. *Sexism and God-talk: toward a feminist theology.* Boston: Beacon Press.

Ralston D. A., Egri C. P., Reynaud E. et al. 2011. A Twenty-First Century Assessment of Values Across the Global Workforce. *Journal of Business Ethics,* November 2011, Volume 104, Issue 1, pp 1-31.

RAND Corporation (Benard C.) 2005. *Civil Democratic Islam. Partners, Resources, and Strategies',* freely available on the Internet at http://www.rand.org/pubs/monograph_ reports/MR1716/.

RAND Corporation (Rabasa A. M. et al.) 2006. *The Muslim World after 9/11. Prepared for the United States Air Force.* Rand Corporation, Santa Monica, California, Rand Project Air Force, available at: http: //www.rand.org/pubs/research_briefs/2005/RAND_RB151.pdf.

Rashed R. 1994. *The development of Arabic mathematics: between arithmetic and algebra. Roshdi Rashed; translated by A.F.W.* Armstrong. Boston: Kluwer Academic, Boston studies in the philosophy of science.

Ray J. J. 1985. Authoritarianism of the left revisited. *Personality and Individual Differences*, 6 (2), pp. 271-272.

Ray J. J., and Lovejoy F. H. 1990. Does attitude to authority exist*? Personality and Individual Differences,* 11 (8), pp. 765-769.

Reeskens T. Hooghe M. 2008. Cross-Cultural Measurement Equivalence of Generalized Trust. Evidence from the European Social Survey (2002 and 2004). *Social Indicators Research,* 85 (3): 515-532.

Reher S. 2011. *Peace Through Integration or Peace Through Separation? World Values Research,* 4 (4): 97-119. Available at: http://www.worldvaluessurvey.org/the_paperseries.

Rehm P. 2005. *Citizen Support for the Welfare State: Determinants of Preferences for Income Redistribution.* Wissenschaftszentrum Berlin.

Rennstich J. 2007. Is Globalization self-organizing?. In: *Globalization as Evolutionary Process. Modeling Global Change.* (G. Modelski, T. C. Devezas, W. R. Thompson Eds.) pp. 74-107. London and New York: Routledge, 74-107.

Rescher N. 1967. *Studies in Arabic philosophy.* Pittsburgh: University of Pittsburgh Press.

Richardson S. 2011. *PIIGS 'R' Us? The Coming U. S. Debt Crisis and What Can Be Done About It* (July 25, 2011). University of Calgary – School of Public Policy Communiqué, Vol. 3, Issue 2, July 2011. URL: Available at SSRN: http://ssrn.com/abstract=1894927.

Ringdal K. 2008. Economic Morality. In: *Nordic social attitudes in a European perspective,* Cheltenham: Edward Elgar Publishing,.

Rizzo H, Abdel-Latif A, and Meyer K. 2007. The Relationship Between Gender Equality and Democracy: A Comparison of Arab Versus Non-Arab Muslim Societies. *Sociology,* 41 (6): 1151-70.

Rock M. T. 2009. Had democracy slowed growth in Asia?, *World Development,* 37 (5), 941-952.

Rodrik D. 2006. Goodbye Washington Consensus, Hello Washington Confusion? A Review of the World Bank's"Economic Growth in the 1990s: Learning from a Decade of Reform" *Journal of Economic Literature,* 44 (4): 973-987.

Rodrik D., Subramanian A. and Trebbi F. 2004. Institutions Rule: The Primacy of Institutions over Geography and Integration in Economic Development. *Journal of Economic Growth,* 9 (2): 131-165.

Röhrich W. 2004. Die Macht der Religionen. *Glaubenskonflikte in der Weltpolitik.* München: C. H. Beck.

Romano O. 2012. How to rebuild democracy, re-thinking degrowth. *Futures,* 44, 582-589.

Roose J. 2010. *Vergesellschaftung an Europas Binnengrenzen. Eine vergleichende Studie zu den Bedingungen sozialer Integration.* Wiesbaden: VS Verlag für Sozialwissenschaften.

Rozman G. 1991. *The East Asian region: Confucian heritage and its modern adaptation.* Princeton: Princeton University Press.

Rubin B. (Ed.) 2007. Introduction to political Islam: ideas and key issues. London and New York: Routledge, *Critical concepts in Islamic studies,* Political Islam; v. 1.

Rubinstein G. 1995. Right-wing authoritarianism, political affiliation, religiosity, and their relation to psychological androgyny. *Sex Roles,* 33 (7-8), pp. 569-586.

Rummel R. J. 1970. Applied factor analysis. Evanston: Ill.: Northwestern University Press.

Rusby J. S. M. 2010. Long-term influences of childhood history and educational level on authoritarianism. *Psychoanalytic Psychology,* 27 (3), pp. 330-354.

Russell L. M. and Clarkson J. S. 1996. *Dictionary of feminist theologies.* Louisville, Ky.: Westminster John Knox Press.

Rydgren J. 2008. Immigration sceptics, xenophobes or racists? Radical right-wing voting in six West European countries. *European Journal of Political Research,* 47 (6): 737-765.

Sacchi S. 1998. The dimensionality of postmaterialism: An application of factor analysis to ranked preference data. *European Sociological Review* 14 (2), pp. 151-175.

Sade D. A. F. (Donatien Alphonse François) Marquis de 1996/67. *Œuvres complètes.* Gilbert Lély. Paris: Cercle du livre précieux.

Safi M. 2009. Immigrants' Life Satisfaction in Europe: Between Assimilation and Discrimination. *European Sociological Review,* 26 (2): 159-176.

Saha S. and R. Gounder R. 2013. Corruption and economic development nexus: Variations across income levels in a non-linear framework. *Economic Modelling,* 31, 70-79.

Sala-I-Martin X. 1997. I Just Ran Two Million Regressions. The American Economic Review, 87 (2) *Papers and Proceedings of the Hundred and Fourth Annual Meeting of the American Economic Association* (May, 1997): 178-183.

Sala-i-Martin X., Doppelhofer, G. and Miller, R. I. 2004. Determinants of Long-Term Growth: A Bayesian Averaging of Classical Estimates (BACE) Approach. *The American Economic Review,* 94 (4): 813-835.

Sandovici M. E. Listhaug O. 2010. Ethnic and Linguistic Minorities and Political Participation in Europe. *International Journal of Comparative Sociology,* 51 (1): 111-136.

Sardar Z. and Davies M. W. 2002. *Why Do People Hate America?* New York: Disinformation Books.

Savage T. M. 2004. *Europe and Islam: Crescent Waxing, Cultures Clashing.* The Washington Quarterly (Summer 2004. 27, 3: 25 – 50. Available at: http: //www.twq.com/04summer/docs/04summer_savage.pdf#search=%22savage%20washington%20crescent%20waxing%22.

Savoia A., Easaw J. and McKay A. 2010. Inequality, democracy, and institutions: A critical review of recent research. *World Development,* 38 (2), 142-154.

Schatz M. S. 2004. Social Tolerance and Social Civility: Key Elements for Transitioning Democratic Countries. *Journal of Comparative Social Welfare,* 20 (1): 23-31.

Scheuer A. 2006. *Integration und Partizipation. In: Report Lebensqualität* 2006, Mannheim: Mannheimer Zentrum für Europäische Sozialforschung (MZES): 118-140.

Scheuer A. 2008. Gesellschaftliche Integration und Partizipation. In: *Datenreport 2008. Ein Sozialbericht der Bundesrepublik Deutschland,* Bonn: Bundeszentrale für Politische Bildung: 376-382.

Scheuregger D. and Spier T. 2007. Working-Class Authoritarianism und die Wahl rechtspopulitsischer Parteien. Eine empirische Untersuchung für fünf westeuropäische Staaten. *Kölner Zeitschrift für Soziologie und Sozialpsychologie,* 59 (1): 59-80.

Schlueter E. Wagner U. 2008. Regional Differences Matter: Examining the Dual Influence of the Regional Size of the Immigrant Population on Derogation of Immigrants in Europe. *International Journal of Comparative Sociology,* 49 (2): 153-173.

Schneider F. 2005. Shadow economies around the world: what do we really know? *European Journal of Political Economy,* 21, 598-642.

Schneider F. 2012. The Shadow Economy and Work in the Shadow: What Do We (Not) Know? *IZA Discussion Papers* 6423, Institute for the Study of Labor (IZA). Available at: http://ideas.repec.org/p/iza/izadps/dp6423.html

Schneider S. L. 2007. Anti-Immigrant Attitudes in Europe: Outgroup Size and Perceived Ethnic Threat. *European Sociological Review,* 24 (1): 53-67.

Schuessler-Fiorenza E. 1983. *In memory of her: a feminist theological reconstruction of Christian origins.* New York: Crossroad.

Schumpeter J. A. 1939. *Business cycles. A theoretical, historical, and statistical analysis of the capitalist process.* New York, London: McGraw-Hill Book.

Schumpeter J. A. 1950. *Capitalism, Socialism and Democracy.* New York & London: Harper & Row.

Schwartz S. H. 2006a. A Theory of Cultural Value Orientations: Explication and Applications. *Comparative Sociology,* 5 (2): 137-182.

Schwartz S. H. 2006b. Les valeurs de base de la personne: Théorie, mesures et applications. *Revue française de sociologie,* 47 (4): 929-968.

Schwartz S. H. 2007a. Universalism Values and the Inclusiveness of our Moral Universe. *Journal of Cross-Cultural Psychology,* 38 (6): 711-728.

Schwartz S. H. 2007b. Value orientations: Measurement, antecedents and consequences across nations. In: *Measuring Attitudes Cross-Nationally: Lessons from the European Social Survey,* London: Sage Publications: 161-193.

Schwartz S. H. 2009. *Cultural Value Orientations: Nature & Implications of National Differences.* The Hebrew University of Jerusalem, Israel Science Foundation Grant No. 921/02, available at http://blogs.helsinki.fi/valuesandmorality/files/2009/09/Schwartz-Monograph-Cultural-Value-Orientations.pdf

Scola A. 2005. *The nuptial mystery. Grand Rapids*, Mich.: William B. Eerdmans Pub. Co.

Sedgwick M. 2010. *Muhammad Abduh.* Oxford: Oneworld.

Seleny A. 2006. Tradition, Modernity, and Democracy: The Many Promises of Islam. *Perspectives on Politics,* 4 (3): 481-494.

Semyonov M., Glikman A., and Krysan M. 2007. Europeans' Preference for Ethnic Residential Homogeneity: Cross-National Analysis of Response to Neighborhood Ethnic Composition. *Social problems,* 54 (4): 434-453.

Semyonov M., Raijman R., and Gorodzeisky A. 2008. Foreigners' Impact on European Societies: Public Views and Perceptions in a Cross-National Comparative Perspective. *International Journal of Comparative Sociology,* 49 (1): 5-29.

Senik C., Stichnoth H. , and Straeten K. Van der 2009. Immigration and Natives' Attitudes towards the Welfare State: Evidence from the European Social Survey. *Social Indicators Research,* 91 (3): 345-370.

Shah T. S. and Duffy M. 2006. Why God Is Winning. *Foreign Policy,* 155 (Jul. - Aug., 2006): 38-43.

Sheikhzadegan A. 2007. Islam in the World System. In: Herkenrath, Mark (ed.). *The Regional and Local Shaping of World Society (*151-175), Münster [etc.]: LIT Verlag.

Sherman S. 2003. The Attacks of September 11 in Three Temporalities. *Journal of World-Systems Research, ix,* 1, winter 2003, 141–169, http: //jwsr. ucr. edu, issn 1076–156x.

Sidaway J. D. 2012. *Geographies of Development: New Maps,* New Visions? Professional Geographer. 64 (1): 49-62.

Sides J. M. Citrin J. 2007. European Opinion About Immigration: The Role of Identities, Interests and Information. *British Journal of Political Science,* 37 (3): 477-504.

Sigelman L. 2006. *Top Twenty Commentaries. American Political Science Review,* November, 100 (4): 667-687.

Silverberg G. 2006. Long Waves in Global Warfare and Maritime Hegeomony? A Complex System Perspective. In: *Kondratieff waves, warfare and world security: [proceedings of the NATO Advanced Research Workshop on The Influence of Chance Events and Socioeconomic Long Waves in the New Arena of Asymmetric Warfare, Covilhã, Portugal, 14 - 18 February 2005].* (Devezas T. C. (Ed.)), 154-164. Amsterdam: IOS Press.

Simiand F. 1932. *Les fluctuations économiques à longue période et la crise mondiale.* Paris: Lib. Felix Alcan.

Skocpol Th. 1979. *States and Social Revolutions.* Princeton: Princeton University Press.

Smelser N. J. 1984. *Sociology.* New Jersey, Prentice-Hall, Inc..

Smith R. J. and Taylor A. M. R. 2001. Recursive and rolling regression-based tests of the seasonal unit root hypothesis. *Journal of Econometrics.* 105 (2): 309-336.

Sobrino J. SJ. 1990. *Companions of Jesus: the Jesuit martyrs of El Salvador.* Maryknoll, N.Y.: Orbis Books.

Söderlund P. Kestilä E. 2009. Dark Side of Party Identification? An Empirical Study of Political Trust among Radical Right-Wing Voters. *Journal of Elections,* Public Opinion & Parties, 19 (2): 159-181.

Solomou S. N. 2013. Kondratieff cycles. The New Palgrave Dictionary of Economics. Second Edition. (Eds. Steven N. Durlauf and Lawrence E. Blume). Palgrave Macmillan, 2008. *The New Palgrave Dictionary of Economics Online.* Basingstoke, U. K.: Palgrave Macmillan. 04 June 2013 <http: //www. dictionaryofeconomics. com/article?id= pde2008_K000037> doi: 10. 1057/9780230226203. 0902.

Solow R. M. 1956. A contribution to the theory of economic growth. *Quarterly Journal of Economics. 70* (1): 65-94.

Soysa I. De and Nordas R. 2006. *Islam's Bloody Innards? Religion and Political Terror, 1980 – 2000.* Conference Paper, International Peace Research Institute, Oslo (PRIO. available at: http: //www. prio. no/files/file48349_desoysa_nordas_wg3meeting. pdf.

Spencer H. 1996 (posthumously). *Herbert Spencer, collected writings.* Herbert Spencer; William Henry Hudson; David Duncan; M W Taylor. London: Routledge/Thoemmes Press.

SPSS, 2007. *Statistical Package for the Social Sciences, User Guide.* Version 14, August 2007.

Stack S. and Kposowa A. 2006. The Effect of Religiosity on Tax Fraud Acceptability: A Cross-National Analysis. *Journal for the Scientific Study of Religion,* 45 (3): 325-351.

Stark R. 2006. *Sociology.* Belmont, CA: Wadsworth, 9th edition.

Stathopoulou T. 2010. Faith and Trust. Tracking patterns of religious and civic commitment in Greece and Europe. An empirical approach. In: *Orthodox Christianity in 21st century Greece. The role of Religion in Culture, Ethnicity and Politics.* London: Ashgate: 193-217.

Steenkamp J. -B. E. M., and Geyskens I. 2012. Transaction cost economics and the roles of national culture: A test of hypotheses based on Inglehart and Hofstede. *Journal of the Academy of Marketing* Science 40 (2), pp. 252-270.

Steindl J. 1952. *Maturity and Stagnation in American Capitalism.* Oxford: Basil Blackwell.

Steindl J. 1979. Stagnation theory and stagnation policy. *Cambridge Journal of Economics,* 1979 (3), 1-14.

Steindl J. 1988. Diskussionsbeitrag zur EG-Frage. *Kurswechsel,* 4 (3), pp. 3–7.

Steindl J. 1990. *Economic Papers* 1941-88. Basingstoke: MacMillan.

Stemplowski R. 2002. *Transnational Terrorism in the World System Perspective (*edited by Ryszard Stemplowski). Warsaw: The Polish Institute of International Affairs, Collections PISM http://www. pism.pl/en.

Sterman J. D. 1985. *The Economic Long Wave: Theory and Evidence.* System Dynamics Group, Sloan School of Management, Massachusetts Institute of Technology, Cambridge, MA 02139 U. S. A., March 1985, Revised September 1985, Invited paper for the Conference on Long Terra Fluctuations in Economic Growth: Their Causes and Consequences, Weimar, GDR, 10-14 Jiine 1985 (URL: http://www.archive.org/stream/economiclongwave00ster/economiclongwave00ster_djvu.txt).

Sterman J. D. 1986. The economic long wave: theory and evidence. *System Dynamics* Review. 2 (2): 87-125.

Stiglitz J. 2002. *Globalization and its discontents,* New York: W. W. Norton.

Stouffer S. 1955. *Communism, Conformity and Civil Liberties.* Garden City, NJ: Doubleday, 1955.

Stroemsnes K. 2008. The Importance of Church Attendance and Membership of Religious Voluntary Organizations for the Formation of Social Capital. *Social Compass,* Dec 2008 (55): 478-496.

Sullins D. P. 2006. Gender and Religion: Deconstructing Universality, Constructing Complexity. *American Journal of Sociology,* 112 (3): 838-880.

Sullivan J. and Transue J. 1999. The Psychological Underpinnings of Democracy: A Selective Review of Research on Political Tolerance, Interpersonal Trust, and Social Capital. *Annual Reviews in Psychology,* 50 (1): 625-650.

Sullivan J. L., Marcus G. E., Feldman S., Piereson J. E. 1981. The Sources of Political Tolerance: A Multivariate Analysis. *The American Political Science Review,* 75 (1): 92-106.

Taagepera R. 2003. *Prospects for Democracy in Islamic Countries.* Working Paper, UC Irvine: Center for the Study of Democracy.

Tabachnick B. G. and Fidell L. S. 2001. *Using multivariate statistics.* Needham Heights, MA: Allyn & Bacon.

Tang C. F. 2010. The Money-prices Nexus for Malaysia: New Empirical Evidence from the Time-varying Cointegration and Causality Tests. *Global Economic Review.* 39 (4): 383-403.

Tanwir F. 2002. Religious Parties and Politics in Pakistan. *International Journal of Comparative Sociology,* 43 (3-5): 250-268.

Tanzi V. 1999. Uses and abuses of estimates of the underground economy. *Economic Journal,* 109, 338-340.

Tanzi V. and Schuknecht L. 1997. Reconsidering the fiscal role of government: the international perspective. *American Economic Review,* 87, 164-168.

Tausch A. 2003. Европейская перспектива: по пути к созданию «общего средиземноморского дома» и интегрированию положительного потенциала общественного развития исламс *Evropa,* 4 (9), 2003: 87 – 109, Warsaw, Polish Institute for International Affairs.

Tausch A. 2004. Fuer immer ausgegrenzt? Neue Studien zur sozialen Lage der Roma in Europa. *Das Juedische Echo,* October 2004, Vol. 53, Tischiri 5765: 147-150, and *Die Zukunft/Akzente,* 06/2004: 35 – 41.

Tausch A. 2005. Europe, the Muslim Mediterranean and the End of the era of Global Confrontation. Alternatives. *Turkish Journal of International Relations,* Volume 3, Number 4, Winter 2004, available at: http://www.alternativesjournal.net/ volume3/ number4/arno3.pdf.

Tausch A. 2006a. Fuer Rueckkehr der Vernunft in der Tuerkei-Politik. *Europaeische Rundschau,* 34, 1: 121 – 132.

Tausch A. 2006b. Towards a New EU Policy for the Mediterranean South? *Current Politics and Economics of Europe,* Volume 17 Issue 3. pp. 373-417.

Tausch A. 2007a. *Against Islamophobia. Quantitative analyses of global terrorism, world political cycles and center periphery structures.* Hauppauge, N. Y.: Nova Science Publishers.

Tausch A. 2007b. Armutsrisiko und Migrationshintergrund in Europa – vergleichende Daten. *Europaeische Rundschau,* 4, 2007: 107 – 120.

Tausch A. 2007c. Global Terrorism and World Political Cycles. *History and mathematics (special issue): Analyzing and Modeling Global Development* (Volgograd, Russia) Grinin L., Munck V. C. de, Korotayev A. (Moscow) (Ed.). 'Anthropology of the East' Center, Russian State University for the Humanities, Moscow Russia: 99 – 126.

Tausch A. 2007d. Quantitative World System Studies Contradict Current Islamophobia: World Political Cycles, Global Terrorism, and World Development. *Alternatives. Turkish Journal of International Relations,* 6, 1&2, Spring and Summer 2007, p. 15 - 81. available at: http: //www. alternativesjournal. net/ (journal is available electronically).

Tausch A. 2007e. War Cycles. *Social Evolution and History (Volgograd, Russia),* Vol. 6, No. 2, September 2007, pp. 39 – 74.

Tausch A. 2007f. Zu einer Lissabon-Prozess-bezogenen Perspektive ueber die muslimische Gemeinschaft in Europa. Zusammenstellung und Wertung europaweit vorliegender Materialien, insbesondere des 'European Social Survey'. *OeMZ. Oesterreichische Militaerische Zeitschrift,* 1, 2007: 53 – 57.

Tausch A. 2007g. Европейский союз: 'град на холме' и Лиссабонская стратегия. *Mirovaia ekonomika i mezhdunarodnye otnosheniia,* RU (0131-2227), Vyp. 50, no. 3 (2007), s. 65-72.

Tausch A. 2008a. Nikolai Kondratiev und die Zukunft der Konflikte in der Weltgesellschaft. Bewertung eines Neuansatzes des 'NATO Institute for Advanced Studies'. *OeMZ. Oesterreichische Militaerische Zeitschrift,* 2008 (3): 289 – 299.

Tausch A. 2008b. Разрушительное созидание?' (Рассуждения в духе Шумпетера о некоторых трендах и Лиссабонском процессе в Европе). *Mirovaia ekonomika i mezhdunarodnye otnosheniia,* RU (0131-2227), Contents # 10, 2008, 34 – 41.

Tausch A. 2009a. Armut von Muslimen und politisches Radikalitätspotential in Europa. Raison, *Zeitschrift fuer gesellschaftliche Entwicklung,* 3, 12, 2009: 29 – 43.

Tausch A. 2009b. Multikulturalität und die Armee der Zukunft in Europa - ein erster Datenbefund basierend auf dem 'World Values Survey' und dem 'European Social Survey'. OeMZ. *Oesterreichische Militaerische Zeitschrift,* 5, 2009: 557 – 564.

Tausch A. 2009c. *Towards a Global Tolerance Index. The Vienna Institute for International Economic Studies Monthly Report,* 1: 9 - 12. Vienna: The Vienna Institute for International Economic Studies (wiiw).

Tausch A. 2009d. Евроислам. миф или реальность? Количественный анализ по данным «всемирной программы изучения ценностей. *Evropa,* 1 (26), 2008: 121 – 151. Warsaw, Polish Institute for International Affairs.

Tausch A. 2010a. *Armut und Radikalität? Soziologische Perspektiven zur Integration der Muslime in Europa, basierend auf dem 'World Values Survey' und dem 'European Social Survey'* Bremen: Europäischer Hochschulverlag, Serie , Schriftenreihe: Studien zu vergleichender Sozialpädagogik und internationaler Sozialarbeit und Sozialpolitik', Bd. 12.

Tausch A. 2010b. Globalisation and development: the relevance of classical "dependency" theory for the world today. *International Social Science Journal* (UNESCO, Paris), 61 (202), December 2010: 467-488.

Tausch A. 2010c. Integrationspolitik in Europa – welche konkreten statistischen Kennziffern stehen der Sozialpolitik ueberhaupt zur Verfuegung? Soziale Sicherheit. *Fachzeitschrift der Oesterreichischen Sozialversicherung, April* 2010: 224 – 230.

Tausch A. 2010d. Passive Globalization and the Failure of the European Union's Lisbon Strategy, 2000-2010: Some New Cross-National Evidence. *Alternatives: Turkish Journal of International Relations,* Vol. 9, No. 1, Spring 2010: 1-91.

Tausch A. 2010e. Paul Boccara's analysis of global capitalism. the return of the Bourbons, and the breakdown of the Brussels / Paris neo-liberal consensus. *Entelequia. Revista Interdisciplinar,* 12, Fall 2010. 105-147. Available from Internet: http://www.eumed.net/entelequia/en.art. php?a=12a06.

Tausch A. 2010f. Towards an Index of global tolerance: a quantitative analysis, based on the 'World Values Survey' data. *Islamic Perspective,* (4)2010 (Center for Sociological Studies, IPCSS, London Academy of Iranian Studies, SOAS London): 263-279, available at: http: //iranianstudies. org/journals/islamic-perspective-journal-number-4-2010/.

Tausch A. 2010g. Towards yet another age of creative destruction? *Journal of Globalization Studies (Moscow),* 1, 1, 2010: 104 – 130 (also available at: http: //www. socionauki. ru/authors/tausch_arn/?sphrase_id=14791).

Tausch A. 2011a. Islamismus durch die Hintertür? *Der Standard,* 1. Juli 2011. Available at: http: //derstandard. at/1308680115372/Wahl-der-Islamischen-Glaubensgemeinschaft-Islamismus-durch-die-Hintertuer.

Tausch A. 2011b. On The Global Political And Economic Environment Of The Current Al-Jazeera Revolution. *Middle East Studies Online Journal-* ISSN 2109-9618- Issue n°5. Volume 2 (2011), *Etudes du Moyen-Orient*. N°5. Volume 2. 2011, دراسات الشرق الأوسط، مجلة فكرية محكمة. العدد الخامس. العدد الثاني 2011 (available at: http: //www. middle-east-studies. net/?p=9765).

Tausch A. 2011c. Vom Behagen und Unbehagen mit der Migration. *Ökonomenstimme. KOF Konjunkturforschungsstelle der ETH Zürich,* available at: http: //www. oekonomenstimme. rg/.

Tausch A. 2012a. *Globalisierung und die Zukunft der EU-2020-Strategie.* Hrsg. Abteilung Wirtschaftswissenschaft d. AK Wien, Wien: Kammer f. Arbeiter u. Angestellte f. Wien, 2011, Serie: Materialien zu Wirtschaft und Gesellschaft; 110, ISBN: 978-3-7063-0422-1.

Also available eletcronically from http: //www. arbeiterkammer. at/online/page. php?P=2968.

Tausch A. 2012b. Planet Happy. *Ökonomenstimme. KOF Konjunkturforschungsstelle der ETH Zürich,* available at: http: //www. oekonomenstimme. org/.

Tausch A. 2012c. Геостратегические соображения по поводу циклов ондратьева, глобализации и войн' [Geostrategic considerations on Kondratiev cycles, globalization and war] *Mirovaia ekonomika i mezhdunarodnye otnosheniia,* 55 (10): 105–114.

Tausch A. and Ghymers C. 2011. *El Papa ¿Cuántas Divisiones Tiene? Sondeo Global del Catolicismo Mundial Según el 'World Values Survey' y el 'European Social Survey'* E-Book N° 49 Centro Argentino de Estudios Internacionales (in Spanish) [English Title: 'The Pope - How Many Divisions Does He Have?' a First Global Survey of World Catholicism Based on the 'World Values Survey' and the 'European Social Survey'], available at: http: //www. caei. com. ar/es/irebooks. htm (English language executive summary available).

Tausch A. and Ghymers Ch. 2007. *From the 'Washington' towards a 'Vienna Consensus'? A quantitative analysis on globalization, development and global governance.* Hauppauge, N. Y.: Nova Science Publishers.

Tausch A. and Herrmann P. (Eds.) 2005. *Dar al Islam. The Mediterranean, the World System and the Wider Europe.* Vol. 1: The 'Cultural Enlargement' of the EU and Europe's Identity; Vol. 2: The Chain of Peripheries and the New Wider Europe. Hauppauge, New York: Nova Science Publishers. Abridged paperback editions, 2006, under the title: 'The West, Europe and the Muslim World' (Vol. 1) and 'Towards a Wider Europe' (Vol. 2).

Tausch A. and Heshmati A. 2004. Turkey and the Lisbon process. A short research note on the position of Turkey on a new 'Lisbon Strategy Index' (LSI).' *Ankara Institute for Turkish Policy Studies, ANKAM, Insight Turkey,* 8, 2, 2006: 7 – 18.

Tausch A. and Heshmati A. 2009. Asabiyya: Re-Interpreting Value Change in Globalized Societies. *IZA (Institute for the Study of Labour, Bonn, FRG) Discussion Papers* 4459, Downloadable at http: //ideas. repec. org/e/pta132. html and http: //papers. ssrn. com/sol3/papers. cfm?abstract_id=1489282. In addition, the title is available from the Social Science Research Network, New York http: //www. ssrn. com/ The study is also available from Middle Eastern Values Studies, http: //mevs. org/pubs/alpha. html at the Michigan Population Studies Center of the University of Michigan, 426 Thompson St., Ann Arbor, MI 48106.

Tausch A. and Jourdon Ph. 2011. *Trois essais pour une économie politique du 21e siècle: Mondialisation, gouvernance mondiale, marginalisation.* Paris: Harmattan.

Tausch A. and Karoui H. 2011. *Les Musulmans: Un cauchemar ou une force pour l'Europe?* Paris: Harmattan.

Tausch A. and Moaddel M. 2009. *What 1.3 billion Muslims really think. An answer to a recent Gallup study, based on the 'World Values Survey'.* Hauppauge, N. Y.: Nova Science Publishers.

Tausch A., Bischof C., and Mueller K. 2010. *'Muslim Calvinism', internal security and the Lisbon process in Europe.* Rozenberg Publishers, Amsterdam.

Tausch A., Bischof C., Kastrun T. and Mueller K. 2007. *Against Islamophobia: Muslim Communities, Social Exclusion and the Lisbon Process in Europe.* Hauppauge, N. Y.: Nova Science Publishers.

Tausch A., Heshmati A. and Brand U. 2012. *Globalization, the Human Condition and Sustainable Development in the Twenty-first Century. Cross-national Perspectives and European Implications.* London, New York and Delhi: Anthem Press.

Taylor Ch. 1992. *Sources of the Self: The Making of the Modern Identity.* Cambridge, M. A.: Harvard University Press.

Telhami S. 2002. *The Stakes: America and the Middle East*, Boulder: Westview Press.

Teorell J. and Hadenius A. 2006. Democracy without Democratic Values: A Rejoinder to Welzel and Inglehart. *Studies in Comparative International Development,* Fall 2006, Vol. 41, No. 3, pp. 95-111.

Tessler M. 2002. Do Islamic orientations influence attitudes toward democracy in the Arab world? Evidence from Egypt, Jordan, Morocco, and Algeria. *International Journal of Comparative Sociology,* 43 (3-5): 229-249.

Tessler M. 2003. Arab and Muslim Political Attitudes: Stereotypes and Evidence from Survey Research. *International Studies Perspectives,* 4: 175-80.

Tessler M. 2004. The view from the street: The attitudes and values of ordinary Algerians. *Journal of North African Studies,* 9 (2), pp. 184-201.

Tessler M. and Gao E. 2009. Democracy and the political culture orientations of ordinary citizens. *International Social Science Journal,* 192: 197-205.

Tessler M. and Jamal A. 2006. Political Attitude Research in the Arab World: Emerging Opportunities. *PS: Political Science and Politics,* 39 (3): 433-437.

Tessler M. and Robbins M. D. H. 2007. What Leads Some Ordinary Arab Men and Women to Approve of Terrorist Acts Against the United States? *Journal of Conflict Resolution,* 51 (2): 305-328.

Tessler M., Moaddel M. and Inglehart R. F. 2006a. What Do Iraqis Want? *Journal of Democracy,* 17 (1): 38-50.

Tessler M., Moaddel M., and Inglehart, R. F. 2006b. Getting to Arab Democracy: What Do Iraqis Want? *Journal of Democracy,* 17 (1): 38-50.

Thompson B. 2004. Exploratory and confirmatory factor analysis: understanding concepts and applications. Washington, D. C.: American Psychological Association; London: Eurospan.

Thompson W. R. and Zuk L. G. 1982. War, Inflation, and the Kondratieff Long Wave. *Journal of Conflict Resolution,* 26 (4): 621-644.

Thorisdottir H., Jost J. T., Liviatan I., Shrout P. E. 2007. Psychological Needs and Values Underlying Left-Right Political Orientation: Cross-National Evidence from Eastern and Western Europe. *Public Opinion Quarterly,* 71 (2): 175-203.

Tibi B. 1990. *Arab nationalism: a critical enquiry.* New York, N. Y.: St. Martin's Press.

Tibi B. 1995a. Islamic Dream of Semi-modernity. *India International Centre Quarterly,* 22 (1): 79-87.

Tibi B. 1995b. *Krieg der Zivilisationen: Politik und Religion zwischen Vernunft und Fundamentalismus.* Hamburg: Hoffmann und Campe.

Tibi B. 1996. *Der Wahre Imam: der Islam von Mohammed bis zur Gegenwart.* München: Piper, 1996.

Tibi B. 1997a. *Arab nationalism: between Islam and the nation-state.* New York: St. Martin's Press.

Tibi B. 1997b. *Challenge of fundamentalism: political Islam and the new world disorder.* Berkeley: University of California Press.

Tibi B. 1997c. *Conflict and war in the Middle East, 1967-91: regional dynamic and the superpowers.* Translated by Clare Krojzl. Houndmills, Basingstoke, Hampshire: Macmillan in association with the Center for International Affairs, Harvard University.

Tibi B. 1998a. *Conflict and war in the Middle East: from interstate war to new security.* New York: St. Martin's Press.

Tibi B. 1998b. *Crisis of modern Islam: a preindustrial culture in the scientific-technological age.* Translated by Judith von Sivers; foreword by Peter von Sivers. Salt Lake City: University of Utah Press.

Tibi B. 2001a. *Islam between culture and politics.* New York: Palgrave, in association with the Weatherhead Center for International Affairs, Harvard University.

Tibi B. 2001b. *Kreuzzug und Djihad. Der Islam und die christliche Welt.* München: Goldmann Taschenbuchausgabe.

Tibi B. 2002. *Islamische Zuwanderung: die gescheiterte Integration.* München: Deutsche Verlags-Anstalt.

Tibi B. 2005. *Mit dem Kopftuch nach Europa?* Die Türkei auf dem Weg in die Europäische Union. Darmstadt: Primus-Verlag.

Tibi B. 2007. The Totalitarianism of Jihadist Islamism and its Challenge to Europe and to Islam. *Totalitarian Movements and Political Religions,* 8, 1, March: 35 – 54.

Tope D, and Meyer K. 2005. Religion and Democratic Support: Evidence from the World Values Survey. *Democracy and Society,* 3 (1): 10-12.

Torgier B., and Schneider, F. 2007. What shapes attitudes toward paying taxes? Evidence from multicultural European countries. *Social Science Quarterly,* 88 (2), pp. 443-470.

Torgler B. and Schneider F. 2009. The impact of tax morale and institutional quality on the shadow economy. *Journal of Economic Psychology,* 30, 228-245.

Torgler B., and García-Valiñas M. A. 2007. The determinants of individuals' attitudes towards preventing environmental damage. *Ecological Economics,* 63 (2-3), pp. 536-552.

Tosini, D. 2007. Sociology of Terrorism and Counterterrorism: A Social Science Understanding of Terrorist Threat. *Sociology Compass*, 1: 664–681.

Tripp Ch. 2006. *Islam and the Moral Economy.* Cambridge: Cambridge University Press.

Troll Chr. W. SJ. 2001a. *Der europäische Islam:* eine reale Perspektive? Berlin: Morus.

Troll Chr. W. SJ. 2001b. Muslime in Deutschland. Ziele, Strömungen, Ogranisationen/ Strukturen. http://www. jesuiten.org/aktuell/jubilaeum/files/jahresthema_2001_troll_1.pdf.

Troll Chr. W. SJ. 2003. *Muslime fragen, Christen antworten.* Dortmund: Topos Plus.

Troll Chr. W. SJ. 2004. *Als Christ dem Islam begegnen.* Würzburg: Echter.

Troll Chr. W. SJ. 2007. *Progressives Denken im zeitgenössischen Islam. Friedrich Ebert Stiftung,* Islam und Gesellschaft, 4 (available at http://www.sankt-georgen.de/leseraum/troll27.pdf).

Troll Chr. W. SJ. and Bsteh A. (Eds.) 1997. *One world for all: foundations of a socio-political and cultural pluralism from Christian and Muslim perspectives.* New Delhi: Vikas Pub. House.

Troll Chr. W. SJ. and Donohue J. 1998. *Faith, power, and violence: Muslims and Christians in a plural society, past and present.* Roma: Pontificio Istituto orientale.

Troll Chr. W. SJ. and Vahiduddin S. 1986. *Islamic experience in contemporary thought.* Delhi: Chanakya Publications.

Tsui A. S., Nifadkar S. S. and Ou A. Y. 2007. Cross-National, Cross-Cultural Organizational Behavior Research: Advances, Gaps, and Recommendations. *Journal of Management,* Jun 2007; 33, 426-478.

Tuch, S. A. 1987. Urbanism, Region, and Tolerance Revisited: The Case of Racial Prejudice. *American Sociological Review*, (52 (4 Aug., 1987: 504-510.

Turner B. S. (Ed.) 2003. *Islam as religion and law.* London and New York: Routledge, Critical concepts in sociology. Islam; v. 1.

Ueberla K. 1970. *Faktorenanalyse. Eine systematische Einfuehrung fuer Psychologen,* Mediziner, Wirtschafts- und Sozialwissenschaftler. Berlin: Springer.

UNDP 2013. *Human Development Report 2013*. The rise of the South: Human progress in a diverse world, United Nations Development Program.

United Kingdom Foreign and Commonwealth Office 2004. *Draft Report on Young Muslims and Extremism.* UK Foreign and Commonwealth Office/Home Office, available at: http://www. globalsecurity. org/security/library/report/2004/muslimext-uk. htm.

Vala J. and Costa-Lopes R. 2010. Youth attitudes toward difference and diversity: a cross-national analysis. *Análise Social,* 45 (195): 255-275.

Vala J., Pereira C., Ramos A. 2006. Racial prejudice, threat perception and opposition to immigration: A comparative analysis. *Portuguese Journal of Social Science,* 5 (2): 119-140.

Valentová M., Šmídová I., and Katrňák T. 2007. Gender Segregation in the Labour Market Placed in the Context of Educational Segregation: Cross-National Comparison. *Sociológia - Slovak Sociological Review,* 39 (3): 214-244.

Van de Vijver F. J. R., and Poortinga Y. H. 2002. Structural equivalence in multilevel research. *Journal of Cross-Cultural Psychology* 33 (2), pp. 141-156.

van der Meer T. van Ingen E. 2008. Schools of democracy? Disentangling the relationship between civic participation and political action in 17 European countries. *European Journal of Political Research,* 48 (2): 281-308.

Van Ewijk C. 1982. A Spectral Analysis of the Kondratieff Cycle. *Kyklos.* 35 (3): 468-499.

Van Hiel A., Cornelis I., and Roets A. 2010. To Have or To Be? A Comparison of Materialism-Based Theories and Self-Determination Theory as Explanatory Frameworks of Prejudice. *Journal of Personality* 78 (3), pp. 1037-1070.

van Tubergen F. and Sindradottir J. 2011. The Religiosity of Immigrants in Europe: A Cross-National Study. *Journal for the Scientific Study of Religion,* 50 (2): 272-288.

Velleman P. F., and R. E. Welsch R. E. 1981. Efficient computing of regression diagnostics. *American Statistician.* 35, 234–242.

Vergolini L. 2011. Social cohesion in Europe: How do the different dimensions of inequality affect social cohesion?. *International Journal of Comparative Sociology,* 52 (3): 197-214.

Vernon R. 1966. International Investment and International Trade in the Product Cycle. *Quarterly Journal of Economics.* 80, (2): 190-207.

Voas D. 2008. The Rise and Fall of Fuzzy Fidelity in Europe. *European Sociological Review,* 25 (2): 155-168.

Vogelsang T. J. 2013. Spectral analysis. The New Palgrave Dictionary of Economics. Second Edition. (Eds. Steven N. Durlauf and Lawrence E. Blume). Palgrave Macmillan, 2008. *The New Palgrave Dictionary of Economics Online*. Basingstoke, U. K.: Palgrave Macmillan. 04 June 2013 <http: //www. dictionaryofeconomics. com/article?id= pde2008_S000201> doi: 10. 1057/9780230226203. 1588.

Voulgaris Y. 2006. Globalization and national identity: Monitoring Greek culture today. *Portuguese Journal of Social Science,* 5 (2): 141-153.

Wagar W. 1996. Toward a Praxis of World Integration. Prepared for presentation at the 90th Annual Meeting of the American Sociological Association, Washington, D. C., August 19-23, 1995. *Journal of World Systems Research,* 2, 2, 1996 http: //jwsr. ucr. edu/archive/vol2/v2_n2. php.

Warner C. M. and Wenner M. W. 2002. *Organizing Islam for Politics in Western Europe.* Hoover Institution, Stanford University, available at http://faculty.washington. edu/tgill/544%20Warner%20Islam%20Europe.pdf.

Warner C. M. and Wenner M. W. 2006. Religion and the Political Organization of Muslims in Europe. *Perspectives on Politics,* 4 (3): 457-479.

Weber M. 1922/1964 (posthumously) *The sociology of Religion,* (E. Fischoff trans.) Boston: Beacon Press.

Weber R. P. 1981. Society and Economy in the Western World System. *Social Forces.* 59 (4): 1130-1148.

Weede E. 2006. *Is there a contradiction between freedom and Islam?* Paper, Department of Sociology, University of Bonn (FRG); available at: http://admin.fnst.org/uploads/ 896/Weede_engl.pdf.

WEF 2013. *The global competiveness report* 2012-2013. World Economic Forum.

Weldon S. A. 2006. The Institutional Context of Tolerance for Ethnic Minorities: A Comparative, Multilevel Analysis of Western Europe. *American Journal of Political Science,* 50 (2): 331-349.

Welzel A. 2007. Are levels of Democracy Affected by Mass Attitudes? Testing Attainment and Sustainment Effects on Democracy. *International Political Science Review,* 28 (4): 397-424.

Welzel C. and Alexander A. C. 2008. Measuring Effective Democracy: The Human Empowerment Approach. *World Values Research,* 1 (1): 1-35. Available at: http://www. worldvaluessurvey.org/the_paperseries.

Welzel C. and Klingemann H. -D. 2008. Evidencing and Explaining Democratic Congruence: The Perspective of "Substantive" Democracy. *World Values Research,* 1 (3): 57-90. Available at: http: //www. worldvaluessurvey.org/wvs/articles/folder_published/ paperseries_25/files/WVR_01_03_2008.pdf.

Wind S. L. 2011. *Eurozone Sovereign Debt Crisis* (November 25, 2011). URL: Available at SSRN: http: //ssrn.com/abstract=1966315 or http: //dx. doi.org/10.2139/ssrn.1966315).

Woolcock M. and Narayan D. 2000. Social Capital: Implications for Development Theory, Research, and Policy. *The World Bank Research Observer,* 15 (2): 225-249.

Wooldridge J. M. 2009. *Introductory Econometrics: A Modern Approach.* 4th edition, Michigan State University: South-Western.

Yazbeck-Haddad Y.; Esposito J. L. 2001. *Daughters of Abraham: feminist thought in Judaism, Christianity, and Islam.* Gainesville, FL: University Press of Florida.

Yesilada B. and Noordijk P. 2010. Changing Values in Turkey: Religiosity and Tolerance in Comparative Perspective. *Turkish Studies,* 11 (1): 9–27.

You J. 2005. *Corruption and Inequality as Correlates of Social Trust: Fairness Matters More Than Similarity.* The Hauser Center for Nonprofit Organizations and The John F. Kennedy School of Government Harvard University Working Paper No. 29.

You J. -S. and Khagram S. 2005. A Comparative Study of Inequality and Corruption. *American Sociological Review,* 70 (1): 136-157.

Zaimeche S. 2002, Islam and Science, *FSTC Limited* 2002, 2003 Manchester, UK.

Zak P. J. and Knack S. 2001. Trust and Growth. *The Economic Journal,* 111 (470): 295-321.

Zakaria F. 2003. *The Future of Freedom: Illiberal Democracy at Home and Abroad,* New York: WW Norton & Company.

Zarigil Z. 2009. Deconstructing the Turkish Military's Popularity. *Armed Forces & Society,* 35 (4): 709-727.

Zivot E. and Wang J. 2006. *Modeling Financial Time Series with S-PLUS,* Second Edition, with Jiahui Wang. Heidelberg, Berlin and New York: Springer.

INDEX

2

A

B

C

D

E

F

G

H

I

J

K

L

M

N

O

P

Q

R

S

T

U

V

W

X

Y

Z